Advances in Plant Breeding

Advances in Plant Breeding

Edited by Bridget Harrington

SYRAWOOD
PUBLISHING HOUSE

New York

Published by Syrawood Publishing House,
750 Third Avenue, 9ᵗʰ Floor,
New York, NY 10017, USA
www.syrawoodpublishinghouse.com

Advances in Plant Breeding
Edited by Bridget Harrington

International Standard Book Number: 978-1-68286-787-7 (Hardback)

Cataloging-in-Publication Data

Advances in plant breeding / edited by Bridget Harrington.
    p. cm.
Includes bibliographical references and index.
ISBN 978-1-68286-787-7
1. Plant breeding. 2. Breeding. 3. Agriculture. I. Harrington, Bridget.
SB123 .A38 2019
631.53--dc23

TABLE OF CONTENTS

PREFACE

This book has been an outcome of determined endeavour from a group of educationists in the field. The primary objective was to involve a broad spectrum of professionals from diverse cultural background involved in the field for developing new researches. The book not only targets students but also scholars pursuing higher research for further enhancement of the theoretical and practical applications of the subject.

Plant breeding is the science concerned with the modification of the traits of plants for the generation of improved characteristics. These include high yield, drought tolerance, disease resistance and adaptability. Classical plant breeding is developed around the practices of selection and interbreeding. It relies on the technique of homologous recombination between chromosomes for generating genetic diversity. The in vitro techniques of embryo rescue and protoplast fusion are used for creating hybrid plants. Radiation, chemical mutagens and transposons can also be used to create mutated plants. This book provides comprehensive insights into the diverse areas of plant breeding. It traces the progress of this field and highlights some of its key concepts and techniques. A number of latest researches have been included to keep the readers up-to-date with the global concepts in this area of study.

It was an honour to edit such a profound book and also a challenging task to compile and examine all the relevant data for accuracy and originality. I wish to acknowledge the efforts of the contributors for submitting such brilliant and diverse chapters in the field and for endlessly working for the completion of the book. Last, but not the least; I thank my family for being a constant source of support in all my research endeavours.

Editor

Silencing the expression of the salivary sheath protein causes transgenerational feeding suppression in the aphid *Sitobion avenae*

Eltayb Abdellatef[1,†], Torsten Will[1,†], Aline Koch[1], Jafargholi Imani[1], Andreas Vilcinskas[1,2] and Karl-Heinz Kogel[1,*]

[1]*Centre for BioSystems, Land Use and Nutrition, Institute of Phytopathology and Applied Zoology, Justus Liebig University, Giessen, Germany*
[2]*Project Group 'Bioresources', Fraunhofer Institute of Molecular Biology and Applied Ecology IME, Giessen, Germany*

*Correspondence

email karl-heinz.kogel@agrar.uni-giessen.de
†Contributed equally to this work.

Keywords: host-induced gene silencing, RNAi, aphid, salivary sheath protein, *Sitobion avenae*, transgenerational silencing.

Summary

Aphids produce gel saliva during feeding which forms a sheath around the stylet as it penetrates through the apoplast. The sheath is required for the sustained ingestion of phloem sap from sieve elements and is thought to form when the structural sheath protein (SHP) is cross-linked by intermolecular disulphide bridges. We investigated the possibility of controlling aphid infestation by host-induced gene silencing (HIGS) targeting *shp* expression in the grain aphid *Sitobion avenae*. When aphids were fed on transgenic barley expressing *shp* double-stranded RNA (*shp*-dsRNA), they produced significantly lower levels of *shp* mRNA compared to aphids feeding on wild-type plants, suggesting that the transfer of inhibitory RNA from the plant to the insect was successful. *shp* expression remained low when aphids were transferred from transgenic plants and fed for 1 or 2 weeks, respectively, on wild-type plants, confirming that silencing had a prolonged impact. Reduced *shp* expression correlated with a decline in growth, reproduction and survival rates. Remarkably, morphological and physiological aberrations such as winged adults and delayed maturation were maintained over seven aphid generations feeding on wild-type plants. Targeting *shp* expression therefore appears to cause strong transgenerational effects on feeding, development and survival in *S. avenae*, suggesting that the HIGS technology has a realistic potential for the control of aphid pests in agriculture.

Introduction

The family Aphididae encompasses more than 4300 aphid species, all of which are specialized in phloem sap feeding (Blackman and Eastop, 1994). Aphid populations proliferate rapidly reflecting their parthenogenetic and viviparous mode of reproduction during the asexual life cycle (International Aphid Genomics Consortium, 2010). The evolution of telescoping generations, wherein embryonic development begins before the mother is born, allows nymphs to reach maturity a few days after birth (Dixon, 1987). Aphids have a severe impact on their host because they deprive plants of nutrition and act as vectors for phytopathogenic viruses; annual crop losses due to aphids therefore cost hundreds of millions of dollars (Blackman and Eastop, 2000). The control of aphid pests is becoming more challenging due to the spread of insecticide-resistant populations, so more effective control measures are urgently required (Elzen and Hardee, 2003).

RNA interference (RNAi) has been developed as a novel method for pathogen and pest control (Price and Gatehouse, 2008; Huvenne and Smagghe, 2010; El-Shesheny et al., 2013; Wang et al., 2013; Zhang et al., 2013; Koch et al., 2013; for review see Koch and Kogel, 2014). RNAi is an ancient and highly conserved form of gene regulation present in almost all eukaryotes (Fire, 2007; Fire et al., 1998), which involves two pathways: post-transcriptional gene silencing (PTGS) and transcriptional gene silencing (TGS) (Vaucheret, 2006; Vaucheret and Fagard, 2001). PTGS occurs in the cytoplasm acting at the mRNA level and begins with the cleavage of a precursor double-stranded RNA (dsRNA) by RNase III-like enzymes of the Dicer family, producing 21–25 nt short interfering RNA (siRNA) duplexes (Baulcombe, 2004). These siRNAs assemble with proteins to form an RNA-induced silencing complex (RISC) that subsequently binds to a complementary mRNA, degrading it and thereby inhibiting translation (Brodersen and Voinnet, 2006; Vaucheret et al., 2004).

More recently, systemic RNAi methods have been developed by studying dsRNA uptake in model organisms such as the beetle *Tribolium castaneum* (Tomoyasu et al., 2008). Several protocols have been developed for the delivery of dsRNA to insects, including dietary supplements with purified dsRNA (Baum et al., 2007), microinjection (Arakane et al., 2004; Suzuki et al., 2008) and delivery by feeding on plants expressing dsRNA transgenes (Will and Vilcinskas, 2013). The latter strategy is known as host-induced gene silencing (HIGS) (Nowara et al., 2010) and has been proposed as a method to engineer crops for resistance to pathogens and pests such as aphids, including the prevalent pest species *Sitobion avenae* and *Myzus persicae* (Bhatia et al., 2012; Pitino et al., 2011; Xu et al., 2014).

Aphids feed on phloem sap from sieve tubes, the nutrient transport channels of the vascular bundles in higher plants (van Bel, 2003). The aphids penetrate the cuticle with specialized mouthparts known as stylets and push the stylet through the apoplast, probing and penetrating cells along its pathway, possibly in response to plant signals (Hewer et al., 2010; Tjallingii, 2006). Stylet movement is accompanied by the secretion of gel saliva, which forms a salivary flange on the epidermis and an enveloping salivary sheath in the apoplast, both of which may provide stability, lubrication and protection during feeding, while the latter also seals the plasma membrane at stylet penetration sites (Miles, 1999; Will and van Bel, 2006; Will et al., 2012). Arnaud (1918) compared the salivary sheath with the padding of an oil borehole and suggested that the sheath prevents sap leaking from the stylet as well as the ingress of undesirable plant material into the feeding channel. Most phytophagous hemiptera form a salivary sheath during feeding, emphasizing its biological relevance (Morgan et al., 2013; Will et al., 2013).

The hardening of the salivary sheath is thought to be caused by the oxidation of cysteine sulfhydryl groups to form intermolecular disulphide bonds (Miles, 1965; Tjallingii, 2006). This view is supported by the observation that salivary sheath formation is inhibited under anoxic and reducing conditions (Will et al., 2012). The A. pisum sheath protein (SHP) contains the highest number of cysteine residues in the aphid saliva proteome and is a pivotal component of the sheath hardening process (Carolan et al., 2009), making it a promising candidate for the disruption of sheath hardening by RNAi. Here, we show that silencing shp expression in the grain aphid S. avenae by HIGS strongly inhibits feeding and reproductive behaviour of the aphid and negatively impacts its survival. Remarkably, we found that shp silencing was transmitted to offspring across several generations of aphids even after switching to wild-type plants.

Results

Barley plants expressing shp-dsRNA induce shp silencing in feeding aphids

Barley plants (cultivar Golden Promise) were transformed with either the silencing vector p7i-Ubi-shp-dsRNA or the empty-vector p7i-Ubi-RNAi2. The former contains two inverted ubiquitin promoters (ubi) driving the constitutive expression of sense and antisense copies of a 491-bp fragment of shp to generate the corresponding dsRNA, while, in the latter, this fragment is replaced by a fragment of the ß-glucuronidase (GUS) gene (Figure S1). Six transgenic lines with independent transformation events were assessed by quantitative real-time PCR (qRT-PCR) analysis for expression of shp-dsRNA. Strongest expression of the inhibitory RNA was found in lines L26, L28 and L33 (Figure S2).

Next, we assessed whether the expression of the inhibitory RNA affects target gene expression in aphids fed on transgenic barley lines. To this end, grain aphids (Sitobion avenae) were fed for 2 weeks on L26 which showed the highest level shp-dsRNA expression. The relative expression level of the aphid's shp gene was reduced 10-fold compared to aphids that fed on wild-type and empty-vector controls (Figure 1a). To determine whether the aphids recovered from silencing, those fed on L26 and empty-vector control plants for 2 weeks were subsequently transferred to wild-type plants, and shp mRNA levels were measured after 1 and 2 further weeks of feeding. In aphids that had previously fed on L26 plants, the shp gene remained strongly silenced. After 1 week of recovery, shp mRNA levels had risen from 10% to only 34% of the control level, and after 2 weeks, they had risen further but only to 38% of the control level (Figure 1b).

HIGS-mediated silencing of shp impairs the fitness of adult aphids

We investigated the impact of shp silencing on sheath formation and various fitness parameters. The aphids were fed for 2 weeks on either L26 or empty-vector/wild-type control lines and were subsequently transferred to an artificial diet on parafilm sachets. After 2 further days, 30 salivary sheaths formed by gel saliva secreted onto the underside of the parafilm were inspected using an inverse microscope. The salivary sheaths deposited by the control aphids showed a typical necklace-like structure (Figure 2a–b), whereas sheath formation was largely inhibited in aphids that had initially fed on L26 (Figure 2c). The salivary sheath of shp silenced aphids showed a significant size reduction of approx. 81% (0.21 ± 0.2 mm^2; $P < 0.001$) while there was no impact on sheath size on empty-vector plants (1.104 ± 0.3 mm^2) compared with wild type (1.143 ± 0.4 mm^2; $P < 0.74$).

The impact of shp silencing on reproduction was determined by placing synchronous 8-day-old adults onto shp-dsRNA or control plants and scoring the number of offspring. The number of nymphs was significantly ($P < 0.001$) lower when aphids were fed on the shp-dsRNA lines (28, 37 and 39 nymphs per adult on L26, L28 and L33, respectively) compared to 63 and 58 nymphs per adult on the wild-type and empty-vector controls, respectively (Figures 3a; S3a). Reproduction ceased after 41 days on the empty-vector controls and after 42 days on the wild-type controls, but after 31 days on L26. The average daily reproduction rate rose to a maximum of ~4 nymphs per day on control lines and ~2.8 nymphs per day on L26 (Figure 3b). Comparable results were obtained with transgenic lines L28 and L33 (Figure S3b, c). The overall survival of aphids feeding on L26 was significantly ($P < 0.05$) lower than aphids fed on the control lines (Figure 3c). Comparable results were obtained with transgenic lines L28 and L33 (Figure S3d).

Aphids feeding on shp-dsRNA lines experience a developmental delay

Ten synchronous one-day-old nymphs were taken from wild-type control lines and placed individually inside clip cages, which were placed on the surface of L26 or control leaves. Development was monitored until maturity. The nymphs feeding on control plants remained at developmental stages 2 and 3 for 4 days and developed normally, reaching maturation within 8 days. By contrast, the nymphs feeding on L26 remained at developmental stages 2 and 3 for 9 days and reached maturation only after 12–14 days (Figure 4a). We also measured the offspring body plan area (BPA) at maturity and found that those feeding on L26 was approximately 50% smaller than those feeding on control lines (Figure 4b).

Feeding on shp-dsRNA lines induces transgenerational silencing in aphids

Silencing in insects can be vertically transmitted, that is parents subjected to RNAi and can transmit the effect to their progeny (Bucher et al., 2002). We investigated the potential transgenerational effects of shp silencing by feeding aphids for 2 weeks on L26 and allowing them to reproduce on wild-type plants for 24 h. We then monitored shp expression in successive generations of aphids fed solely on wild-type plants. Remarkably, we found that shp expression was not only inhibited in the parental generation that had fed on L26, but remained significantly reduced in the six subsequent generations that fed only on wild-type plants (Figure 5). Aphids recovered only slowly from shp silencing over successive generations, with relative expression levels peaking at 2.5%, 4.2%, 21%, 19%, 44%, 77% and 94% of control levels in generations 1–7, respectively). The offspring were also characterized by prolonged development (15.5, 15, 13.7, 14.2, 12.75, 11 and 9.2 days in generations 1–7, respectively) compared to

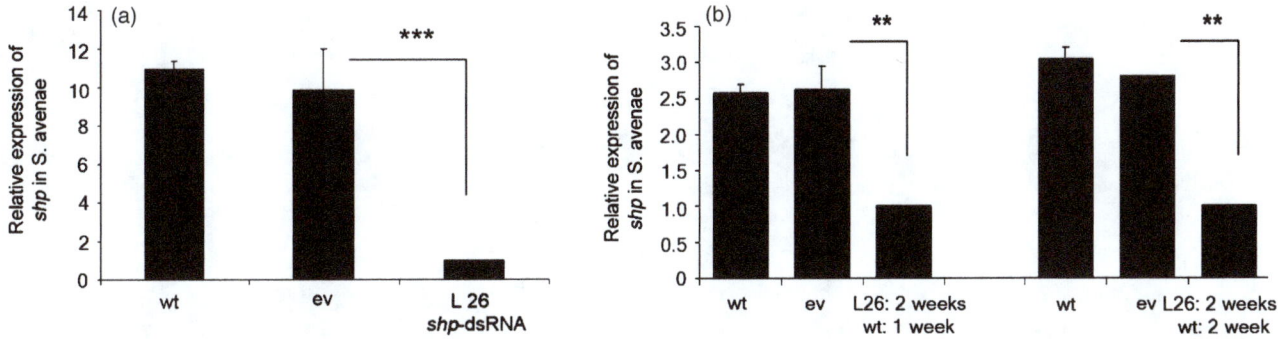

Figure 1 Quantification of *shp* transcripts in *Sitobion avenae* by qRT-PCR. The cDNA was generated from total RNA extracted from aphids after feeding on the *shp*-dsRNA line L26 as well as empty-vector (ev) and wild-type (wt) control plants. (a) The abundance of *shp* mRNA after 2 weeks feeding on L26, empty-vector and wild-type control lines. The lower *shp* mRNA levels in aphids fed on L26 was statistically significant (***$P < 0.001$, *t*-test). (b) Abundance of *shp* mRNA after 2 weeks feeding on L26 before transfer to wild-type plants for 1 or 2 weeks. The lower *shp* mRNA levels in aphids fed on L26 was statistically significant (**$P < 0.01$, *t*-test). Bars represent mean values ± SDs for three independent samples with 10 aphids each.

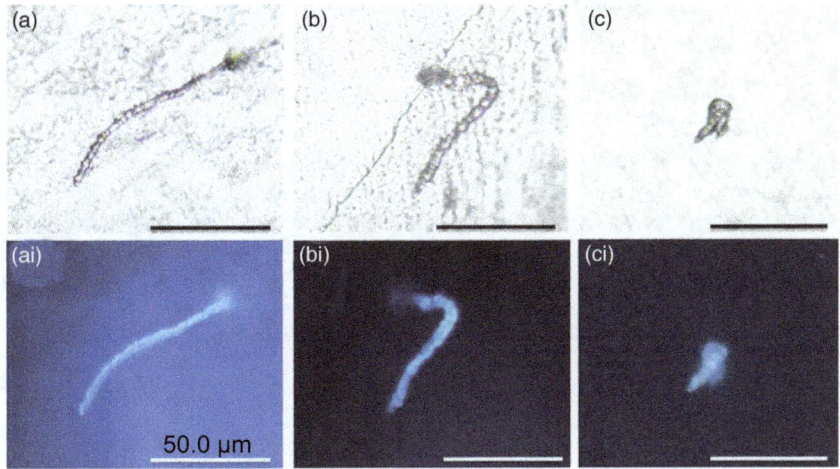

Figure 2 Influence of *shp* silencing on salivary sheath formation. (a) and (b) salivary sheath formation by aphids maintained on wild-type (wt) control (a) or empty-vector (ev) control (b) plants after transfer to artificial diet and incubation for 48 h, showing typical necklace structure. (c) Salivary sheath formation by aphids maintained on L26 before transfer to artificial diet for 48 h, showing that sheath formation on the artificial diet is disrupted. (a), (b) and (c) were observed by bright-field microscopy, (ai), (bi) and (ci) by fluorescence microscopy. Thirty randomly selected salivary sheaths from a total of 10 aphids were observed for each treatment, and images show typical examples.

8 days in controls (Table 1) and a higher percentage of winged adults (80%, 70%, 68%, 69%, 70%, 50% and 30% more than controls in generations 1–7, respectively; Table 2). These data show that aphid feeding can induce long-lasting *shp* silencing via HIGS in successive aphid generations.

Discussion

RNAi is a prospective tool for pest control that has been experimentally validated in a large number of studies (e.g. Baum *et al.*, 2007; Huvenne and Smagghe, 2010; Mao *et al.*, 2007; Zhang *et al.*, 2013). Although RNAi has been used to silence targets in several aphid organs (Jaubert-Possamai *et al.*, 2007; Sapountzis *et al.*, 2014; Zhang *et al.*, 2013), the salivary glands appear to be a promising target because aphid saliva (and particularly the proteins in it) are indispensable for the infestation of plants (Will *et al.*, 2013). Previous work has suggested that the efficacy of gene silencing against salivary effectors is species/method dependent. For example, the silencing of salivary protein C002 in the green peach aphid *Myzus persicae* by HIGS reduced

the reproduction rate but did not affect survival (Pitino *et al.*, 2011). By contrast, Mutti *et al.* (2006, 2008) achieved a reduction in feeding, reproduction and survival by targeting the same protein by microinjection in the pea aphid *Acyrthosiphon pisum*, suggesting that targets must be explored on a case-by-case basis.

The composition of salivary proteins differs among aphid species even if they feed on similar host plants (Cooper *et al.*, 2011; Rao *et al.*, 2013; Vandermoten *et al.*, 2014). Furthermore, even similar aphid salivary effectors are species restricted and are adapted to particular hosts (Pitino and Hogenhout, 2013). By contrast, the highly conserved sheath protein SHP (Rao *et al.*, 2013) is present in diverse aphid species such as *A. pisum* and *S. avenae* (Carolan *et al.*, 2009, 2011) and therefore represents an ideal target to control different aphid pests infesting various crops. SHP is necessary for sheath formation and is found not only in aphids but also in other hemipteran pests such as white flies, making it a candidate target with broad applicability (Freeman *et al.*, 2001). SHP is not found in nontarget and beneficial insects, which reduces the risk of off-target effects (Carolan *et al.*, 2009).

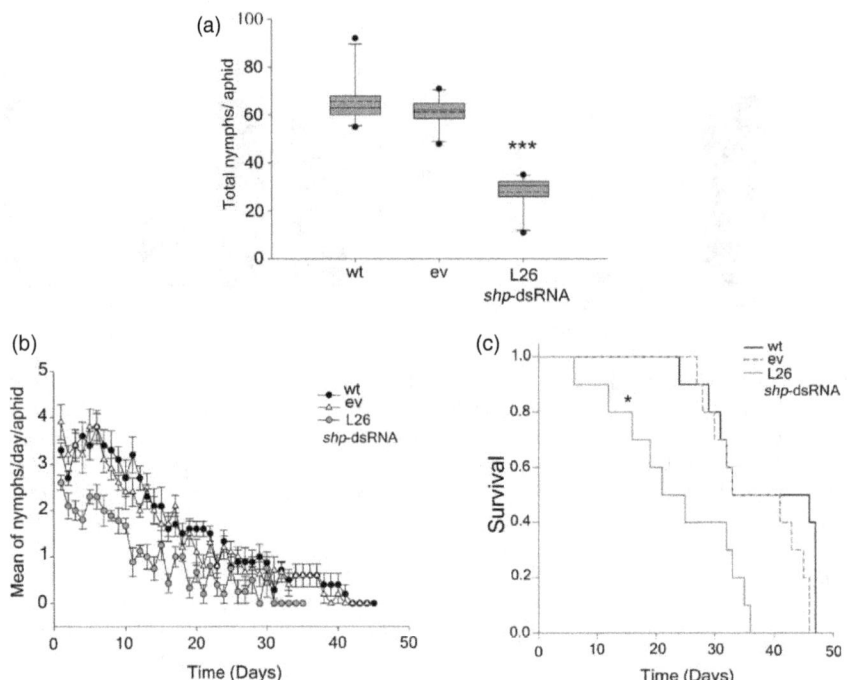

Figure 3 Fitness analysis of aphids fed on *shp*-dsRNA plants. (a) The reproduction of aphids fed on *shp*-dsRNA line L26 was significantly lower than that of aphids fed on empty-vector (ev) and wild-type (wt) control lines. Synchronous eight-day-old nymphs were placed either on *shp*-dsRNA or on control lines. (***$P < 0.001$, *t*-test). (b) Aphids fed on *shp*-dsRNA L26 show a significant decline in reproduction rate and a shorter overall duration of reproduction than aphids fed on ev and wt control plants. (c) Mortality was recorded for 48 days. Aphids fed on *shp*-dsRNA lines died earlier than those fed on ev and wt control lines, and the total survival rate was significantly lower in the *shp*-dsRNA line L26 compared to the controls (*$P < 0.05$, *t*-test). All experiments were repeated three times, with 15 aphids, and each replicate was similar to the results shown here. Bars represent mean values ± SDs of 15 single animals. Whiskers of box plots defining the 10th and 90th percentiles.

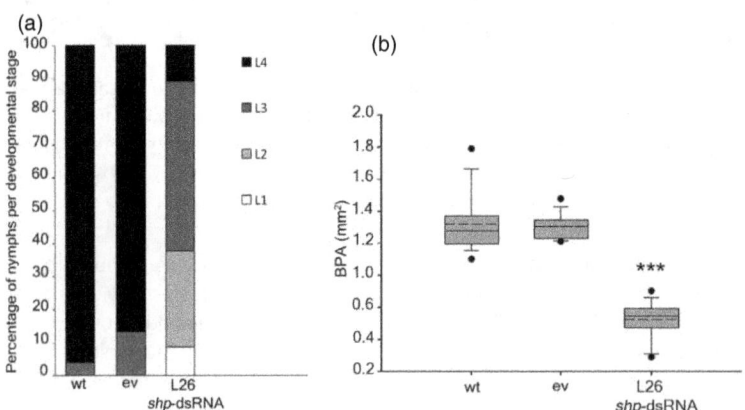

Figure 4 Aphid feeding on *shp*-dsRNA line disrupts offspring nymph development and reduces the adult size. (a) Aphid feeding on *shp*-dsRNA line L26 delays nymph maturation compared to nymphs fed on empty-vector (ev) and wild-type (wt) controls after 8 days. The experiment was carried out three times with similar results. (b) Adults were smaller when the nymphs were fed on *shp*-dsRNA L26 compared to nymphs fed on ev and wt controls. The body plan area of aphids fed on the *shp*-dsRNA was significantly smaller than aphids fed on ev and wt control plants (***$P < 0.001$, *t*-test). All experiments were repeated three times, with 10 aphids, and each replicate was similar to the results shown here. Bars represent mean values ± SDs of 10 single animals.

We have demonstrated that aphids feeding on transgenic barley plants expressing *shp*-dsRNA experienced potent *shp* silencing, which prevented the formation of normal sheathes and thus confirmed the pivotal role of SHP in the structural integrity of the salivary sheath (Carolan *et al.*, 2009). The disruption of sheath formation prevented normal food uptake, and thus prolonged development, produced aphids with a smaller body plan area and reduced fecundity and survival. In addition to these direct effects of *shp* silencing, we observed indirect yet potent transgenerational effects lasting at least seven generations. These effects included prolonged development and a higher percentage of winged aphids (alataes), a morph type that correlates with reduced parental nutrition (Braendle *et al.*, 2006).

Table 1 Transgenerational effects of *shp* silencing on seven successive aphid generations.

Line	Maturation time (mean)/days						
	1st generation	2nd generation	3rd generation	4th generation	5th generation	6th generation	7th generation
Wild type	8.5 ± 0.0	8.5 ± 0.0	8.5 ± 0.0	8.5 ± 0.0	8.5 ± 0.0	8.5 ± 0.0	8.5 ± 0.0
Empty vector	8.5 ± 0.0	8.5 ± 0.0	8.5 ± 0.0	8.5 ± 0.0	8.5 ± 0.0	8.5 ± 0.0	8.5 ± 0.0
shp-dsRNA L26	15.5 ± 0.7**	15 ± 0.7**	13.75 ± 1.7**	14.25 ± 2.4*	12.75 ± 2.0	11 ± 1.4	9.25 ± 0.3

Maturation time: delayed maturation (mean values ± SDs) was observed in 50 nymphs fed on *shp*-dsRNA L26 through seven generations. Until the sixth-generation nymphs matured after 11–15 day compared to 8–9 days for nymphs fed on wt plants. Maturation time in the aphid with background feeding on *shp*-dsRNA compared to feeding on empty-vector and wild-type controls was statistically significant (**$P < 0.01$, *$P < 0.05$ Student's *t*-test).

Table 2 Transgenerational effects of *shp* silencing on seven successive aphid generations.

Line	Winged adult (proportion and percentage)						
	1st generation (%)	2nd generation (%)	3rd generation (%)	4th generation (%)	5th generation (%)	6th generation (%)	7th generation (%)
Wild type	3.5 ± 1/50 (7)	4.5 ± 2/50 (9)	4.0 ± 2/50 (8)	2.0 ± 2/50 (4)	2.0 ± 1/50 (4)	7.5 ± 2/50 (15)	4.5 ± 2/50 (9)
Empty vector	5.0 ± 1/50 (10)	3.5 ± 0.7/50 (7)	4.5 ± 3/50 (9)	7.5 ± 3/50 (15)	9.0 ± 1/50 (18)	5.5 ± 2/50 (11)	8.0 ± 1/50 (16)
shp-dsRNA L26	40 ± 1/50** (80)	35 ± 2/50** (70)	34 ± 4/50* (68)	34.5 ± 4/50* (69)	35 ± 4/50* (70)	25 ± 4/50* (50)	16 ± 4/50 (32)

Winged adults: wing development (mean values ± SDs) was observed in 50 nymphs until maturation through seven generations. Until the sixth generation, 80–50% of nymphs fed on *shp*-dsRNA L26 developed wings compared to 7–15% for nymphs fed on wild-type control plants. There was a statistically significant difference in the development of wings when the first generation of aphids were initially fed on the *shp*-dsRNA line compared to empty-vector and wild-type controls (**$P < 0.01$, *$P < 0.05$, *t*-test).

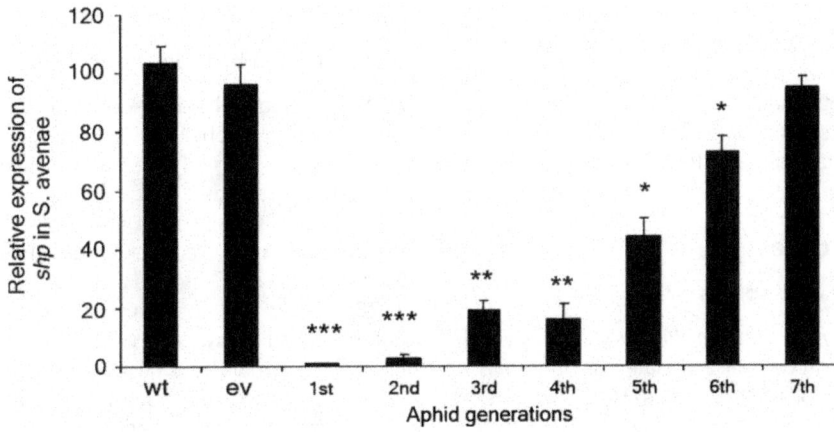

Figure 5 Transgenerational effect of *shp* silencing. Aphids were fed for two weeks on *shp*-dsRNA L26 and empty-vector (ev) and wild-type (wt) controls and subsequently allowed to reproduce on wt. The *shp* transcript level was determined in seven successive aphid generations. The *shp* transcript levels in aphids initially fed on the *shp*-dsRNA lines were significantly lower than those fed on ev and wt controls (***$P < 0.001$, **$P < 0.01$, *$P < 0.05$, *t*-test). Each generation included 50 aphids. Bars represent mean values ± SDs of 10 single animals, and three replicates were conducted.

The induction of RNAi by the microinjection of siRNA or dsRNA was shown to cause silencing that lasted up to 7 days in *A. pisum* (Jaubert-Possamai *et al.*, 2007), 25 days in the honeybee *Apis mellifera* (Amdam *et al.*, 2003) and more than 6 months in the beetle *Tribolium castaneum* (Miller *et al.*, 2012). In the latter case, there was a positive correlation between the amount of dsRNA and the duration of the silencing effect. Silencing could be achieved by continuous feeding with either single-stranded or double-stranded siRNAs (Bhatia *et al.*, 2012). We found that *shp* silencing by HIGS lasts for more than 2 weeks when aphids were transferred onto wild-type plants after feeding on *shp*-dsRNA

lines. This suggests that long-term feeding could potentially cause the accumulation of new inhibitory RNA species, as observed for *T. castaneum* (Miller *et al.*, 2012).

In insects, parental RNAi was first reported as the delivery of dsRNA to the mother and the induction of RNAi in the offspring in *T. castaneum* (Bucher *et al.*, 2002) and *Locusta migratoria* (He *et al.*, 2006). We found that the parental transfer of RNAi to offspring also occurs in *S. avenae* but lasts for seven generations, at least when the target is *shp*. We observed a slow but continuous weakening of *shp* silencing over successive generations, accompanied by a decreasing percentage of alataes and prolonged development in the offspring, which appears to be directly induced by reduced nutritional uptake in the nymphs.

Bucher *et al.* (2002) speculated that parental RNAi could be caused either by a specific cellular uptake mechanism for dsRNA or secondary amplification of small amounts of incidentally incorporated dsRNA. The transfer of siRNA/dsRNA to following generations could in aphids be facilitated by the phenomenon of telescoping generations, meaning that a parthenogenetically adult already carries its developing grandchildren in it. The long-lasting silencing of *shp* in *S. avenae* cannot adequately be explained by the uptake of dsRNA into the ovaries resulting in the delivery of dsRNA or siRNA to the developing embryos. More likely, the observed transgenerational effects reflect an attenuated amplification process, in which the total amount of siRNA must decline in successive generations. However, we cannot exclude the possibility that long-term transgenerational silencing of *shp* reflects a direct interaction between the delivered inhibitory RNA and the aphid's transcriptional machinery, resulting in an epigenetic modification directed by small RNAs (Castel and Martienssen, 2013).

We have demonstrated that HIGS targeting the structural sheath protein in aphids is a powerful strategy for aphid control. Direct effects on adults and indirect effects on the offspring are induced by the inhibition of sheath formation as well as a long-term parental RNAi effect, and this should lead to a significant reduction of plant infestation. The high percentage of alataes among the offspring could result in the emigration of a large proportion of the offspring in subsequent generations, but the parental RNAi effect will limit the effectiveness of infestations caused by those migrating aphids, that is population growth will be delayed compared to aphids fed on wild-type plants. The presence of SHP in all aphid species studied thus far (Rao *et al.*, 2013) makes it unlikely that population decline in one aphid species will lead to the settlement of a new aphid species. Our results demonstrate the substantial potential of HIGS as an alternative or complementary strategy for insect pest control in agriculture.

Experimental procedures

Maintenance of plants and aphids

Barley (*Hordeum vulgare*) cv. Golden Promise (GP) and all transgenic GP plants were grown in a climate chamber with a 16-h photoperiod (260 µmol/m^2/s) at 18 °C/14 °C (light/dark) with 65% relative humidity. Grain aphids (*S. avenae*) were reared on three-week-old barley plants in a climate chamber. To obtain synchronized insects, reproductive mature aphids were placed in clip cages (one aphid per cage) on GP plants for 24 h. The adults were then removed, and the offspring were used for experiments as previously described (Gaupels *et al.*, 2008; Schmitz *et al.*, 2012). All experiments were conducted in a climate cabinet under the conditions stated above.

Construction of *shp* templates and generation of transgenic barley plants

For the constitutive overexpression of *shp*-dsRNA in barley, a 491-bp cDNA template fragment (Figure S1) from the 3621-bp *Acyrthosiphon pisum* (Ap) *shp* cDNA (XM_001943863, ACYPI009881) was amplified using primers Ap-HindII_F and Ap-AflII_R (Table S1) and inserted into the binary RNAi vector p7i-Ubi-RNAi2 (DNA Cloning Service, Hamburg, Germany) to replace

the *GUS* gene. The resulting vector (p7i-Ubi-*shp*-RNAi), containing the *shp* fragment under the control of inverted plant ubiquitin (*ubi*) promoters, was transferred by electroporation into *Agrobacterium tumefaciens* strain AGL1 (Lazo *et al.*, 1991), which was used for the transformation of immature barley embryos as described (Imani *et al.*, 2011; Kogel *et al.*, 2010). PCR analysis was used to confirm integration of the transferred DNA using primers *shp*-RNA_F and *shp*-RNA_R. Primers 35sP_F and Hyg_R were used to identify the empty-vector (ev) lines that contained p7i-Ubi-RNAi2 (Table S1).

RNA extraction and qRT-PCR analysis

The expression of *shp*-dsRNA was analysed by qRT-PCR following the isolation of RNA from barley leaves using TRIzol$^{®}$ reagent (Ambion, Carlsbad, CA, USA) according to the manufacturer's instructions. Freshly extracted mRNA from 10 aphids was converted into cDNA using the QuantiTect$^{®}$ Reverse Transcription Kit (Qiagen, Hilden, Germany) and 40 ng of cDNA was used as the template for qRT-PCR in an Applied Biosystems 7500 FAST real-time PCR system. Each reaction comprised 7.5 µL SYBER Green JumpStart Taq ReadyMix (Sigma-Aldrich, Steinheim, Germany) and 0.5 pmol of the gene-specific primers *shp*-RNA-qpcr-F1 and *shp*-RNA-qpcr-R1 (Table S1). After initial heating to 95 °C for 5 min, the target was amplified by 40 cycles at 95 °C for 30 s, 52 °C for 30 s and 72 °C for 30 s. Ct values were determined using 7500 Fast software (Applied Biosystems, Darmstadt, Germany). The *shp*-dsRNA transcript levels were determined using the $2^{-\Delta\Delta Ct}$ method (Livak and Schmittgen, 2001) by normalizing against barley ubiquitin mRNA (GenBank M60175). The *shp* transcript levels in aphids were determined using the same primers but normalizing against 18S ribosomal RNA (GenBank APU27819).

Electrical penetration graph (EPG) technique

Aphid feeding behaviour was monitored using the electrical penetration graph (EPG) technique (Tjallingii, 1988). A 1 cm × 20 µm gold wire was attached to the dorsal abdomen of apterous aphids, using electrically conductive silver glue (Electrolube; Swadlincote, Derbyshire, UK) and a vacuum device for immobilization (van Helden and Tjallingii, 2000). The aphid electrode was connected to a DC EPG Giga-8 amplifier (Tjallingii, 1978, 1988), and the EPG output was recorded with stylet hardware and software (EPG Systems, Wageningen, the Netherlands). A second electrode was inserted into the soil of potted plants. The experimental set-up was placed in a Faraday cage to shield it from electromagnetic interference. Aphids starved for 3–4 h were placed onto the abaxial side of a mature leaf of a 3-week-old barley plant. EPG recordings commenced immediately and continued for 1 h while behaviour was observed continuously. Only aphids that secreted gel saliva during the observation period were used for *shp* expression analysis by qRT-PCR. If the aphid behaviour differed from gel saliva secretion during the observation period, the aphids were forced to withdraw their stylets penetrate the plant again. EPG waveforms were interpreted as described by Prado and Tjallingii (1994).

Aphid salivary sheath formation on the artificial diet

An artificial diet that mimics the plant cell wall milieu (20 mM KCl, 1 mM CaCl$_2$, 10 mM MES, pH 5.5) as described by Cosgrove and Cleland (1983) was used to stimulate the secretion of gel saliva. The food was filter sterilized through a 0.45-µm PVDF membrane, and 150 µL was placed in a parafilm sachet (Verheggen *et al.*, 2009) sterilized with 30% H_2O_2 for approximately 30 min prior to use. After feeding for 3 weeks on the shp-dsRNA and control lines, a single aphid was placed on a diet sachet and 10 single aphids were observed per plant line. The diet sachet was placed downwards on a small aphid cage, and aphids were allowed to feed for 48 h. Parafilm foils pierced by the aphids were placed facing upwards in a Petri dish, and bright-field/fluorescence microscopy using an inverse microscope (Leica DMLB; Leica Microsystems, Mannheim, Germany) was used to locate the salivary sheaths. Three replicates (10 aphids each) were prepared for each treatment, and 10 randomly chosen salivary sheaths were observed in each replicate. ImageJ v1.42q (Wayne Rosband; National Institute of Health, Bethesda, Maryland) was used to measure the area of the randomly selected salivary sheaths on parafilm, and data of aphids from shp-dsRNA plants were compared with controls using t-test.

Evaluation of aphid fitness parameters after feeding on shp-dsRNA lines

Aphid survival ($n = 3$) and fecundity assays ($n = 3$) used for each replicate and a total of 10 synchronized mature aphids in individual clip cages placed on the abaxial side of leaves (one cage per leaf) from intact barley plants. Parameters were checked daily throughout the aphid lifespan. Survival was analysed using the Kaplan–Meier log-rank test (Kaplan and Meier, 1985) in Sigma Plot v11. For the analysis of nymph development and growth ($n = 3$), 10 synchronous one-day-old nymphs born on wild-type barley plants were segregated into individual clip cages on shp-dsRNA and control lines. Nymph development was recorded after 8 days and scored from 1 to 5 according to the developmental stages listed by Chau and Mackauer (2000). Growth was measured by determining the body plan area (Daniels et al., 2009) after nymphs reached maturity. For this purpose, images of individual aphids were captured using a Leica MZ16FA (Leica Microsystems) and analysed using ImageJ v1.42q. Obtained data for reproduction and body plan area were compared with t-test.

Permanency and transgenerational analysis of shp silencing

Recovery from shp silencing was observed in 10 aphids that fed for 2 weeks on shp-dsRNA plants before switching to wild-type plants for a further 1–2 weeks. The t-test was applied for the comparison of normalized $2^{-\Delta\Delta Ct}$ values obtained from aphids that fed on shp-dsRNA plants with aphids that were feeding on control plant lines. Transgenerational shp silencing was investigated by feeding 50 synchronized aphids on shp-dsRNA plants for 2 weeks and again transferring them to wild-type plants. Offspring produced over the next 24 h were maintained on wild-type plants until maturity and the birth of the next-generation nymphs. The 50 adults were then used for transcript analysis. Each time, aphids were allocated to 3 individual plants. Controls were treated in a similar way without incubation on shp-dsRNA plants. We assessed the maturation time, frequency of winged aphids and shp expression levels until the seventh generation whereas each single aphid was considered as an experimental unit. Data from shp-dsRNA were compared with control lines for each individual generation by t-test. Both experiments were repeated three times.

Acknowledgements

We thank Dr. Edgar Schliephake (Julius Kuehn Institute—Federal Research Centre for Cultivated Plants, Institute for Resistance Research and Stress Tolerance) for the starter colony of S. avenae. We thank Marlène Sophie Birk for her assistance with experiments and data acquisition. This work was supported by the Landes-Offensive zur Entwicklung Wissenschaftlich-ökonomischer Exzellenz (LOEWE) programme (project Insect Biotechnology) of the Hessian State, Germany, to T.W., A.V. and K.-H.K., the German Research Council (Deutsche Forschungsgemeinschaft, DFG) (to K.-H.K.) and Deutscher Akademischer Austauschdienst (DAAD) to E.A. The authors thank Richard M. Twyman for valuable comments concerning manuscript editing.

References

Amdam, G.V., Simoes, Z.L.P., Guidugli, K.R., Noraberg, K. and Omholt, S.W. (2003) Disruption of vitellogenin gene function in adult honeybees by intra-abdominal infection of double straned RNA. BMC Biotech. **3**, 1–8.

Arakane, Y., Hogenkamp, D.G., Zhu, Y.C., Kramer, K.J., Specht, C.A., Beeman, R.W., Kanost, M.R. and Muthukrishnan, S. (2004) Characterization of two chitin synthase genes of the red flour beetle, Tribolium castaneum, and alternate exon usage in one of the genes during development. Insect Biochem. Mol. Biol. **34**, 291–304.

Arnaud, G. (1918) Les Astérinées, pp. 52–65. Montpelier: Annales de l'Ecole d'Agriculture.

Baulcombe, D. (2004) RNA silencing in plants. Nature, **431**, 356–363.

Baum, J.A., Bogaert, T., Clinton, W., Heck, G.R., Feldmann, P., Ilagan, O., Johnson, S., Plaetinck, G., Munyikwa, T., Pleau, M., Vaughn, T. and Roberts, J. (2007) Control of coleopteran insect pests through RNA interference. Nat. Biotechnol. **25**, 1322–1326.

van Bel, A.J.E. (2003) The phloem, a miracle of ingenuity. Plant, Cell Environ. **26**, 125–149.

Bhatia, V., Bhattacharya, R., Uniyal, P.L., Singh, R. and Niranjan, R.S. (2012) Host generated siRNAs attenuate expression of serine protease gene in Myzus persicae. PLoS ONE, **7**, e46343.

Blackman, R.L. and Eastop, V.F. (1994) Aphids on the World's Trees: An Identification and Information Guide. Wallingford, UK: Cambridge University Press.

Blackman, R.L. and Eastop, V.F. (2000) Aphids on the World's Crops: An Identification and Information Guide, 2nd edn. Chichester, UK: John Wiley and Sons, Ltd.

Braendle, C., Davis, G.K., Brisson, J.A. and Stern, D.L. (2006) Wing dimorphism in aphids. Heredity, **97**, 192–199.

Brodersen, P. and Voinnet, O. (2006) The diversity of RNA-silencing pathways in plants. Trends Genet. **22**, 268–280.

Bucher, G., Scholten, J.S. and Klingler, M. (2002) Parental RNAi in Tribolium (Coleoptera). Curr. Biol. **12**, R85–R86.

Carolan, J.C., Fitzroy, C.I.J., Ashton, P.D., Douglas, A.E. and Wilkinson, T.L. (2009) The secreted salivary proteome of the pea aphid Acyrthosiphon pisum characterised by mass spectrometry. Proteomics, **9**, 2457–2467.

Carolan, J.C., Caragea, D., Reardon, K.T., Mutti, N.S., Dittmer, N., Pappan, K., Cui, F., Castaneto, M., Poulain, J. Dossat, C., Tagu, D., Reese, J.C., Reeck, G.R., Wilkinson, T.L. and Edwards, O.R. (2011) Predicted effector molecules in the salivary secretome of the pea aphid (Acyrthosiphon pisum): a dual transcriptomic protemic approach. J. Proteome Res. **10**, 1505–1518.

Castel, S.E. and Martienssen, R.A. (2013) RNA interference in the nucleus: roles for small RNAs in transcription, epigenetics and beyond. Nat. Rev. Genet. **14**, 100–112.

Chau, A. and Mackauer, M. (2000) Host instar selection in the aphid parasitoid Monoctonus paulensis(Hymenoptera: Braconidae, Aphidiinae): a preference for small pea aphids. Eur. J. Entomol. **97**, 347–353.

Cooper, W.R., Dillwith, J.W. and Puterka, G.J. (2011) Comparison of salivary proteins from five aphid (Hemiptera: Aphididae) species. Envrion. Entomol. **40**, 151–156.

Cosgrove, D.J. and Cleland, R.E. (1983) Solutes in the free space of growing stem tissues. Plant Physiol. **72**, 326–331.

Daniels, M., Bale, J.S., Newbury, H.J., Lind, R.J. and Pritchard, J. (2009) A sub lethal dose of thiamethoxam causes a reduction in xylem feeding by the bird cherry-oat aphid (*Rhopalosiphum padi*), which is associated with dehydration and reduced performance. *J. Insect Physiol.* **55**, 758–765.

Dixon, A.F.G. (1987) Parthenogenetic reproduction and the rate of increase in aphids. In *Aphids Their Biology, Natural Enemies and Control. Vol. A* (Minks, A.K. and Harrewijn, P., eds), pp. 269–287. Amsterdam: Elsevier.

El-Shesheny, I., Hajeri, S., El-Hawary, I., Gowda, S. and Killiny, N. (2013) Silencing abnormal wing disc gene of the asian citrus psyllid, *Diaphorina citri*, disrupts adult wing development and increases nymph mortality. *PLoS ONE*, **8**, e65392.

Elzen, G.W. and Hardee, D.D. (2003) United States department of agriculture-agricultural research service research on managing insect resistance to insecticides. *Pest Manag. Sci.* **59**, 770–776.

Fire, A.Z. (2007) Gene silencing by double-stranded RNA. *Cell Death Differ.* **14**, 1998–2012.

Fire, A., Xu, S., Montgomery, M.K., Kostas, S.A., Driver, S.E. and Mello, C.C. (1998) Potent and specific genetic interference by double-stranded RNA in *Caenorhabditis elegans. Nature*, **391**, 806–811.

Freeman, T.P., Buckner, J.S., Nelson, D.R., Chu, C.C. and Henneberry, T.J. (2001) Stylet penetration by *Bemisia argentifolii* (Homoptera: Aleyrodidae) into host leaf tissue. *Ann. Entomol. Soc. Am.* **94**, 761–768.

Gaupels, F., Buhtz, A., Knauer, T., Deshmukh, S., Waller, F., van Bel, A.J.E., Kogel, K.H. and Kehr, J. (2008) Adaptation of aphid stylectomy for analyses of proteins and mRNAs in barley phloem sap. *J. Exp. Bot.* **59**, 3297–3306.

He, Z.B., Cao, Y.Q., Yin, Y.P., Wang, Z.K., Chen, B., Peng, G.X. and Xia, Y.X. (2006) Role of hunchback in segment patterning of *Locusta migratoria manilensis* revealed by parental RNAi. *Dev. Growth Differ.* **48**, 439–445.

vanHelden, M. and Tjallingii, W.F. (2000) Experimental design and analysis in EPG experiments with emphasis on plant resistance research. In *Principles and Applications of Electronic Monitoring and Other Techniques in the Study of Homopteran Feeding Behavior* (Walker, G.P. and Backus, E.A., eds), pp. 144–171. Lanham, MD: Entomological Society of America.

Hewer, A., Will, T. and van Bel, A.J.E. (2010) Plant cues for aphid navigation in vascular tissues. *J. Exp. Biol.* **213**, 4030–4042.

Huvenne, H. and Smagghe, G. (2010) Mechanisms of dsRNA uptake in insects and potential of RNAi for pest control. *J. Insect Physiol.* **56**, 227–235.

Imani, J., Li, L., Schäfer, P. and Kogel, K.H. (2011) STARTS-a stable root transformation system for rapid functional analyses of proteins of the monocot model plant barley. *Plant J.* **67**, 726–735.

International Aphid Genomics Consortium. (2010) Genome sequence of the pea aphid *Acyrthosiphon pisum. PLoS Biol.* **8**, e1000313.

Jaubert-Possamai, S., Le Trionnaire, G., Bonhomme, J., Christophides, G.K., Rispe, C. and Tagu, D. (2007) Gene knockdown by RNAi in the pea aphid *Acyrthosiphon pisum. BMC Biotechnol.* **7**, 63.

Kaplan, E.L. and Meier, P. (1985) Non parametric estimation from incomplete observations. *J. Am. Stat. Assoc.* **53**, 457–481.

Koch, A. and Kogel, K.H. (2014) New wind in the sails: improving the agronomic value of crop plants through RNAi-mediated gene silencing. *Plant Biotechnol. J.* **12**, 821–831.

Koch, A., Kumara, N., Weber, L., Keller, L., Imani, J. and Kogel, K.H. (2013) Host-induced gene silencing of cytochrome P450 lanosterol C14α-demethylase–encoding genes confers strong resistance to *Fusarium* species. *Proc. Natl Acad. Sci. USA*, **110**, 19324–19329.

Kogel, K.H., Voll, L.M., Schäfer, P., Jansen, C., Wuc, Y., Langen, G., Imani, J., Hofmann, J., Schmiedl, A., Sonnewald, S., von Wettstein, D., Cook, J. and Sonnewald, U. (2010) Transcriptome and metabolome profiling of fieldgrown transgenic barley lack induced differences but show cultivar-specific variances. *Proc. Natl Acad. Sci. USA*, **107**, 6198–6203.

Lazo, G.R., Stein, P.A. and Ludwig, R.A. (1991) A DNA transformation-competent Arabidopsis genomic library in Agrobacterium. *Biotechnology*, **9**, 963–967.

Livak, K.J. and Schmittgen, T.D. (2001) Analysis of relative gene ex pression data using real-time quantitative PCR and the $2^{-\Delta\Delta Ct}$ method. *Methods*, **25**, 402–408.

Mao, Y.B., Cai, W.J., Wang, J.W., Hong, G.J., Tao, X.Y., Wang, L.J., Huang, Y.P. and Chen, X.Y. (2007) Silencing a cotton bollworm P450 mono

oxygenase gene by plant-mediated RNAi impairs larval tolerance of gossypol. *Nat. Biotechnol.* **25**, 1307–1313.

Miles, P.W. (1965) Studies on the salivary physiology of plant-bugs: the salivary secretions of aphids. *J. Insect Physiol.* **11**, 1261–1268.

Miles, P.W. (1999) Aphid saliva. *Biol. Rev.* **74**, 41–85.

Miller, S.C., Miyata, K., Brown, S.J. and Tomoyasu, Y. (2012) Dissecting systemic RNA interference in the Red Flour Beetle *Tribolium castaneum*. Parameters affecting the efficiency of RNAi. *PLoS ONE*, **7**, e47431.

Morgan, J.K., Luzio, G.A., Ammar, E.D., Hunter, W.B., Hall, D.G. and Shatters, R.G. (2013) Formation of stylet sheaths in aere (in air) from eight species of phytophagous hemipterans from six families (suborders: Auchenorrhyncha and Sternorrhyncha). *PLoS ONE*, **8**, e62444.

Mutti, N.S., Park, Y., Reese, J.C. and Reeck, G.R. (2006) RNAi knockdown of a salivary transcript leading to lethality in the pea aphid, *Acyrthosiphon pisum. J. Insect Sci.* **6**, 1–7.

Mutti, N.S., Louis, J., Pappan, L.K., Pappan, K., Begum, K., Chen, M.S., et al. (2008) A protein from the salivary glands of the pea aphid, *Acyrthosiphon pisum*, is essential in feeding on a host plant. *Proc. Natl Acad. Sci. USA*, **105**, 9965–9969.

Nowara, D., Gay, A., Lacomme, C., Shaw, J., Ridout, C., Douchkov, D., Hensel, G., Kumlehn, J. and Schweizer, P. (2010) HIGS: host-induced gene silencing l the obligate biotrophic fungal pathogen *Blumeria graminis. Plant Cell*, **22**, 3130–3141.

Pitino, M. and Hogenhout, S.A. (2013) Aphid protein effectors promote aphid colonization in a plant species-specific manner. *Mol. Plant Microbe Interact.* **26**, 130–139.

Pitino, M., Coleman, A.D., Maffei, M.E., Ridout, C.J. and Hogenhout, S.A. (2011) Silencing of aphid genes by dsRNA feeding from plants. *PLoS ONE*, **6**, e25709.

Prado, E. and Tjallingii, W.F. (1994) Aphid activities during sieve element punctures. *Entomol. Exp. Appl.* **72**, 157–165.

Price, D.R.G. and Gatehouse, J.A. (2008) RNAi-mediated crop protection against insects. *Trends Biotechnol.* **26**, 393–400.

Rao, S.A., Carolan, J.C. and Wilkinson, T.L. (2013) Proteomic profiling of cereal aphid saliva reveals both ubiquitous and adaptive secreted proteins. *PLoS ONE*, **8**, e57413.

Sapountzis, P., Duport, G., Balmand, S., Gaget, K., Jaubert-Possamai, S., Febvay, G., Charles, H., Rahbé, Y., Colella, S. and Calevro, F. (2014) New insight into the RNA interference response against cathepsin-L gene in the pea aphid, *Acyrthosiphon pisum*: Molting or gut phenotypes specifically induced by injection or feeding treatments. *Insect Biochem. Mol. Biol.* **51**, 20–32.

Schmitz, A., Anselme, C., Ravallec, M., Rebuf, C., Simon, J.C., Gatti, J.L. and Poirié, M. (2012) The cellular immune response of the pea aphid to foreign intrusion and symbiotic challenge. *PLoS ONE*, **7**, e42114.

Suzuki, Y., Truman, J.W. and Riddiford, L.M. (2008) The role of Broad in the development of *Tribolium castaneum*: implications for the evolution of the holometabolous insect pupa. *Development*, **135**, 569–577.

Tjallingii, W.F. (1978) Electronic recording of penetration behavior by aphids. *Entomol. Exp. Appl.* **24**, 721–730.

Tjallingii, W.F. (1988) Electrical recording of stylet penetration activities. In *Aphids: Their Biology, Natural Enemies and Control, Vol. 2B. World Crop Pests* (Minks, A.K. and Harrewijn, P., eds), pp. 95–108. Amsterdam: Elsevier.

Tjallingii, W.F. (2006) Salivary secretions by aphids interacting with proteins of phloem wound responses. *J. Exp. Bot.* **57**, 739–745.

Tomoyasu, Y., Miller, S.C., Tomita, S., Schoppmeier, M., Grossmann, D. and Bucher, G. (2008) Exploring systemic RNA interference in insects: a genome-wide survey for RNAi genes in Tribolium. *Genome Biol.* **9** (1): R10.1 R10.22, doi:10.1186/gb-2008-9-1-r10.

Vandermoten, S., Harmel, N., Mazzucchelli, G., De Pauw, E., Haubruge, E. and Francis, F. (2014) Comparative analyses of salivary proteins from three aphid species. *Insect Mol. Biol.* **23**, 67–77.

Vaucheret, H. (2006) Post-transcriptional small RNA pathways in plants: mechanisms and regulations. *Genes Dev.* **20**, 759–771.

Vaucheret, H. and Fagard, M. (2001) Transcriptional gene silencing in plants: targets, inducers and regulators. *Trends Genet.* **17**, 29–35.

Vaucheret, H., Vazquez, F., Crété, P. and Bartel, D.P. (2004) The action of ARGONAUTE1 in the miRNA pathway and its regulation by the miRNA pathway are crucial for plant development. *Genes Dev.* **15**, 1187–1197.

Verheggen, F.J., Haubruge, E., De Moraes, C.M. and Mescher, M.C. (2009) Social environment influences aphid production of alarm pheromone. *Behav. Ecol.* **20**, 283–288.

Wang, Z., Dong, Y., Desneux, N. and Niu, C. (2013) RNAi silencing of the HaHMG-CoA reductase gene inhibits oviposition in the *Helicoverpa armigera* cotton bollworm. *PLoS ONE,* **8**, e67732.

Will, T. and van Bel, A.J.E. (2006) Physical and chemical interactions between aphids and plants. *J. Exp. Bot.* **57**, 729–737.

Will, T. and Vilcinskas, A. (2013) Aphid proof plants: biotechnical approaches for aphid control. *Adv. Biochem. Eng. Biotech.* **136**, 179–204.

Will, T., Steckbauer, K., Hardt, M. and van Bel, A.J.E. (2012) Aphid gel saliva: sheath structure, protein composition and secretory dependence on Stylet Tip Milieu. *PLoS ONE,* **7**, e46903.

Will, T., Furch, A.C.U. and Zimmermann, M.R. (2013) How phloem-feeding insects face the challenge of phloem-located defenses. *Front. Plant Sci.* **4**, 336.

Xu, L., Xiaoliang, D., Yanhua, L., Xiaohua, Z., Zhansheng, N., Chaojie, X., Zhongfu, N. and Rongqi, L. (2014) Silencing of an aphid carboxylesterase gene by use of plant mediated RNAi impairs *Sitobion avenae* tolerance of Phoxim insecticides. *Transgenic Res.* **23**, 389–396.

Zhang, M., Yuwen, Z., Hui, W., Huw, D.J., Qiang, G., Dahai, W., Youzhi, M. and Lanqin, X. (2013) Identifying potential RNAi targets in grain aphid (*Sitobion avenae* F.) based on transcriptome profiling of its alimentary canal after feeding on wheat plants. *BMC Genom.* **14**, 560.

A fusion protein derived from plants holds promising potential as a new oral therapy for type 2 diabetes

Jeehye Choi[1,†], Hong Diao[2,†], Zhi-Chao Feng[2,3], Arthur Lau[4], Rennian Wang[2,3], Anthony M. Jevnikar[2,5] and Shengwu Ma[1,2,5,*]

[1]Department of Biology, University of Western Ontario, London, ON, Canada

[2]Lawson Health Research Institute, London, ON, Canada

[3]Department of Physiology & Pharmacology, University of Western Ontario, London, ON, Canada

[4]Department of Pathology, University of Western Ontario, London, ON, Canada

[5]Plantigen Inc., London, ON, Canada

*Correspondence

email sma@uwo.ca

[†]These authors contributed equally to this work.

Keywords: plant green bioreactor, incretin hormone, exendin-4, transferrin, fusion protein, novel antidiabetic therapy.

Summary

The incretin hormone glucagon-like peptide-1 (GLP-1) is recognized as a promising candidate for the treatment of type 2 diabetes (T2D), with one of its mimetics, exenatide (synthetic exendin-4) having already been licensed for clinical use. We seek to further improve the therapeutic efficacy of exendin-4 (Ex-4) using innovative fusion protein technology. Here, we report the production in plants a fusion protein containing Ex-4 coupled with human transferrin (Ex-4-Tf) and its characterization. We demonstrated that plant-made Ex-4-Tf retained the activity of both proteins. In particular, the fusion protein stimulated insulin release from pancreatic β-cells, promoted β-cell proliferation, stimulated differentiation of pancreatic precursor cells into insulin-producing cells, retained the ability to internalize into human intestinal cells and resisted stomach acid and proteolytic enzymes. Importantly, oral administration of partially purified Ex-4-Tf significantly improved glucose tolerance, whereas commercial Ex-4 administered by the same oral route failed to show any significant improvement in glucose tolerance in mice. Furthermore, intraperitoneal (IP) injection of Ex-4-Tf showed a beneficial effect in mice similar to IP-injected Ex-4. We also showed that plants provide a robust system for the expression of Ex-4-Tf, producing up to 37 μg prEx-4-Tf/g fresh leaf weight in transgenic tobacco and 137 μg prEx-4-Tf/g freshweight in transiently transformed leaves of N. benthamiana. These results indicate that Ex-4-Tf holds substantial promise as a new oral therapy for type 2 diabetes. The production of prEx-4-Tf in plants may offer a convenient and cost-effective method to deliver the antidiabetic medicine in partially processed plant food products.

Introduction

Diabetes is a global epidemic that is expected to affect approximately 350 million people by 2030 (Collins et al., 2011). Type 2 is the most common form of diabetes, accounting for about 90%–95% of all cases of the disease. Type 2 diabetes is a costly disease, associated with substantial morbidity and mortality, particularly from its cardiovascular complications (Potenza et al., 2011; Zhang P et al., 2010). Type 2 diabetes is characterized by insulin resistance, impaired glucose-induced insulin secretion and inappropriately regulated glucagon secretion which in combination eventually leads to the development of hyperglycaemia (Alonso-Magdalena et al.,2011). Although several classes of antidiabetic drugs are available, achieving and maintaining long-term glycaemic control is often challenging, and many current agents have treatment-limiting side effects (Molitch, 2013). Therefore, the development of new drugs to treat diabetes and to prevent associated complications is necessary and urgent.

Incretin-based therapies represent an important and major step forward in the development of new treatments for type 2 diabetes (Koliaki and Doupis, 2011; Schwartz and Defronzo, 2013). Incretins are hormones released from the gut in response to nutrient ingestion that potentiate glucose-stimulated insulin secretion. The predominant incretin hormone is glucagon-like peptide-1 (GLP-1). In addition to stimulating insulin secretion,

GLP-1 suppresses glucagon release, slows gastric emptying, improves insulin sensitivity and reduces food intake (Mudaliar and Henry, 2012). However, native GLP-1 has a very short physiological half-life due to degradation by dipeptidyl peptidase-4 (DPP-4). GLP-1 survives <90 s following intravenous infusion and 1 h after subcutaneous injection (Holst and Gromada, 2004). Therefore, to be clinically effective as a diabetes treatment, native GLP-1 would have to be given as a continuous infusion. For this reason, the clinical focus for potential treatments for T2D has shifted towards long-acting GLP-1 receptor agonists and DPP-4 inhibitors.

Exendin-4 (Ex-4), a 39 amino acid peptide isolated from salivary secretions of Gila monster, displays 52% identity to mammalian GLP-1 and is a potent and long-acting agonist of GLP-1 receptor (Lovshin and Drucker, 2009). Ex-4 is resistant to degradation by DPP-IV and consequently has a longer half-life compared with GLP-1 (Giorgino et al., 2007). Ex-4 increased insulin production by β-cells in response to glucose, promoted β-cell regeneration and protected against apoptosis (Hadjiyanni et al., 2008; Lovshin and Drucker, 2009). The synthetic version of Ex-4, exenatide, was approved by the Food and Drug Administration (FDA) in 2005 for the treatment of T2D and was the first marketed GLP-1 receptor agonist (Kyriacou and Ahmed, 2010; Robles and Singh-Franco, 2009). While exenatide has demonstrated its value in treating patients with T2D, its therapeutic utility is limited due to the frequent injections required (twice daily), thus making it an inconvenient and expensive treatment (Garber, 2011). If an effective oral incretin mimetic is developed, it could lead to numerous benefits including convenience, greater patient satisfaction and compliance, and fewer side effects compared with the same drug that is administered systemically.

We are keenly interested in developing transferrin fusion protein technology as a novel strategy to improve the therapeutic efficacy of protein and peptide drugs and as well to achieve the preferred noninvasive oral delivery of biopharmaceuticals. Transferrin (Tf) is an abundant, naturally occurring serum protein with the capacity to bind and transport iron to cells through Tf receptor-mediated endocytosis. Tf receptor (TfR) is present on the surface of most proliferating higher eukaryotic cells. TfRs are also highly expressed in human GI epithelial cells. Furthermore, Tf is stable and resistant to proteolytic enzymes, which leads to a tremendously long plasma half-life (14–17 days) (Brandsma et al., 2011; Li and Qian, 2002; Melanie, 2005). These properties of human Tf make it a very valuable tool in developing Tf-based novel fusion protein technology to enhance protein expression, extend the serum half-lives of protein/peptide drugs and to achieve active targeted (smart) drug delivery. Being an endogenous protein, Tf as a drug delivery system is nonimmunogenic and nontoxic in humans (Jiang et al., 2007).

In recent years, plants have emerged as an attractive alternative expression platform for the production of recombinant pharmaceutical proteins and offer several advantages. Given that plant cells are eukaryotic cells, the cells are able to perform many of the post-translational protein modifications (glycosylation, disulphide bond formation and protein processing and folding) necessary for the biological functions of many mammalian proteins. Moreover, plant cells do not harbour human or zoonotic pathogens, making them a safe host for the production of biopharmaceuticals. Also, plant production of foreign proteins can be targeted into edible parts of the plant, which allows for the possibility of oral administration of therapeutic proteins in partially processed plant food products, reduces the cost of manufacturing and therefore decreases the cost of therapy for the patient (Ma et al., 2004; Ma and Wang, 2012; Tremblay et al., 2010). Recently, Kwon et al. (2012) reported the production of Ex-4 as a fusion protein with cholera toxin B subunit (CTB) in tobacco chloroplasts and demonstrated that oral delivery of plant-made Ex-4-CTB lowers blood glucose levels in mice. While CTB has proven to be an effective carrier system for transmucosal delivery of biopharmaceuticals, CTB is in itself a potent immunogen and adjuvant that elicit strong systemic and mucosal antibody responses (Kim et al., 1998). Accordingly, CTB is mostly employed as a carrier system for the delivery of oral vaccine antigens, exploiting the intrinsic immunogenic and adjuvant properties of CTB to increase immunogenicity of a linked vaccine. However, the intrinsic adjuvanticity and immunogenicity of CTB could become detrimental in cases where CTB is being used to deliver an oral

protein/peptide drug to which its therapeutic effect is not related to or dependent on the induction of a specific immune response. The development of unwanted immune response against a therapeutic protein can have serious clinical consequences (Chirino et al., 2004).

In this study, we report the production of Ex-4-Tf in plants using stable and transient systems and the characterization of plant-derived Ex-4-Tf (prEx-4-Tf). We show that prEx-4-Tf retain functions of both fusion partners. In particular, the fusion protein stimulated insulin secretion from pancreatic β-cells, promoted proliferation of β-cells, stimulated differentiation of pancreatic precursor cells into insulin-producing β-cells, retained the ability to internalize into human intestinal epithelial cells and resisted stomach acid and proteolytic enzymes. Importantly, oral administration of partially purified prEx-4-Tf significantly improved glucose tolerance, whereas administration of commercial Ex-4 via the same route failed to show any significant improvement in glucose tolerance in mice. Furthermore, intraperitoneal (IP) injection of prEx-4-Tf showed a beneficial effect in mice similar to IP-injected Ex-4. We also demonstrated that plants provide a robust system for the expression of Ex-4-Tf, producing up to 37 µg prEx-4-Tf/g fresh leaf weight in T_0 transgenic tobacco plants and 137 µg prEx-4-Tf/g freshweight in transiently transformed leaves of N. benthamiana. These results indicate that Ex-4-Tf holds substantial promise as a new oral therapy for T2D. The production of Ex-4-Tf in plants may offer a convenient and cost-effective method to deliver the antidiabetic medicine in partially processed plant food products.

Results

Vector construction

The plant transformation vector pBI101.1-Ex-4-Tf was constructed as described under Experimental Procedures. As shown in Figure 1, Ex-4 was fused in-frame to the N-terminal end of the mature form of human Tf through a flexible linker to minimize the steric hindrance between the two fusion partners. To ensure correct N-terminal processing of the fusion protein in plant cells to obtain a final chimeric product containing histidine at the N-terminus, a barley alpha amylase signal peptide (SP) was added to the fusion protein. The tobacco etch virus (TEV) RNA 5' untranslated leader (UTL) sequence as well as a 3'endoplasmic reticulum (ER) retention signal motif KDEL was incorporated onto pBI101.1-Ex-4-Tf to maximize fusion protein accumulation (Brandsma et al., 2010; Wang et al., 2008). A 6xHis tag was included to facilitate fusion protein purification.

Stable and transient expression of Ex-4-hTf fusion protein in plants

Low-alkaloid tobacco (cultivar 81V9) was stably transformed using Agrobacterium-mediated transformation. Over 25 kanamycin-resistant primary transgenic lines (T_0 generation) were

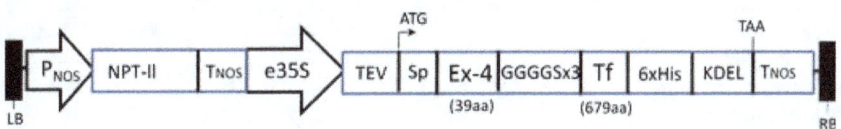

Figure 1 Diagrammatic representation of the T-DNA region of vector pBI101.1-Ex-4-Tf. RB, right border; LB, left boarder; PNOS, nopaline synthase promoter; npt II, neomycin phosphotransferase gene; TNOS, nopaline synthase terminator; e35S, enhanced double CaMV 35S promoter; TEV, tobacco etch virus 5'leader sequence; SP, barley α-amylase signal peptide; Ex-4, a synthetic gene encoding 39 amino acids (aa); GGGGSx3, linker sequence; Tf, Tf cDNA encoding a mature protein of 679 aa; 6xHis, six consecutive histidine residues; KDEL, ER-retention signal.

generated. The expression of Ex-4-Tf fusion protein in putative T_0 transgenic tobacco plants was analysed by Western blot using a commercial anti-human Tf antibody. As shown in Figure 2a, the anti-Tf antibody detected a protein band of 76 kDa, the expected size of the fusion protein. No protein of similar size was detected in extracts from wild-type plants. The yields of prEx-4-Tf in the T_0 stable transgenic tobacco reached up to 37 µg/g fresh leaf weight.

Transient expression of Ex-4-Tf in *N. benthamiana* was achieved with the co-infiltration of *Agrobacterium* harbouring the gene for p19, a viral protein that inhibits RNA silencing and is known to boost the yield of transiently expressed proteins (Lakatos *et al.*,2004; Tremblay *et al.*, 2011). As shown in Figure 2b, the expression of Ex-4-Tf in infiltrated leaves reached the highest level at day 6 postinfiltration (dpi), with yields upto 137 µg prEx-4-Tf/g freshweight.

Plant-derived Ex-4-Tf-stimulated glucose-dependent insulin secretion from pancreatic beta cells

Ex-4 is known to stimulate the release of insulin from pancreatic beta cells in a glucose-dependent fashion (Holst and Gromada, 2004). We therefore investigated whether prEx-4-Tf retained the ability of Ex-4 to stimulate glucose-dependent insulin secretion in the mouse β-cell line MIN6. To test its effect, the fusion protein was partially purified from leaves of stably transformed tobacco plants using a HiTrap Chelating HP column (around 50% recovery) and the purity confirmed by SDS-PAGE gel/coomassie blue staining (Figure 3). Chen *et al.* (2012) demonstrated that Ex-4 at 10–15 nM stimulated significant insulin

secretion from MIN6 cells. Thus, we tested the effect of prEx-4-Tf on insulin secretion from MIN6 cells under different concentrations (15, 5 and 1 nM). As shown in Figure 4, in the presence of 10 mM glucose, prEx-4-Tf stimulated insulin secretion from MIN6 cells in a dose-dependent manner. Commercial Ex-4 was used as a positive control. As expected, incubation of MIN6 cells with Ex-4 resulted in a dose-dependent increase in insulin secretion. Interestingly, Ex-4-Tf appears to stimulate insulin secretion from pancreatic MIN6 cells significantly more than the commercial Ex-4 of the same concentration ($P < 0.05$). No stimulatory effect on insulin secretion was observed in MIN6 cells cultured with plant-derived Tf alone. Both prEx-4-Tf and commercial Ex-4 had no effect on insulin secretion in the absence of glucose in MIN6 cells (data not shown), suggesting their glucose-dependent action.

Plant-derived Ex-4-Tf-stimulated proliferation of pancreatic beta cells

We further examined whether prEx-4-Tf retained the ability to stimulate pancreatic β-cells to proliferate using INS-1 cells. INS-1 cells were chosen as a β-cell line model because unlike slow-growing glucose-responsive β-cell lines such as MIN6, INS-1 cells grow at a rate that allow sensitive measurement of rate of DNA synthesis using a [³H]thymidine incorporation assay (Asfari *et al.*, 1992; Gahr *et al.*, 2002; Hügl *et al.*, 1998). We tested the effect of prEx-4-Tf on glucose-dependent INS-1 cell proliferation at several concentrations (10, 1, 0.1, 0.01, 0.001 and 0.0001 nM). A previous study indicated that the threshold concentration for GLP-1 to induce INS-1 cell proliferation was in the range of

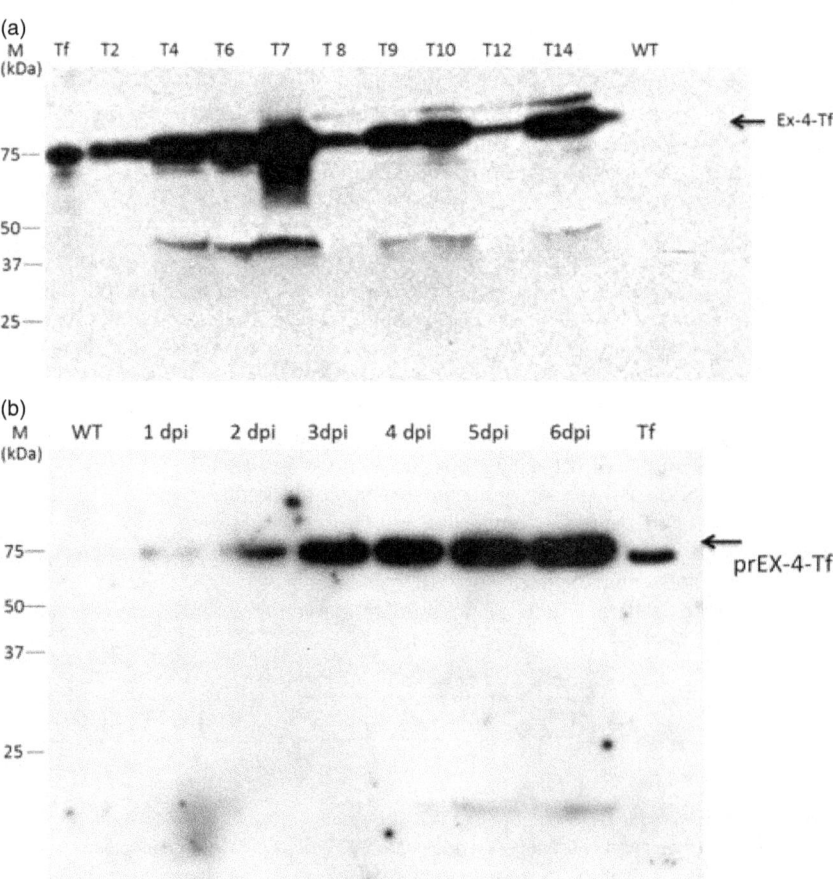

Figure 2 Western blot analysis of prEx-4-Tf expression in transgenic tobacco (a) and in infiltrated *N. benthamiana* leaves (b). Protein samples (40 µg per well) were separated by 10% SDS-PAGE, transferred onto PVDF membrane and probed with anti-hTf antibodies. T2 to T14, representatives of the individual T0 transgenic lines; Tf, commercial Tf (13 ng/well); dpi, days of postinfiltration. The position of prEx-4-Tf is indicated by the arrow.

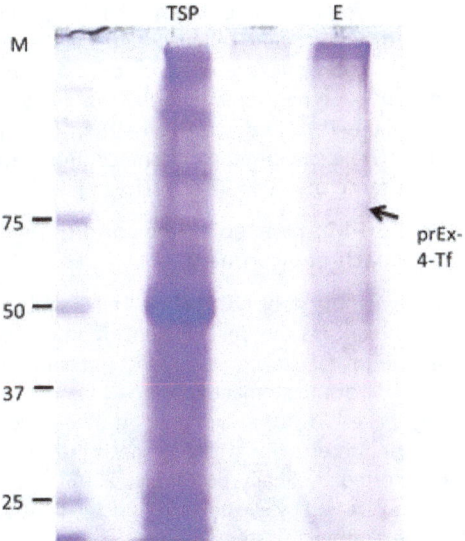

Figure 3 Coomassie-stained 10% SDS-PAGE gel showing the partial purification of prEx-4-Tf. The fusion protein was partially purified by affinity chromatography. M, protein markers; TSP, total soluble protein (100 μg loaded); E, eluate from HiTrap™ Chelating HP Column (750 ng loaded).

INS-1 cells towards the impurities present in partially purified prEx-4-Tf (Figure 3). Interestingly, prEx-4-Tf displayed a stimulatory effect on INS-1 cell proliferation even in the absence of glucose. prEx-4-Tf at 0.001 nM, without glucose, induced ~2.5-fold increase in INS-1 cell proliferation. There was little effect of plant-derived Tf (prTf) alone on INS-1 cell proliferation in the presence or absence of glucose (data not shown), suggesting that Tf itself has no effect on β-cell proliferation. We also tested the effect of commercial Ex-4 on INS-1 cell proliferation using the same concentrations. As shown in Figure 5b, Ex-4 enhanced glucose-induced proliferation of INS-1 cells in a dose-dependent manner as expected. However, prEx-4-Tf appears to be a more potent stimulator of INS-1 cell proliferation than Ex-4, as only prEx-4-Tf demonstrated considerable stimulatory effect on INS-1 cell proliferation at concentrations below 0.1 nM (Figure 5a,b).

Plant-derived Ex-4-Tf induced pancreatic precursor cells to differentiate into insulin-producing beta-like cells

We next examined whether prEx-4-Tf retained the ability to promote differentiation of pancreatic precursor cells into insulin-secreting cells using rat pancreatic ductal cells (ARIP) as a model. Previous work has shown that GLP-1 treatment can induce the differentiation of ARIP cells into insulin-synthesizing

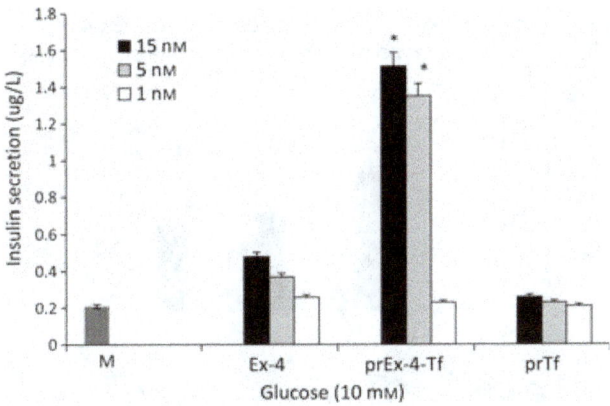

Figure 4 Effect of prEx-4-Tf on insulin secretion from MIN6 cells. MIN6 cells were treated with prEx-4-Tf or commercial Ex-4 at different concentrations in the presence of 10 mM glucose. Supernatants from treated cells were collected and assayed for insulin concentration using insulin ELISA Kits. The assay was performed in triplicates and repeated. Data were expressed as means ± SEM. *, statistically significant difference ($P < 0.05$) in insulin release from MIN 6 cells induced by prEx-4-Tf vs. commercial Ex-4.

10^{-12}–10^{-11}M (Buteau et al., 1999). As shown in Figure 5a, glucose by itself was able to stimulate INS-1 cell proliferation in a dose-dependent manner, and the addition of prEx-4-Tf further increased this glucose-induced INS-1 cell proliferation. For example, glucose at 8 mM increased INS-1 cell proliferation by ~9-fold compared with the control (i.e. medium without glucose). Upon addition of 0.001 nM or 0.0001 nM prEx-4-Tf, INS-1 cell proliferation increased ~8- and 5-fold, respectively, above the proliferation observed at 8 mM glucose alone, suggesting a dose-dependent effect of prEx-4-Tf. The addition of higher concentrations of partially purified prEx-4-Tf (≥0.1 nM) was found to be toxic causing considerable cell death due to the high sensitivity of

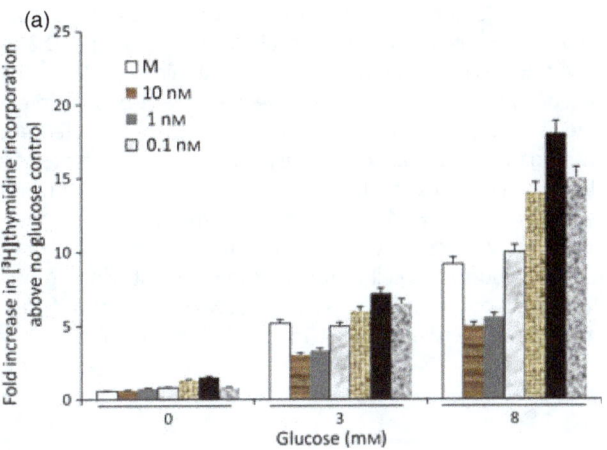

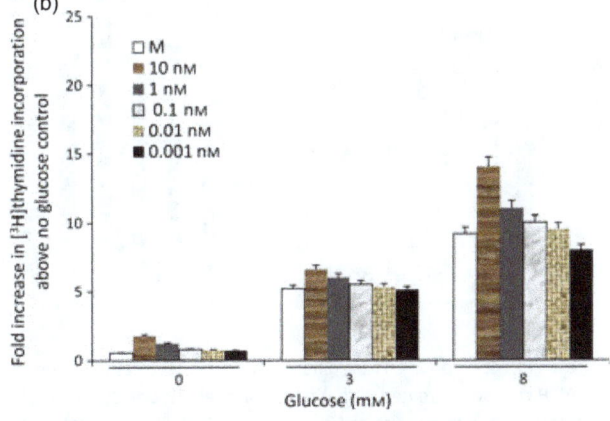

Figure 5 Effect of prEx-4-Tf on pancreatic β-cell proliferation. INS-1 cells were incubated with different concentrations of prEx-4-Tf (a) or commercial Ex-4 (b) in the presence of different glucose concentrations. Cell proliferation was determined by [³H]thymidine incorporation. Data were expressed as means ± SEM. The assay was performed in triplicate and repeated. M, medium only.

β-cells (Hui *et al.* (2001). ARIP cells were cultured with different concentrations of prEx-4-Tf (10, 1, 0.1, 0.01, 0.001 and 0.0001 nᴍ) for 48 h in the presence of 12 mᴍ glucose. The supernatant was assayed for insulin secretion by ELISA. As shown in Figure 6, insulin levels in the supernatants of ARIP cells treated with 0.01, 0.001 nᴍ or 0.0001 nᴍ Ex-4-Tf were all significantly elevated compared with the basal insulin level measured in supernatants of ARIP cells grown in 12 mᴍ glucose without prEx-4-Tf ($P < 0.05$). Partially purified prEx-4-Tf at higher concentrations (≥0.1 nᴍ) was toxic to ARIP cells. The treatment of ARIP cells with Ex-4 of the same concentrations were used as controls. As expected, Ex-4 induced insulin secretion in a dose-dependent manner (Figure 6). As prEx-4-Tf, but not Ex-4, at concentrations below 0.01 nᴍ were still capable of inducing insulin secretion, prEx-4-Tf appears to be more potent than Ex-4 in promoting the differentiation of ARIP cells into insulin-secreting cells.

Plant-derived Ex-4-Tf retained the ability to be internalized into human intestinal epithelial cells

Internalization into mammalian cells is an important property of Tf (Li and Qian, 2002). Therefore, we investigated whether prEx-4-Tf would carry the ability of Tf to get internalized using the human intestinal epithelial cell line Caco-2, an *in vitro* cell model widely used to study cellular uptake, transport, metabolism processes or oral bioavailability of drug candidates (Angelis and Turco, 2011). Caco-2 cells were incubated with 1.0 μg or 0.2 μg/mL of prEx-4-Tf, prTf or Tf standard for 60 min. Cells were then treated with acid buffer to remove membrane-bound foreign proteins. Following cell lysis, the supernatant was collected after centrifugation and analysed for the presence of target proteins by ELISA. The ELISA results shown in Figure 7a, clearly demonstrate the presence of prEx-4-Tf, prTf or Tf in the lysate supernatant from the treated cells. A correlation between the amount of the protein added to the cell culture and the amount of the protein internalized into cells was also observed. No significant difference was observed between prEx-4-Tf and prTf in their ability to get

internalized. However, compared with Tf standard, both prEx-4-Tf and prTf had a significant reduction in their ability to get internalized ($P < 0.05$). The presence of prEX-4Tf, prTf or Tf standard in the supernatant was additionally visualized by Western blot (Figure 7b). No detectable levels of endogenous Tf were found in lysate supernatants from control Caco-2 cells grown in the absence of prEx-4-Tf or prTf.

Plant-derived Ex-4-Tf resisted acidic conditions of simulated stomach environment

We also investigated the stability of prEx-4-Tf in a simulated acidic stomach environment, as the development of an oral version of Ex-4 requires the protein to be stable and remain intact in the acidic environment of the stomach (pH 1.5–3.5). This was tested by subjecting prEx-4-Tf to a synthetic gastric fluid solution containing pepsin, an enzyme in the stomach that breaks down proteins, as reported by Shaji and Patole (2008) and Tremblay *et al.* (2010). As revealed by Westen blot, prEx-4-Tf was found to be stable in the synthetic gastric fluid solution (Figure 8). Plant-derived Tf was used as a control. There was no difference observed in the stability against digestion by the synthetic gastric fluid between the two proteins.

Oral and intraperitoneal routes of administration of partially purified Ex-4-Tf improved glucose tolerance in mice

We first investigated whether oral delivery of prEx-4-Tf could enhance glucose metabolism, as one of our major goals was to

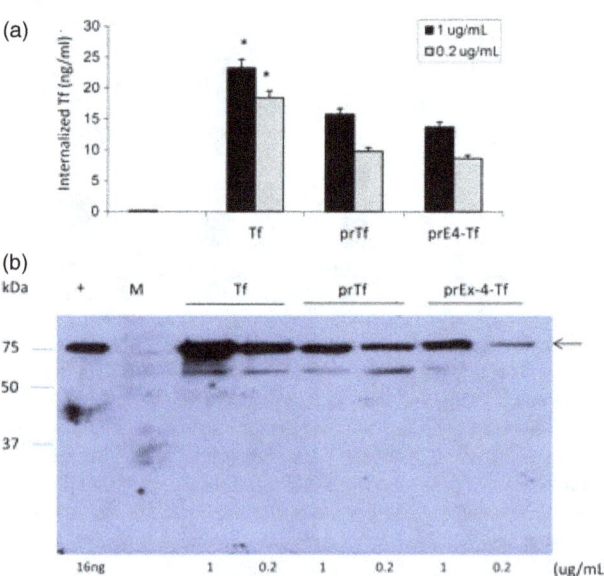

Figure 7 Analysis of cellular uptake of prEx-4-Tf. Caco-2 cells were incubated with 1.0 μg/mL (1) or 0.2 μg/mL (0.2) of prEx-4-Tf, prTf or Tf. Following incubation, the cell membrane-bound foreign protein was removed. Cells were lysed, and supernatants were collected after centrifugation and assayed for the concentrations of internalized target proteins by ELISA (a) and Western blot (b). The ELISA was performed in triplicate and repeated. Data were expressed as the mean with standard deviation. For Western blot, 6 μg of total protein was separated on a 10% SDS-PAGE. +, Tf standard (16 ng); *, statistically significant difference ($P < 0.05$) in concentrations of internalized Tf standard in lysate supernatants from Tf treated cells compared with concentrations of internalized prEx-4-Tf or prTf in lysates from prEx-4-Tf or prTf-treated cells. The position of internalized proteins is indicated by the arrow.

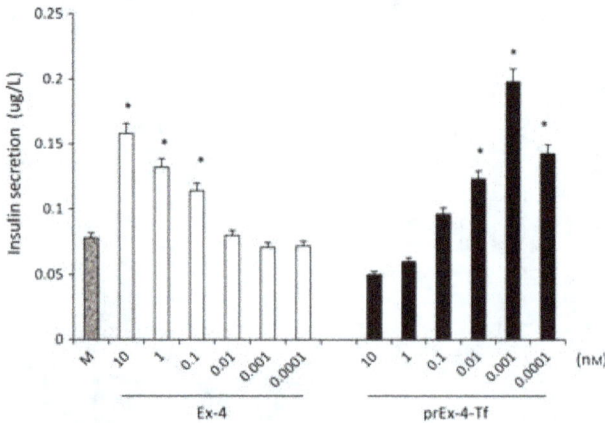

Figure 6 Effect of prEx-4-Tf on differentiation of pancreatic precursor cells. ARIP cells were treated with different concentrations of prEx-4-Tf or Ex-4 for 48 h in the presence of 12 mᴍ glucose (detailed in the section of Experimental Procedures). Supernatants from treated cells were collected and assayed for insulin concentrations using a commercial insulin ELISA Kit. M, medium containing 12 mᴍ glucose only. *, statistically significant difference ($P < 0.05$) in levels of insulin in supernatants from cells treated with prEx-4-Tf or Ex-4 compared with basal insulin levels measured in the supernatant from untreated control cells.

develop novel Ex-4 derivatives that can be used as safe and effective oral antidiabetic medicines. To determine its effect, fasting plasma glucose levels of C57BL/6J male mice before and after receiving oral prEx-4-Tf were measured and compared with those in the control and Ex-4 group. In the Ex-4-Tf group, the plasma glucose level 1 h after oral treatment tended to be lower when compared with basal levels before treatment ($P = 0.07$, Figure 9a), indicating that oral administration of prEx-4-Tf at 1.5 µg/g did not significantly affect basal plasma glucose levels. This data suggest that unlike many conventional antidiabetic medicines, no side effect risks such as hypoglycaemia are associated with the use of prEx-4-Tf in rodent. No differences were observed in fasting plasma glucose levels in mice from both control and Ex-4 groups before and after oral treatment. While IPGTT results revealed that all experimental groups exhibited a plasma glucose peak at 30 min after glucose loading followed by a gradual return to basal levels by 120 min (Figure 9b), the prEx-4-Tf-treated mice displayed significantly improved glucose tolerance, as demonstrated by decreased area under the curve (AUC) compared with the control ($P < 0.01$, Figure 9c) or Ex-4 group

($P < 0.05$, Figure 9c). No significant difference in glucose clearance was observed between the control group and the Ex-4 group.

We also studied the effect of the intraperitoneally delivered prEx-4-Tf on glucose metabolism in mice. No changes in fasting plasma glucose levels was observed among the experimental groups, although both Ex-4-Tf and Ex-4 groups tended to have lower fasting plasma glucose levels when measured at 30 min post-treatment (Figure 10a). However, IPGTT results showed a significant enhancement in glucose clearance, occuring at 15 min post-glucose loading, in both Ex-4-Tf and Ex-4 groups, as evidenced by lower AUC compared with control group (Ex-4-Tf group vs. control, $P < 0.01$; Ex-4 group vs. control, $P < 0.001$) (Figure 10b,c). No significant difference in IPGTT was observed between the prEx-4-Tf group and the Ex-4 group.

Discussion

In this work, we produced in plants Ex-4 as a fusion protein with Tf as a strategy for achieving novel effective oral delivery of prEx-

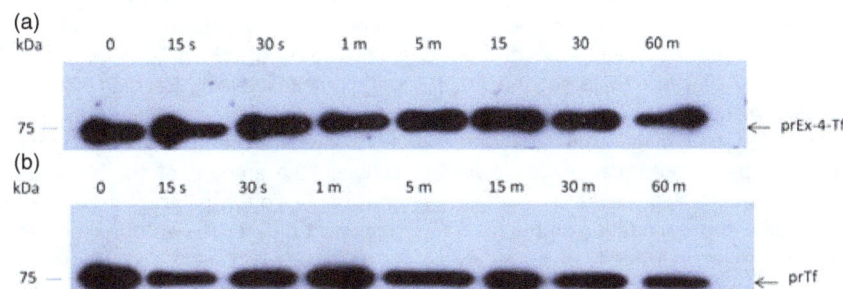

Figure 8 Digestion of partially purified prEx-4-Tf in synthetic gastric fluid (pH 2.5). Partially purified prEx-4-Tf (a) or prTf (b) was digested in the synthetic gastric fluid for different lengths of time. Digestion stopped by adding neutralization buffer. The digested samples were separated by 10% SDS-PAGE, transfered to membrane and probed with anti-Tf antibodies. Each lane contains the starting equivalent of approximately 100 ng purified protein.

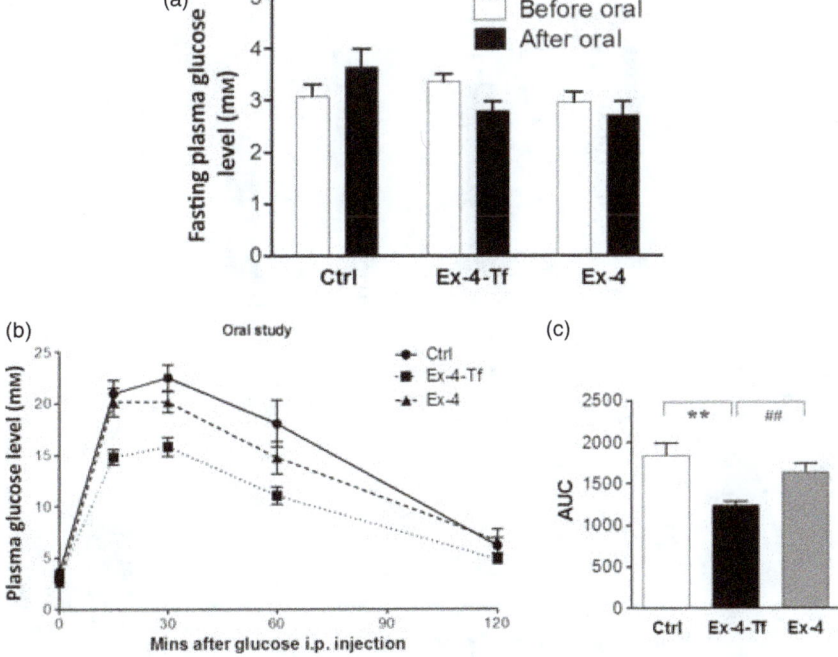

Figure 9 Effect of oral administration of partially purified prEx-4-Tf on plasma glucose homoeostasis in mice. (a) fasting plasma glucose and (b) IPGTT of mice treated with sterile saline (Ctrl), partially purified prEx-4-Tf (Ex-4-Tf) or commercial Ex-4 (Ex-4) via an oral route. (c) Glucose responsiveness of the corresponding experimental groups is shown as a measurement of area under the curve (AUC) of the IPGTT graphs with units of mM/min. Data are expressed as means ± SEM ($n = 5–6$). ** $P < 0.01$ (One-way ANOVA) and ## $P < 0.05$ (unpaired Student's t-test) versus the ctrl group.

4-Tf for the treatment of T2D. Currently, exenatide (synthetic exendin-4) is only available in injectable form and moreover, twice daily injections are required to achieve near-normal glucose control in T2D patients. This makes the incretin-based treatment to be inconvenient, expensive and stressful. An oral version of exenatide would be more advantageous, enhancing ease in administration, reducing side effects and improving patient acceptance and compliance (Brandsma et al., 2011). We demonatrated that prEx-4-Tf exhibited the activity of both EX-4 and Tf. We also showed that the antidiabetic activity of Ex-4 was also enhanced by fusion to Tf. More importantly, we demonstrated that oral administration of partially purified prEx-4-Tf significantly improved glucose tolerance, whereas commercial Ex-4 administered by the same oral route failed to show any significant improvement in glucose tolerance in mice. These findings have important implications. One implication is that prEx-4-Tf has good potential to be developed into a new effective oral antidiabetic medicine. Another implication is that the production of Ex-4-Tf in plants may offer a cost-effective method to deliver oral prEx-4-Tf in partially processed food products. Indeed, preliminary results from our newest animal feeding trial to test whether oral consumption of unprocessed prEx-4-Tf tobacco leaves can have an effect indicated that mice fed raw leaf material for short time showed some improvement in glucose tolerance compared with mice fed wild-type leaf material or normal feed, although the difference did not reach statistical significance (data not shown), supporting the feasibility of using plants as a vehicle to deliver oral prEx-4-Tf. It should be mentioned that in this new trial, the amount of prEx-4-Tf delivered per mouse daily was lower compared with the amount of prEx-4-Tf delivered as partially purified protein by oral gavage, due to the limitation on the plant material that can be added into animal feed as a dietary supplement.

Our in vitro cellular assay data showed that the antidiabetic activity of prEx-4-Tf was improved compared with commercial Ex-4. This is highly likely to have resulted from increased stability of Ex-4 acquired through its fusion to Tf. Human Tf is very stable, possessing a t1/2 in excess of 14–17 days (Brandsma et al., 2011;

Li and Qian, 2002). Therefore, if a fusion molecule between Ex-4 or any peptide and Tf can be created, the stabilizing properties of Tf will likely be imbued to the peptide. The results of our in vitro stability study showed that prEx-4-hTf or prTf was stable and resistant to synthetic digestive fluids (Figure 8), further supporting this assertion. Correct N-terminal processing of prEx-4-Tf may also play a major role in assuring the maximal antidiabetic activity of the fusion protein. The N-terminal histidine (His) of GLP-1 or Ex-4 has been shown to be critical for pancreatic receptor binding and insulinotropic activity. Removal of this N-terminal His from GLP-1or Ex-4 or its replacement with other amino acids was shown to result in drastic loss of its receptor binding and insulinotropic activity (Xiao et al., 2001; Kieffer and Habener, 1999). Previously, we expressed GLP-1as a large multimeric protein (GLP-1 × 10) in plants (Brandsma et al., 2009). As the N-terminus of GLP-1 × 10 contains two additional amino acids (Met and Gly) serving as part of a translational start site, such an N-terminal extension had a significantly negative impact on GLP-1 activity. To ensure that plant-derived Ex-4-Tf has His at its N-terminus, Ex-4-hTf was fused with barley α-amylase signal peptide for effective processing. We have previously demonstrated that barley α-amylase signal peptide is capable of directing efficient and accurate processing of human interleukin-4 when expressed in plants (Ma et al., 2005).

The ability of prEx-4-Tf to get internalized was demonstrated using human intestinal Caco-2 cells. The level of internalized prEx-4-Tf in Caco-2 cell lysates was quantified by ELISA and confirmed by Western blotting (Figure 7). There was good correlation between the amount of the protein added to the cell culture and the amount of the protein internalized into cells. No significant difference was observed between prEx-4-Tf and prTf; however, both prEx-4-Tf and prTf had significant reduction in their ability to get internalized compared with Tf standard (P < 0.05). Although further studies are needed, the difference between Tf standard and prEx-4-Tf or prTf may be due to difference in protein quality. Tf standard used in this study was in highly purified form, whereas prEX-4-Tf or prTf was only in partially purified form. We (Brandsma et al., 2010) and others (Zhang D et al., 2010) have

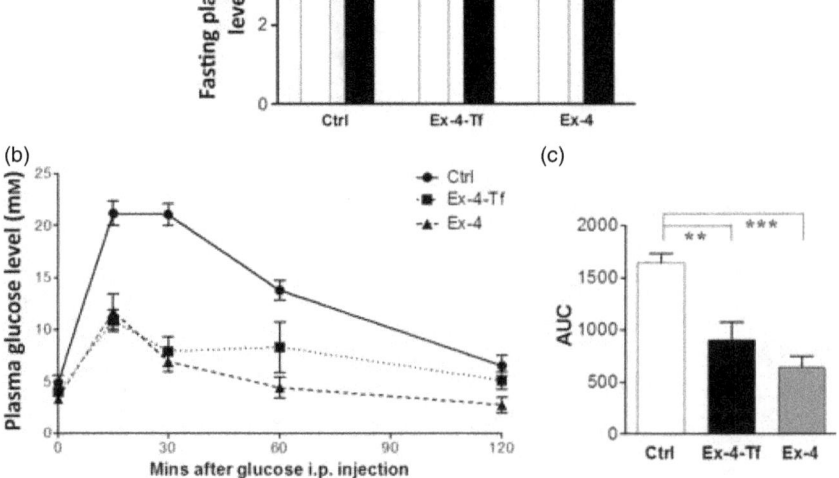

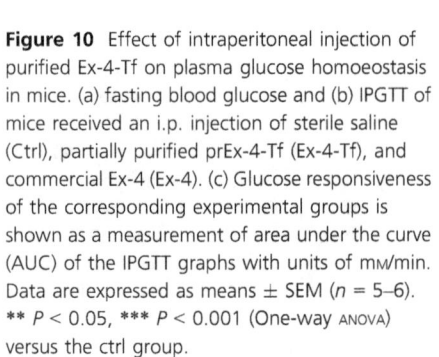

Figure 10 Effect of intraperitoneal injection of purified Ex-4-Tf on plasma glucose homoeostasis in mice. (a) fasting blood glucose and (b) IPGTT of mice received an i.p. injection of sterile saline (Ctrl), partially purified prEx-4-Tf (Ex-4-Tf), and commercial Ex-4 (Ex-4). (c) Glucose responsiveness of the corresponding experimental groups is shown as a measurement of area under the curve (AUC) of the IPGTT graphs with units of mM/min. Data are expressed as means ± SEM (n = 5–6). ** P < 0.05, *** P < 0.001 (One-way ANOVA) versus the ctrl group.

shown previously that plant-derived human Tf is not glycosylated, though native Tf is a glycoprotein. However, the glycosylation on the Tf has no known influence on receptor binding or any other biological function (Huebers and Finch, 1987). With Ex-4, the fusion protein only had a 4-kDa increase in molecular weight. Therefore, the size of the fusion protein is not likely to be the factor affecting its cellular uptake. Chen et al. (2011) investigated the possible effect of different linkers connecting Tf to growth hormone (GH) or granulocyte colony-stimulating factor (G-CSF) on receptor binding. They showed that the linkers can exert positive or negative effects on the binding affinities of the fusion protein to both protein receptors, thereby affecting subsequent endocytosis of Tf. It would be worthwhile to test other linkers than GGGGSx3 to see if the cellular uptake of prEx-4-Tf can be enhanced.

The data from animal feeding trials revealed that oral administration of partially purified Ex-4-Tf (1.5 μg/g body weight) significantly improved glucose tolerance but had little effect on the basal plasma glucose level in mice (Figure 9a–c). These results suggest that Ex-4-Tf as an oral antidiabetic medicine is not only effective but also carries no or low risk of hypoglycaemia that can lead to seizures, coma and even death. The use of conventional antidiabetic medicines such as insulin is often associated with increased risk of developing hypoglycaemia. The results also suggest that Tf fusion protein technology can be used as an effective strategy to achieve oral delivery of Ex-4 and probably many other therapeutic proteins. In recent years, there has been increasing interest in exploiting Tf fusion protein technology for the development of orally effective peptide and protein drugs. Xia et al. (2000) reported that oral administration of insulin-Tf conjugate (In-Tf) to diabetic rats lowered blood sugar levels. Bai et al. (2005) reported that oral administration of G-CSF and Tf fusion protein promoted proliferation of neutrophils in BDF1 mice. Amet et al. (2010) showed that rats receiving oral human growth hormone hGH-Tf fusion protein (GHT) gained body weight. Recently, Bobst et al. (2012) investigated possible mechanisms underlying the successful oral delivery of GHT through analysis of the fusion protein's stability to proteolysis and its binding affinity for Tf receptors, the key factors in overcoming the primary barriers to successful oral delivery. They showed that in addition to the anticipated monomeric form (GHT1), a significant population of GHT exists in an oligomeric form (GHTx), a form proving to be exceptional stable in gut environment. On the other hand, oligomerized GHT did not affect its binding to Tf receptor.

The relative high-level accumulation of prEx-4-Tf in plants (up to 37 μg/g fresh leaf weight in transgenic tobacco and 140 μg/g freshweight in infiltrated leaves) may be attributed to several factors acting in synergy, including the stable nature of Tf itself, the use of a strong CaMV 35S promoter and the targeting of prEx-4-Tf into the lumen of the ER. Moreover, the use of a flexible linker (GGGGS)x3 between Ex-4 and Tf may also contribute to its higher expression. Trinh et al. (2004) reported a 30-fold increase in the yield of a hybrid protein consisting of a single-chain Fv antibody specific for HER2/neu and the CH3 region of human anti-rat transferrin receptor IgG3 heavy chain when they were connected through the (GGGGS)x3 linker and expressed in a mammalian transient expression system. To achieve higher and more stable expression, we have begun to produce transgenic tobacco lines homozygous for the introduced Ex-4-Tf gene.

Recently, Kim et al. (2010) reported the expression of Ex-4 fused to Tf in yeast and demonstrated that injection of yeast-derived Ex-4-Tf lowered blood glucose levels in mice. While yeast may provide a useful system for the expression of Ex-4-Tf or other therapeutic proteins, a major disadvantage of yeast-based expression system is its comparatively low-expression yield. The vacuoles of yeast cells, the homologue to lysosomes of higher cells, are filled with highly active proteases that can cause rapid degradation of recombinant proteins during cell breakage (Holkeri and Makarow, 1998). Moreover, like bacterial and mammalian cells, yeast requires complex cell culture facilities that are not only expensive but also make the scaling up of pharmaceutical protein production difficult. However, these problems can likely be solved with the use of plants as an alternative expression host.

In summary, we have produced a fusion protein consisting of Ex-4 coupled to Tf in plants. The in vitro analysis showed that prEx-4-Tf retained the activity of both proteins. Oral administration of partially purified prEx-4-Tf significantly improved glucose tolerance but had little effect on basal plasma glucose level, whereas administration of Ex-4 by the same oral route had no effect in mice. Furthermore, IP injection of partially purified prEx-4-Tf showed a beneficial effect similar to IP-injected Ex-4 in mice. These results indicate that prEx-4-Tf holds promising potential as a new oral therapy for type 2 diabetes. The expression of Ex-4-Tf in plants may offer a convenient and cost-effective method to deliver the antidiabetic medicine in partially processed plant food products. This study also suggests that Tf fusion protein technology may offer a powerful new method for improving the therapeutic efficacy of peptide and protein drugs.

Experimental procedures

Construction of Ex-4-Tf fusion protein expression vector

Ex-4 is a 39–amino acid peptide. A synthetic gene encoding Ex-4 was assembled based on published cDNA sequence (Pohl and Wank, 1998). A cDNA clone encoding human Tf was obtained from OriGene (Rockville, MD). Standard PCR and recombinant DNA techniques were used to assemble Ex-4 and Tf as a fusion gene. Briefly, the synthetic Ex-4 gene was modified by adding a DNA sequence encoding barley α-amylase signal peptide (SP) (Rogers and Milliman, 1983) to its 5′ end, whereas the Tf gene was modified by deleting its native signal peptide sequence (Yang et al., 1984) combined with adding a 6xHis-tag and an ER-retention signal (KDEL) sequence to its 3′ end. The modified Ex-4 gene was fused in-frame to the N-terminal end of Tf gene through a (GGGGS)3 linker (Trinh et al., 2004). The resulting Ex-4-Tf chimeric gene was cloned into plasmid pRTL-GUS (Carrington and Freed, 1990) by replacing the GUS gene. The Ex-4-Tf expression cassette, consisting of 35S promoter, SP-Ex-4-Tf -6xHis-KDEL and NOS terminator, was released from pRTL-Ex-4-Tf and cloned into pBI101.1 (Brandsma et al., 2010) to obtain the final construct pBI101.1-Ex-4-Tf.

Plant transformation

Prior to plant transformation, pBI101.1-Ex-4-Tf was transferred into Agrobacterium tumefaciens LBA4404 by tri-parental mating (Ma et al., 2005). For stable transformation, leaf discs of tobacco (cv.81V9) were transformed with Agrobacterium containing pBI101.1-Ex-4-Tf using standard techniques (Horsch et al.,1985). For transient transformation, 6–8-week-old leaves of N. benthamiana were co-infiltrated with two strains of Agrobacterium harbouring pBI101.1-Ex-4-Tf and the p19 silencing suppressor as previously described by Tremblay et al. (2011). Leaf tissues were harvested at 1–7 days post-infection (dpi).

Accumulation of Ex-4-Tf fusion protein in plants

The expression of Ex-4-Tf in plants was analysed by Western blot using commercial anti-human Tf antibodies as described by Brandsma et al. (2010).

The level of prEx-4-Tf accumulation was quantified by ELISA for Tf as described by Brandsma et al. (2010), compared with known quantities of Tf standard (Sigma-Aldrich Canada Co., Oakville, Ontario, Canada).

His-purification of plant-derived Ex-4-Tf fusion protein

Hs-tagged prEx-4-Tf was purified from leaf extracts of stable transgenic tobacco plants using HiTrap™ Chelating HP Columns (GE Healthcare Life Sciences, Baie d'Urfe, Quebec, Canada). Eluted Ex-4-Tf fractions were dialysed extensively against PBS and concentrated using a speed vacuum at 4 °C.

Effect of plant-derived Ex-4-Tf fusion protein on insulin secretion from pancreatic β-cells

The effect of prEx-4-Tf on insulin secretion was evaluated in the mouse beta cells MIN6 as described by Brandsma et al. (2009). In brief, MIN6 cells were cultured in DMEM high glucose with 50 μM 2-mercaptoethanol and 10% (v/v) foetal calf serum (FCS). Once the cells reached about 80% confluence, they were seeded into 96-well (flat-bottomed) microtiter plates at a density of 3×10^4 cells per well. Following incubation for 3 days, cells were washed twice with Earle's balanced salt solution (Sigma) containing 0.1% BSA. After starvation in EBSS plus 0.1% BSA for 1 h, cells were incubated with prEx-4-Tf, commercial Ex-4 or prTf in the presence or absence of glucose. After incubation for 135 min, cell culture supernatants were collected, and insulin content was measured using mouse insulin ELISA Kits (Crystal Chem Inc., Downers Grove, IL) according to the manufacturer's instructions.

Effect of plant-derived Ex-4-Tf fusion protein on proliferation of pancreatic beta cells

The effect of prEx-4-Tf on beta cell proliferation was evaluated using the beta cell line INS-1 according to Buteau et al. (2001). Briefly, INS-1 cells were grown in RPMI 1640 medium. Two days before the experiment, INS-1 cells were seeded in 96-well plates (8×10^4 cells/well) and grown in RPMI medium. Cells were washed with PBS and preincubated for 24 h in minimal RPMI medium, that is, without serum and glucose but with 0.1% BSA. Cells were then grown in fresh minimal RPMI medium containing prEx-4-Tf, prTf or Ex-4 in the presence of glucose. Proliferation was determined by incorporation of [^{3}H]thymidine during the final 4 h of the 24-h incubation period. Cells were harvested with a PHD cell harvester (Cambridge technology) and the radioactivity retained on the dried glass fibre filters was measured.

Effect of plant-derived Ex-4-Tf fusion protein on the differentiation of pancreatic precursor cells

The differentiating effects of prEx-4-Tf on pancreatic precursor cells into insulin-producing cells was evaluated in the rat pancreatic ductal (ARIP) cell according to the method of Hui et al. (2001). In brief, ARIP cells were grown to 80% confluence in F12K medium (Gibco-BRL) with 10% FCS, washed with serum-free F12K followed by 'wash-out' incubation for 6 h with F12K medium. Cells were then incubated with fresh F12K containing prEx-4-Tf or Ex-4 at 12 mM glucose. After incubation for 48 h, the supernatants of the treated cells were collected and analysed for insulin concentration content using mouse insulin ELISA Kits.

Assay of cellular internalization ability of plant-derived Ex-4-Tf fusion protein

Cellular internalization capacity of Ex-4-Tf was assessed using the human intestinal epithelial cells Caco-2. Caco-2 cells were maintained in tissue culture flasks containing RPMI-1640 (Invitrogen) with 10% FCS. To assay cellular uptake, cells were plated to 60×15-mm petri plates. Following incubation for 24 h, cells were washed twice with serum-free F12 medium and incubated in fresh serum-free F12 medium for 1 h at 370 °C to starve the cells of serum. Following the addition of prEx-4-Tf, prTf or Tf standard, plates were incubated for a further 60 min at 37 °C. The reaction was stopped by placing plates on ice, followed by washes with cold PBS to remove excess and unbound proteins. The cell membrane–bound foreign protein was removed by treating plates with acid buffer (500 mM NaCl, 200 mM acetic acid, pH 4.5) for 5 min on ice followed by washes with PBS as described by Karin and Mintz (1981). Cells were then lysed with lysis buffer (0.5% (v/v) Triton X-100, 10 mM Hepes, 10 mM KCl, 1 mM EDTA, 0.1 mM EGTA, 0.1% NP40, 1 mM DTT, 0.5 mM PMSF) and centrifuged at 10 000 g at 4 °C for 10 min to remove cellular debris. The supernatant (containing cytoplastic and nuclear proteins) was collected and assayed for released prEx-4-Tf by ELISA using anti-human Tf antibodies and visualized on 10% (w/v) SDS–PAGE gels followed by Western blot as desctibed by Brandsma et al. (2010).

Digestion of Plant-derived Ex-4–hTF in synthetic gastric fluid

The stability of prEx-4-Tf in an acidic stomach environment was assessed in synthetic gastric fluid according to Tremblay et al. (2011) with minor modifications. Briefly, the His-purified protein sample was incubated in synthetic gastric fluid (0.2 g NaCl, 0.32 g pepsin, 700 μL HCl, in 100 mL dH2O, pH 2.5) at 37 °C. The digestion was stopped by the addition of neutralization buffer (3.4 g Na2CO3 in 100 mL dH2O) at times 0, 15 s, 30 s, 1 min, 5 min, 15 min, 30 min and 1 h. The neutralized samples were boiled for 10 min and analysed by 10% SDS-PAGE followed by Western blot using antibody against human Tf.

Administration in mice of partially purified Ex-4-Tf to mice by oral and intraperitoneal routes

The C57BL/6J male mice at 8 weeks of age (Charles River, Senneville, Quebec, Canada) were used. All mice had free access to standard diet and water. The animal protocol used was approved by Animal User Subcommittee at the University of Western Ontario in accordance with the guiddlines of the Canadian Council of Animal Care.

Mice with similar body weight underwent overnight fasting (12 h) was randomly divided into three experimental groups: Control, Ex-4-Tf and Ex-4. For the study of oral delivery, the control group received an oral gavage of sterile saline, while the Ex-4-Tf and Ex-4 group received an oral gavage of partially purified prEx-4-Tf and commercial Ex-4 (1.5 μg/g body weight), respectively, 1 h prior to intraperitoneal glucose tolerance test (IPGTT). For the study of intraperitoneal delivery, the control group received sterile saline by injection, while the Ex-4-Tf and Ex-4 group received an injection of partially purified prEx-4-Tf or commercial Ex-4 (10 ng/g body weight) 30 min prior to IPGTT. Plasma glucose level before and after the treatment was measured.

For IPGTT, mice were given an intraperitoneal injection of glucose (2 mg/g; D-(+)-glucose; Sigma), and plasma glucose levels were then measured at 0, 15, 30, 60 and 120 min after injection. Area under curve (AUC) was used to quantify responsiveness (Feng et al., 2012a,b, 2013). Five to six mice were used per experimental group ($n = 5$–6).

Statistics

Data are expressed as mean±SE (standard error). Statistically significant differences between groups were analysed by using the Student's t-test or an ANOVA followed by the Bonferroni post hoc test. Differences were considered to be statistically significant when $P < 0.05$.

Acknowlegements

INS-1 cells were received as a gift from Dr. Savita Dhanvantari, Department of Medical Biophysics at the University of Western Ontario. This research was supported in part by NSERC. The authors have no conflicts of interest to declare.

References

Alonso-Magdalena, P., Quesada, I. and Nadal, A. (2011) Endocrine disruptors in the etiology of type 2 diabetes mellitus. Nat. Rev. Endocrinol. 7, 346–353.

Amet, N., Wang, W. and Shen, W.C. (2010) Human growth hormone-transferrin fusion protein for oral delivery in hypophysectomized rats. J. Control. Release, 141, 177–182.

Angelis, I.D. and Turco, L. (2011) Caco-2 cells as a model for intestinal absorption. Curr. Protoc. Toxicol., Chapter 20, Unit20.6.

Asfari, M., Janjic, D., Meda, P., Li, G., Halban, P.A. and Wollheim, C.B. (1992) Establishment of 2-mercaptoethanol-dependent differentiated insulin-secreting cell lines. Endocrinology, 130, 167–178.

Bai, Y., Ann, D.K. and Shen, W.C. (2005) Recombinant granulocyte colony-stimulating factor transferrin fusion protein as an oral myelopoietic agent. Proc. Natl Acad. Sci. USA, 102, 7292–7296.

Bobst, C.E., Wang, S., Shen, W.C. and Kaltashov, I.A. (2012) Mass spectrometry study of a transferrin-based protein drug reveals the key role of protein aggregation for successful oral delivery. Proc. Natl Acad. Sci. USA, 109, 13544–13548.

Brandsma, M., Wang, X., Diao, H., Kohalmi, S.E., Jevnikar, A.M. and Ma, S. (2009) A proficient approach to the production of therapeutic glucagon-like peptide-1 (GLP-1) in transgenic plants. Open Biotechnol. J. 3, 50–56.

Brandsma, M., Diao, H., Wang, X., Kohalmi, S.E., Jevnikar, A.M. and Ma, S. (2010) Plant-derived human serum transferrin demonstrates multiple functions. Plant Biotechnol. J. 8, 489–505.

Brandsma, M., Jevnikar, A.M. and Ma, S. (2011) Recombinant human transferrin: beyond iron binding and transport. Biotechnol. Adv. 29, 230–238.

Buteau, J., Roduit, R., Susini, S. and Prentki, M. (1999) Glucagon-like peptide-1 promotes DNA synthesis, activates phosphatidylinositol 3-kinase and increases transcription factor pancreatic and duodenal homeobox gene 1 (PDX-1) DNA binding activity in beta (INS-1)-cells. Diabetologia, 42, 856–864.

Buteau, J., Foisy, S., Rhodes, C.J., Carpenter, L., Biden, T.J. and Prentki, M. (2001) Protein kinase Czeta activation mediates glucagon-like peptide-1-induced pancreatic beta-cell proliferation. Diabetes, 50, 2237–2243.

Carrington, J.C. and Freed, D.D. (1990) Cap-independent enhancement of translation by a plant potyvirus 5'nontranslated region. J. Virol. 64, 1590–1597.

Chen, X., Lee, H.F., Zaro, J.L. and Shen, W.C. (2011) Effects of receptor binding on plasma half-life of bifunctional transferrin fusion proteins. Mol. Pharm. 8, 457–465.

Chen, W., Wang, Lin, Wang, Y., Chen, Z., Liu, X., Liu, X.H. and Liu, L. (2012) Exendin-4 Protects MIN6 cells from t-BHP-induced Apoptosis via IRE1-JNK-Caspase-3 signaling. Int. J. Endocrinol. 2012, 54908.

Chirino, A.J., Ary, M.L. and Marshall, S.A. (2004) Minimizing the immunogenicity of protein therapeutics. Drug Discovery Today, 9, 82–90.

Collins, G.S., Mallett, S., Omar, O. and Yu, L.M. (2011) Developing risk prediction models for type 2 diabetes: a systematic review of methodology and reporting. BMC Med. 9, 103.

Feng, Z.C., Li, J., Turco, B.A., Riopel, M., Yee, S.P. and Wang, R. (2012a) Critical role of c-Kit in beta cell function: increased insulin secretion and protection against diabetes in a mouse model. Diabetologia, 55, 2214–2225.

Feng, Z.C., Donnelly, L., Li, J., Krishnamurthy, M., Riopel, M. and Wang, R. (2012b) Inhibition of Gsk3b activity improves b-cell function in c-KitWv/+ male mice. Lab. Invest. 92, 543–555.

Feng, Z.C., Riopel, M., Li, J., Donnelly, L. and Wang, R. (2013) Down-regulation of Fas activity could rescue early onset of diabetes in c-KitWv/+ mice. Am. J. Physiol. Endocrinol. Metab. 67, E557–E565.

Gahr, S., Merger, M., Bollheimer, L.C., Hammerschmied, C.G., Scholmerich, J. and Hugl, S.R. (2002) Hepatocyte growth factor stimulates proliferation of pancreatic beta-cells particularly in the presence of subphysiological glucose concentrations. J. Mol. Endocrinol. 28, 99–110.

Garber, A.J. (2011) Long-acting glucagon-like peptide 1 receptor agonists: a review of their efficacy and tolerability. Diabetes Care, 34, S279–S284.

Giorgino, F., Natalicchio, A., Leonardini, A. and Laviola, L. (2007) Exploiting the pleiotropic actions of GLP-1 for the management of type 2 diabetes mellitus and its complications. Diabetes Res. Clin. Pract. 78, S59–S67.

Hadjiyanni, L., Baggio, L.L., Poussier, P. and Drucker, D.J. (2008) Exendin-4 modulates diabetes onset in nonobese diabetic mice. Endocrinology, 149, 1338–1349.

Holkeri, H. and Makarow, M. (1998) Different degradation pathways for heterologous glycoproteins in yeast. FEBS Lett. 429, 162–166.

Holst, J.J. and Gromada, J. (2004) Role of incretin hormones in the regulation of insulin secretion in diabetic and nondiabetic humans. Am. J. Physiol. Endocrinol. Metab. 287, E199–E206.

Horsch, R.B., Fry, J.E., Hoffmann, N.L., Eicholtz, D., Rogers, S.G. and Fraley, R.T. (1985) A simple and general method for transferring genes into plants. Science, 227, 1229–1231.

Huebers, H.A. and Finch, C.A. (1987) The physiology of transferrin and transferrin receptors. Physiol. Rev. 67, 520–582.

Hügl, S.R., White, M.F. and Rhodes, C.J. (1998) Insulin-like growth factor I (IGF-I)-stimulated pancreatic beta-cell growth is glucosedependent. Synergistic activation of insulin receptor substratemediated signal transduction pathways by glucose and IGF-I in INS-1 cells. J. Biol. Chem. 273, 17771–17779.

Hui, H., Wright, C. and Perfetti, R. (2001) Glucagon-like peptide 1 induces differentiation of islet duodenal homeobox-1-positive pancreatic ductal cells into insulin-secreting cells. Diabetes, 50, 785–796.

Jiang, Y.Y., Liu, C., Hong, M.H., Zhu, S.J. and Pei, Y.Y. (2007) Tumor cell targeting of transferrin-PEG-TNF-alpha conjugate via a receptor-mediated delivery system: design, synthesis, and biological evaluation. Bioconjug. Chem. 18, 41–49.

Karin, M. and Mintz, B. (1981) Receptor-mediated endocytosis of transferrin in developmentally totipotent mouse teratocarcinoma stem cells. J. Biol. Chem. 256, 3245–3252.

Kieffer, T.M. and Habener, J.F. (1999) The glucagon-like peptides. Endocr. Rev. 6, 876–913.

Kim, P.H., Eckmann, L., Lee, W.J., Han, W. and Kagnoff, M.F. (1998) Cholera toxin and cholera toxin B subunit induce IgA switching through the action of TGF-beta 1. J. Immunol. 160, 1198–1203.

Kim, B.J., Zhou, J., Martin, B., Carlson, O.D., Maudsley, S., Greig, N.H., Mattson, M.P., Ladenheim, E.E., Wustner, J., Turner, A., Sadeghi, H. and Egan, J.M. (2010) Transferrin fusion technology: a novel approach to prolonging biological half-life of insulinotropic peptides. J. Pharmacol. Exp. Ther. 334, 682–692.

Koliaki, C. and Doupis, J. (2011) Incretin-based therapy: a powerful and promising weapon in the treatment of type 2 diabetes mellitus. Diabetes Ther. 2, 101–121.

Kwon, K.C., Nityanandam, R., New, J.S. and Daniell, H. (2012) Oral delivery of bioencapsulated exendin-4 expressed in chloroplasts lowers blood glucose level in mice and stimulates insulin secretion in beta-TC6 cells. *Plant Biotechnol. J.* **11**, 66–76.

Kyriacou, A. and Ahmed, A.B. (2010) Exenatide use in the management of type 2 diabetes mellitus. *Pharmaceuticals*, **3**, 2554–2567.

Lakatos, L., Szittya, G., Silhavy, D. and Burgyan, J. (2004) Molecular mechanism of RNA silencing suppression mediated by p19 protein of tombusviruses. *EMBO J.* **23**, 876–884.

Li, H. and Qian, Z.M. (2002) Transferrin/transferrin receptor-mediated drug delivery. *Med. Res. Rev.* **22**, 225–250.

Lovshin, J.A. and Drucker, D.J. (2009) Incretin-based therapies for type 2 diabetes mellitus. *Nat. Rev. Endocrinol.* **5**, 262–269.

Ma, S. and Wang, A.M. (2012) Molecular farming in plants: host systems and technologies for expression and downstream processing. In *Molecular Farming in Plants: Recent Advances and Future Prospects*(Wang, A.M. and Ma, S., eds), pp. 1–20. Berlin: Springer.

Ma, S., Huang, Y., Yin, Z., Menassa, R., Brandle, J.E. and Jevnikar, A.M. (2004) Induction of oral tolerance to prevent diabetes with transgenic plants requires glutamic acid decarboxylase (GAD) and IL-4. *Proc. Natl Acad. Sci. USA*, **101**, 5680–5685.

Ma, S., Huang, Y., Davis, A., Yin, Z.Q., Mi, Q., Menassa, R., Brandle, J. and Jevnikar, A.M. (2005) Production of biologically active human interleukin-4 in transgenic tobacco and potato. *Plant Biotech. J.* **3**, 309–318.

Melanie, B. (2005) Transferrin' the load. *Nat. Rev. Drug Discov.* **4**, 537.

Molitch, M.E. (2013) Current state of type 2 diabetes management. *Am. J. Manag. Care*, **19**, s136–s142.

Mudaliar, S. and Henry, R.R. (2012) The incretin hormones: from scientific discovery to practical therapeutics. *Diabetologia*, **55**, 1865–1868.

Pohl, M. and Wank, S.A. (1998) Molecular cloning of the helodermin and exendin-4 cDNAs in the lizard. Relationship to vasoactive intestinal polypeptide/pituitary adenylate cyclase activating polypeptide and glucagon-like peptide 1 and evidence against the existence of mammalian homologues. *J. Biol. Chem.* **273**, 9778–9784.

Potenza, M.A., Nacci, C., Gagliardi, S. and Montagnani, M. (2011) Cardio-vascular complications in diabetes: lessons from animal models. *Curr. Med. Chem.* **18**, 1806–1819.

Robles, G.I. and Singh-Franco, D. (2009) A review of exenatide as adjunctive therapy in patients with type 2 diabetes. *Drug Des. Devel. Ther.* **3**, 219–240.

Rogers, J.C. and Milliman, C. (1983) Isolation and sequence analysis of a barley alpha-amylase cDNA clone. *J. Biol. Chem.* **258**, 8169–8174.

Schwartz, S. and Defronzo, R.A. (2013) Is incretin-based therapy ready for the care of hospitalized patients with type 2 diabetes? *Diabetes Care*, **36**, 2107–2111.

Shaji, J. and Patole, V. (2008) Protein and peptide drug delivery: oral approaches. *Indian J. Pharm. Sci.* **70**, 269–277.

Tremblay, R., Wang, D., Jevnikar, A.M. and Ma, S. (2010) Tobacco, a highly efficient green bioreactor for production of therapeutic proteins. *Biotechnol. Adv.* **28**, 214–221.

Tremblay, R., Feng, M., Menassa, R., Huner, N.P., Jevnikar, A.M. and Ma, S. (2011) High-yield expression of recombinant soybean agglutinin in plants using transient and stable systems. *Transgenic Res.* **20**, 345–356.

Trinh, R., Gurbaxani, B., Morrison, S.L. and Seyfzadeh, M. (2004) Optimization of codon pair use within the (GGGGS)3 linker sequence results in enhanced protein expression. *Mol. Immunol.* **40**, 717–722.

Wang, D.J., Brandsma, M., Yin, Z., Wang, A., Jevnikar, A.M. and Ma, S.W. (2008) A novel platform for biologically active recombinant human interleukin-13 production. *Plant Biotechnol. J.* **6**, 504–515.

Xia, C.Q., Wang, J. and Shen, W.C. (2000) Hypoglycemic effect of insulin-transferrin conjugate in streptozotocin-induced diabetic rats. *J. Pharmacol. Exp. Ther.* **295**, 594–600.

Xiao, Q., Giguere, J., Parisien, M., Jeng, W., St-Pierre, S.A., Brubaker, P.L. and Wheeler, M.B. (2001) Biological activities of glucagon-like peptide-1 analogues *in vitro* and *in vivo*. *Biochemistry*, **40**, 2860–2869.

Yang, F., Lum, J.B., McGill, J.R., Moore, C.M., Naylor, S.L., van Bragt, P.H., Baldwin, W.D. and Bowman, B.H. (1984) Human transferrin: cDNA characterization and chromosomal localization. *Proc. Natl Acad. Sci. USA*, **81**, 2752–2756.

Zhang, P., Zhang, X., Brown, J., Vistisen, D., Sicree, R., Shaw, J. and Nichols, G. (2010) Global healthcare expenditure on diabetes for 2010 and 2030. *Diabetes Res. Clin. Pract.* **87**, 293–301.

Zhang, D., Nandi, S., Bryan, P., Pettit, S., Nguyen, D., Santos, M.A. and Huang, N. (2010) Expression, purification, and characterization of recombinant human transferrin from rice (*Oryza sativa* L.). *Protein Expr. Purif.* **74**, 69–79.

3

The production of recombinant cationic α-helical antimicrobial peptides in plant cells induces the formation of protein bodies derived from the endoplasmic reticulum

Nuri Company[1], Anna Nadal[1], José-Luis La Paz[2], Sílvia Martínez[2], Stefan Rasche[3], Stefan Schillberg[3], Emilio Montesinos[1] and Maria Pla[1],*

[1]Institute for Food and Agricultural Technology (INTEA), University of Girona, Girona, Spain
[2]Center for Research in Agricultural Genomics (CRAG), Barcelona, Spain
[3]Fraunhofer Institute for Molecular Biology and Applied Ecology (IME), Aachen, Germany

*Correspondence

email maria.pla@udg.edu

Keywords: antimicrobial peptide, BP100, cationic peptide, molecular farming, protein body, transgenic plant.

Summary

Synthetic linear antimicrobial peptides with cationic α-helical structures, such as BP100, are valuable as novel therapeutics and preservatives. However, they tend to be toxic when expressed at high levels as recombinant peptides in plants, and they can be difficult to detect and isolate from complex plant tissues because they are strongly cationic and display low extinction coefficient and extremely limited immunogenicity. We therefore expressed BP100 with a C-terminal tag which preserved its antimicrobial activity and demonstrated significant accumulation in plant cells. We used a fluorescent tag to trace BP100 following transiently expression in *Nicotiana benthamiana* leaves and showed that it accumulated in large vesicles derived from the endoplasmic reticulum (ER) along with typical ER luminal proteins. Interestingly, the formation of these vesicles was induced by BP100. Similar vesicles formed in stably transformed *Arabidopsis thaliana* seedlings, but the recombinant peptide was toxic to the host during latter developmental stages. This was avoided by selecting active BP100 derivatives based on their low haemolytic activity even though the selected peptides remained toxic to plant cells when applied exogenously at high doses. Using this strategy, we generated transgenic rice lines producing active BP100 derivatives with a yield of up to 0.5% total soluble protein.

Introduction

Antimicrobial peptides (AMPs) are key components of innate immunity in plants and animals and are also produced by microbes in antibiosis processes (Bulet and Stocklin, 2005; Cooter et al., 2010; Degenkolb et al., 2003; Ganz, 2003; Hancock, 2001; Jack and Jung, 2000; Lay and Anderson, 2005; Ng, 2004; Raaijmakers et al., 2006; Tincu and Taylor, 2004; Toke, 2005; Zasloff, 2002). They are generally short peptides, which can have linear structures that often adopt an amphipathic α-helical conformation that binds to the phospholipid membranes of target microbes before the hydrophobic face is inserted into the membrane bilayer (Bechinger, 2004; Boman, 2003; Brogden, 2005; Ferré et al., 2009; Huang, 2006; Marcos and Gandía, 2009; Tossi et al., 2000). This unique mode of action helps to avoid the emergence of resistance in target pathogens (Brogden, 2005; Peschel and Sahl, 2006; Yeaman and Yount, 2003; Yount and Yeaman, 2005).

Antimicrobial peptides are valuable as novel therapeutic agents (Hancock, 2001; Marcos et al., 2008; Montesinos, 2007; Moreno et al., 2006; Rajasekaran et al., 2012; Zasloff, 2002) because of their broad-spectrum activity against bacteria, fungi, viruses, parasites and tumour cells (Ajesh and Sreejith, 2009; Broekaert et al., 1997; Brogden et al., 2003; Bulet et al., 2004; Jenssen et al., 2006; Otvos, 2000; Torrent et al., 2012; Zasloff, 2002). Natural AMPs have been optimized to increase their potency against selected pathogens while protecting nontarget organisms and

enhancing stability (Badosa et al., 2007; Cavallarín et al., 1998; López-García et al., 2002; Marcos et al., 2008; Monroc et al., 2006).

The synthetic CECMEL11 peptide library, a 125-member cationic α-helical undecapeptide library designed using a combinatorial approach, contains groups of sequences with potent and selective activity against a number of bacterial and fungal phytopathogenic reporter strains (Badosa et al., 2007, 2009; Bardají, 2006). The synthetic peptide BP100 (KKLFKKILKYL-NH$_2$) was found to be effective at micromolar concentrations against *Xanthomonas axonopodis* pv. *vesicatoria* in pepper, *Erwinia amylovora* in apple and *Pseudomonas syringae* pv. *syringae* in pear (Badosa et al., 2007). The efficacy was comparable to standard antibiotics, and BP100 is also highly biocompatible, as determined by acute oral toxicity tests in mice (Montesinos and Bardají, 2008).

Genetically modified (GM) plants with moderate resistance to pathogenic bacteria and fungi have been developed by expressing AMPs either constitutively or induced by pathogens (reviewed by Marcos et al., 2008; Montesinos, 2007). Many different recombinant proteins have been expressed in plants (Fischer et al., 2004; Hoja and Sonnewald, 2013; Twyman et al., 2003), and the applications envisaged for CECMEL11 peptides suggest that plant-based expression would be preferable, to develop plants either that are disease resistant or that express AMPs for medical and industrial applications. Large-scale chemical synthesis of peptides above around six amino acids is only economically viable for applications of very high added value. Rice is a suitable platform for large-scale production, and the target product can be easily stored in the kernels for a long time allowing to decouple the production and the processing step.

BP100 was recently used as a proof-of-concept to show that the constitutive expression of short cationic α-helical synthetic peptides can have a strong negative impact on the fitness of transgenic rice plants (Nadal et al., 2012). Transformation with five sequences encoding BP100 derivatives (BP100der) resulted in a transformation efficiency more than 100-fold less efficient than a control transgene. These BP100der peptides contained endoplasmic reticulum (ER) retention motifs to prevent peptide degradation in the cytosol and minimize toxicity to the host plant. This did not affect the antimicrobial activity of the products in vitro using bacterial growth inhibition tests. However, when applied at high doses ($>10^2$-fold of the minimal inhibitory concentration, MIC), the peptides exerted toxic effects such as erythrocyte lysis, leaf damage in Nicotiana benthamiana and the inhibition of rice seedling development. The extreme physicochemical properties and low immunogenicity of the recombinant peptides prevented their direct detection in GM rice, but low levels of transgene mRNA were detected and the plants were more resistant to oxidative stress and pathogens such as Dickeya chrysanthemi and Fusarium verticillioides. We speculated that BP100der toxicity was associated with constitutive transgene expression but was ameliorated by the conferred stress tolerance.

Here, we assessed the production of recombinant BP100 derivatives in transformed N. benthamiana and Arabidopsis thaliana plants directly, by expressing the peptides as fusions with the fluorescent marker protein DsRed. We also screened a series of BP100 derivatives elongated with sequences from natural AMPs and found that peptides with lower haemolytic activity achieved greater transformation efficiency. We were therefore able to confirm that peptides with potent antimicrobial and low haemolytic activity can accumulate in stably transformed rice plants at levels of up to 0.5% total soluble protein (TSP).

Results

BP100 derivatives can be produced by transient expression in N. benthamiana cells

We previously described transgenic rice plants constitutively expressing transgenes that encoded three BP100 derivatives: bp100.1, bp100.2i and bp100.2mi (Nadal et al., 2012). The synthesis of BP100der in these plants could only be confirmed indirectly, that is, by demonstrating the presence of transgene mRNA, a resistant phenotype and ultrastructural changes in the plant cell. To confirm that active cationic α-helical antimicrobial peptides can be made to accumulate in transgenic plants, we developed a strategy based on the fusion of BP100 to the Discosoma spp. red fluorescent reporter protein DsRed (Matz et al., 1999) and the epitope tag54 sequence for the detection of recombinant proteins using the specific antibody mAb54 (Rasche et al., 2011).

Chimeric genes encoding BP100-DsRed-tag54-KDEL (hereafter described as BP100-DsRed-tag54) and a DsRed-tag54-KDEL control lacking AMP sequences (hereafter described as DsRed-tag54) were placed under the control of the constitutive Cauliflower mosaic virus 35S promoter for constitutive expression and product accumulation in the ER. The constructs were agroinfiltrated into N. benthamiana leaves together with HC-Pro silencing suppressor, and the accumulation of each recombinant protein was monitored by tracking the DsRed

fluorescence by confocal microscopy. Three days postinfiltration (dpi), epidermal cells from N. benthamiana leaves transformed with the AMP and control constructs were both found to produce strong DsRed fluorescence signals in three agroinfiltrated fields representing the upper, medial and basal portions of different leaves. Cells expressing the BP100-DsRed-tag54 and DsRed-tag54 constructs had fluorescence intensities of 1004 ± 574 and 817 ± 189 fluorescence units per field, respectively, representing statistically similar expression levels as determined by one-way ANOVA ($P = 0.395$).

Western blot analysis using the tag54-specific monoclonal antibody revealed single bands of the anticipated sizes for DsRed-tag54 (29 kDa) and BP100-DsRed-tag54 (~33 kDa) as shown in Figure 1. This confirmed the presence of the BP100 peptide in the BP100-DsRed-tag54 fusion protein. The lack of a 29-kDa band in BP100-DsRed extracts showed that the BP100 peptide was not specifically cleaved from the fusion partner. There were no visible bands of the size expected for the tag54 alone, indicating the tag was not cleaved from the fusion protein either (data not shown). The observed differences in band intensity are likely to reflect the challenging extraction procedure for BP100 derivatives when dealing with complex matrices such as leaf samples. However, our data show clearly that BP100 derivatives can be synthesized in plants and directed to accumulate in plant cells.

BP100 induces the formation of ER-derived vesicles

The analysis of N. benthamiana leaves by confocal microscopy showed different patterns of fluorescence between the BP100

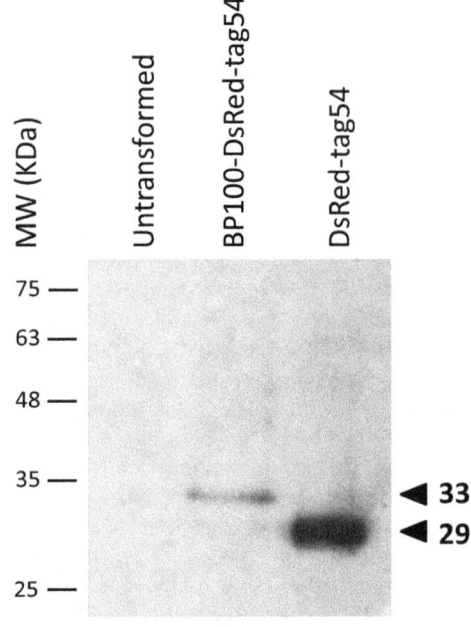

Figure 1 Transient expression of BP100-DsRed-tag54 and DsRed-tag54 in Nicotiana benthamiana leaves. Western blot of proteins extracted from N. benthamiana leaves agroinfiltrated with bp100-dsred-tag54 or dsred-tag54 as a control. Leaf tissue was collected at 3 dpi, and 1 µg of total protein was separated by SDS-PAGE before transfer to nitrocellulose filters. The recombinant proteins were detected using antibody mAb54k (diluted 1 : 1500) followed by the horseradish peroxidase-labelled anti-mouse IgG secondary antibody (diluted 1 : 10000) and ECL chemiluminescent detection. Arrowheads indicate BP100-DsRed-tag54 (32.9 kDa) and DsRed-tag54 control (28.5 kDa) proteins.

and control constructs. As expected, DsRed fluorescence in the DsRed-tag54 control leaves displayed a characteristically reticulate pattern 2 days after infiltration, consistent with typical ER morphology in the plant epidermal cells. Although some cells transformed with BP100-DsRed-tag54 displayed a similar fluorescence pattern at 2 dpi, indicating that BP100 does not inhibit protein transit and retention in the ER, the majority contained numerous and widely distributed spherical structures ~1–2 μm in diameter showing intense fluorescence, obscuring the normal ER network. After 3–5 dpi, most cells expressing BP100-DsRed-tag54 contained larger (up to 15 μm) highly fluorescent and irregular structures accompanied by smaller punctate fluorescence.

To gain insight into the origin of these structures and the integrity of the ER in cells accumulating BP100-DsRed-tag54, we co-transformed N. benthamiana leaves with BP100-DsRed-tag54 and another construct encoding ER-localized cyan fluorescent protein (eCFP) (Joseph et al., 2012). The merged fluorescence images in double-transformed cells at 3 dpi showed that BP100-DsRed-tag54 and eCFP co-localized in the induced vesicles described above while eCFP was also visible in the ER (Figure 2a–d). This confirmed that BP100-DsRed resides transiently in the ER lumen but rapidly accumulates in novel ER-derived vesicles that contain other luminal ER proteins such as eCFP. As expected, control cells co-transformed with the DsRed-tag54 and eCFP constructs showed co-localization of the proteins within the normal ER (Figure 2e–h). The observed changes in ER structure therefore appear to be driven by the BP100 component of the BP100-DsRed-tag54 fusion protein.

Stably transformed Arabidopsis thaliana seedlings accumulate BP100-DsRed-tag54 in vesicles but fail to develop into mature plants

Having confirmed the accumulation of BP100-DsRed-tag54 in large ER-derived vesicles following transient expression in plant cells, we set out to determine whether this could also be achieved in A. thaliana plants stably transformed with the same constructs using the floral dip method. Two weeks after in vitro germination, the radicles of hygromycin-resistant seedlings expressing the BP100 and control constructs were analysed by confocal microscopy. All analysed transgenic plants displayed red fluorescence, and those expressing BP100-DsRed-tag54 revealed fluorescent spots 1–10 μm in diameter, ranging from 1–3 μm in root tip cells to 2–10 μm in elongating epidermal and lateral root cells (Figure 3a,d). The morphology and distribution of the fluorescent vesicles were similar to the pattern observed in N. benthamiana epithelial cells.

Although we were able to recover fertile transgenic seeds producing each of the constructs, most of the seedlings expressing BP100-DsRed-tag54 did not survive acclimation, whereas those expressing the control construct developed normally. This emphasized that BP100 is phytotoxic at high concentrations even when it accumulates in vesicles. We further investigated BP100-DsRed-tag54 toxicity by incubating the transgenic radicles with SYTOX, a nucleic acid stain that can only cross damaged cell membranes. We found that DsRed and SYTOX were not co-localized at this developmental stage (Figure 3). Most of the cells expressing BP100-DsRed-tag54 and the control construct DsRed-tag54 were viable, and only a few epidermal cells above the lateral root cap revealed both SYTOX and DsRed fluorescence, and the distribution of SYTOX staining was similar in wild-type A. thaliana cells (data not shown), suggesting the cell damage reflected normal physiological processes rather than a BP100-specific effect (Truernit and Haseloff, 2008).

Haemolytic activity can be used to predict the feasibility of constitutive BP100 expression in stably transformed plants

To improve the accumulation of BP100 derivatives (BP100der) in transgenic plants, we screened a recently described synthetic

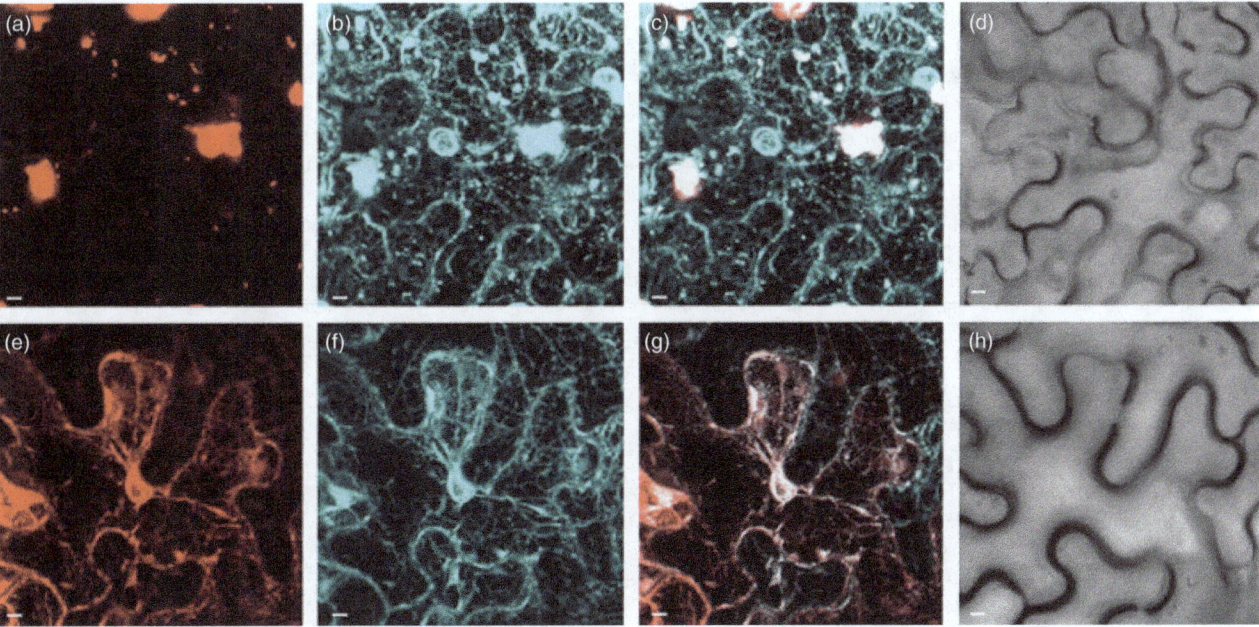

Figure 2 Confocal micrographs of Nicotiana benthamiana epidermal cells transiently expressing BP100der-DsRed-tag54 or the control DsRed-tag54. Nicotiana benthamiana leaves were co-transformed with constructs encoding ER-targeted eCFP and BP100-DsRed-tag54 (a–d) or DsRed-tag54 (e–h). (a,d), DsRed red fluorescence; (b,f) eCFP cyan fluorescence; (c,g), cyan and red merged images showing co-localization in white; (d,h), bright field. Scale bars, 5 μm.

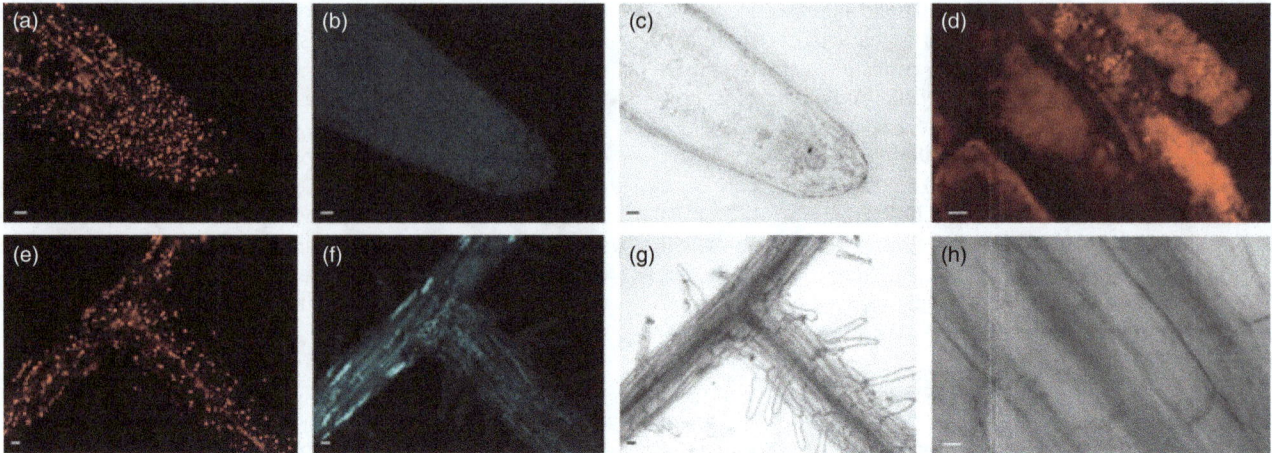

Figure 3 Confocal micrographs of transgenic *Arabidopsis thaliana* radicles expressing BP100-DsRed-tag54. *Arabidopsis thaliana* seeds obtained by floral dip transformation with recombinant bacteria carrying the BP100-DsRed-tag54 (a–c and e–g) or the DsRed-tag54 (d,h) constructs were germinated *in vitro* for a total of 16 days and observed by confocal microscopy following SYTOX staining. (a–c), radicle tip; (d–h), lateral root. (a,d,e), DsRed fluorescence; (b,f), SYTOX fluorescence; (c,g,h) bright field. Scale bars, 5 μm (d,h), 10 μm (a–c) and 20 μm (e–g).

library of BP100 derivatives to identify less phytotoxic peptides that maintained their potent antimicrobial activity (Badosa et al., in preparation). The synthetic library of 40 BP100 derivatives included 20 carrying a C-terminal KDEL ER retention signal, and their haemolytic activity at 150 μM ranged from 0 to 100%. Because haemolytic activity has been widely used to determine the toxicity of peptides towards eukaryotic cells, we used this property as a tentative predictor of phytotoxicity and, by extension, of transformation efficiency in plants.

We therefore selected four BP100 derivatives with potent antimicrobial activity but the lowest haemolytic activity: BP100.m (15%), BP100.g (4%), BP100.g2 (4%) and BP100.c (1%), as shown in Table 1. All four peptides also showed stronger activity against the model plant pathogen *Xanthomonas axonopodis* pv. *vesicatoria* (*Xav*) than the original BP100 based on *in vitro* growth inhibition assays, and their activities against *Erwinia amylovora* (*Ea*) and *Pseudomonas syringae* pv. *syringae* (*Pss*) were similar to BP100. The modifications involved elongation with sequences derived from natural AMPs: mellitin in BP100.m, magainin in BP100.g and BP100.g2, and cecropine A in BP100.c. We also produced the additional construct BP100.gtag, which was equivalent to BP100.g but contained three copies of the tag54

epitope (tag54×3) at the C-terminal end. This reduced the haemolytic activity of the peptide to 0% at 150 μM and increased its antimicrobial potency, even though the tag sequence alone had no antimicrobial activity (Table 1).

Constructs encoding the five BP100 derivatives (BP100.m, BP100.g, BP100.gtag, BP100.g2 and BP100.c, all including a C-terminal KDEL ER retention sequence) were introduced in-frame with the sequence encoding the *N. tabacum* pathogenesis-related protein PR1a signal peptide, under the control of the *ubi* constitutive promoter and *nos* terminator. All constructs were introduced in *Agrobacterium tumefaciens* and used to transform rice plants along with the *hpt*II marker for hygromycin selection. Each construct yielded hygromycin-resistant plants, most of which were shown to contain the BP100der transgene by testing leaf genomic DNA by PCR (data not shown). The number of transgenic events achieved with the BP100der constructs was approximately 35–50% of the number achieved using the selectable marker alone, specifically 48% for BP100.g and BP100.gtag, 41% for BP100.m, 36% for BP100.g2 and 34% for BP100.c, much better than reported for other BP100 derivatives in the same expression cassette introduced as well into rice using the same method (Nadal et al., 2012).

Table 1 Sequence, and antibacterial and haemolytic activity of selected BP100 derivatives

AMP code	Sequence	MIC (μM)			Haemolytic activity vs. BP100 (150 μM)
		Xav	Pss	Ea	
BP100	KKLFKKILKYL	5–10	2.5–5	2.5–5	22.0 ± 2.8
BP100.m	KKLFKKILKYL **AGPA** TTGLPALISW <u>KDEL</u>	5–7.5	2.5–5	5–7.5	0.7
BP100.g	KKLFKKILKYL **AGPA** KFLHSAK <u>KDEL</u>	5–7.5	7.5–10	2.5–5	0.2
BP100.gtag	KKLFKKILKYL **AGPA** KFLHSAK **AGPA** *KDWEHLKDWEHLKDWEHL* <u>KDEL</u>	2.5–5	2.5–5	2.5–5	0.0
BP100.g2	KKLFKKILKYL **AGPA** GIGKFLHSAK <u>KDEL</u>	2.5–5	2.5–5	2.5–5	0.2
BP100.C	KKLFKKILKYL **AGPA** VAVVGQATQIAK <u>KDEL</u>	1.25–2.5	2.5–5	2.5–5	0.1
Tag54	*KDWEHLKDWEHLKDWEHL* <u>KDEL</u>	>100	>100	>100	0.0

The KDEL ER retention sequence is underlined, the AGPA linker sequence is highlighted in bold, and the tag54 sequence in italics. Antibacterial activity was determined against *Erwinia amylovora* (*Ea*), *Pseudomonas syringae* pv. *syryngae* (*Pss*) and *Xanthomonas axonopodis* pv. *vesicatoria* (*Xv*) as reporter species. Minimal inhibitory concentrations (MIC) are shown in μM and were calculated with 10^8 bacterial CFU/mL. Haemolytic activity is shown as the ratio between each peptide and the reference peptide BP100, calculated at 150 μM and presented as a percentage value with confidence interval for $\alpha = 0.05$ (Badosa et al., 2007).

The transgene copy number and expression profile were assessed in three independent events representing each of the five bp100der transgenes. The ratio of bp100der and hptII DNA to the endogenous actin sequence was 0.5, as determined by quantitative PCR (qPCR) using leaf genomic DNA from T_0 plants, suggesting single-copy insertions. Transgene expression was verified by quantitative RT-PCR (RT-qPCR) in the same T_0 leaf samples, revealing that both hptII and the bp100der transgene were expressed in all events, with mRNA levels ranging from 0.01 through to 350-fold the level of actin mRNA used for normalization (GeNorm M < 0.5 in these samples; Figure S1). The transformation efficiency and expression level of the bp100.gtag transgene were statistically similar to bp100.g lacking the tag54 sequence.

BP100.gtag accumulates in stably transformed rice seedlings

Five independent transgenic events expressing bp100.gtag (S-bp100.gtag) were selected for further analysis. qPCR characterization confirmed that all five events contained single transgene copies (Table 2). They were self-crossed to obtain homozygous T_2 lines. Transgene expression was monitored by RT-qPCR using stage V2 leaves (two-leaf stage) from plants grown under controlled-environment conditions. The level of bp100.gtag mRNA varied from 2 through to 67-fold the level of actin mRNA (Table 2).

We used the tag54×3 sequence to measure the accumulation of the peptide in transgenic rice plants by Western blot. A ~10.1-kDa band was detected in total soluble protein (TSP) extracts from one-week-old seedlings in four of five S-bp100.gtag events (Figure 4). Chemically synthesized BP100.gtag migrates at ~5.7 kDa when mixed with nontransformed rice extracts, which is commensurate with its theoretical mass, and a minor ~10.1-kDa band is visible at higher peptide concentrations. Serial dilutions of chemically synthesized BP100.gtag mixed with 20 μg of nontransformed rice extracts (0.72–0.045% of rice TSP) were used as a reference in Western blots to determine peptide yields in the transgenic plants. Accordingly, biologically produced BP100.gtag was shown to accumulate at levels up to 0.5% TSP in some events (e.g. S-bp100.gtag-11). There was also a positive correlation between the levels of bp100.gtag mRNA and the peptide detected in the different S-bp100.gtag events (Pearson correlation coefficient = 0.9).

We further assessed the activity of the recombinant BP100.gtag by an in vitro bacterial growth inhibition test. E. amylovora growth was reduced in the presence of protein extracts obtained from S-bp100.gtag-11 homozygous T_2 seedlings in these assays, compared with protein extraction buffer and extracts of control seedlings harbouring the hptII selection gene. Inhibition levels were 63 ± 8% (one-way ANOVA $P = 0.000$) when compared with the extraction buffer and 51 ± 7% (one-way ANOVA $P = 0.000$) when compared with control seedlings. This demonstrated the antibacterial activity of the recombinant BP100 derivative.

Ultrastructural changes in rice cells induced by the accumulation of BP100.gtag

We investigated the ultrastructure of rice cells producing BP100.gtag by transmission electron microscopy (TEM) at stage V2. We found that the ER in S-BP100.gtag cells was morphologically distinct in the collar region (Figure 5a). Compared with nontransformed rice cells, there was some ER fragmentation and an increase in the abundance of dictyosome structures. As shown in Figure 5b, several cells also featured strongly dilated intracisternal spaces and ribosome-decorated ER-derived vesicles that were in some cases comparable to mitochondria in size. Dictyosome vesicles presented as a compact mass of vesicles in the cytosol of these cells.

The exogenous application of high doses (i.e. 50 μM) of chemically synthesized BP100.gtag also induced significant morphological changes in rice seedlings at the ultrastructural level. The ER, with swollen cisternae, was distributed in the cytosol near the cell wall. The nuclear envelope was more dilated than that of control cells, and there were numerous irregular vesicles containing electron-dense granules. No visible injury was observed in other cellular components (Figure 5c). However, in some cells, there was substantial disruption to the ER, including numerous irregular vesicles up to 1 μm across, containing electron-dense granules, as well as irregular mitochondria with altered cristae (Figure 5d). Membrane invagination and plasmolysis were observed in the worst-affected cells.

Discussion

Several synthetic linear undecapeptides from the CECMEL11 library are useful leads for the development of novel antimicrobial

Table 2 Transgene copy numbers and mRNA expression in S-bp100.gtag rice

GMO event code	Copy numbers per qPCR			Normalized transgene DNA levels		Normalized transgene mRNA expression (RT-qPCR)	
	bp100.gtag	hptII	actin	bp100.gtag	hptII	bp100.gtag	hptII
S-bp100.gtag-2	2.17E + 04	1.59E + 04	7.88E + 04	0.27	0.20	41.1 ± 2.48	83.85 ± 11.18
S-bp100.gtag-5	1.29E + 04	2.05E + 04	2.86E + 04	0.45	0.72	1.03 ± 0.22	101.98 ± 7.90
S-bp100.gtag-6	5.05E + 03	4.92E + 03	1.71E + 04	0.30	0.29	68.82 ± 2.28	153.84 ± 2.93
S-bp100.gtag-7	8.96E + 03	6.04E + 03	3.27E + 04	0.27	0.18	33.69 ± 3.96	81.52 ± 15.09
S-bp100.gtag-11	5.02E + 03	5.62E + 03	1.68E + 04	0.30	0.33	66.60 ± 10.97	189.97 ± 33.85

The transgene copy number was determined by qPCR using genomic DNA extracted from T_0 leaf tissue. Means of 10 qPCR experimental replicates are shown with relative standard deviation (RSD) values consistently below 2.5%. Transgene copy numbers were normalized against β-actin. Considering that the T_0 plants are hemizygous concerning the transgene, ratios close to 0.5 indicate single transgene copies. Transgene mRNA expression was assessed by RT-qPCR and normalized against β-actin (GeNorm M values below 0.5) using RNA extracted from leaves of homozygous plants. For each line, means and SDs are shown for three biological replicates of 10 plants per biological replicate.

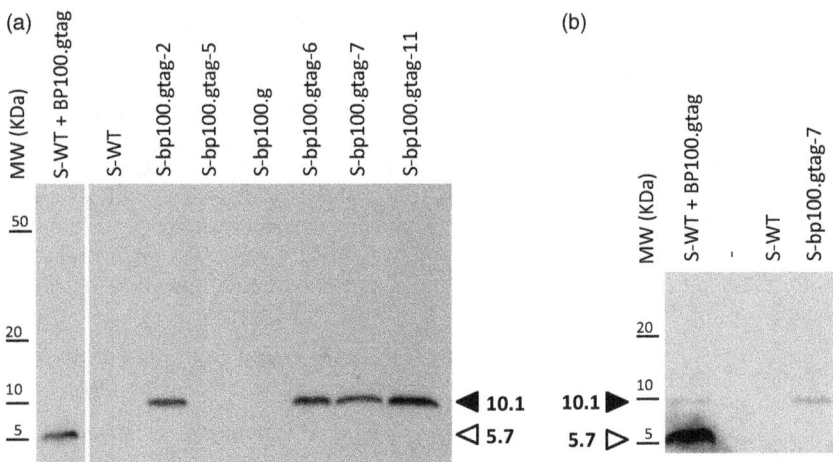

Figure 4 BP100.gtag accumulation in transgenic rice seedlings. Western blot analysis of proteins from five homozygous T_2 S-bp100.gtag rice events. Total soluble protein was extracted from rice seedling samples (five plants per event), and 20 μg (a) or 4 μg (b) of protein per lane was separated by SDS-PAGE before transfer to nitrocellulose filters. Recombinant proteins were detected using antibody mAb54k (diluted 1 : 1500) followed by the horseradish peroxidase-labelled anti-mouse IgG secondary antibody (diluted 1 : 10000) and ECL chemiluminescent detection with 30 ng (a) or 400 ng (b) chemically synthesized BP100.gtag mixed with wild-type rice protein extract as a control (S-WT). S-WT and S-bp100.g (i.e. transgenic rice with the same *bp100der* except for the tag54 sequence) were used as additional controls. Open arrowhead indicates chemically synthesized BP100.gtag (Mw = 5.7 kDa). Closed arrowhead indicates BP100.gtag produced in transgenic plants (Mw = 10.1 kDa).

agents, especially for the protection of plants against pathogens (Montesinos and Bardají, 2008; Montesinos *et al.*, 2012). These cationic, amphipathic α-helical peptides show potent activity against both microbial pathogens and tumour cells, but its toxicity at high concentrations towards eukaryotic cells makes them difficult to produce at high levels in plants (Nadal *et al.*, 2012). They are also difficult to detect and isolate due to their extreme physicochemical properties (highly cationic, pI = 11.5), low extinction coefficient (low contents or absence of aromatic amino acids) and lack of immunogenicity as determined by in silico prediction (OptimumAntigen™ Design Tool, GenScript, NJ) and the low antibody titres observed after immunization of rabbit with a BP100-KLH conjugate (data not shown).

BP100 is a member of the CECMEL11 library with potent activity against the major bacterial pathogens of plants (Badosa *et al.*, 2007). We expressed this peptide as a fusion with the fluorescent marker protein DsRed and the epitope tag54 to confirm that such peptides can be expressed transiently in *N. benthamiana* cells and can accumulate to similar levels as the nontoxic marker protein DsRed alone, when controlled by a strong constitutive promoter such as CaMV 35S (Piotrzkowski *et al.*, 2012).

We also showed that targeting BP100-DsRed-tag54 to the ER using a C-terminal KDEL tag resulted in the induction and rapid accumulation of ER-derived vesicles or protein bodies (PB) that merged to form structures up to 15 μm in diameter by 4 dpi. In contrast, the control protein DsRed behaved as a typical ER luminal protein, showing that the novel protein vesicles were induced by the BP100 component of BP100-DsRed-tag54. The BP100 peptide therefore appears to function in a similar manner to the highly tandemly repeated VPGXG elastin-like motif (ELP) and hydrophobins, which also induce the formation of novel ER-derived protein bodies when joined to ER-targeted fusion partners (Conley *et al.*, 2009; Joensuu *et al.*, 2010). The 93-residue Zera polypeptide derived from the maize seed storage protein γ-zein can also induce the formation of ER-derived protein bodies although in this case the capability appears to be intrinsic and

does not require an additional KDEL sequence. This property has been used to facilitate purification of the recombinant protein (Torrent *et al.*, 2009). Fusions with peptides or polypeptides that induce vesicle formation may be suitable for hard-to-express and toxic protein candidates (Conley *et al.*, 2011).

When we expressed an ER-targeted cyan fluorescent protein along with the DsRed constructs, this not only accumulated in the typical ER along with DsRed in the control experiments, but also co-localized with BP100-DsRed-tag54 in the novel vesicles confirming that the vesicles are derived from the ER compartment and contain typical ER-resident proteins. The proteome of Zera-DsRed protein bodies induced by transient expression in tobacco leaves has recently been characterized (Joseph *et al.*, 2012), revealing the presence of nearly 200 additional proteins including typical ER-trafficking and ER-resident proteins that appear to be recruited to the new vesicles as if they represent extensions of the typical ER. The vesicles induced by BP100 appear to behave in a similar manner. BP100, ELP and Zera share an amphipathic structure that facilitates self-assembly (Llop-Tous *et al.*, 2010) and membrane interactions (Alves *et al.*, 2010). Therefore, although BP100 is much smaller and has a pI of 11.5, its propensity for self-assembly may promote its ability to induce the formation of novel vesicles.

BP100-derived peptides are phytotoxic at concentrations 100-fold higher than the MIC for target microbes, and this was confirmed by the expression of BP100-DsRed-tag54 in stably transformed *A. thaliana* plants. Two-week-old plants demonstrated the same vesicular properties as *N. benthamiana* leaves transiently expressing the same construct, but they were unable to develop fully suggesting that long-term exposure to the peptide has a deleterious impact. Confocal micrographs of radicles expressing BP100-DsRed-tag54 and stained with SYTOX showed the same pattern of cell death as control radicles expressing DsRed alone and wild-type plants (Truernit and Haseloff, 2008). It is therefore clear that BP100 can be expressed transiently in plants without harmful effects but that long-term exposure interferes with normal growth and development.

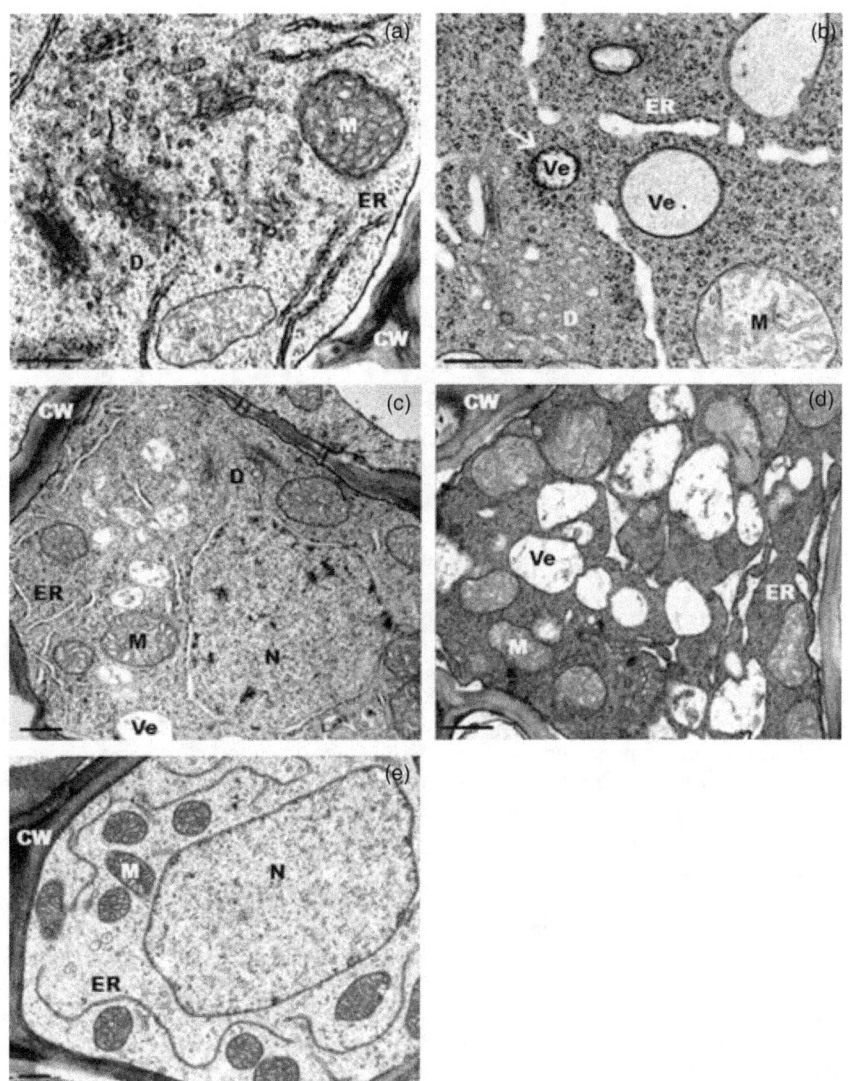

Figure 5 Transmission electron micrographs showing the effects of recombinant or exogenous BP100.gtag on the ultrastructure of rice cells. Surrounding vascular cells of the crown region of S-bp100.gtag.11 seedlings producing BP100.gtag (a,b) and untransformed rice seedlings either grown in the presence of 50 μM chemically synthesized BP100.gtag for 7 days (c,d) or in control conditions (e). A, detail showing increased abundance of dictyosomes; (b), detail showing dilation of ER cisternae, ER-derived vesicles and more abundant dictyosome vesicles; (c), placement of the ER near the cell wall, with some dilation of flattened cisternae, swelling of the nuclear envelope and appearance of vesicles containing electron-dense granules; (d), altered ER morphology, numerous large vesicles with electron-dense granules, altered mitochondria. CW, cell wall; (d), dictyosome; ER, endoplasmic reticulum; M, mitochondria; N, nucleus; Ne, nuclear envelope; Ve, vesicle; Arrow, ribosomes decorating vesicles. Scale bars, 0.5 μm.

The long-term toxicity of BP100 means that it is only suitable for short-term applications such as transient expression and local delivery, for example BP100 has been developed as an efficient cell-penetrating agent to deliver functional cargoes such as the actin-binding Lifeact peptide (MGVADLIKKFESISKEE) into tobacco cells (Eggenberger et al., 2011). However, longer-term applications such as stable expression in plants for molecular farming or to provide pathogen resistance in crops would benefit from rational modification to achieve a targeted reduction in phytotoxicity without affecting antimicrobial potency. We have previously noted that modified BP100 peptides (BP100der) with potent antimicrobial activity against E. amylovora, X. axonopodis pv. vesicatoria and P. syringae pv. syringae can vary significantly in terms of their haemolytic activity (Badosa et al., in preparation). Therefore, we hypothesized that this property could be used as a marker for phytotoxicity, allowing the selection of potent

derivatives suitable for stable expression in plants based on our previous experiments in rice (Nadal et al., 2012).

We found an inverse correlation between the haemolytic activity and transformation efficiency of nine BP100der constructs, with a Pearson coefficient of −0.951 and a bilateral significance of 0.000. One-way ANOVA distinguished two groups of sequences based on their haemolytic activity: those with haemolytic activities up to 15% at 150 μM achieved transformation efficiencies of 30–50% that of the control plasmid, whereas those with haemolytic activities in the 68–99% range at 150 μM achieved only minimal transformation efficiencies (Figure 6). BP100 derivatives selected on the basis of low haemolytic activities therefore appear to provide the greatest likelihood of successful expression in stably transformed plants.

Five BP100 derivatives with low haemolytic activities were expressed in stably transformed rice, including an epitope tag to

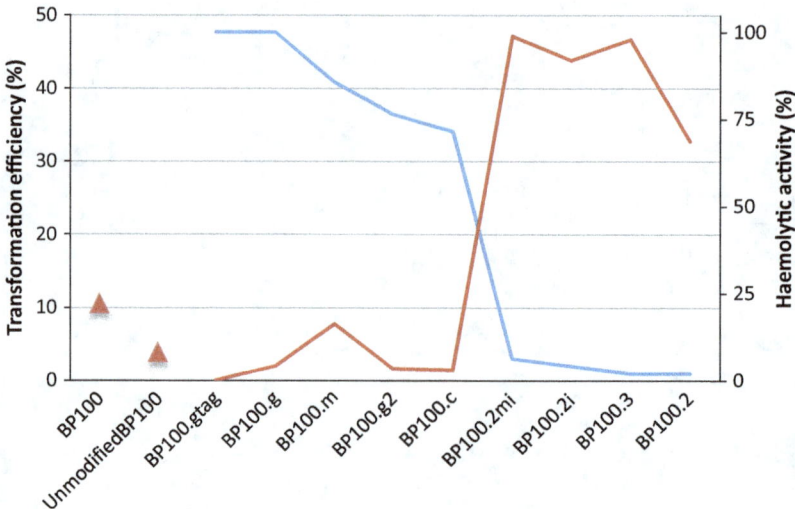

Figure 6 Haemolytic activity of BP100 derivatives and corresponding transformation efficiencies. Transformation efficiency is expressed as a percentage compared to the empty vector (*hpt*II, blue line). Haemolytic activity is expressed as a percentage at 150 μM peptide (red line). Triangles represent BP100 (amidated) and unmodified BP100, for which no transformation efficiency data are available.

facilitate detection. A direct comparison of BP100.g and BP100.gtag confirmed that the tag did not affect the antimicrobial profile or transformation efficiency of the peptide. Although the expression levels varied considerably among independent transformants with the same construct, we found a good correlation between transgene mRNA and recombinant BP100.gtag peptide levels in extracts from the transgenic plants and that the best performing lines produced up to 0.5% BP100der as a proportion of TSP and, importantly, had antibacterial activity. Similar yields of recombinant proteins have been reported using constitutive promoters in *Triticum aestivum* and *N. tabacum* (Khan *et al.*, 2012). We cannot completely exclude the possibility that our yield estimates were affected by the unusual properties of the BP100der peptides, which result in nonspecific interactions with many components of the plant cell, thus reducing the efficiency of isolation from complex matrices. BP100 yield decreases due to unspecific adherence to plant-derived molecules and surfaces of materials used during extraction and purification have been observed (L. Montesinos, personal communication). The recombinant BP100.gtag was larger than its chemically synthesized counterpart used in the control lanes (10.1 vs 5.7 kDa), a phenomenon that has been described for other AMPs such as Sarcotoxin IA and Cecropin A when targeted to the ER in tobacco and rice, respectively (Mitsuhara *et al.*, 2000; Coca *et al.*, 2006). This gain in electrophoretic mobility did not affect the activity of the peptide and may potentially reflect the proposed higher stability of these peptides as dimer.

BP100.gtag was produced in stably transformed rice plants without significantly impairing fitness, suggesting lower toxicity to the host plant than other BP100 derivatives such as BP100.2 that could not be expressed in rice plants using the same approach (Nadal *et al.*, 2012). However, the exogenous exposure of plant tissues to high concentrations of chemically synthesized BP100.gtag did have a significant deleterious impact, as demonstrated in *N. benthamiana* leaf microinfiltration and rice seedling development assays. Indeed, all nine chemically synthesized BP100der peptides with available transformation efficiency data produced similar results in plant toxicity tests, statistical analyses (ANOVA and Tukey B post-test) clustering most peptides together

(Figure S2). There was no correlation between the *N. benthamiana* leaf lesion size or inhibition of seedling development and transformation efficiency (Pearson coefficient of −0.488 and 0.586, respectively). The assays discussed above therefore should not be relied upon to predict plant transformation efficiencies when using constitutive expression constructs and ER-targeted peptides. It may be possible to extrapolate the data obtained with BP100.gtag to other cationic α-helical antimicrobial peptides that show some toxicity towards plant cells when applied at high concentrations.

The exogenous application of BP100.gtag to the roots of rice plants for 7 days resulted in ultrastructural changes similar to those observed in other plants exposed to heavy metals, for example modified ER and dictyosome morphology, modified nuclear envelope structure, the appearance of numerous vesicles containing electron-dense granules and in some cases the degeneration of other organelles and plasmolysis, perhaps reflecting a common response to strongly cationic molecules (Fan *et al.*, 2011; Jiang and Liu, 2010; Liu *et al.*, 2009). Conversely, the ultrastructure of rice cells stably transformed with *bp100.gtag* was less severely affected, and the dictyosome and ER modifications appeared similar to those associated with a highly active endomembrane system or the morphological changes observed during very short exposure to metal ions. The accumulation of recombinant peptides in ER-derived vesicles probably protects the cell from the worst effects, thus minimizing the residual toxicity of these BP100-derived peptides. Similarly, the antimicrobial peptide Cecropine A could be expressed in rice when targeted for accumulation in the ER but not when targeted to the apoplast (Coca *et al.*, 2006).

In conclusion, cationic α-helical peptides such as BP100 can be produced as recombinant proteins in plant cells, using an approach based on constitutive expression and ER targeting. Using a fluorescent fusion partner, we demonstrated that BP100 derivatives accumulate in ER-derived vesicles when transiently expressed in *N. benthamiana* and when stably expressed in *A. thaliana* and that the BP100 component induces the formation of these vesicles. BP100 expression in transgenic plants does not cause cell death in young tissues, but long-term expression nevertheless impairs normal growth and development. We demonstrated that antimicrobial peptides such as BP100 could be produced in a transient-expression plant-based platform. We also showed that BP100 derivatives with low haemolytic activity but high antimicrobial potency retain a degree of phytotoxicity at high doses but can nevertheless be produced in stably transformed rice plants at levels of up to 0.5% TSP when targeted to the ER.

Experimental procedures

Chemical synthesis of BP100-derived peptides and *in vitro* characterization

The BP100-derived peptides were synthesized using solid-phase Fmoc-type chemistry as previously described (Badosa *et al.*, 2007) and assessed for purity by HPLC (the purity was >90% in all cases). Peptide identity was confirmed by electrospray ionization mass spectrometry. Peptides were solubilized in sterile Milli-Q H_2O to a concentration of 1 mM and filter-sterilized (0.22-μm pore filter).

Antimicrobial activity tests were carried out as previously described (Nadal *et al.*, 2012) using the plant pathogens *Erwinia amylovora* PMV6076 (INRA, Angers, France), *Pseudomonas syringae* pv. *syringae* EPS94 (UdG, Girona, Spain) and *Xanthomonas axonopodis* pv. *vesicatoria* 2133-2. Positive (water instead of peptide) and negative (water instead of microbial suspension) controls were included in each experiment. Three biological replicates were performed, each comprising two experimental replicates. The lowest peptide concentration inhibiting microbial growth at the end of the experiment was established as the minimal inhibitory concentration (MIC). Haemolytic activity was assessed by measuring the release of haemoglobin from erythrocyte suspensions prepared from fresh human blood as previously described (Badosa *et al.*, 2007). The percentage haemolysis was calculated relative to mellitin and Tris buffer. Each experiment was carried out three times.

Phytotoxic activity was determined using a rice seedling development test and a *N. benthamiana* leaf inoculation assay as previously described (Nadal *et al.*, 2012). Toxicity in the first assay was expressed as the inverse shoot length of 12 plantlets per treatment. Toxicity in the second assay was expressed as the mean of the lesion diameters of six replicates infiltrated with 100 μl of 50 μM peptide.

Construction of vectors

The *bp100:dsred:tag54* and *dsred:tag54* constructs included the *Petrosilinum hortense* chalcone synthase 5' untranslated region and a sequence encoding the codon optimized signal peptide from the heavy chain of monoclonal antibody 24 (Vaquero *et al.*, 1999) upstream of BP100-DsRed or DsRed (Matz *et al.*, 1999), and the epitope tag 12-tag54 (Rasche *et al.*, 2011) and KDEL ER retention motif downstream, all placed under the control of the double-enhanced *Cauliflower mosaic virus* 35S promoter and terminator (Piotrzkowski *et al.*, 2012). The constructs were directionally inserted into the KpnI and SbfI sites of pCAMBIA1300 to obtain pCbp100-dsred-tag54 and pCdsred-tag54. After sequencing the whole insert, these plasmids were transferred into *A. tumefaciens* strain GV2260 by electroporation and infiltrated into *N. benthamiana* leaves or used to transform *A. thaliana* plants by floral dipping. Subcloning procedures and the transformation of *Escherichia coli* strain XL1Blue were carried out using standard techniques (Sambrook and Russell, 2001).

Sequences encoding the five BP100 derivatives fused in-frame to the *N. tabacum* pathogenesis-related protein PR1a signal peptide sequence were designed according to rice codon usage. Synthetic *bp100.c*, *bp100.g*, *bp100.g2*, *bp100.gtag* and *bp100.m* genes were prepared by GenScript (Piscataway NJ) and included terminal BamHI restriction sites to facilitate insertion into pAHC17 (Oh *et al.*, 2000). The constructs were flanked by the promoter, first exon and first intron of the maize *ubiquitin-1* gene (Huang, 2000) and the *A. tumefaciens* nopaline synthase (*nos*) terminator. After verification by sequencing, the complete cassettes were introduced into the KpnI site of pCAMBIA1300 in the opposite orientation to the *hptII* gene to generate five pCBP100der vectors for the stable transformation of rice. The resulting binary vectors were transferred into *A. tumefaciens* strain EHA105 by cold shock (Sambrook and Russell, 2001).

Agroinfiltration of *N. benthamiana* leaves

Wild-type *N. benthamiana* plants were cultivated for 4–5 weeks in a greenhouse at 18–28 °C with a long-day (16-h) photoperiod of light. And the bacteria cultures containing the expression vectors discussed above as well as the one encoding eCFP (pCSPECFPKDEL, Joseph *et al.*, 2012; kindly provided by D. Ludevid, CRAG) were mixed with cultures carrying the HC-Pro suppressor of silencing (Goytia *et al.*, 2006). Agroinfiltration of the abaxial side of the upper leaves was carried out using a syringe without needle.

Stable transformation of *A. thaliana*

Arabidopsis thaliana plants were transformed with *A. tumefaciens* cultures carrying the expression vectors discussed above by floral dipping (Clough and Bent, 1998). Seeds were surface-sterilized and allowed to germinate on MS medium with 20 mg/L hygromycin B in culture chambers for 3 days in the dark at 4 °C (stratification) and 13 days under long-day conditions (16-h photoperiod) at 22 °C. Plantlets were analysed by confocal microscopy or allowed to grow to maturity under standard conditions.

Stable transformation of rice

Embryonic rice callus (*Oryza sativa* L., ssp *japonica*, cv Senia) derived from mature embryos as described (Pons *et al.*, 2000) was transformed with the constructs discussed above and pCAMBIA 1300 (containing *hptII*) as a control. We transformed 500 callus pieces with the constructs containing the *bp100.gtag*, *bp100.m*, *bp100.g* and *bp100.c* genes and 1000 callus pieces with the constructs containing the *bp100.c2* and *bp100.g2* genes. Hygromycin-resistant, fertile T_0 plants were grown to maturity, and leaves were tested for transgene insertion by qPCR. The efficiency of transformation was calculated by comparing the number of transformants obtained with each BP100der construct compared with the pCAMBIA 1300 control. S-bp100.gtag plants were cultured under standard greenhouse conditions to obtain homozygous transgenic lines in the T_2 generation.

Protein extraction and Western blot analysis

Total soluble protein was extracted from three *N. benthamiana* leaf sections (~1 cm^2 agroinfiltrated tissue) or shoots from six rice seedlings germinated in a culture chamber at 25 ± 1 °C with a photoperiod of 16 h light/8 h dark under fluorescent Sylvania Cool White lamps. Tissue was extracted in lysis buffer (10 mM Tris-HCl pH 6.2, 50 mM KCl, 6 mM $MgCl_2$, 0.4 M NaCl, 1% (v/v) Triton X-100 and 10 mM EDTA) and centrifuged at 16 000 **g** for 15 min at 4 °C, and the protein concentration in the supernatant was determined using the Bradford method. The extracted protein was separated by 12% or 15% (w/v) SDS polyacrylamide gel electrophoresis (20 or 4 μg per lane) and transferred to nitrocellulose filters, and the tag54 epitope was detected with monoclonal antibody mAb54k (Rasche et al., 2011) diluted 1 : 1500, overnight at 4 °C. Binding was detected using a horseradish peroxidase-conjugated anti-mouse IgG secondary antibody (GE Healthcare Life sciences) diluted 1 : 10000 for 1 h at room temperature. The signal was detected by ECL chemiluminescence (Supersignal® West Femto, Thermo Scientific, Waltham, MA) and quantified using Multi Gauge v3.0 software based on 25.0, 12.5, 6.25, 3.125 and 1.5625 pmol standards corresponding to 143.4–9.0 ng of chemically synthesized BP100.gtag mixed with protein extracts from wild-type rice, or 0.72–0.045% rice TSP, run on the same gel.

Antimicrobial activity of recombinant peptides

Protein extracts were obtained from 30 rice seedlings germinated for 4 days in a culture chamber at 25 ± 1 °C with a photoperiod of 16 h light/8 h dark. Tissue was ground in liquid nitrogen and treated with 10 mM phosphate buffer pH 7.5 and 0.6 M sucrose buffer, previously sterilized through a 0.2-μm pore filter. The extract was centrifuged at 200 g for 10 min at 4 °C, and the supernatant was re-centrifuged at 2000 g for 10 min at 4 °C to precipitate a fraction enriched in vesicles and protein bodies, which was re-suspended in 10 mM phosphate buffer pH 7.5 with protease inhibitor cocktail (Sigma-Aldrich, Munich, Germany). Protein concentrations were determined using the Bradford method and were consistently in the 12–15 mg/mL range. Preliminary experiments showed that the protein extraction buffer had no effect on E. amylovora growth, but protein extracts from control rice samples inhibited bacterial growth. Dilution of protein extracts to 1/3 in the same buffer permitted growth of the indicator bacteria after a short lag phase. To evaluate the expected inhibition of the recombinant BP100.gtag, controls were carried out with extracts from rice seedlings harbouring only the hptII selection gene. Three biological replicates were performed, each comprising three experimental replicates. The area under the growth curve (AUC) was used to calculate the inhibition effects of the extracts from transgenic seedlings expressing BP100.gtag compared with control seedlings or the buffer control: (Ac–At)*100/Ac, where Ac and At are the AUC of control and test seedlings, respectively.

Imaging

Confocal microscopy

Nicotiana benthamiana leaves and A. thaliana leaves/radicles were analysed by confocal microscopy using an FV1000 Olympus microscope. Red fluorescence images were collected at 559 nm excitation and 570–670 nm emission, whereas cyan fluorescence images were collected at 405 nm excitation and 460–500 nm emission. ImageJ software (http://rsb.info.nih.gov/ij/) was used to calculate the number and size of fluorescent spots or aggregates. SYTOX staining was carried out by incubating plantlets with 2.5 μM SYTOX (Invitrogen Life Technologies, Carlsbad, CA) for 24 h prior to analysis at 559 nm excitation and 570–670 nm emission. A total of three agroinfiltrated fields in the upper, medial and basal portions of different leaves were analysed using Olympus Fluoview v3.1 software with standard parameters. IBM SPSS Statistics19 software was used to compare means by one-way ANOVA and Tukey B post-test.

Transmission electron microscopy

Rice seeds were allowed to germinate for 7 days in distilled water under controlled conditions, with or without 50 μM chemically synthesized BP100.gtag. Sections of the collar region were excised and fixed in 1% (v/v) glutealdehyde and 2.5% (v/v) paraformaldehyde in 0.1 M phosphate buffer, pH 7.4, overnight at room temperature. TEM sections were prepared by the Service for Microscopy at the Autonomous University of Barcelona, and images were collected on a GEM-1400 TEM. Wild-type rice and plants transformed with the pCAMBIA 1300 vector (containing hptII) were used as controls.

Nucleic acid extraction, qPCR and RT-qPCR

To assess transgene copy number, genomic DNA was extracted from 1 g of mature hemizygous T_0 leaf tissue using a CTAB method (Coll et al., 2008) and analysed by qPCR targeting conserved sequences in each construct, with 10 experimental replicates (Nadal et al., 2012) and the rice β-actin gene as a standard for normalization (Montero et al., 2011). All oligonucleotides were purchased from MWG Biotech AG (Germany).

Transgene expression was measured in leaf samples from three independent T_0 transgenic rice plants representing each construct and in five homozygous lines expressing bp100.gtag. Leaves of the homozygous T_2 plants were sampled at the two-leaf vegetative stage, and three biological replicates of 10 plants (i.e. a total of 30 plants) were analysed per line. Total RNA was extracted from 400-mg samples using Trizol reagent (Invitrogen Life Technologies) with DNase I treatment (Ambion, Grand Island, NY), according to the manufacturer's instructions. The concentration and quality of the RNA were confirmed by UV absorption at 260 and 280 nm using a NanoDrop ND1000 spectrophotometer (Nanodrop technologies, Wilmington, DE). RT-qPCR was carried out as previously described (Nadal et al., 2012). For each sample, cDNA was prepared with random primers in duplicate, and the reactions were performed in triplicate. The absence of DNA targets was demonstrated using DNase-treated samples. All reactions had linearity coefficients exceeding 0.99 and efficiency values above 0.95. The β-actin gene was used for normalization, its suitability having been confirmed using the geNORM v3.4 statistical algorithm (Vandesompele et al., 2002; M values below 0.5 in our samples).

Acknowledgements

This work was supported by the Spanish Ministerio de Ciencia e Innovación (projects AGL2010-17181/AGR and PLANT-KBBE EUI2008-03769) and BMBF PLANT-KBBE FKZ0315456B. We thank J. Messeguer and E. Melé (IRTA, Cabrils) for valuable suggestions; D. Ludevid (CRAG, Barcelona) for providing plasmid material; M. Amenós (CRAG) and the staff of the Microscopy Service (Universitat Autonoma de Barcelona) for technical assistance; E. Badosa and L. Montesinos for assessing the antimicrobial activity of chemically synthesized peptides; and the scientific writer R.M. Twyman for revision of the manuscript.

References

Ajesh, K. and Sreejith, K. (2009) Peptide antibiotics: an alternative and effective antimicrobial strategy to circumvent fungal infections. Peptides, **30**, 999–1006.

Alves, C.S., Melo, M.N., Franquelim, H.G., Ferré, R., Planas, M., Feliu, L., Bardají, E., Kowalczyk, W., Andreu, D., Santos, N.C., Fernandes, M.X. and Castanho, M.A. (2010) *Escherichia coli* cell surface perturbation and disruption induced by antimicrobial peptides BP100 and pepR. *J. Biol. Chem.* **285**, 27536–27544.

Badosa, E., Ferré, R., Planas, M., Feliu, L., Besalú, E., Cabrefiga, J., Bardají, E. and Montesinos, E. (2007) A library of linear undecapeptides with bactericidal activity against phytopathogenic bacteria. *Peptides*, **28**, 2276–2285.

Badosa, E., Ferre, R., Frances, J., Bardaji, E., Feliu, L., Planas, M. and Montesinos, E. (2009) Sporicidal activity of synthetic antifungal undecapeptides and control of Penicillium rot of apples. *Appl. Environ. Microbiol.* **75**, 5563–5569.

Bardají, E. *Antimicrobial linear peptides*. (Patent WO/2007/125142 A1). (2006).

Bechinger, B. (2004) Structure and function of membrane-lytic peptides. *Crit. Rev. Plant Sci.* **23**, 271–292.

Boman, H.G. (2003) Antibacterial peptides: basic facts and emerging concepts. *J. Intern. Med.* **254**, 197–215.

Broekaert, W., Cammue, B., De Bolle, M., Thevissen, K., De Samblanx, G. and Osborn, R. (1997) Antimicrobial peptides from plants. *Crit. Rev. Plant Sci.* **16**, 297–323.

Brogden, K.A. (2005) Antimicrobial peptides: pore formers or metabolic inhibitors in bacteria? *Nat. Rev. Microbiol.* **3**, 238–250.

Brogden, K.A., Ackermann, M., McCray, P.B. Jr and Tack, B.F. (2003) Antimicrobial peptides in animals and their role in host defences. *Int. J. Antimicrob. Agents*, **22**, 465–478.

Bulet, P. and Stocklin, R. (2005) Insect antimicrobial peptides: structures, properties and gene regulation. *Protein Pept. Lett.* **12**, 3–11.

Bulet, P., Stocklin, R. and Menin, L. (2004) Anti-microbial peptides: from invertebrates to vertebrates. *Immunol. Rev.* **198**, 169–184.

Cavallarín, L., Andreu, D. and San Segundo, B. (1998) Cecropin A-derived peptides are potent inhibitors of fungal plant pathogens. *Mol. Plant Microbe Interact.* **11**, 218–227.

Clough, S.J. and Bent, A.F. (1998) Floral dip: a simplified method for Agrobacterium-mediated transformation of *Arabidopsis thaliana*. *Plant J.* **16**, 735–743.

Coca, M., Peñas, G., Gómez, J., Campo, S., Bortolotti, C., Messeguer, J. and San Segundo, B. (2006) Enhanced resistance to the rice blast fungus *Magnaporthe grisea* conferred by expression of a cecropin A gene in transgenic rice. *Planta*, **223**, 392–406.

Coll, A., Nadal, A., Palaudelmàs, M., Messeguer, J., Melé, E., Puigdomènech, P. and Pla, M. (2008) Lack of repeatable differential expression patterns between MON810 and comparable commercial varieties of maize. *Plant Mol. Biol.* **68**, 105–117.

Conley, A.J., Joensuu, J.J., Menassa, R. and Brandle, J.E. (2009) Induction of protein body formation in plant leaves by elastin-like polypeptide fusions. *BMC Biol.* **7**, 48.

Conley, A.J., Joensuu, J.J., Richman, A. and Menassa, R. (2011) Protein body-inducing fusions for high-level production and purification of recombinant proteins in plants. *Plant Biotechnol. J.* **9**, 419–433.

Cooter, P.D., Hill, C. and Ross, P. (2010) Bacterial lantibiotics: strategies to improve therapeutic potential. *Curr. Prot. Pept. Sci.* **6**, 61–75.

Degenkolb, T., Berg, A., Gams, W., Schlegel, B. and Grafe, U. (2003) The occurrence of peptaibols and structurally related peptaibiotics in fungi and their mass spectrometric identification via diagnostic fragment ions. *J. Pept. Sci.* **9**, 666–678.

Eggenberger, K., Mink, C., Wadhwani, P., Ulrich, A.S. and Nick, P. (2011) Using the peptide Bp100 as a cell-penetrating tool for the chemical engineering of actin filaments within living plant cells. *ChemBioChem*, **12**, 132–137.

Fan, J.L., Wei, X.Z., Wan, L.C., Zhang, L.Y., Zhao, X.Q., Liu, W.Z., Hao, H.Q. and Zhang, H.Y. (2011) Disarrangement of actin filaments and Ca^{2+} gradient by $CdCl_2$ alters cell wall construction in *Arabidopsis thaliana* root hairs by inhibiting vesicular trafficking. *J. Plant Physiol.* **168**, 1157–1167.

Ferré, R., Melo, M.N., Correira, A.D., Feliu, L., Bardají, E., Planas, M. and Castanho, M. (2009) Synergistic effects of the membrane actions of cecropin-melittin antimicrobial hybrid peptide BP100. *Biophys. J.* **96**, 1815–1827.

Fischer, R., Stoger, E., Schillberg, S., Christou, P. and Twyman, R.M. (2004) Plant-based production of biopharmaceuticals. *Curr. Opin. Plant Biol.* **7**, 152–158.

Ganz, T. (2003) Defensins: antimicrobial peptides of innate immunity. *Nat. Rev. Immunol.* **3**, 710–720.

Goytia, E., Fernández-Calvino, L., Martínez-García, B., López-Abella, D. and López-Moya, J.J. (2006) Production of plum pox virus HC-Pro functionally active for aphid transmission in a transient-expression system. *J. Gen. Virol.* **87**, 3413–3423.

Hancock, R.E. (2001) Cationic peptides: effectors in innate immunity and novel antimicrobials. *Lancet Infect. Dis.* **1**, 156–164.

Hoja, U. and Sonnewald, U. (2013) *Molecular Farming in Plants, Encyclopedia of Life Sciences*. Chichester: John Wiley & Sons Ltd.

Huang, H.W. (2000) Action of antimicrobial peptides: two-state model. *Biochemistry*, **39**, 8347–8352.

Huang, H.W. (2006) Molecular mechanism of antimicrobial peptides: the origin of cooperativity. *Biochim. Biophys. Acta*, **1758**, 1292–1302.

Jack, R.W. and Jung, G. (2000) Lantibiotics and microcins: polypeptides with unusual chemical diversity. *Curr. Opin. Chem. Biol.* **4**, 310–317.

Jenssen, H., Hamill, P. and Hancock, R.E. (2006) Peptide antimicrobial agents. *Clin. Microbiol. Rev.* **19**, 491–511.

Jiang, W. and Liu, D. (2010) Pb-induced cellular defense system in the root meristematic cells of *Allium sativum* L. *BMC Plant Biol.* **10**, 40.

Joensuu, J.J., Conley, A.J., Lienemann, M., Brandle, J.E., Linder, M.B. and Menassa, R. (2010) Hydrophobin fusions for high-level transient protein expression and purification in *Nicotiana benthamiana*. *Plant Physiol.* **152**, 622–633.

Joseph, M., Ludevid, M.D., Torrent, M., Rofidal, V., Tauzin, M., Rossignol, M. and Peltier, J.B. (2012) Proteomic characterisation of endoplasmic reticulum-derived protein bodies in tobacco leaves. *BMC Plant Biol.* **12**, 36.

Khan, I., Twyman, R.M., Arcalis, E. and Stoger, E. (2012) Using storage organelles for the accumulation and encapsulation of recombinant proteins. *Biotechnol. J.* **7**, 1099–1108.

Lay, F.T. and Anderson, M.A. (2005) Defensins–components of the innate immune system in plants. *Curr. Protein Pept. Sci.* **6**, 85–101.

Liu, D., Jiang, W., Meng, Q., Zou, J., Gu, J. and Zeng, M. (2009) Cytogenetical and ultrastructural effects of copper on root meristem cells of *Allium sativum* L. *Biocell*, **33**, 25–32.

Llop-Tous, I., Madurga, S., Giralt, E., Marzabal, P., Torrent, M. and Ludevid, M.D. (2010) Relevant elements of a maize gamma-zein domain involved in protein body biogenesis. *J. Biol. Chem.* **285**, 35633–35644.

López-García, B., Pérez-Paya, E. and Marcos, J.F. (2002) Identification of novel hexapeptides bioactive against phytopathogenic fungi through screening of a synthetic peptide combinatorial library. *Appl. Environ. Microbiol.* **68**, 2453–2460.

Marcos, J.F. and Gandía, M. (2009) Antimicrobial peptides: to membranes and beyond. *Expert Opin. Drug Discov.* **4**, 659–671.

Marcos, J.F., Muñoz, A., Pérez-Paya, E., Misra, S. and López-García, B. (2008) Identification and rational design of novel antimicrobial peptides for plant protection. *Annu. Rev. Phytopathol.* **46**, 273–301.

Matz, M.V., Fradkov, A.F., Labas, Y.A., Savitsky, A.P., Zaraisky, A.G., Markelov, M.L. and Lukyanov, S.A. (1999) Fluorescent proteins from nonbioluminescent *Anthozoa* species. *Nat. Biotechnol.* **17**, 969–973.

Mitsuhara, I., Matsufuru, H., Ohshima, M., Kaku, H., Nakajima, Y., Murai, N., Natori, S. and Ohashi, Y. (2000) Induced expression of sarcotoxin IA enhanced host resistance against both bacterial and fungal pathogens in transgenic tobacco. *Mol. Plant Microbe Interact.* **13**, 860–868.

Monroc, S., Badosa, E., Besalú, E., Planas, M., Bardají, E., Montesinos, E. and Feliu, L. (2006) Improvement of cyclic decapeptides against plant pathogenic bacteria using a combinatorial chemistry approach. *Peptides*, **27**, 2575–2584.

Montero, M., Coll, A., Nadal, A., Messeguer, J. and Pla, M. (2011) Only half the transcriptomic differences between resistant genetically modified and conventional rice are associated with the transgene. *Plant Biotechnol. J.* **9**, 693–702.

Montesinos, E. (2007) Antimicrobial peptides and plant disease control. *FEMS Microbiol. Lett.* **270**, 1–11.

Montesinos, E. and Bardají, E. (2008) Synthetic antimicrobial peptides as agricultural pesticides for plant-disease control. *Chem. Biodivers.* **5**, 1225–1237.

Montesinos, E., Badosa, E., Cabrefiga, J., Planas, M., Feliu, L. and Bardají, E. (2012) Antimicrobial peptides for plant disease control. From discovery to application. In *Small Wonders: Peptides for Disease Control* (Rajasekaran, K., Cary, J., Jaynes, J. and Montesinos, E., eds.), pp. 235–261. Washington, DC: Oxford University Press.

Moreno, A.B., Martínez, D.P. and San Segundo, B. (2006) Biotechnologically relevant enzymes and proteins. Antifungal mechanism of the *Aspergillus giganteus* AFP against the rice blast fungus *Magnaporthe grisea*. *Appl. Microbiol. Biotechnol.* **72**, 883–895.

Nadal, A., Montero, M., Company, N., Badosa, E., Messeguer, J., Montesinos, L., Montesinos, E. and Pla, M. (2012) Constitutive expression of transgenes encoding derivatives of the synthetic antimicrobial peptide BP100: impact on rice host plant fitness. *BMC Plant Biol.* **12**, 159.

Ng, T.B. (2004) Peptides and proteins from fungi. *Peptides*, **25**, 1055–1073.

Oh, D., Shin, S.Y., Lee, S., Kang, J.H., Kim, S.D., Ryu, P.D., Hahm, K.S. and Kim, Y. (2000) Role of the hinge region and the tryptophan residue in the synthetic antimicrobial peptides, cecropin A(1-8)-magainin 2(1-12) and its analogues, on their antibiotic activities and structures. *Biochemistry*, **39**, 11855–11864.

Otvos, L. Jr. (2000) Antibacterial peptides isolated from insects. *J. Pept. Sci.* **6**, 497–511.

Peschel, A. and Sahl, H.G. (2006) The co-evolution of host cationic antimicrobial peptides and microbial resistance. *Nat. Rev. Microbiol.* **4**, 529–536.

Piotrzkowski, N., Schillberg, S. and Rasche, S. (2012) Tackling heterogeneity: a leaf disc-based assay for the high-throughput screening of transient gene expression in tobacco. *PLoS ONE*, **7**, e45803.

Pons, M.J., Marfà, V., Melé, E. and Messeguer, J. (2000) Regeneration and genetic transformation of Spanish rice cultivars using mature embryos. *Euphytica*, **114**, 117–122.

Raaijmakers, J.M., De Bruijn, I. and de Kock, M.J. (2006) Cyclic lipopeptide production by plant-associated *Pseudomonas* spp.: diversity, activity, biosynthesis, and regulation. *Mol. Plant Microbe Interact.* **19**, 699–710.

Rajasekaran, K., Cary, J., Jaynes, J. and Montesinos, E. (2012) *Small Wonders: Peptides for Disease Control*. American Chemical Society: Oxford University Press.

Rasche, S., Martin, A., Holzem, A., Fischer, R., Schinkel, H. and Schillberg, S. (2011) One-step protein purification: Use of a novel epitope tag for highly efficient detection and purification of recombinant proteins. *Open Biotechnol. J.* **5**, 1–6.

Sambrook, J. and Russell, D. (2001) *Molecular Cloning: A Laboratory Manual*, 3rd ed. Cold Spring Harbor: Cold Spring Harbor Laboratory Press.

Tincu, J.A. and Taylor, S.W. (2004) Antimicrobial peptides from marine invertebrates. *Antimicrob. Agents Chemother.* **48**, 3645–3654.

Toke, O. (2005) Antimicrobial peptides: new candidates in the fight against bacterial infections. *Biopolymers*, **80**, 717–735.

Torrent, M., Llompart, B., Lasserre-Ramassamy, S., Llop-Tous, I., Bastida, M., Marzabal, P., Westerholm-Parvinen, A., Saloheimo, M., Heifetz, P.B. and Ludevid, M.D. (2009) Eukaryotic protein production in designed storage organelles. *BMC Biol.* **7**, 5.

Torrent, M., Pulido, D., Rivas, L. and Andreu, D. (2012) Antimicrobial peptide action on parasites. *Curr. Drug Targets*, **13**, 1138–1147.

Tossi, A., Sandri, L. and Giangaspero, A. (2000) Amphipathic, alpha-helical antimicrobial peptides. *Biopolymers*, **55**, 4–30.

Truernit, E. and Haseloff, J. (2008) A simple way to identify non-viable cells within living plant tissue using confocal microscopy. *Plant Methods*, **4**, 15.

Twyman, R.M., Stoger, E., Schillberg, S., Christou, P. and Fischer, R. (2003) Molecular farming in plants: host systems and expression technology. *Trends Biotechnol.* **21**, 570–578.

Vandesompele, J., De Preter, K., Pattyn, F., Poppe, B., Van Roy, N., De Paepe, A. and Speleman, F. (2002) Accurate normalization of real-time quantitative RT-PCR data by geometric averaging of multiple internal control genes. *Genome biology*, **3**, Research 0034.

Vaquero, C., Sack, M., Chandler, J., Drossard, J., Schuster, F., Monecke, M., Schillberg, S. and Fischer, R. (1999) Transient expression of a tumor-specific single-chain fragment and a chimeric antibody in tobacco leaves. *Proc. Natl Acad. Sci. USA*, **96**, 11128–11133.

Yeaman, M.R. and Yount, N.Y. (2003) Mechanisms of antimicrobial peptide action and resistance. *Pharmacol. Rev.* **55**, 27–55.

Yount, N.Y. and Yeaman, M.R. (2005) Immunocontinuum: perspectives in antimicrobial peptide mechanisms of action and resistance. *Protein Pept. Lett.* **12**, 49–67.

Zasloff, M. (2002) Antimicrobial peptides of multicellular organisms. *Nature*, **415**, 389–395.

4

Integrated analysis of seed proteome and mRNA oxidation reveals distinct post-transcriptional features regulating dormancy in wheat (*Triticum aestivum* L.)

Feng Gao[1], Christof Rampitsch[2], Vijaya R. Chitnis[1], Gavin D. Humphreys[2], Mark C. Jordan[2] and Belay T. Ayele[1,*]

[1]Department of Plant Science, University of Manitoba, Winnipeg, MB, Canada
[2]Cereal Research Centre, Agriculture and Agri-Food Canada, Winnipeg, MB, Canada

*Correspondence
email b_ayele@umanitoba.ca

Keywords: after-ripening, mRNA oxidation, proteomics, seed dormancy, seed imbibition, wheat.

Summary

Wheat seeds can be released from a dormant state by after-ripening; however, the underlying molecular mechanisms are still mostly unknown. We previously identified transcriptional programmes involved in the regulation of after-ripening-mediated seed dormancy decay in wheat (*Triticum aestivum* L.). Here, we show that seed dormancy maintenance and its release by dry after-ripening in wheat is associated with oxidative modification of distinct seed-stored mRNAs that mainly correspond to oxidative phosphorylation, ribosome biogenesis, nutrient reservoir and α-amylase inhibitor activities, suggesting the significance of post-transcriptional repression of these biological processes in regulating seed dormancy. We further show that after-ripening induced seed dormancy release in wheat is mediated by differential expression of specific proteins in both dry and hydrated states, including those involved in proteolysis, cellular signalling, translation and energy metabolism. Among the genes corresponding to these proteins, the expression of those encoding α-amylase/trypsin inhibitor and starch synthase appears to be regulated by mRNA oxidation. Co-expression analysis of the probesets differentially expressed and oxidized during dry after-ripening along with those corresponding to proteins differentially regulated between dormant and after-ripened seeds produced three co-expressed gene clusters containing more candidate genes potentially involved in the regulation of seed dormancy in wheat. Two of the three clusters are enriched with elements that are either abscisic acid (ABA) responsive or recognized by ABA-regulated transcription factors, indicating the association between wheat seed dormancy and ABA sensitivity.

Introduction

Dormancy is an adaptive trait that confers seeds a mechanism to remain quiescent until conditions are optimal for germination (Finkelstein et al., 2008). Modern cereal cultivars, however, are characterized by reduced seed dormancy due to selection by breeders for rapid and uniform germination and seedling establishment (Simpson, 1990). As a result, they are susceptible to preharvest sprouting (PHS), which causes substantial loss in grain yield and quality (Gubler et al., 2005), especially in years when wet conditions occur along with harvest maturity. Given that wheat (*Triticum aestivum* L.) is one of the most important crops worldwide, incorporation of a certain degree of dormancy into commercial cultivars is essential to mitigate the economic losses associated with sprouting damage. However, this requires detailed dissection of the molecular mechanisms underlying dormancy in wheat seeds.

During the late maturation phase, seeds of many plant species acquire primary dormancy, and its decay is regulated by a number of endogenous and exogenous cues that exert both synergistic and competing effects (Finkelstein et al., 2008). Seed dormancy decay by after-ripening, a period of dry seed storage, has been demonstrated in many plant species (Iglesias-Fernandez et al., 2011); however, the underlying molecular features are still not well known. Thermodynamic analyses of water sorption isotherms have shown that seed dormancy release by dry after-ripening is associated with changes in molecular mobility and seed water status (Bazin et al., 2011a). It has been proposed by the same authors that the change in seed water status is triggered by nonenzymatic oxidative processes and can induce active metabolic reactions including gene transcription. Consistently, changes in transcript abundance appear to occur during dry after-ripening of seeds in various species (Bazin et al., 2011b; Bove, 2005; Gao et al., 2012; Liu et al., 2013), although de novo transcription is not indispensable for seed germination (Rajjou et al., 2004). After-ripening also induces imbibition-mediated transcriptional changes related to various biological processes. For example, transcripts of genes involved in cell organization and biogenesis, proteolysis and those associated with the protein synthesis machinery are over-represented in imbibing after-ripened (AR) seeds of *Nicotiana plumbaginifolia* (Bove, 2005), Arabidopsis (*Arabidopsis thaliana*, Cadman et al., 2006) and wheat (Gao et al., 2012), whereas the expressions of genes involved in abscisic acid (ABA) biosynthesis, gibberellin (GA) catabolism and stress response are enriched in the corresponding dormant (D) seeds. Furthermore, the transcription of genes associated with the protein synthesis machinery was shown to be repressed in imbibing D seeds. These results highlight the significance of transcriptional changes in controlling de novo protein synthesis, which is essential for seed germination (Rajjou et al., 2004).

Gene expression is also regulated by a variety of post-transcriptional mechanisms, and pictures about the role of such mechanisms in regulating seed dormancy are emerging (Bazin et al., 2011b). Oxygen permeates through a glassy matrix such as vitreous cytoplasm (Andersen et al., 2000), indicating its ability to diffuse into dry seeds during storage and thereby acts as a major source of reactive oxygen species (ROS). This is because respiratory and enzymatic activities are repressed when the seeds are in the dry state. In agreement with this, after-ripening-induced seed dormancy release in sunflower (Helianthus annuus) is associated with the accumulation of ROS (Oracz et al., 2007). Apart from lipids and proteins, ROS are able to oxidize nucleic acids; and RNA molecules are more susceptible to oxidative damage than DNA (Kong and Lin, 2010). It has been found by Bazin et al. (2011b) that mRNAs are more sensitive to after-ripening-induced oxidation than other RNA species in seeds. Oxidized mRNAs mainly derive from the oxidation of base guanine that produces 8-oxo-7, 8-dihydroguanine (8-OHG). The 8-OHG serves as a marker for changes in the degree of mRNA oxidations upon detection with anti-8-OHG antibodies. Seed dormancy release by after-ripening has recently been shown to be tightly associated with the oxidation of specific seed-stored mRNAs, which always leads to decreased protein synthesis (Bazin et al., 2011b). As the anatomical, physiological and metabolic bases of dormancy and germination are distinct between dicot and monocot species, comparative analysis of after-ripening-induced oxidation of seed-stored transcripts between the two species provides valuable insights into unique and conserved post-transcriptional mechanisms regulating dormancy release. To date, however, after-ripening-induced oxidation of seed-stored mRNAs has not been studied in any monocot species.

Imbibed D and AR seeds exhibit similar capacity but distinct pattern of protein synthesis (Chibani et al., 2006), highlighting the importance for differential regulation of translational activity. Accordingly, proteomic studies have identified specific proteins involved in the regulation of seed dormancy in different species. For example, comparative proteomic analysis between dry D and AR seeds of Arabidopsis indicated that after-ripening is associated with differential regulation of seed storage proteins, such as cruciferin precursors, and those related to energy and protein metabolism (Chibani et al., 2006). Other studies with seeds of tree species have also indicated the importance of specific proteins related to signal initiation and transduction, transcription, translation, energy metabolism and cell cycle in regulating seed dormancy (Pawłowski, 2007; Pawłowski, 2009; Pawłowski, 2010). These results demonstrated the importance of selective but not global de novo synthesis of seed proteins in controlling dormancy.

Furthermore, using a candidate protein approach, Schoonheim et al. (2007) have shown that 14-3-3 proteins, which act as regulators of signalling networks (van Hemert et al., 2001), mediate ABA action through their interaction with ABI5 and ABF, important mediators of ABA signalling in seeds (Nambara et al., 2010). This result implicates a role for 14-3-3s in regulating seed dormancy. The germination of AR or nondormant (ND) seeds is characterized by imbibition of dry seeds and subsequent protrusion of the expanding embryo axis through seed covering layers (Finch-Savage and Leubner-Metzger, 2006). Proteomic studies in Arabidopsis (Gallardo et al., 2002), barley (Bønsager et al., 2007), wheat (Mak et al., 2009) and rice (Oryza sativa L., He et al., 2011) have provided valuable insights into the mechanisms underlying the transition of seeds from the state of metabolic quiescence induced by desiccation to its activation during imbibition. However, the specific proteins that underlie after-ripening-induced translational changes during imbibition of wheat seeds are not well unknown.

To gain insights into the gap of knowledge on the molecular features of seed dormancy in wheat, we previously examined transcriptional changes triggered by after-ripening. To extend our understanding of the mechanisms acting at the post-transcriptional levels, this study investigated after-ripening-induced oxidative modification of seed-stored mRNAs in the dry state and changes in seed proteome between AR and D wheat seeds in both dry and imbibed states. Co-expression analysis of the genes differentially expressed (Gao et al., 2012) and oxidized during dry after-ripening, and those representing proteins differentially regulated between D and AR seeds in both dry and imbibed states was performed to identify more candidate genes and transcriptional regulatory mechanisms involved in controlling dormancy in wheat seeds.

Results and discussion

Germination of dormant vs. after-ripened seeds

Comparison between the D and AR seeds of wheat cv. AC Domain for their germination performance revealed that 95% of the AR seeds germinated following 1-day imbibition, when the corresponding D seeds did not show any sign of germination (Figure 1). Only approximately 15% of the D seeds germinated after 5-day imbibition.

Oxidative modification of seed-stored mRNAs during after-ripening

To gain insights into post-transcriptional mechanisms regulating seed dormancy in wheat, we investigated oxidation of seed-stored mRNAs in response to dry after-ripening. Our analysis identified 120 differentially oxidized probesets, which refers to the 8-OHG fraction, between the dry D and AR seeds (at cut-off values of twofold change and $P \leq 0.05$) in which 80 probesets were highly oxidized in dry AR and the remaining 40 probesets in

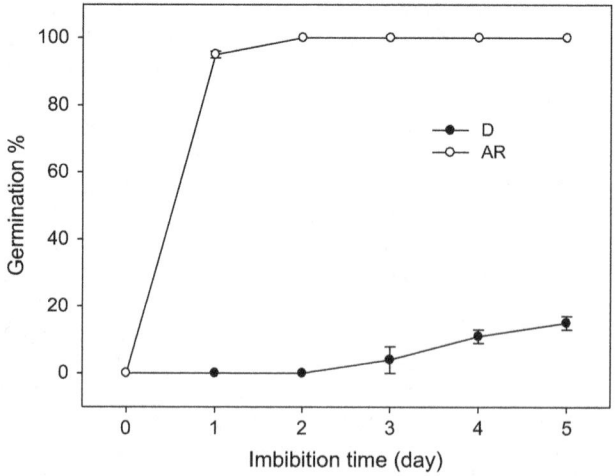

Figure 1 Effect of after-ripening on germination of dormant and after-ripened seeds of wheat cv. AC Domain imbibed in water. Seeds were considered germinated when the coleorhiza was visible beyond the seed coat. Germination was scored daily for 5 day. Data are means ± SD, $n = 3$, where n refers to a batch of 25 seeds.

the corresponding D seeds (Table S1). The presence of oxidized mRNAs in D seeds suggests the initiation of mRNA oxidation during seed maturation. Our results confirm that mRNA oxidation during dry after-ripening is highly selective and is associated with seed dormancy release (Bazin et al., 2011b). These authors have proposed that the selective nature of this oxidative process could be related to the mRNAs' location relative to the site of ROS production in the cell or their functional state within the messenger ribonucleoproteins, as they found no association between oxidized mRNAs and their guanine content or abundance.

The results of our study along with that of Bazin et al. (2011b), who used sunflower seeds as their experimental material, suggest cross-species evolutionary conservation of mRNA oxidation as a post-transcriptional seed dormancy–regulating mechanism. Following annotation by Arabidopsis genes, at least 20% of probesets oxidized in our dry AR whole seed samples are found to be expressed in imbibing wheat embryos (Bassel et al., 2011). However, comparison of mRNAs highly oxidized in our system with those oxidized in dry AR sunflower embryos (Bazin et al., 2011b) identified no wheat–sunflower orthologues, implying that the mRNAs targeted for after-ripening-induced oxidation are distinct between species.

Gene ontology of probesets corresponding to oxidized seed-stored transcripts

As mRNA oxidation always leads to decreased protein synthesis (Bazin et al., 2011b), our results suggest the significance of biological processes represented by the oxidized probesets in regulating seed dormancy. Probesets corresponding to mRNAs highly oxidized in AR seeds are over-represented in nutrient reservoir activity (GO: 0045735, $P = 1.8e-56$; Table S1) and α-amylase inhibitor activity (GO: 0015066, $P = 8.3e-20$; Table S1). Thirty-seven of the nutrient reservoir activity–related probesets represent genes encoding gliadin and glutenin, which account for approximately 80% of seed storage proteins in wheat (Payne et al., 1982). These proteins accumulate during the late stages of seed development (Bewley and Black, 1994) and are proteolytically degraded during germination to serve as a source of amino acids and nitrogen (Spencer, 1984). Thus, it is likely that oxidative modification of the related mRNAs forms a mechanism by which their synthesis is repressed during imbibition. Consistently, probesets over-represented in nutrient reservoir activity exhibited down-regulation during imbibition (Gao et al., 2012). Nine of the probesets highly oxidized in AR seeds are enriched in α-amylase inhibitor activity, suggesting the significance of enhanced hydrolytic starch degradation in regulating dormancy release. In agreement with this hypothesis, after-ripening led to imbibition induced up- and down-regulation of probesets corresponding to α-amylase and its inhibitors, respectively (Gao et al., 2012; Potokina et al., 2002). Furthermore, transcripts of peroxidases and ribosomal protein are also oxidized during after-ripening (Table S1), implying that post-transcriptional control of ROS production and synthesis of metabolically active proteins plays an important role in regulating seed dormancy in wheat.

Probesets highly oxidized in the D seeds are over-represented in oxidative phosphorylation (GO: 0006119, $P = 6.2e-09$; Table S1) and ribosome biogenesis (GO: 0042254, $P = 1.4e-06$; Table S1). Oxidative phosphorylation plays a major role in the synthesis of ATP, which is used as a source of energy for metabolic processes, during seed germination (Moroashi and Sugimoto, 1988). Our data therefore suggest the significance of post-

transcriptional repression of this biological process in regulating the maintenance of seed dormancy. The oxidation of ribosome biogenesis–related mRNAs in dry D seeds is consistent with a recent report by Zhang et al. (2012), who showed the association of inhibition of seed germination with transcriptional repression of rRNA that is involved in the biogenesis of protein-synthesizing ribosomes.

Proteins differentially expressed during dry after-ripening

Our proteomic analysis revealed after-ripening mediated differential abundance of specific storage proteins, proteases/enzyme inhibitors and those involved in cellular signalling (Table 1; Figure S1). After-ripening induced a decrease in the abundance of storage protein triticin (7.0-fold, $P \leq 0.05$; Table 1), leading to dormancy release. In contrast, the level of another storage protein designated as 27K exhibited an increase (over twofold, $P \leq 0.05$), although such proteins mainly accumulate in the later stages of seed development (Bewley and Black, 1994).

Protease inhibitor cystatin decreased in abundance during after-ripening (3.0-fold, $P \leq 0.05$; Table 1). Cystatin is involved in controlling the activity of gliadain, a cysteine protease associated with wheat seed germination (Kiyosaki et al., 2007). Thus, it is likely that after-ripening renders a mechanism to degrade cystatin and in turn activate gliadain-mediated proteolysis. Consistently, repression of cystatin in barley aleurone treated with the germination promoting GA is associated with enhanced degradation of storage proteins (Martinez et al., 2009). The level of trypsin/α-amylase inhibitor designated as CM16, because of its solubility in chloroform/methanol, was also reduced by after-ripening (5.0-fold, $P \leq 0.05$). These results, overall, reflect the association of enhanced proteolysis and hydrolysis of storage reserves with seed dormancy release.

After-ripening also decreased the abundance of proteins related to cellular signalling, including 14-3-3 proteins and superoxide dismutase (SOD) (over twofold, $P \leq 0.05$; Table 1). The 14-3-3s are involved in the activation of seed ABA response (Schoonheim et al., 2007); the decrease in their level likely leads to loss of seed ABA sensitivity and dormancy. Whereas the reduction in the level of SOD, one of the antioxidative enzymes, might form a mechanism to maintain cellular homoeostasis of ROS that acts as a signal in governing embryo transition from a D to ND status (Oracz et al., 2007, 2009).

Regulation of proteins differentially expressed during after-ripening

After-ripening is associated with accumulation of ROS derived from nonenzymatic reactions (Oracz et al., 2007), and oxidation of storage proteins occurs in dry seeds (Job et al., 2005). This along with the unlikely occurrence of transcriptional and translational activities in dry D seeds (Bazin et al., 2011a) might suggest that the decreased abundance of triticin, cystatin, SOD and 14-3-3 during after-ripening (Table 1) results from their oxidative degradation. No differential expression and oxidation of probesets annotated as cystatin, SOD and 14-3-3 was evident in response to after-ripening; however, the transcript abundance of specific triticin probesets was decreased (over threefold, $P \leq 0.05$; Figure 2), implying the decay of triticin mRNAs. The reduction in the level of CM16 during after-ripening (Table 1) is accompanied by oxidation of its corresponding probesets (Table S1), and this might suggest its translational repression initiated possibly by the change in seed water status due to nonenzymatic

Table 1 Proteins differentially expressed between dry after-ripened and dry dormant seeds

Spot*	Identification	ID number[†]	MS/MS MASCOT peptides (score)[‡]	AR-0/D-0[§] (P-value[¶])
a1	Triticin	ACB41345.1	R.LLAEALGTSGK.I (90)	0.14 (0.020)
			R.CTGVFAIR.R (59)	
a2	Unnamed protein product cystatin	CAI84619.1	R.NEQAGIVGHKDVTDVTGDR.G (64)	0.33 (0.036)
a3	α-amylase/trypsin inhibitor CM16	CAA35596.1	K.SRPDQSGLMELPGCPR.E + Oxidation (M) (64)	0.20 (0.014)
a4	Manganese superoxide dismutase	AAB68036.1	K.ALEQLDAAVSK.G (81)	0.36 (0.009)
			K.NLKPISEGGGEPPHGK.L (70)	
a5	Hv14-3-3b	CAA63658.1	K.AAQEIALAELPPTHPIR.L (77)	0.05 (0.000)
			K.TVDSEELTVEER.N (73)	
a6	Hv14-3-3b	CAA63658.1	K.TVDSEELTVEER.N (98)	0.09 (0.003)
			R.IISSIEQKEESR.G (77)	
a7	14-3-3 protein homologue	CAA44259.1	K.QAFDEAIAELDSLGEESYK.D (94)	0.21 (0.006)
			K.SAQDIALADLPTTHPIR.L (93)	
a8**	Aspartate aminotransferase	CV777781	K.EYLPITGLADFNKLSAK.L (39)	0.06 (0.017)
			R.LEGGVGGGGGGR.V (26)	
a9	No good hit	CL1Contig1401	-.VATVSIPR.T (65)	0.11 (0.040)
b1	27K protein	BAC76688.1	R.GHNLSLEYGR.Q (55)	2.31 (0.046)

*Protein spot names correspond to the 2D gels in Figure S1.

[†]GenBank IDs of the matching proteins. The ID names starting with 'CL' are from the local wheat EST database.

[‡]Amino acid sequences of the top two peptides matching the MS/MS spectra. Ions score is $-10*Log$ (P), where P is the probability that the observed match is a random event. Individual ions scores >50 indicate identity or extensive homology ($P < 0.05$).

[§]Normalized protein spot volumes in dry after-ripened (AR-0) seeds divided by the corresponding normalized spot volume in dry dormant (D-0) seeds. Data are means of 2–3 independent biological replicates.

[¶]Significant differences between samples were determined using Student's t-test at P-value of ≤ 0.05.

**Protein spot identified as aspartate aminotransferase with less confidence as the matching peptides exhibited relatively less score (<50).

oxidative reactions (Bazin et al., 2011a,b). The accumulation of 27K during after-ripening (Table 1) with no apparent differential expression of its probesets between dry AR and D seeds (Figure 2) can be explained by its enhanced solubility and/or extractability due to oxidative processes that increase protein–water interactions (Cherian and Chinachoti, 1997).

Proteins differentially regulated by imbibition in dormant and after-ripened seeds

To identify proteins that possibly regulate dormancy during imbibition, changes in protein abundance were compared between dry and imbibed seeds in each of the D and AR samples, and also between imbibed D and imbibed AR seeds. Our comparative study was focused on 24-h imbibed seeds, as changes in the proteome of cereal seeds appear to occur mainly at/after this stage (Bønsager et al., 2007; Yang et al., 2007). Furthermore, 95% of the AR but none of the D seeds completed germination at 24 h after imbibition, reflecting that differences in seed proteome at this stage arose from the phase of germination sensu stricto.

Proteins differentially expressed by imbibition specifically in dormant seeds

Imbibition down-regulated 14-3-3, calreticulin (CRT)-like and serpin proteins specifically in D seeds (Table 2, Figure S2a,b). The abundance of a 14-3-3 homologue protein decreased by after-ripening (over fourfold, $P \leq 0.05$; Table 1) and this level was maintained during imbibition (Figure S2c,d). In D seeds, however, this protein was down-regulated only after imbibition (over fivefold, $P \leq 0.05$; Table 2), reinforcing our hypothesis that 14-3-3 protein plays important role in regulating wheat seed dormancy. The level of CRT-like proteins decreased to undetect-

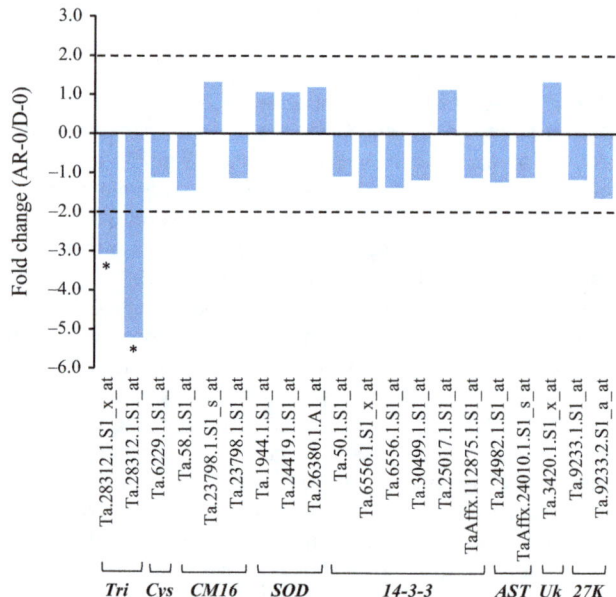

Figure 2 Fold change in the expression of probesets corresponding to proteins differentially expressed between dry after-ripened (AR-0) and dry dormant (D-0) seeds (AR-0/D-0). Asterisks indicate significant difference in expression ($\geq$ twofold change and $P \leq 0.05$). Abbreviations: Tri, triticin; Cys, cystatin; CM16, α-amylase/trypsin inhibitor; SOD, superoxide dismutase; 14-3-3, 14-3-3 proteins; AST, aspartate aminotransferase; Uk, unknown (no good hit); 27K, 27K protein. Fold changes in the expression of these probesets between D and AR seeds in both dry and imbibed states are shown in Table S2.

able level during imbibition in D but not in AR seeds (Table 2). Calreticulin is a major Ca^{2+} sequestering protein implicated in regulating seed ABA response and dormancy (Lorenzo et al.,

2002; Pawłowski, 2007). Our data, thus, imply that maintenance of seed ABA sensitivity and dormancy in wheat requires CRT degradation. Consistent with this hypothesis, treatment of seeds with ABA decreases the level of CRT (Pawłowski, 2007).

While the abundance of a specific serpin in D seeds was repressed to undetectable level during imbibition, the level of another one (serpin 3) was up-regulated (over twofold, $P \leq 0.05$), leading to a nearly threefold more accumulation than that found in AR seeds ($P \leq 0.05$; Table 2). Serpins, a super family of versatile proteins, take part in the regulation of complex proteolytic systems, and those in cereal seeds are likely to inhibit endogenous and exogenous proteases irreversibly (Østergaard et al., 2000). Our data therefore may suggest decreased proteolysis, which might attribute at least partially to the failure in germination of D seeds.

Regulations of proteins differentially expressed by imbibition specifically in dormant seeds

The 14-3-3 protein appeared to be regulated post-transcriptionally as its abundance decreased during imbibition (Table 2) while expression of its probesets either increased or remained unaffected (Figure 3a). Expression of all probesets annotated as CRT remained unaffected during imbibition of D seeds; however, three specific probesets exhibited over twofold down-regulation relative to AR seeds ($P \leq 0.05$; Figure 3a,b). Given that imbibition substantially reduced CRT level in D but not in AR seeds (Table 2), our data imply its transcriptional regulation. Both transcriptional and post-transcriptional mechanisms likely regulate serpins as the two serpins showed opposite pattern in abundance (Table 2), while expression of their probesets either decreased or did not change during imbibition of D seeds or in imbibed D relative to AR samples (Figure 3a,b). Probesets corresponding to all proteins down-regulated by imbibition specifically in D seeds are not among those oxidized in dry D seeds, suggesting that their regulation during imbibition is independent of seed maturation–associated mRNA oxidation.

Proteins differentially expressed by imbibition specifically in after-ripened seeds

Imbibition differentially regulated specific seed storage proteins and protease inhibitors and those involved in energy metabolism, translation, protein folding and amino acid metabolism specifically in AR seeds (Table 3, Figure S2c,d).

Seed storage proteins and protease inhibitors

Storage protein globulin 3 decreased to undetectable level during imbibition of AR seeds, but remained unaffected in D seeds (Table 3), implying the role of after-ripening in inducing imbibition-mediated proteolysis of this protein and in turn dormancy decay. Imbibition of AR seeds also decreased the abundance of storage protein triticin (twofold, $P \leq 0.05$) but increased that of globulin 2 (twofold, $P \leq 0.05$). The up-regulation of globulin 2 in imbibing AR seeds may imply the significance of re-induction of specific seed maturation programmes in enhancing seed vigour (Rajjou et al., 2008), leading to dormancy decay.

Two serpin proteins decreased in abundance during imbibition specifically in AR seeds (over twofold, $P \leq 0.05$), leading to their significantly lower abundance in imbibed AR than in the corresponding D seeds (Table 3). These results may imply the role of after-ripening in repressing the action of serpins in imbibing seeds and thereby activate proteolytic degradation of storage proteins and dormancy release.

Proteins involved in energy metabolism

Granule-bound starch synthase I (GBSSI) was down-regulated during imbibition specifically in AR seeds (2.4-fold, $P \leq 0.05$), leading to a significant reduction in its accumulation relative to that found in the corresponding D seeds (nearly twofold, $P \leq 0.05$; Table 3). These results suggest that after-ripening induces imbibition-mediated suppression of starch synthesis, and thereby dormancy release. Imbibition of AR seeds also led to substantial repression of glucose/ribitol dehydrogenase (GRD;

Table 2 Proteins differentially expressed during imbibition specifically in dormant seeds, and their comparison with those in imbibed after-ripened seeds

Spot*	Identification	ID number[†]	MS/MS MASCOT Amino acids sequence (score)[‡]	D-24/D-0[§] (P-value[¶])	D-24/AR-24** (P-value)
d1	Serpin	CAB52709.1	K.GAWTDQFDSSGTK.N (98)	≤0.01 (NA)	≤0.01 (NA)
			R.VSSVFHQAFVEVNEQGTEAAASTAIK.M (96)		
d2	14-3-3 protein homologue	CAA44259.1	K.QAFDEAIAELDSLGEESYK.D (94)	0.19 (0.03)	0.77 (0.019)
			K.SAQDIALADLPTTHPIR.L (93)		
d3	Calreticulin-like protein	AAW02798.1	K.SGTLFDNILITDDAALAK.T (112)	≤0.01 (NA)	≤0.01 (NA)
			R.FYAISAEYPEFSNK.D (81)		
d4	Calreticulin-like protein	AAW02798.1	R.FYAISAEYPEFSNK.D (105)	≤0.01 (NA)	≤0.01 (NA)
			K.SGTLFDNILITDDAALAK.T (93)		
e1	Serpin 3	ACN59485.1	R.VSSVFHQAFVEVNEQGTEAAASTAIK.M (113)	2.39 (0.001)	2.86 (0.035)
			K.YKAETQSVDFQTK.A (96)		

*Protein spot names correspond to the 2D gels in Figure S2a,b.

[†]GenBank IDs of the matching proteins. The ID names starting with 'CL' are from the local wheat EST database.

[‡]Amino acid sequences of the top two peptides matching the MS/MS spectra. Ions score is $-10*Log (P)$, where P is the probability that the observed match is a random event. Individual ions scores >50 indicate identity or extensive homology ($P < 0.05$).

[§]Normalized protein spot volumes in imbibed dormant (D-24) seeds divided by normalized spot volumes in the corresponding dry (D-0) seeds. Data are means of 2–3 independent biological replicates; ≤0.01 indicates that the abundance of the corresponding protein in D-24 seeds was close to background.

[¶]Significant differences between samples were determined using Student's t-test at $P \leq 0.05$; NA = not available.

**Normalized protein spot volumes in D-24 seeds divided by normalized spot volumes in the corresponding after-ripened (AR-24) seeds. Data are means of 2–3 independent biological replicates; ≤0.01 indicates that the abundance of the corresponding protein in D-24 seeds was close to background.

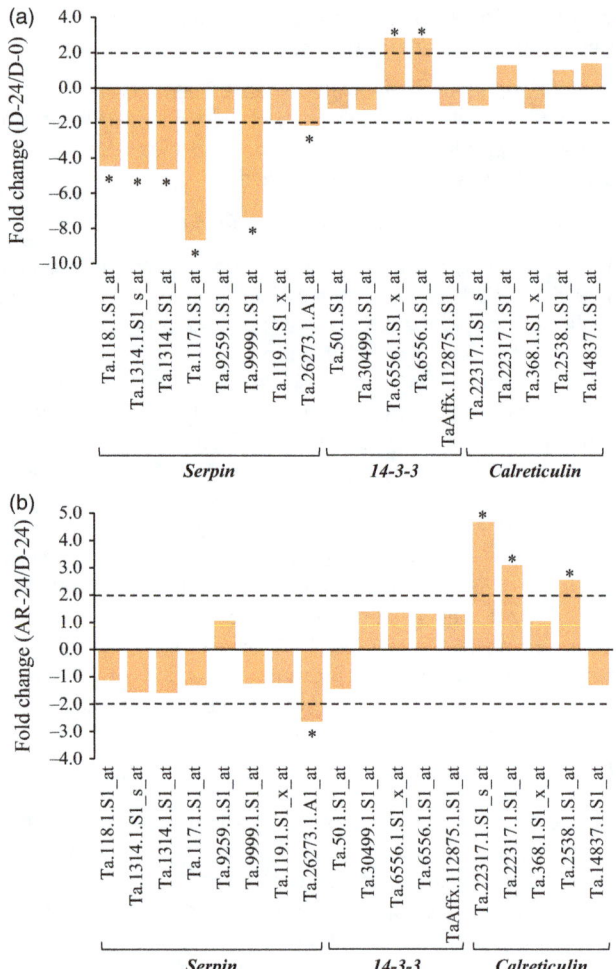

Figure 3 Fold change in the expression of probesets corresponding to proteins differentially expressed during imbibition specifically in dormant (D) seeds (D-24/D-0; a), and their respective comparison between imbibed D and the corresponding after-ripened (AR) seeds (D-24/AR-24; b). Asterisks indicate significant difference in expression ($\geq$ twofold change and $P \leq 0.05$). Fold changes in the expression of these probesets between D and AR seeds in both dry and imbibed states are shown in Table S2.

over fivefold, $P \leq 0.05$). Consistently, *GRD* was down-regulated in imbibing D seeds of *Trollius ledebouri* treated with GA, which stimulates dormancy breaking and germination in this species, but not in the untreated seeds (Bailey *et al.*, 1996), implying the association of after-ripening induced suppression of GRD with dormancy decay.

Proteins involved in translation, protein folding and amino acid metabolism

While maintained at a similar level in D seeds, two eukaryotic translation initiation factors (eIFs), eIF6 and eIF5A1, decreased in their abundance during imbibition of AR seeds (Table 3). The eIF6 protein was virtually undetectable in imbibed AR seeds, while the level of eIF5A1 showed substantial reduction (over 20-fold, $P \leq 0.05$), leading to a significantly lower abundance when compared with that found in imbibed D seeds (over sixfold, $P \leq 0.05$). The eIF6 acts as a ribosome dissociation factor because it can bind to the 60S ribosomal subunit and, in turn,

prevent the association between the 60S and 40S subunits (Russell and Spremulli, 1980). Our data, thus, imply the role of after-ripening in inducing imbibition-mediated activation of specific translational programmes, and thereby dormancy release.

Imbibition of AR seeds also caused suppression of two precursor forms of protein disulfide isomerase (PDI), PDI2 and PDI3, that acts to catalyse protein folding (Ellgaard and Ruddock, 2005), leading to a lower accumulation than that found in D seeds (over twofold, $P \leq 0.05$; Table 3). As increased abundance of PDIs is associated with the accumulation of seed storage proteins (Xia and Kermode, 1999), our data might suggest the significance of after-ripening in repressing PDIs, which promotes proteolysis, and in turn seed dormancy release and germination.

N-Acetyl-gamma-glutamyl-phosphate reductase (AGPR), which is involved in the synthesis of proline, arginine and glutamine (Kishor *et al.*, 2005), was slightly up-regulated during the imbibition of AR seeds, leading to a nearly twofold more accumulation than that detected in D seeds ($P \leq 0.05$; Table 3). This result suggests the significance of seed produced amino acids in dormancy release. Indeed, antisense repression of a proline biosynthetic gene, *Δ1-pyrroline-5-carboxylate synthetase*, in Arabidopsis caused a delay in seed germination (Hare *et al.*, 2003).

Regulation of proteins differentially expressed by imbibition specifically in AR seeds

Seed dormancy release by dry after-ripening is associated with targeted mRNA oxidation, which could lead to translational repression of the corresponding proteins during imbibition and thereby regulate the germination process (Bazin *et al.*, 2011b). However, probesets of proteins down-regulated by imbibition specifically in AR seeds were not oxidized during dry after-ripening, except a specific probeset annotated as GBSSI (Table S1). It is likely that most of the proteins annotated by oxidized probesets are expressed at a level below the detection limit of our 2D gel system.

Although their proteins showed increased or decreased abundance during imbibition of AR seeds or in imbibed AR relative to D seeds (Table 3), probesets corresponding to globulin 2, globulin 3, eIF5A1, eIF6, PDIs and AGPR exhibited either no differential expression or a pattern opposite to that shown by their respective protein (Figures 4a,c,d; S3a,c,d). These data imply post-transcriptional regulation of these proteins by mechanisms other than dry after-ripening-induced mRNA oxidation. Consistent with their protein expression pattern, however, either all or specific probesets of triticin, serpin, GBSSI and GRD were repressed during imbibition of AR seeds (over twofold, $P \leq 0.05$; Figure 4a,b). Furthermore, specific serpin and GBSSI probesets exhibited suppression in imbibed AR relative to D seeds (over twofold, $P \leq 0.05$; Figure S3a,b), suggesting their transcriptional regulation. The decreased abundance of GBSSI in imbibing AR seeds (Table 3) can also be attributed to the oxidation of one of its probesets during dry after-ripening.

Previous proteomic analyses of rice and wheat embryos and different tissues of barley seeds have shown that most storage proteins and proteases/enzyme inhibitors are mainly localized in the endosperm/aleurone tissues, whereas those involved in translation and protein folding such as eIF5A1 and PDIs, and cellular signalling such as 14-3-3s and CRT, in the embryos (Bønsager *et al.*, 2007; Kim *et al.*, 2009; Mak *et al.*, 2009). These results suggest the tissue specificity of the related proteins identified in this study, which involved whole seed samples.

Table 3 Proteins differentially expressed during imbibition specifically in after-ripened seeds, and their comparison with those in imbibed dormant seeds

Spot*	Identification	ID number[†]	MS/MS MASCOT Amino acids sequence (score)[‡]	AR-24/AR-0[§] (P-value[¶])	AR-24/D-24** (P-value)
f1	Triticin	ACB41345.1	R.LLAEALGTSGK.I (93) R.SSQLHSSQNIFSGFDVR.L (59)	0.49 (0.016)	0.66 (0.010)
f2	Globulin 3 (cupin domain contained)	CL1Contig7398	R.QASEGGQGHHWPLPPFR.G (83) R.DTFNLLEQRPK.I (73)	$\leq$ 0.01 (NA)	$\leq$ 0.01 (NA)
f3	Serpin	CAA72274.1	K.AFVEVNETGTEAAATTIAK.V (126) K.AAEVTAQVNSWVEK.V (97)	0.42 (0.007)	0.63 (0.025)
f4	Serpin	CAA72274.1	K.DILPAGSIDNTTR.L (66) K.GAWTDQFDPR.A (56)	$\leq$ 0.01 (NA)	$\leq$ 0.01 (NA)
f5	Serpin 2	ACN59484.1	R.VAFANGVFVDASLSLKPSFQELAVCNYK.S (112) K.GLWTEKFDESK.T (87)	0.38 (0.018)	0.42 (0.044)
f6	Serpin 2	ACN59484.1	K.ISFGFEATNLLK.S (96) K.VVVDQFMLPK.F + Oxidation (M) (82)	0.18 (0.037)	0.12 (0.033)
f7	Glucose and ribitol dehydrogenase homolog	T06212	K.GNATLLDYTATK.G (85) K.VALVTGGDSGIGR.A (82)	0.19 (0.040)	0.74 (0.638)[††]
f8	Granular bound starch synthase I	BAA88511.1	K.GPDVMIAAIPEIVK.E + Oxidation (M) (88) K.EEDVQIVLLGTGK.K (81)	0.42 (0.000)	0.56 (0.032)
f9	Putative eukaryotic translation initiation factor 6	BAC45212.1	K.ATEELIADVLGVEVFR.Q (80) R.IQFENNCEVGVFSK.L (63)	$\leq$ 0.01 (NA)	$\leq$ 0.01 (NA)
f10	Eukaryotic translation initiation factor 5A1	AAZ95171.1	K.LPTDDVLLGQIK.T (94) K.TYPQQAGAIR.K (81)	0.05 (0.000)	0.16 (0.019)
f11	Protein disulfide isomerase 2 precursor	AAK49424.1	K.AYYGAVEEFSGK.D (85) R.KSEPIPEANNEPVK.V (64)	0.41 (0.025)	0.37 (0.014)
f12	Protein disulfide isomerase 3 precursor	AAK49425.1	K.APEDATYLEDGK.I (97) K.LAPILDEAAATLQSEEDVVIAK.M (70)	0.44 (0.039)	0.44 (0.033)
f13	Protein disulfide isomerase 3 precursor	AAK49425.1	R.TADEIVDYIK.K (82) K.VVVADNVHDVVFK.S (82)	0.32 (0.045)	0.40 (0.044)
f14	Unnamed protein product	CAA34060.1	K.TAIAIDTILNQK.Q (101) R.AAELTTLLESR.M (96)	0.34 (0.007)	0.59 (0.017)
g1	N-acetyl-gamma-glutamyl-phosphate reductase	CL187Contig2 CL187Contig3	K.IVDLSADFR.L (69)	1.51 (0.046)	1.82 (0.031)
g2	Globulin 2	CL1Contig7433	R.AQPESVFVAGPQQQR.R (94) K.ALAFPQQAR.E (60)	2.10 (0.026)	1.37 (0.048)

*Protein spot names correspond to the 2D gels in Figure S2c, d.

[†]GenBank IDs of the matching proteins. The ID names starting with 'CL' are from the local wheat EST database.

[‡]Amino acid sequences of the top two peptides matching the MS/MS spectra. Ions score is $-10*Log (P)$, where P is the probability that the observed match is a random event. Individual ions scores >50 indicate identity or extensive homology ($P < 0.05$).

[§]Normalized protein spot volumes in imbibed after-ripened (AR-24) seeds divided by normalized spot volumes in the corresponding dry (AR-0) seeds. Data are means of 2–3 independent biological replicates; $\leq$ 0.01 indicates that the abundance of the corresponding protein in AR-24 seeds was close to background.

[¶]Significant differences between samples was determined using Student's t-test at $P \leq 0.05$; NA = not available.

**Normalized protein spot volumes in AR-24 seeds divided by normalized spot volumes in the corresponding dormant (D-24) seeds. Data are means of 2–3 independent biological replicates; $\leq$ 0.01 indicates that the abundance of the corresponding protein in AR-24 seeds was close to background.

[††]Inconsistent spots between biological replicates of imbibed D seeds. No clear comparison of glucose and ribitol dehydrogenase accumulation between imbibed AR and D samples.

Gene co-expression analysis

To identify more candidate genes involved in the regulation of seed dormancy in wheat, probesets that were found to be differentially expressed (58 probesets; Gao et al., 2012) and oxidized (120 probesets; Table S1) during dry after-ripening and those representing proteins differentially expressed between dry D and AR seed samples (20 unique probesets; Table S2) were subjected to gene co-expression analysis using Web tools in WheatNet (http://aranet.mpimp-golm.mpg.de/wheatnet). This analysis revealed that probesets down-regulated or oxidized during dry after-ripening are enriched in two co-expression clusters designated as cluster 116 and cluster 132 (Figure 5). Whereas those corresponding to the proteins down-regulated by dry after-ripening are enriched exclusively in cluster 132, suggesting that the two co-expression clusters represent further candidate genes regulating after-ripening-mediated seed dormancy release in wheat. Given that change in seed water status capable of inducing active metabolic reactions appear not to occur in dry seeds that maintain dormancy, the genes in the two clusters are more likely to be regulated by nonenzymatic post-transcriptional and post-translational mechanisms (Figure 5).

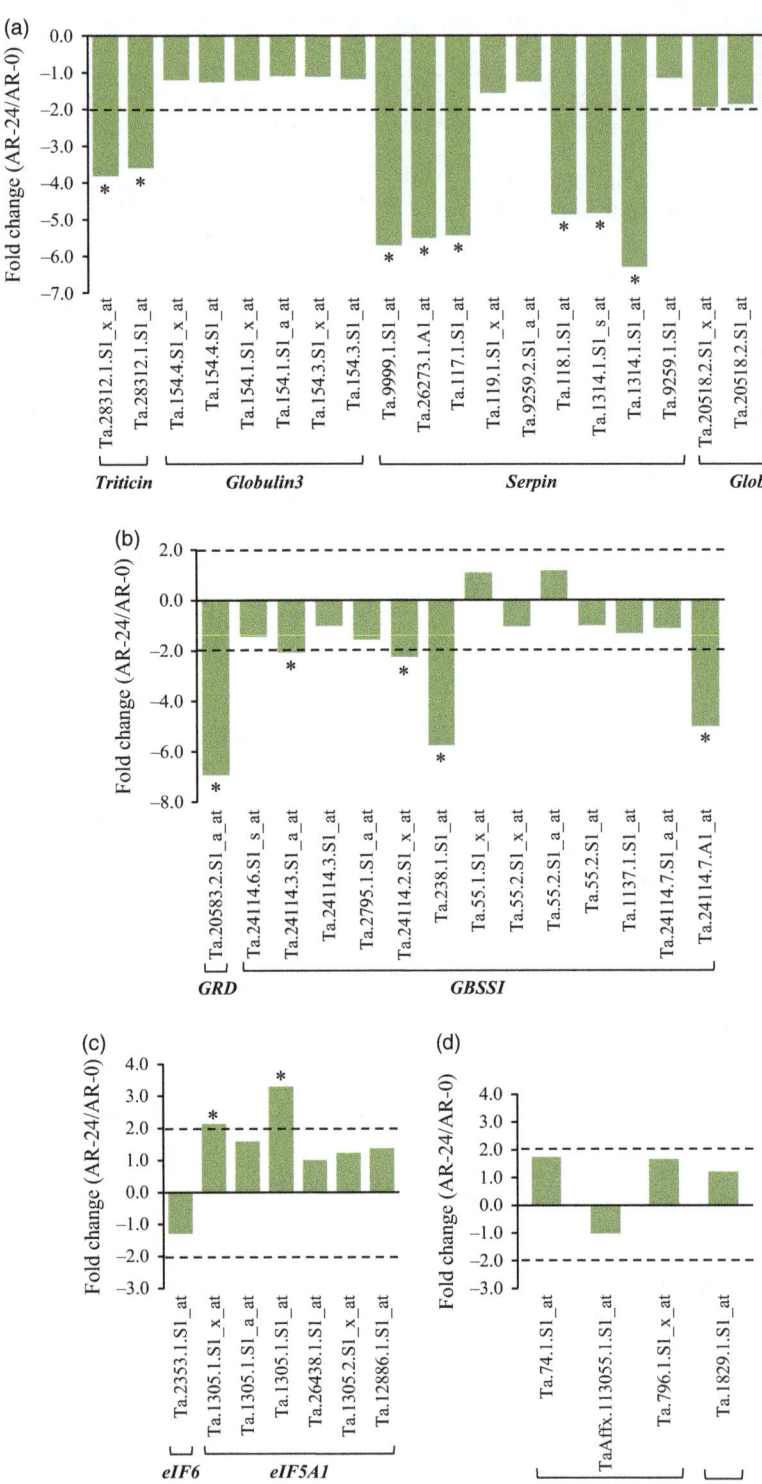

Figure 4 Fold change in the expression of probesets corresponding to the proteins differentially expressed during imbibition specifically in after-ripened (AR) seeds (AR-24/AR-0; a-d). Asterisks indicate significant difference in expression ($\geq$ twofold change and $P \leq 0.05$). Fold changes in the expression of these probesets between dormant and AR seeds in both dry and imbibed states are shown in Table S2. Abbreviations: GRD, glucose/ribitol dehydrogenase; GBSSI, granule-bound starch synthase I; eIF6, eukaryotic translation initiation factor 6; eIF5A1, translation initiation factors 5A1; PDI, protein disulfide isomerase; AGPR, N-acetyl-gamma-glutamyl-phosphate reductase.

Co-expression clusters 116 and 132 contain a total of 82 and 72 probesets (Table S3), respectively. Annotation of these probesets by HarvEST WheatChip and GO analysis revealed that the probesets in both clusters are enriched in nutrient reservoir (GO: 0045735, $P \leq$ 8.6e-34) and enzyme regulatory (GO: 0030234, $P \leq$ 4.5e-15) activities (Table S3). Furthermore, the two clusters consist of probesets corresponding to protease inhibitors, storage proteins and NAC domain containing proteins,

suggesting the linkage of these genes with after-ripening-induced seed dormancy release.

To extend our search for more candidate genes regulating seed dormancy in wheat, we performed co-expression analysis of probesets of proteins differentially expressed between imbibed D and AR seeds (57 unique probesets; Table S2). These probesets are enriched in cluster 95 and cluster 116 (Figure 5). As active metabolic reactions occur during seed imbibition and both similar

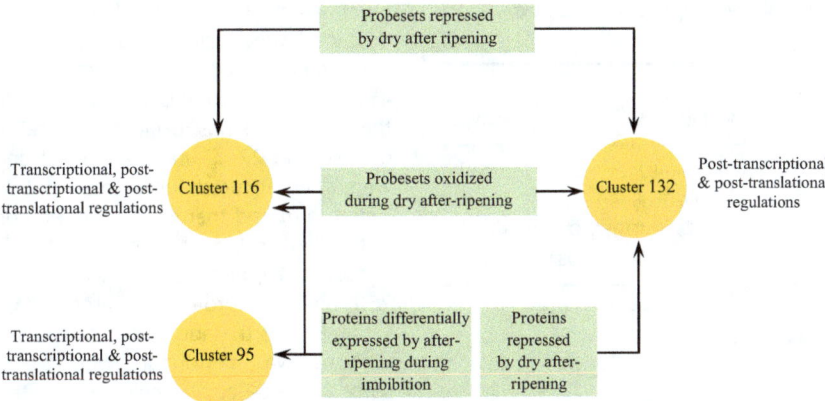

Figure 5 Co-expression analysis of probesets differentially expressed and oxidized by dry after-ripening and those corresponding to proteins that exhibited differential abundance between dormant (D) and after-ripened seeds (AR) in both dry and imbibed states by WheatNet. These probesets appear to be regulated at transcriptional, post-transcriptional and post-translational levels and are enriched in co-expression clusters 95, 116 and 132. It is likely that these co-expression clusters represent further candidate genes regulating seed dormancy in wheat. Probesets in clusters 116 and 132 are enriched in nutrient reservoir and enzyme regulatory activities. No GO category is enriched with probesets contained in cluster 95.

and differential expression patterns were evident between the proteins and their respective probesets, it is likely that the genes in the two clusters are regulated by enzymatic transcriptional, post-transcriptional and post-translational mechanisms (Figure 5). However, we cannot rule out the possibility that nonenzymatic regulatory mechanisms operate during seed imbibition. Cluster 95 contained a total of 82 probesets (Table S3) with no enriched GO category, but consists of probesets corresponding to more candidate genes that potentially involve in the regulation of seed dormancy, including those representing dehydrin family proteins such as responsive to ABA 18 (RAB18), late embryogenesis (LEAs) and XERO1, and caleosins such as responsive to desiccation 20 (RD20). The presence of probesets representing ABI5, an important regulator of ABA signalling and dormancy in seeds (Nambara *et al.*, 2010), in this cluster indicates the validity of this approach for the discovery of dormancy-related genes.

Transcriptional regulation of co-expressed genes

To identify candidate transcription regulators controlling the co-expressed genes, the probesets in each cluster were annotated by rice genes using HarvEST WheatChip. Analysis of the 1-kb upstream region of the identified rice genes by Osiris (http://www.bioinformatics2.wsu.edu/cgi-bin/Osiris/cgi/visualize_select.pl; Morris *et al.*, 2008) revealed that two motifs, ACGTOSGLUB1 and PROLAMINBOXOSGLUB1 (found in the upstream region of rice *GluB-1* gene), are enriched in clusters 95 and 116, respectively (Table 4). The ACGT motif, which is found in the promoter regions of genes encoding storage proteins, acts as a putative binding site for AtbZIP10 and AtbZIP25 transcription factors that interact with ABI3 (Lara *et al.*, 2003). Consistently, the expression of seed storage protein genes exhibits a significant reduction in *abi3* mutants (Parcy *et al.*, 1994). Furthermore, two more ABA-regulated *cis*-elements, ABREOSRAB21 and ACGTABREMO-TIFA2OSEM, are enriched in the co-expression cluster 95 (Table 4). All these data suggest that the genes assigned in co-expression cluster 95 are regulated by ABA and play important roles in controlling seed dormancy. The *cis*-element PROLAMIN-BOX (5′-TGTAAAG-3′) is recognized by DNA binding with One Finger (DOF) transcription factor (Yanagisawa, 2004) that mediate ABA-induced transcriptional changes in its target genes

in germinating seeds (Moreno-Risueno *et al.*, 2007), suggesting that the genes in cluster 116 might also be controlled by ABA and play roles in regulating dormancy release by after-ripening.

In summary, our data show that selective oxidation of seed-stored mRNAs form a mechanism to control seed dormancy in wheat and the changes in patterns of seed transcripts and proteins between D and AR samples before and during imbibition highlight the significance of specific post-transcriptional and post-translational features regulating seed dormancy maintenance and release. These findings contribute to a better understanding of the molecular mechanisms controlling wheat seed dormancy, which is necessary to design strategies for the improvement in PHS tolerance in wheat.

Experimental procedures

Plant materials and growth conditions

Seeds of common wheat cv. AC Domain were used in this study. AC Domain is hard red spring wheat characterized by a high degree of PHS tolerance at harvest and is adapted to the Canadian prairies (Townley-Smith and Czarnecki, 2008). Plant growth conditions, seed harvesting and generation of AR seeds are as described before (Gao *et al.*, 2012). Briefly, mature air-dried seeds were harvested from the middle region of spikes of wheat plants grown in a greenhouse at 18–22 °C/14–18 °C (day/night) under a 16/8-h photoperiod. A portion of the seeds was immediately stored at −80 °C, whereas the remaining portion was first stored at room temperature and ambient relative humidity for 10 months (which gave a seed moisture content of 0.07 g water g DW^{-1}) and then stored at −80 °C until further use.

Seed germination assays

Germination assays were performed as described previously (Gao *et al.*, 2012). Briefly, the D and AR seed samples were surface-sterilized and then imbibed between two layers of filter papers moistened with sterile water in a Petri-plate (25 seeds per plate) at 22 °C in darkness. Seeds were considered germinated when the coleorhiza was visible beyond the seed coat. Unimbibed D and AR seeds were used for analyses of mRNA oxidation and seed proteomes in the dry state.

Table 4 Sequences of *cis*-elements over-represented in co-expression clusters 95 and 132*

Co-expression Cluster #	Motif sequence	Motif ID	P-value[†]
95	ACGTSSSC (S = G/C)	ABREOSRAB21	<10⁻⁴
	GTACGTG	ACGTOSGLUB1	<10⁻⁴
	ACGTGKC (K = G/T)	ACGTABREMOTIFA2OSEM	<10⁻¹⁰
116	TGCAAAG	PROLAMINBOXOSGLUB1	<10⁻⁴

*Cis-elements were identified using the OSIRIS Web-based resource.
[†]Significance was determined at $P < 0.001$.

Immunoprecipitation of oxidized mRNAs

Total and mRNA samples were extracted from three independent biological replicates of the same dry D and AR seeds used for our germination and transcriptomic studies following the protocol described before (Gao et al., 2012). Oxidized mRNAs were immunoprecipitated according to Bazin et al. (2011b) and Shan et al. (2003). Briefly, the mRNA samples (approximately 3 µg) were incubated at 65 °C for 2 min and then on ice to remove secondary structures. Subsequently, the mRNA samples were mixed with monoclonal anti-8-OHG antibody 15A3 (2 µg; QED Bioscience, San Diego, CA) in 200 µL 1× PBS buffer and incubated at room temperature for 2 h. No antibody was added to the negative control experiments. Washed Dynabeads (Invitrogen, Carlsbad, CA) were then added followed by overnight incubation of the mix at 4 °C to capture the antibody. The beads were washed three times with 1× PBS with 0.02% (v/v) Tween-20, and the immunoprecipitated RNA extracted from the pellet by adding 150 µL of 1× PBS, 0.02% (v/v) Tween-20 and 1% (w/v) SDS and incubating the mixture at 70 °C for 10 min. The Dynabeads were then captured by a magnet and the supernatant precipitated with ethanol/NaCl in the presence of 1 µL of 15 mg/mL glycoblue (Invitrogen) as coprecipitant. The pellet containing the mRNA was dissolved in RNase-free H₂O and used for microarray hybridization.

DNA microarray analysis

Labelling and hybridizing the oxidized mRNA samples (100 ng mRNA in 3 µL) to the Affymetrix GeneChip Wheat Genome Array (Affymetrix, Santa Clara, CA), verification of the reproducibility of the data from the three independent biological replicates and subsequent analysis were performed as described previously (Gao et al., 2012). Differentially oxidized probesets were identified using FlexArray software (http://genomequebec.mcgill.ca/Flex--Array; Blazejczyk et al., 2007) by analysis of variance (≥ twofold change and $P \leq 0.05$). Validation of the microarray data was performed with qPCR as described before (Gao et al., 2012) using eight randomly chosen probesets with or with no differential oxidation (Table S1). The microarray data derived from the mRNA oxidation experiment have been deposited in the Gene Expression Omnibus (GEO) database with the accession number GSE41949.

Protein extraction and two-dimensional gel electrophoresis

Total soluble protein was extracted from three independent biological replicates of dry and 24 h imbibed whole seeds (0.5 g fresh weight) of the same D and AR samples used for

germination, mRNA oxidation and transcriptomic studies as described previously (El-Bebany et al., 2010). To perform the two-dimensional electrophoresis, protein samples [500 µg/450 µL isoelectric focusing (IEF) solution] were used to rehydrate the 24-cm IEF strips (pH 4–7, GE Healthcare, Little Chalfont, UK) overnight at 22 °C under a layer of dry strip fluid. The IEF strips were then focused for a total of 58.3 kVh (Multiphor II: GE Healthcare) and then equilibrated before run on second dimension 12% sodium dodecyl sulfate (SDS)-polyacrylamide gels. Subsequently, gels were stained with Coomassie Brilliant Blue R-250 and the protein patterns compared between samples. Estimation of the abundance of spots representing the targeted proteins was performed with Quantity One quantification software (Bio-Rad, Hercules, CA) after taking the background signal into account. Significant differences in protein abundance between samples were determined using Student's t-test at $P \leq 0.05$.

Mass spectrometry

Gel pieces containing the differential protein spots were excised and processed as described by El-Bebany et al. (2010). Following digestion in situ with trypsin (modified sequencing grade: Promega, Madison, WI), the peptide samples were analysed with a linear ion trap mass spectrometer (LTQ XL; ThermoFisher Scientific, Waltham, MA) coupled with a nano-HPLC system (Ultimate 3000; Dionex, Germering, Germany) using the parameters described in El-Bebany et al. (2010). Following conversion to Mascot generic format files, the output files were queried with Mascot (v2.2; Matrix Science, London, UK) against the protein NCBI nonredundant (nr) protein and a local nonredundant wheat EST databases.

Conversion of proteins into probesets and co-expression analysis

To investigate whether the proteins differentially expressed between D and AR seeds are controlled transcriptionally, and to perform gene co-expression analysis, the differential proteins were converted into their respective probesets using WheatNet Blast2probesets tool (http://aranet.mpimp-golm.mpg.de/wheatnet; cut-off value e ≤ 1e-50). The expression profiles of the resulting probesets were extracted from transcriptomic data derived from the same D and AR wheat seed samples (GEO accession number GSE32409). Analysis of the transcriptomic data was performed as described previously (Gao et al., 2012), and probesets were considered to be differentially expressed if they exhibit ≥ twofold change at $P \leq 0.05$.

Probesets corresponding to proteins differentially expressed by after-ripening in both dry and imbibed states (Table S2) and those differentially expressed (Gao et al., 2012) and oxidized (Table S1) during after-ripening were queried to generate co-expression clusters using Web tools in WheatNet (http://aranet.mpimp-golm.mpg.de/wheatnet). Following the co-expression analysis, probesets in each co-expression clusters were annotated by rice genes using HarvEST WheatChip. Analysis of the 1-kb upstream region of the identified rice genes was performed with Osiris (http://www.bioinformatics2.wsu.edu/cgi-bin/Osiris/cgi/visualize_select.pl; Morris et al., 2008).

Acknowledgements

This work was supported by a grant from the Natural Sciences and Engineering Research Council of Canada to BTA. The authors thank Zhen Yao, Brenda Oosterveen and Jacqueling Ching for their technical assistance. The authors have no conflict of interest to declare.

References

Andersen, A.B., Risbo, J., Andersen, M.L. and Skibsted, L.H. (2000) Oxygen permeation through an oil-encapsulating glassy food matrix studied by ESR line broadening using a nitroxyl spin probe. *Food Chem.* **70**, 499–508.

Bailey, P.C., Lycett, G.W. and Roberts, J.A. (1996) A molecular study of dormancy breaking and germination in seeds of *Trollius ledebouri*. *Plant Mol. Biol.* **32**, 559–564.

Bassel, G.W., Lan, H., Glaab, E., Gibbs, D.J., Gerjets, T., Krasnogor, N., Bonner, A.J., Holdsworth, M.J. and Provart, N.J. (2011) Genome-wide network model capturing seed germination reveals coordinated regulation of plant cellular phase transitions. *Proc. Natl Acad. Sci. USA*, **108**, 9709–9714.

Bazin, J., Batlla, D., Dussert, S., El-Maarouf-Bouteau, H. and Bailly, C. (2011a) Role of relative humidity, temperature, and water status in dormancy alleviation of sunflower seeds during dry after-ripening. *J. Exp. Bot.* **62**, 627–640.

Bazin, J., Langlade, N., Vincourt, P., Arribat, S., Balzergue, S., El-Maarouf-Bouteau, H. and Bailly, C. (2011b) Targeted mRNA oxidation regulates sunflower seed dormancy alleviation during dry after-ripening. *Plant Cell*, **23**, 2196–2208.

Bewley, J.D. and Black, M. (1994) *Seeds-Physiology of Development and Germination*. New York, NY: Plenum Press.

Blazejczyk, M., Miron, M. and Nadon, R. (2007) *FlexArray (version 1.6.1): A Statistical Data Analysis Software for Gene Expression Microarrays*. Montreal, QC, Canada: Genome Quebec. http://genomequebec.mcgill.ca/FlexArray.

Bønsager, B., Finnie, C., Roepstorff, P. and Svensson, B. (2007) Spatio-temporal changes in germination and radical elongation of barley seeds tracked by proteome analysis of dissected embryo, aleurone layer, and endosperm tissues. *Proteomics*, **7**, 4528–4540.

Bove, J. (2005) Gene expression analysis by cDNA-AFLP highlights a set of new signalling networks and translational control during seed dormancy breaking in *Nicotiana plumbaginifolia*. *Plant Mol. Biol.* **57**, 593–612.

Cadman, C.S., Toorop, P.E., Hilhorst, H.W. and Finch-Savage, W.E. (2006) Gene expression profiles of Arabidopsis *Cvi* seeds during dormancy cycling indicate a common underlying dormancy control mechanism. *Plant J.* **46**, 805–822.

Cherian, G. and Chinachoti, P. (1997) Action of oxidants on water sorption, 2H nuclear magnetic resonance mobility, and glass transition behavior of gluten. *Cereal Chem.* **74**, 312–317.

Chibani, K., Ali-Rachedi, S., Job, C., Job, D., Jullien, M. and Grappin, P. (2006) Proteomic analysis of seed dormancy in Arabidopsis. *Plant Physiol.* **142**, 1493–1510.

El-Bebany, A., Rampitsch, C. and Daayf, F. (2010) Proteomic analysis of the phytopathogenic soilborne fungus *Verticillium dahliae* reveals differential protein expression in isolates that differ in aggressiveness. *Proteomics*, **10**, 289–303.

Ellgaard, L. and Ruddock, L.W. (2005) The human protein disulphide isomerase family: substrate interactions and functional properties. *EMBO Rep.* **6**, 28–32.

Finch-Savage, W.E. and Leubner-Metzger, G. (2006) Seed dormancy and the control of germination. *New Phytol.* **171**, 501–523.

Finkelstein, R., Reeves, W., Ariizumi, T. and Steber, C. (2008) Molecular aspects of seed dormancy. *Annu. Rev. Plant Biol.* **59**, 387–415.

Gallardo, K., Job, C., Groot, S.P.C., Puype, M., Demol, H., Vandekerckhove, J. and Job, D. (2002) Proteomics of Arabidopsis seed germination: a comparative study of wild-type and gibberellin-deficient seeds. *Plant Physiol.* **129**, 823–837.

Gao, F., Jordan, M.C. and Ayele, B.T. (2012) Transcriptional programs regulating seed dormancy and its release by after-ripening in common wheat (*Triticum aestivum* L.). *Plant Biotechnol. J.* **10**, 465–476.

Gubler, F., Millar, A.A. and Jacobsen, J.V. (2005) Dormancy release, ABA and pre-harvest sprouting. *Curr. Opin. Plant Biol.* **8**, 183–187.

Hare, P.D., Cress, W.A. and van Staden, J. (2003) A regulatory role for proline metabolism in stimulating *Arabidopsis thaliana* seed germination. *Plant Growth Regul.* **39**, 41–50.

He, D., Han, C., Yao, J., Shen, S. and Yang, P. (2011) Constructing the metabolic and regulatory pathways in germinating rice seeds through proteomic approach. *Proteomics*, **11**, 2693–2713.

van Hemert, M.J., Steensma, H.Y. and van Heusden, G.P. (2001) 14-3-3 proteins: key regulators of cell division, signalling and apoptosis. *BioEssays*, **23**, 936–946.

Iglesias-Fernández, R., Rodriguez-Gacio, M.C. and Matilla, A.J. (2011) Progress in research on dry after-ripening. *Seed Sci. Res.* **21**, 69–80.

Job, C., Rajjou, L., Lovigny, Y., Belghazi, M. and Job, D. (2005) Patterns of protein oxidation in Arabidopsis seeds and during germination. *Plant Physiol.* **138**, 790–802.

Kim, S.T., Wang, Y., Kang, S.Y., Kim, S.G., Rakwal, R., Kim, Y.C. and Kang, K.Y. (2009) Developing rice embryo proteomics reveals essential role for embryonic proteins in regulation of seed germination. *J. Proteome Res.* **8**, 3598–3605.

Kishor, P.B.K., Sangam, S., Amrutha, R.N., Laxmi, P.S., Naidu, K.R., Rao, K.R.S.S., Rao, S., Reddy, K.J., Theriappan, P. and Sreenivasulu, N. (2005) Regulation of proline biosynthesis, degradation, uptake and transport in higher plants: its implications in plant growth and abiotic stress tolerance. *Curr. Sci.* **88**, 424–438.

Kiyosaki, T., Matsumoto, I., Asakura, T., Funaki, J., Kuroda, M., Misaka, T., Arai, S. and Abe, K. (2007) Gliadain, a gibberellin-inducible cysteine proteinase occurring in germinating seeds of wheat, *Triticum aestivum* L., specifically digests gliadin and is regulated by intrinsic cystatins. *FEBS J.* **274**, 1908–1917.

Kong, Q.M. and Lin, C.L.G. (2010) Oxidative damage to RNA: mechanisms, consequences, and diseases. *Cell. Mol. Life Sci.* **67**, 1817–1829.

Lara, P., Oñate-Sánchez, L., Abraham, Z., Ferrándiz, C., Díaz, I., Carbonero, P. and Vicente-Carbajosa, J. (2003) Synergistic activation of seed storage protein gene expression in *Arabidopsis* by ABI3 and two bZIPs related to OPAQUE2. *J. Biol. Chem.* **278**, 21003–21011.

Liu, A., Gao, F., Kanno, Y., Jordan, M.C., Kamiya, Y., Seo, M. and Ayele, B.T. (2013) Regulation of wheat seed dormancy by after-ripening is mediated by specific transcriptional switches that induce changes in seed hormone metabolism and signaling. *PLoS ONE*, **8**, e56570.

Lorenzo, O., Nicolás, C., Nicolás, G. and Rodríguez, D. (2002) Molecular cloning of a functional protein phosphatase 2C (FsPP2C2) with unusual features and synergistically up-regulated by ABA and calcium in dormant seeds of *Fagus sylvatica*. *Physiol. Plant.* **114**, 482–490.

Mak, Y.X., Willows, R.D., Roberts, T.H., Wrigley, C.W., Sharp, P.J. and Copeland, L. (2009) Germination of wheat: a functional proteomics analysis of the embryo. *Cereal Chem.* **86**, 281–289.

Martinez, M., Cambra, I., Carrillo, L., Diaz-Mendoza, M. and Diaz, I. (2009) Characterization of the entire cystatin gene family in barley and their target cathepsin L-like cysteine proteases, partners in the hordein mobilization during seed germination. *Plant Physiol.* **151**, 1531–1545.

Moreno-Risueno, M.A., Diaz, I., Carrillo, L., Fuentes, R. and Carbonero, P. (2007) The HvDOF19 transcription factor mediates the abscisic acid dependent repression of hydrolase genes in germinating barley aleurone. *Plant J.* **51**, 352–365.

Moroashi, Y. and Sugimoto, M. (1988) ATP synthesis in cotyledons of cucumber and mung bean seeds during the first hours of imbibition. *Plant Cell Physiol.* **29**, 893–896.

Morris, R.T., O'Connor, T.R. and Wyrick, J.J. (2008) Osiris: an integrated promoter database for *Oryza sativa* L. *Bioinformatics*, **24**, 2915–2917.

Nambara, E., Okamoto, M., Tatematsu, K., Yano, R., Seo, M. and Kamiya, Y. (2010) Abscisic acid and the control of seed dormancy and germination. *Seed Sci. Res.* **20**, 55–67.

Oracz, K., El-Maarouf Bouteau, H., Farrant, J.M., Cooper, K., Belghazi, M., Job, C., Job, D., Corbineau, F. and Bailly, C. (2007) ROS production and protein oxidation as a novel mechanism for seed dormancy alleviation. *Plant J.* **50**, 452–465.

Oracz, K., El-Maarouf Bouteau, H., Kranner, I., Bogatek, R., Corbineau, F. and Bailly, C. (2009) The mechanisms involved in seed dormancy alleviation by hydrogen cyanide unravel the role of reactive oxygen species as key factors of cellular signaling during germination. *Plant Physiol.* **150**, 494–505.

Østergaard, H., Rasmussen, S., Roberts, T. and Hejgaard, J. (2000) Inhibitory serpins from wheat grain with reactive centers resembling glutamine-rich repeats of prolamin storage roteins-cloning and characterization of five major molecular forms. *J. Biol. Chem.* **275**, 33272–33279.

Parcy, F., Valon, C., Raynal, M., Gaubier-Comella, P., Delseny, M. and Giraudat, J. (1994) Regulation of gene expression programs during Arabidopsis seed development: roles of the ABI3 locus and of endogenous abscisic acid. *Plant Cell*, **6**, 1567–1582.

Pawłowski, T.A. (2007) Proteomics of European beech (*Fagus sylvatica* L.) seed dormancy breaking: influence of abscisic and gibberellic acids. *Proteomics*, **7**, 2246–2257.

Pawłowski, T.A. (2009) Proteome analysis of Norway maple (*Acer platanoides* L.) seeds dormancy breaking and germination: influence of abscisic and gibberellic acids. *BMC Plant Biol.* **9**, 48.

Pawłowski, T.A. (2010) Proteomic approach to analyze dormancy breaking of tree seeds. *Plant Mol. Biol.* **73**, 15–25.

Payne, P.I., Holt, L.M., Lawrence, G.J. and Law, C.N. (1982) The genetics of gliadin and glutenin, the major storage proteins of the wheat endosperm. *Plant Food Hum. Nutr.* **31**, 229–241.

Potokina, E., Sreenivasulu, N., Altschmied, L., Michalek, W. and Graner, A. (2002) Differential gene expression during seed germination in barley (*Hordeum vulgare* L.). *Funct. Integr. Genomics*, **2**, 28–39.

Rajjou, L., Gallardo, K., Debeaujon, I., Vandekerckhove, J., Job, C. and Job, D. (2004) The effect of a-amanitin on the Arabidopsis seed proteome highlights the distinct roles of stored and neosynthesized mRNAs during germination. *Plant Physiol.* **134**, 1598–1613.

Rajjou, L., Lovigny, Y., Groot, S.P.C., Belghazi, M., Job, C. and Job, D. (2008) Proteome-wide characterization of seed aging in Arabidopsis: a comparison between artificial and natural aging protocols. *Plant Physiol.* **148**, 620–641.

Russell, D.W. and Spremulli, L.L. (1980) Mechanism of action of the wheat germ ribosome dissociation factor: interaction with the 60S subunit. *Arch. Biochem. Biophys.* **201**, 518–526.

Schoonheim, P.J., Sinnige, M.P., Casaretto, J.A., Veiga, H., Bunney, T.D., Quatrano, R.S. and de Boer, A.H. (2007) 14-3-3 adaptor proteins are intermediates in ABA signal transduction during barley seed germination. *Plant J.* **49**, 289–301.

Shan, X., Tashiro, H. and Lin, C.L.G. (2003) The identification and characterization of oxidized RNAs in Alzheimer's disease. *J. Neurosci.* **23**, 4913–4921.

Simpson, G.M. (1990) *Seed Dormancy in Grasses.* Cambridge, UK: Cambridge University Press.

Spencer, D. (1984) The physiological role of storage proteins in seeds. *Philos. Trans. R. Soc. Lond. B: Biol. Sci.* **304**, 275–785.

Townley-Smith, T.F. and Czarnecki, E.M. (2008) AC Domain hard red spring wheat. *Can. J. Plant Sci.* **88**, 347–350.

Xia, J.H. and Kermode, A.R. (1999) Analyses to determine the role of embryo immaturity in dormancy maintenance of yellow cedar (*Chamaecyparis nootkatensis*) seeds: synthesis and accumulation of storage proteins and proteins implicated in desiccation tolerance. *J. Exp. Bot.* **50**, 107–118.

Yanagisawa, S. (2004) Dof domain proteins: plant-specific transcription factors associated with diverse phenomena unique to plants. *Plant Cell Physiol.* **45**, 386–391.

Yang, P., Li, X., Wang, X., Chen, H., Chen, F. and Shen, S. (2007) Proteomic analysis of rice (*Oryza sativa*) seeds during germination. *Proteomics*, **7**, 3358–3368.

Zhang, L., Hu, Y., Yan, S., Li, H., He, S., Huang, M. and Li, L. (2012) ABA-mediated inhibition of seed germination is associated with ribosomal DNA chromatin condensation, decreased transcription, and ribosomal RNA gene hypoacetylation. *Plant Mol. Biol.* **79**, 285–293.

5

Manipulating corn germplasm to increase recombinant protein accumulation

Elizabeth E. Hood[1],*, Shivakumar P. Devaiah[1], Gina Fake[2], Erin Egelkrout[2], Keat (Thomas) Teoh[1], Deborah Vicuna Requesens[1], Celine Hayden[2], Kendall R. Hood[3], Kameshwari M. Pappu[4], Jennifer Carroll[5] and John A. Howard[2]

[1]*Arkansas State University, Jonesboro AR, USA*
[2]*Applied Biotechnology Institute, San Luis Obispo, CA, USA*
[3]*Infinite Enzymes, Jonesboro, AR, USA*
[4]*University of South Florida, Tampa, FL, USA*
[5]*Cal Poly State University, San Luis Obispo, CA, USA*

Summary

Correspondence
email ehood@astate.edu

Using plants as biofactories for industrial enzymes is a developing technology. The application of this technology to plant biomass conversion for biofuels and biobased products has potential for significantly lowering the cost of these products because of lower enzyme production costs. However, the concentration of the enzymes in plant tissue must be high to realize this goal. We describe the enhancement of the accumulation of cellulases in transgenic maize seed as a part of the process to lower the cost of these dominant enzymes for the bioconversion process. We have used breeding to move these genes into elite and high oil germplasm to enhance protein accumulation in grain. We have also explored processing of the grain to isolate the germ, which preferentially contains the enzymes, to further enhance recovery of enzyme on a dry weight basis of raw materials. The enzymes are active on microcrystalline cellulose to release glucose and cellobiose.

Keywords: cellulase, corn, cellulose digestion, cellobiose, germplasm, high oil.

Introduction

The heavy reliance of western economies on fossil fuels has given rise to energy security concerns. These concerns together with limited petroleum reserves, negative environmental impacts and the rising cost of petroleum have prompted the development of viable alternatives that are cleaner and environmentally neutral. Biofuels and biobased products have emerged as one of the alternatives to address these concerns. They are obtained from renewable biomass, a key long-term component of a sustainable industry. Sustainable, renewable resources are those derived primarily from plant biomass, re-produced with minimal inputs using energy from the sun. Biomass for biofuels and biobased products can include many sources of material: agricultural harvests such as grains, agricultural residues such as stalks and leaves, perennial crops such as hay and trees, building waste wood, municipal solid waste such as paper, and various food industry processing wastes.

Ethanol from corn is the first-generation biofuel produced in the United States and Europe. However, grain-based fuels alone will not be sufficient to address the nation's energy needs. Biofuels must also be produced from renewable cellulosic feedstock materials to reach federally mandated levels of alternative fuels by 2022 (i.e. RFS2). The 'Billion Ton Study' published by the US Department of Energy and United States Department of Agriculture (USDA) estimated that sufficient renewable cellulosic resources are available from forest and agricultural lands to sustainably produce enough biofuel to displace 30% or more of US transportation fuel needs annually (Perlack *et al.*, 2005). This could significantly reduce the US dependence on imported oil.

However, plant biomass is a very complex matrix of polymers comprising the polysaccharides cellulose and hemicellulose in addition to lignin as the major structural components. Although cellulose is a simple, linear polymer of glucose, its semi-crystalline structure is notoriously resistant to hydrolysis by both enzymatic and chemical means. Thus, a strategy designed to use cellulosic material for biobased products must include the ability to efficiently and inexpensively convert the polysaccharide components of plant cell walls into simple sugars.

The preferred method for deconstructing plant cell walls is to pretreat the biomass with steam pressure often in conjunction with an acid or base catalyst (Saville, 2011). The cellulose in this pretreated biomass is then digested with enzymes (cellulases) to release the glucose for fermentation. Pretreatment and saccharification technologies are two of the three cost-limiting factors for cellulosic ethanol industry deployment, the third being delivered cost of feedstock materials. A number of enzymes are required to break down cellulose and work synergistically including endocellulase, exocellulase and glucosidase (Wood and Ingram, 1992). Other proteins, such as expansin, may also be helpful in facilitating this process (Cosgrove, 1999).

Cellulases are a subset of the glycosyl hydrolase superfamily of enzymes that have been grouped into at least 115 families based on amino acid sequence similarity, enzyme reaction mechanism and protein fold motif (Beguin and Aubert, 1994; Hayashi *et al.*, 2005). Two specific cellulases, endocellulase E1, from *Acidothermus cellulolyticus* and cellobiohydrolase I, or CBH I, from *Trichoderma reesei* have been shown to be effective and synergistic partners in the hydrolysis of cellulose from biomass (Baker *et al.*, 1998).

The current trend to use biofuels to offset the dependency on petroleum has catalysed intensive efforts to find new enzymes, feedstocks and microorganisms that will be more efficient and cost-effective in the breakdown of cellulose. This is challenging because the complex nature of crystalline cellulose, as well as the particular source of feedstock, offers unique structures and contaminants to the enzymes, affecting their efficiency (Beguin and Aubert, 1994; Kabel et al., 2007). An alternative approach to finding more unique enzymes is to find an inexpensive way to produce the huge volumes of enzymes required for the industrialization of biomass conversion.

Two hundred and thirty-five (235) million dry tonnes of biomass would be needed to produce 20 billion gallons of ethanol, assuming 40% cellulose and 85 gallons per tonne (considering the glucose fraction only) at 80% cellulose conversion into glucose (Howard et al., 2011). The amount of enzyme needed per gallon of ethanol to convert a feedstock is a moving target based on the continued improvement in enzyme mixes and the types of feedstocks. Early estimates by the National Renewable Energy Lab (NREL; 119 g/gallon with a predicted threefold increase in activity) and current enzyme manufacturing recommendations indicate that a reasonable assumption would be in the range of 30 g/gallon. With this assumption, the total amount of enzyme required for this volume of ethanol would be approximately 0.6 million tonnes. If these enzymes were to be produced by conventional fungal fermentation, the tanks and instrumentation alone for this much enzyme would cost nearly $30 billion before enzyme production processes even begin (based on calculations in Howard et al., 2011). The agricultural bioproduction system offers the potential for a viable and scalable alternative to lower the cost of enzymes for biomass deconstruction.

A plant-based system for the production of industrial enzymes has numerous advantages such as low capital investment and easily scalable large volume production. The ability to achieve high levels of accumulation of the enzymes in grain allows for their production at extremely low cost in a form that is easily harvested, stored and transported (Howard et al., 2011). Plant production systems that accumulate cellulase in the normally unused or low value portion of the plant are now approaching competitive cost structure with microbial systems. As expression technology improves and cellulase reaches 4% of the dry weight of the plant organ of choice, direct delivery of plant tissue can be the enzyme delivery system of choice and will easily reach cost targets set by NREL (Howard et al., 2011).

The practical aspects of producing enzymes in commodity crops and the associated logistics have also been demonstrated, particularly as they apply to cellulosic ethanol (Howard and Hood, 2005, 2007). We demonstrate in this study the use of germplasm and genetics to increase cellulase expression to generate very high accumulation levels of enzyme production in the field and the utility of these enzymes in biomass deconstruction experiments. Increases in target protein accumulation on a dry weight basis decrease the cost of production of enzymes for industrial applications. In order for the plant production system to be competitive with other production systems, the cost on a weight basis must be low, and the way to achieve this is to create plant lines with very high protein accumulation amounts. Thus, the selection for increased enzyme amount in the seed of these transgenic corn plants is critical to commercialization success. These plant-made enzymes can be a major enabler for the biomass to bioproducts industry.

Results

Enhancement of cellulase activity using germplasm

An independent transgenic event for endocellulase E1, referred to here as BCH01, and an independent transgenic event for cellobiohydrolase, or CBH I, referred to here as BCC02, were backcrossed for six generations into each of two elite inbred germplasm lines, a Lancaster (SP122) and a Stiff Stalk (SP114), to improve the agronomic performance of the transgenic lines in the field. The progeny of these crosses were concomitantly selected for increased target protein accumulation in seed at each generation.

The breeding scheme is illustrated in Figure 1. Tissue culture-derived plants are termed T0 or transformed first-generation nonseed-derived plants. Up to 10 plants per independent transgenic event were recovered, grown in the greenhouse, and pollinated with elite inbred pollen of the Lancaster parent (SP122) only. Six individual seeds from the ears of those plants were analysed for cellulase content by enzyme assay (Hood et al., 2007). The three highest-expressing ears per event, based on the individual seed assays, were chosen for continued propagation. Approximately 60 of the remaining seeds of these high-expressing individuals, 20 from each ear, were planted in the greenhouse and leaf-painted with Liberty™ herbicide to identify the segregating transgenic and nontransgenic progeny. The plants showed 1 : 1 segregation when treated with glufosinate ammonium, or Liberty™. Resistance to this herbicide is provided through the phosphinothricin acetyl transferase (pat) gene (Hood et al., 2007) that is linked to the cellulase gene cassette.

A fifty seed pool from each ear harvested from the greenhouse-grown plants was ground and analysed for total extractable protein and enzyme activity level. Enzyme values are expressed per unit dry weight of grain flour. Based on this analysis, up to 15 high-expressing T2 ears [nomenclature described in (Meksem and Kahl, 2010)] were selected from progeny derived from T1 plants pollinated with each inbred parent (SP122 or SP114) to continue in the backcross programme in the field. At this point, SP122 (Lancaster) lines had two doses of inbred, and SP114 (Stiff Stalk) lines had one dose of SP122 and one of SP114.

A single copy of the insert is present in each transgenic line (Hood et al., 2007). Thus, dosage of the linked pat and cellulase genes described above is one copy in half the seeds (see Table 1) because nontransgenic inbred pollen is used to produce seed on the hemizygous transgenic plants. This condition remains throughout the backcross programme until the plants are self-pollinated in the later generations. When homozygous plants are recovered in T8, they contain a pair of transgenic loci in every individual, and thus, the gene dosage increases from the original hemizygous ½ N to a final 2N. The relative expression could be as much as four times greater than the hemizygous ears assuming a linear correlation between the gene dosage and cellulase levels.

During each planting season, one to several rows of 20 seeds from each selected ear was planted, and all were treated with Liberty™ herbicide. Surviving plants were manually crossed with the appropriate recurrent parent and individually harvested, yielding from 75 to 200 ears per inbred parent. After analysis, ranking and selection based on high protein accumulation, seed from chosen individuals was then replanted in the following season's nursery (Figure 1).

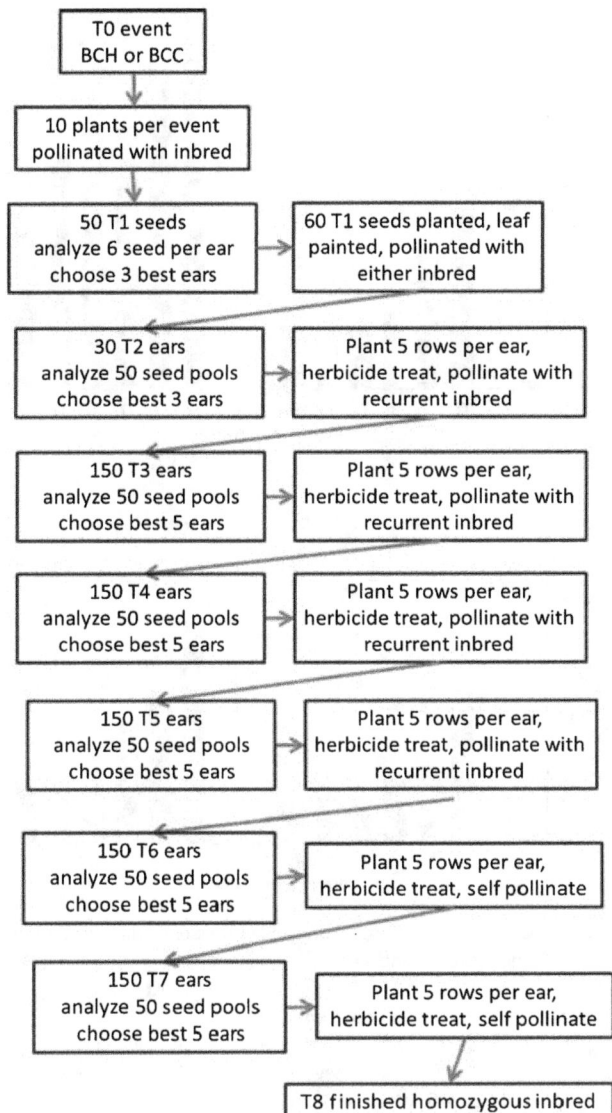

Figure 1 Breeding scheme for introducing the transgenic cellulase trait into elite germplasm. The 'T' nomenclature for transgenic backcrosses is standard to refer to each generation beyond transformation and tissue culture. T0 ears are pollinated with the SP122 (Lancaster) parent only in the greenhouse. T1 ears are pollinated with both inbred parents, half the ears with each parent. Each generation (T1 through T8) is selected based on field performance, and transgenic protein accumulation in seed. Fifty to 150 individual ears are screened at each generation, and using the top five ears, one to several rows of 20 plants each are propagated in the following field season, treated with herbicide to remove nontransgenic segregants and backcrossed to the recurrent parent. T6 and T7 plants are self-pollinated to ensure homozygosity of the inbred and the transgene.

Data for all ears from a single original plant were compared across generations to assess the overall increase in target protein accumulation. Using this approach, the total amount of target enzyme recovered per ear per generation can be ranked, and the highest-expressing transgenic lines selected. After three backcross generations, progeny from one specific plant line derived from a single transgenic event expressing each enzyme, BCH0101 (E1) or BCC0206 (CBH I), were continued in the

Table 1 Increase in transgene copy number with self-pollination of backcrossed inbreds and hybrid seed for production

Germplasm pool	Generation	Zygosity	Gene dosage	Potential expression
Backcrossed inbreds	T1-T6	Heterozygous segregating	1/2 N*	1
Self-pollinated inbreds	T7-T8	Approaching homozygous	N-2N†	2 to 4
Hybrids for production	T8 and beyond	Homozygous	2N	4
Hybrid grain	T8 and beyond	Homozygous	2N	4

*Individuals in the population either have 1 copy of the gene (hemizygous) or are null.

†In the first selfed populations, ½ are still hemizygous, ¼ are homozygous and ¼ are null, generating the apparent equivalent of 1 copy of the transgene in each plant (or N). In the second selfed populations, the hemizygous plants will produce the same ratio of progeny, but the homozygous plants will produce 2N progeny. Homozygous individuals are selected by herbicide screening for the inclusion in hybrid production.

breeding programme because they showed the largest increases in expression and the best agronomic phenotype. Although abiotic factors likely influence protein accumulation on an individual plant basis, cumulative data for progeny from a single original line show statistically significant increases in protein accumulation over generations.

For E1-expressing lines using elite germplasm as the recurrent male parent, cellulase increased to approximately 0.10% of dry weight in the highest lines harvested (Figure 2). Although significant differences in target gene expression level were not observed in the entire population of ears in generations T4-T6, they were all significantly different than the T1 generation at the 95% confidence level. The highest-expressing ears were sevenfold higher than the average protein accumulation level of 0.014% dry weight in first-generation (T1) seed. Because this T6 seed was still segregating 1 : 1, when the lines were self-pollinated to complete the breeding programme, up to four times the gene dose would be expected in the final homozygous hybrid, which should promote enzyme accumulation even further (Table 1). We have not observed gene silencing in seed-expressed transgenes in homozygous condition previously or in this project. Indeed, the first selfed generation (T7) of the Lancaster (SP122) parent increased the gene dosage to ~1 N and the amount of E1 enzyme to 0.15–0.2% DW of grain (Figure 2), significantly different (95% confidence level) than the hemizygous seed in the T6 generation.

Similar experiments were conducted with the CBH I lines. The parental inbreds each achieved accumulation of enzyme to approximately 0.2% DW of grain flour, a 20-fold increase over the original T1 amount of approximately 0.01% DW (Figure 3). T5 and T6 generations were not significantly different from one another, but in both inbred backgrounds, were significantly different than T1. The Lancaster (SP122) inbred is one breeding generation ahead of the Stiff Stalk (SP114) lines because it was the original male parent of greenhouse plants from tissue culture. T7 plants are the progeny of selfs of T6 plants (Figure 3a), and the enzyme accumulation levels were as high as 0.45% of dry weight of grain flour or twice the levels observed in the

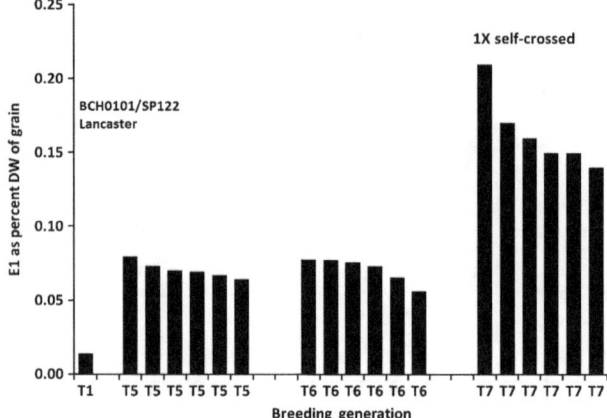

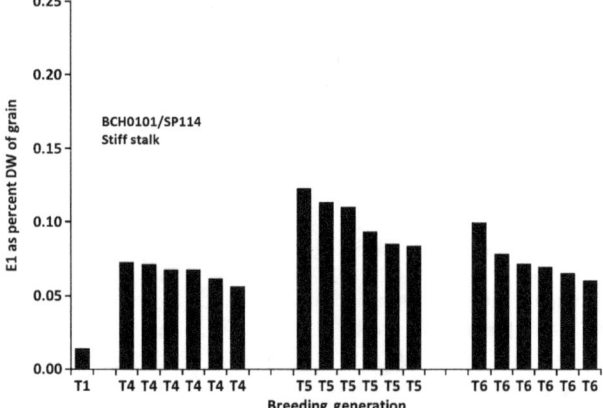

Figure 2 Selection of high-expressing E1 lines with elite germplasm. T1, T4, T5 and T6 represent seed generations post-transformation (T0). Each bar on the graph for T4–T7 is the result of the extraction and quantification of a 50-seed pool from a single plant in the backcross programme and is the average of three replicates. An average of 0.014% DW was obtained for the six T1 seed analysed (positive and negative) for the plant BCH0101, whose progeny are represented in the figure. Using statistical analysis described in Methods, the mean of T1 seed was lower when compared to the means of later generations at the 95% confidence level. For E1, T7 seed was also significantly higher than T5 and T6 seed. The top expressing lines (not the means) of each generation are shown in the graph.

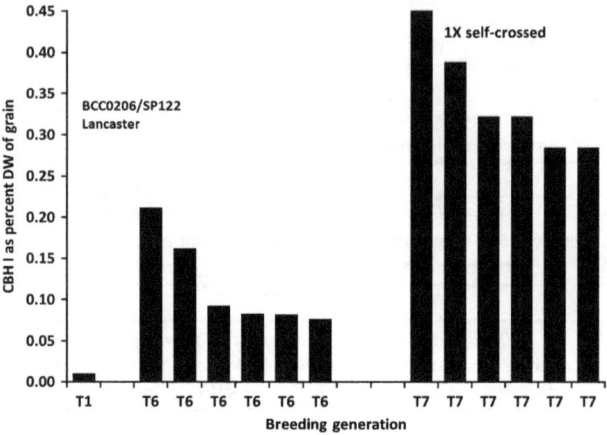

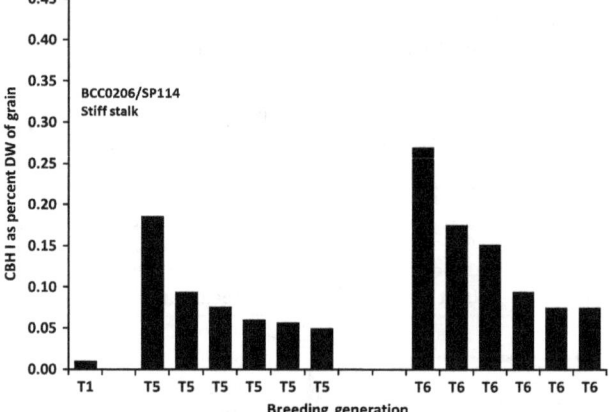

Figure 3 Selection of high-expressing CBH I lines with elite germplasm. T1, T5, T6 and T7 represent seed generations post-transformation (T0). Each bar on the graph for T5–T7 is the result of the extraction and quantification of a 50-seed pool from a single plant in the backcross programme and is the average of three replicates. An average of 0.01% DW was obtained for the six T1 seed analysed (positive and negative) for the plant BCC0206, whose progeny are represented in the figure. The mean of T1 seed was lower when compared to the means of later generations at the 95% confidence level. For CBH I, T7 seed was also significantly higher than T6 seed. The top expressing lines (not the means) of each generation are shown in the graph.

previous generation and significantly different from the T6 generation at the 95% confidence level. This is expected because the gene dosage is increasing (Table 1).

Alternative germplasm can also increase yields of protein from transgenic maize, as has been demonstrated for high oil corn (HOC) (Hood et al., 2002). Therefore, we tested the effect of high oil germplasm on the accumulation of cellulases in our transgenic lines. Each of the four transgene/inbred combinations was crossed to the HOC line—those yielding the most ears are shown (Figure 4). The effect of high oil germplasm was dramatic in a single cross using the high oil line as the male parent for both cellulase enzymes—E1 and CBH I (Figure 4). E1 accumulated to approximately 50% higher levels, and CBH I accumulated to nearly 2.5- to 5-fold higher levels compared with elite crosses of the same generation (significantly different at the 95% confidence level). High oil germplasm generally produces seed with large embryos that have increased oil

content. Because our transgene product accumulates in embryos, one might argue that the increased size alone may account for the increase in protein. However, in previous experiments, we have determined that factors other than embryo size alone account for this phenomenon (Hood et al., 2002).

Hybrid production and fractionation of grain

The purpose of the backcross programme is to produce transgenic inbred lines that can be used to generate transgenic hybrids with good agronomic performance that yield at parity with current hybrids. These hybrids should also contain high concentrations of the target proteins in grain. A hybrid was prepared at the T5 generation for each transgene [BCC0206 (CBH I) Stiff Stalk by Lancaster; BCH0101 (E1) Stiff Stalk by Lancaster] and used for production on an Arkansas farm in summer 2009. The amount of enzyme recovered on a grain dry weight basis in this trial production was extremely encouraging (Table 2)—at this rate, E1 (0.086% DW) would be recovered at approximately 1 kg

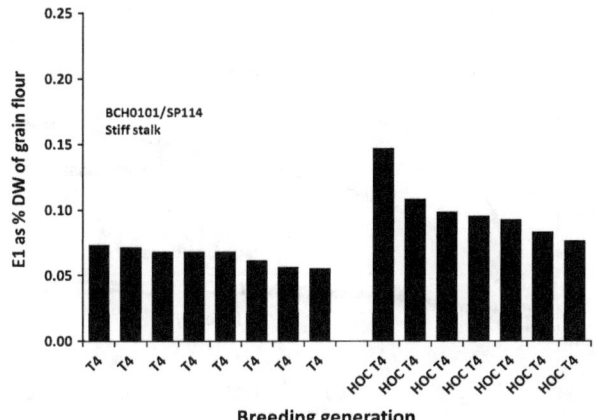

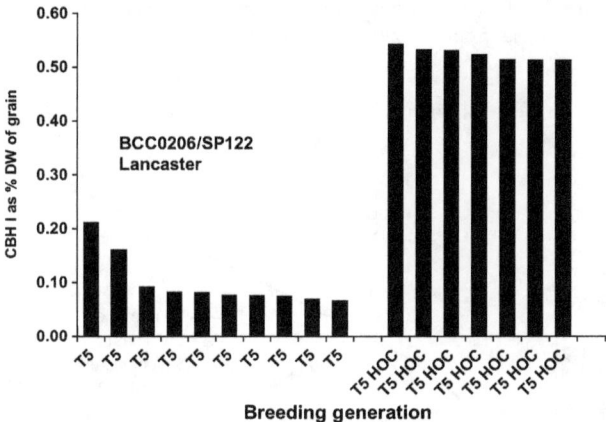

Figure 4 High oil germplasm increases accumulation of E1 and CBH I in seed. Each transgenic line was crossed onto high oil, or high oil pollen was crossed onto the transgenic line and F1 seed analysed in 50-seed bulks. Each bar on the graph is the result of the extraction and quantification of a 50-seed pool from a single plant in the backcross programme and is the average of three replicates. The high oil (HOC) lines were significantly higher compared to their elite counterparts using the median for all lines tested (n = 6–20 per generation). The top expressing lines (not the means) of each generation are shown in the graph.

Table 2 Yields of E1 and CBH I in hybrid germplasm and fractions therefrom

	E1 endocellulase	CBH I exocellulase
Transgenic grain fraction	Mean % DW ± standard deviation	
Whole flour	0.086 ± 0.001	0.176 ± 0.05
Endosperm flour (grits)	0.01 ± 0.002	0 ± 0
Mixed embryo fractions*	0.175 ± 0.02	1.143 ± 0.143
Hand-picked embryos	0.463 ± 0.021	1.5 ± 0.576

*The milled fraction from Satake maize de-germinator prior to separation of germ.

per tonne of grain and CBH I (0.176% DW) would be nearly 1.8 kg per tonne. These hybrid production levels are approximately equivalent to those seen in the highest T5 ears as shown in Figures 2 and 3, clearly showing that the expression levels can

be maintained under nonselective production conditions. Also, from the point of view of land-use efficiency, assuming 150 bu/acre grain yield, these expression levels translate to approximately 3.2 kg/ac for E1 and 6.6 kg/ac for CBH I.

Because the enzymes are predominantly confined to the embryo (germ) of the kernel, we tested dry mill fractionation of the grain to enrich the concentration of the enzymes in dry matter. Approximately 500 kg of grain for each hybrid was processed at 20% moisture content through a Satake De-Germinator at their facility in Stafford, TX. Two fractions were recovered—one with primarily endosperm (grits) that is mostly starch and a second that contains the protein-rich germ/oil fraction also containing small endosperm pieces and powdered starch (mixed embryo fractions; Table 2). Each fraction was analysed for extractable cellulase activity. The endosperm had no appreciable activity of either enzyme. The mixed embryo fractions from the de-germer amounted to only 0.25 tonnes each and thus were not a large enough volume to process on a gravity table for germ separation. Therefore, as a first step to assess germ-specific enzyme activity, embryos were hand-picked from the mixed fractions and analysed. On a dry weight basis, E1 embryos contained approximately five times more activity than the whole grain (Table 2), whereas CBH I germ exhibited a ninefold increase in activity over the whole grain. It is not clear whether these differences in germ enrichment of enzyme are because of differences in handling during processing or to impurities in the germ that may differentially impact enzyme activity assays.

Quantity of enzyme in grain

Total enzyme available in the grain has an impact on commercial value of the material. To determine the absolute quantity of the enzymes in grain, an exhaustive sequential extraction experiment was conducted. CBH I and E1 from corn meal were extracted five successive times, and protein detection was performed by enzyme assay and Western blot analysis on each separate extract (Figure 5). Sixty-four per cent of the CBH I activity was recovered in the first extraction and subsequent extractions resulted in 26%, 8% and 2% of the total (Figure 3a). The fifth extraction did not show any CBH I enzyme activity. A similar trend was observed in the case of E1, 63% of the E1 activity was recovered in the first extraction, and subsequent extractions resulted in 23%, 9%, 4% and 1% of the total E1 activity. Results of Western blot analysis showed a strong correlation with the enzyme activity results (Figure 5b).

Purification of enzymes

To characterize the enzyme activity on cellulosic substrates, enzyme was purified from grain samples. Purified, concentrated protein fractions were analysed by SDS–PAGE. Each protein showed >95% purity by Coomassie blue staining (Figure 6). The size of E1 (42 kDa) indicates that it is the catalytic subunit without the cellulose binding domain. CBH I at 53 kDa is intact with (apparent) multiple glycosylated forms.

Activity of enzyme on a cellulosic substrate

The data presented above assume that the activity of the enzymes on soluble substrates (MUC) is similar to the activity that catalyses cleavage of beta-1,4-glycosidic linkages in cellulose chains from cellulosic substrates. To verify this critical assumption, we first tested the ability of the plant-produced cellulases to release fermentable sugars from cellulose as the

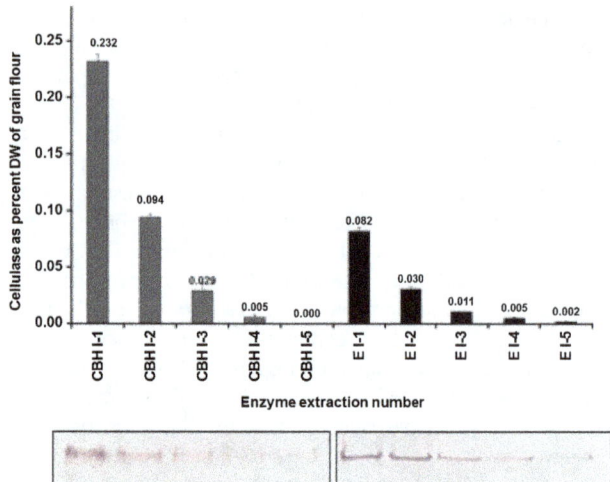

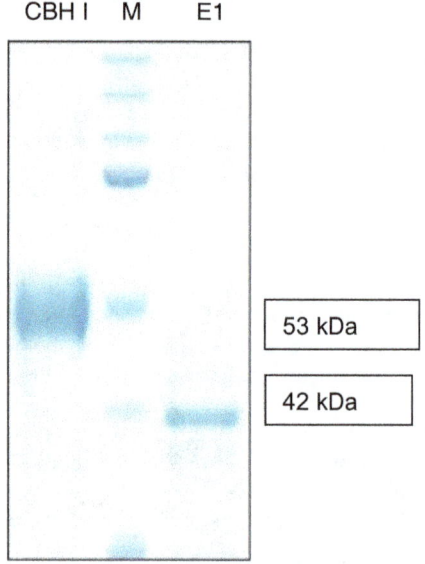

Figure 5 Successive extraction of E1 and CBH I from corn meal was performed to determine total activity recoverable. (a) (top) CBH I-1 through CBH I-5 represent the five successive extractions of cellobiohydrolase. E1-1 through E1-5 represent the five successive extractions of β-1,4-endoglucanase. (b) (bottom) Western blot of CBH I (left) and E1 (right) with protein-specific antibodies raised in rabbits against each protein. 10-g protein was loaded in each lane.

Figure 6 Coomassie blue–stained acrylamide gel of purified E1 and CBH I. These enzymes were used in deconstruction experiments described below. M = molecular weight markers.

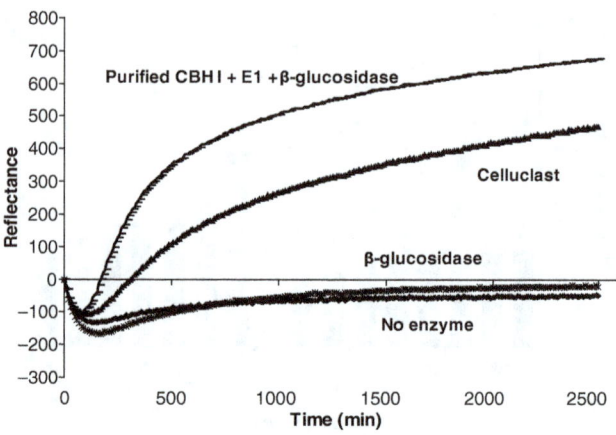

Figure 7 Celluclast and purified plant-produced enzymes' release of free sugars from Sigmacell as measured by microbial assay (BacT).

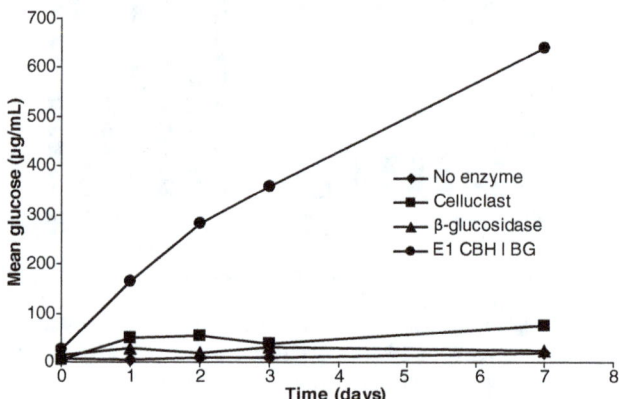

Figure 8 Release of free sugars using purified plant-produced enzymes as measured biochemically (GLOX). E1 = endoglucanase; CBH I = cellobiohydrolase; BG = β-glucosidase.

sole carbon source using the microbial growth assay as described earlier (Jimenez-Flores *et al.*, 2010). This experiment also provides evidence that in the process of releasing free sugars, no interfering compounds are released that would significantly interfere with the growth of an ethanol-producing microbe. The release of free sugars from cellulosic material requires at least three enzymes: an endocellulase (E1), an exocellulase (CBH I) and β-glucosidase to hydrolyse cellobiose into monomer glucose molecules. Therefore, we added β-glucosidase to our purified, plant-produced E1 and CBH I to ensure that fermentable sugars were available for the yeast in the microbial

assay. The results in Figure 7 illustrate that yeast can grow on the commercial preparation of cellulase (Celluclast, which is known to contain beta-glucosidase activity) as well as the plant-produced preparations. Cultures with no enzyme and cultures with β-glucosidase alone acted as controls to identify background fermentable sugars in the assay.

While the microbial growth assay confirms the ability of these enzyme preparations to generate fermentable sugars from cellulose, it does not positively identify which sugars are present. Glucose oxidase (GLOX) is a convenient assay that is specific for glucose (http://www.worthington-biochem.com/gop/default. html). Therefore, this assay was used in experiments similar to those in Figure 7 to measure the amount of glucose released from cellulose by the enzymes. The results in Figure 8 illustrate that both the plant-produced cellulase mixture and the commercial preparation of cellulase were capable of releasing glucose in amounts that were significantly greater than the controls. No attempt was made to make these samples equal in protein concentration. The combination of proteins in commercial cellulase is complex and not well-defined, and the corn-derived enzymes do not have all the factors present in the commercial cellulase.

In a final confirming experiment, free sugars released by the enzyme treatments were quantified by conventional HPLC analysis. The results in Figure 9 demonstrate that the commercial cellulase provides a significant amount of glucose (minimum detectable limit = 0.065 mg/mL) after 24 h and accumulating to over 0.1 mg/mL by 72 h and beyond. Cellobiose concentrations released by the commercial cellulase were much higher, accumulating to over 0.4 mg/mL. The plant-produced cellulase (E1 + CBHI) that also contained microbially produced β-glucosidase showed approximately the same amount of free sugar released but in this case it is almost all in the form of glucose rather than cellobiose. The abundance of additional β-glucosidase that converts cellobiose to glucose included with the plant-produced enzymes is a likely cause for this result. This result also explains why the plant-produced enzyme appears to be much more active than the commercial cellulase in the previous test because the GLOX assay only detects glucose and not cellobiose.

Discussion

With the current state of technology for biomass conversion, the overwhelming enzyme requirement is for cellulases: endocellulase, exocellulase and β-glucosidase (Merino and Cherry, 2007). The specific activity of commercially available cellulases is quite low (Jorgensen et al., 2007; Sticklen, 2008), and considerable effort has focused on increasing their activity levels. However, even with improved enzymes and improved methods of production, the amount of cellulase required to deconstruct the

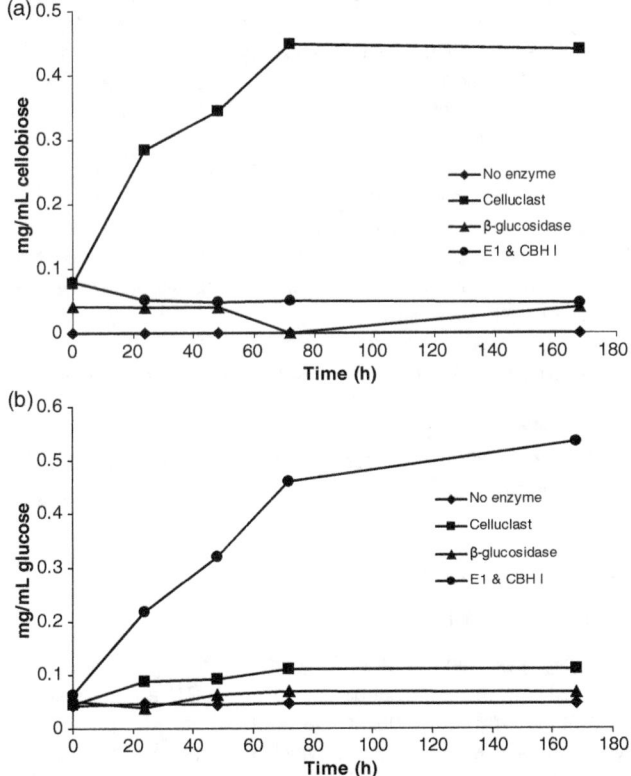

Figure 9 Release of free sugars as measured chemically (HPLC) showing cellobiose (a) and glucose (b). Treatments included: no enzyme; celluclast; β-glucosidase; plant-produced E1 + CBHI + microbial beta-glucosidase.

volumes of biomass necessary for 30% replacement of transportation fuel (the 2022 RFS2 target) is in the millions of tons per year.

This is an unprecedented challenge in terms of the amount and the extremely low cost of enzymes that is required for competitively priced biofuels and biobased products. Moreover, to produce the enzymes through conventional fungal fermentation, the amount of upfront capital investment required for fermenter capacity is daunting. This situation has led many groups to investigate ways to reduce this cost burden. Reports continually appear of improvements from many groups on different enzyme cost reduction technologies. One solution described here is through plant-produced enzymes.

Plant bioproduction of industrial enzymes offers an alternative to fungal fermentation. The advantages of the plant system include high expression levels in seed (Clough et al., 2006; Hood et al., 2007; Miles et al., 2007), established infrastructure for growing and processing the crops on a commodity scale, a stable and easily transportable and processable production package (i.e. seed), easy scaling up or down to meet market demand, and no capital investment for production. An additional advantage for corn is the ability to enhance the levels of target protein accumulation to a high degree in production lines through breeding and selection, which will help to minimize the footprint of this technology on productive land.

Maize genetics is a powerful tool with which to manipulate the expression of input and output traits engineered through biotechnology. For the transformation-competent germplasm of maize (Armstrong et al., 1991), field performance characteristics are poor. Nevertheless, grain yields similar to those of commercial hybrids are imperative for the production of genetically engineered lines and must be improved through transfer of transgenic traits into elite, high-yielding germplasm. We conducted a breeding programme to improve agronomic performance, enhance grain yields and increase cellulase protein accumulation in grain. We have also demonstrated the quality of activity of the cellulases produced in the grain. The quality and phenotype of the inbreds and hybrids produced from these crosses matched the wild-type germplasm as would be expected from an increase of 98% elite inbred genes into the transgenic lines. Moreover, parallel increases in transgenic protein accumulation could also be selected.

In this study, the increase in enzyme accumulation through backcrossing to elite inbreds was sevenfold for E1 and approximately 20-fold for CBH I. In each case, the potential increase with self-pollination of the inbred lines is an additional two to fourfold, and we have demonstrated up to a twofold with the first self-pollinated generation (Table 3; Figures 2 and 3). Thus, the potential increase over T1 expression in elite grain produced from hybrid seed is greater than 14-fold for E1 and >40-fold for CBH I. We have observed this phenomenon numerous times with various transgenes (Streatfield et al., 2002; Hood et al., 2003; Hood, 2004; Clough et al., 2006). Each of these proteins has different characteristics, and the germplasm used in the crosses vary; thus, the phenomenon appears to be a general one. Although it is empirically reproducible, the mechanism is unknown. We have constructed isogenic lines of high and low expression at advanced generations and are exploring the mechanism of this phenomenon.

Efficient recovery of enzyme from the production raw material plays an important role in reducing the cost of industrial enzymes. The bottom line for industrial enzymes is how much

Table 3 Summary of increase in protein accumulation per dry weight at various steps in breeding and processing of mature grain over T1 (first-generation seed from tissue culture plants) selected lines

Enzyme	Increase from T1–T6	Increase with self-pollination	Increase with hybrid grain	Increase with high oil germplasm	Potential increase in elite grain*	Processing increase in germ flour	Total possible increase with processing[†]
E1	Sevenfold	two to fourfold	Equal	0.33-fold	14- to 28-fold	fivefold	~70- to 100-fold
CBH I	20-fold	two to fourfold	Equal	fivefold	40- to 80-fold	ninefold	~360- to 720-fold

*Potential increase in elite grain is derived from multiplying the T6 increase by the self-pollination increase.
[†]Final number is derived from multiplying elite grain increase by processing increase.

enzyme can be recovered from production material on a dry weight basis. Separating the germ fraction from the grain and removing the oil using standard process conditions provide a defatted germ that is often used in animal feed and in ethanol production (Duensing et al., 2003). Because the cellulase is expressed from an embryo-preferred promoter, the germ should contain as much as 10-fold more cellulase than the whole grain, which provides a potential formulation for adding concentrated enzymes to industrial processes. However, processing is not a 100% efficient procedure, and losses occur at each step. Thus, it was critical to assess the enzyme concentration and activity in the germ fractions. Thus, dry-milled, processed grain, producing the embryo (germ) and the endosperm (grits) fractions, was analysed to determine the recovery of enzyme from the germ fraction. In the recovered germ, the enzyme shows as much as a fivefold increase in accumulation on a dry weight basis for E1 and as much as a ninefold increase for CBH I (Table 3). When all breeding and processing steps available for corn production are combined, the resulting recovery of enzyme per dry weight will allow highly cost-effective production of E1 and CBH I from the corn seed biofactory.

We previously demonstrated that high oil germplasm can be used to enhance the yield of target proteins in maize (Hood, 2004). In the current study, we tested the effect of a single high oil cross on the accumulation of E1 and CBH I. The increase in CBH I was fivefold, and a 50% increase in total E1 protein was observed. These results suggest a production method—that high oil germplasm could be used as a component of a production field with out-crossing of the transgenic line pollen onto a high oil line for production of high accumulation of enzymes and high yields of grain. Thus, using a variety of methods, the quantity of enzyme in transgenic lines can be increased 70- to 100-fold or 360- to 720-fold for E1 or CBH I, respectively, over the original transgenic lines recovered from tissue culture (Table 3). This strategy is powerful for promoting industrial enzyme accumulation in maize and has impressive potential for lowering production costs associated with the system.

High accumulation is only useful if the enzymes so accumulated are active in their target applications. Thus, enzymes were analysed for their ability to release free sugars by a variety of methods all of which indicated their ability to deconstruct microcrystalline cellulose. E1 and CBH I purified from corn flour were active in deconstructing cellulose when combined with β-glucosidase. The primary product of this reaction (over 24 h) was glucose in contrast to purchased control mixtures of T. reesei enzymes for which cellobiose was the primary product. Further experiments to optimize plant-made enzyme load alone and in combination with microbial enzymes are in progress.

In the near term, the accepted production method for cellulase enzymes for biomass conversion is likely to be microbial fermentation. However, because of the potential in cost savings for large-scale production using plants, in the longer term, the plant process could enhance the use of microbial enzymes, lowering the capital investment necessary for this cost-sensitive industry.

To grow these corn lines with enzymes at the current accumulation levels, an approximately equal number of acres would be required to deconstruct the harvestable stover (Howard and Hood, 2007). However, the technology is in its infancy and will produce new lines with more than 10-fold greater amounts of enzyme than are currently available. In addition, because the system does not require capital infrastructure, the cost of production is far less than for microbial enzyme fermentation (Howard et al., 2011). Nevertheless, because they are genetically engineered, deregulation of the lines through USDA APHIS will be required for commercial production. However, even though the cost of this regulatory approval process could be as high as $30 million (McElroy, 2003), it is far less than the cost of building the fermentation infrastructure required for the microbial production systems.

The plant biofactory is thus a viable system for biomass conversion enzyme production. To scale these enzymes to industrial production, the next steps include to:
1. determine the efficacy of biomass conversion using these enzymes in industry relevant conditions, i.e. biomass substrates and process conditions.
2. scale-up reactions to pilot plant scale
3. use enzymes in combination with microbial enzymes at pilot scale.

This excellent system can demonstrate the power of plant production to reduce the cost of effective enzymes for biomass conversion processes.

Experimental procedures

Transgenic plant lines

Each cellulase gene is driven by the embryo-preferred maize globulin-1 promoter (Belanger and Kriz, 1991) and is linked to a CaMV 35S promoter-driven maize-optimized phosphinothricin acetyl transferase (pat) gene, which confers resistance to the herbicide Bialaphos (Hood et al., 2007). Each independent transgenic event [BCC02 (CBH I) and BCH01 (E1)] has a single insertion and a single copy of the respective cellulase gene. Breeding experiments described below were carried out with individual plants derived from these two events.

Germplasm sources

Transgenic plants (from Hi II germplasm) were generated using *Agrobacterium tumefaciens*-mediated transformation experiments described earlier (Hood *et al.*, 2007). The plants include two independent events expressing the E1 endocellulase (E.C. 3.2.1.4 or Cel5A; GenBank Accession #U33212) from *Acidothermus cellulolyticus* and the cellobiohydrolase I gene (E.C.3.2.1.91 or Cel7A; GenBank Accession #X69976) from *T. reesei* (*Hypocrea jecorina*). Each gene is expressed using the globulin I promoter from maize. Up to 10 individual plants were recovered for each independent transgenic event and pollinated with the SP122 (Lancaster) inbred in the greenhouse. The resulting ears were individually analysed, and 10–15 of the ears showing the highest-expressing seed were chosen for subsequent agronomic performance improvement through successive backcrosses to elite germplasm.

In each crossing experiment, multiple rows (typically five rows, 20 seeds per row) of transgenic seed were planted from each line described above. Nontransgenic pollen donors were planted in intervals at 5 days prior to transgenic plantings, on the same day as the transgenic plantings and at 5 days post-transgenic plantings to ensure sufficient pollen for crossing to the transgenic lines. Transgenic rows were sprayed at the 5-leaf stage with Liberty™ herbicide (1% active ingredient) to select against nontransgenic segregating plants. Selected lines were crossed separately to either an elite Stiff Stalk (SP114) or an elite Lancaster (SP122) inbred line to generate both parents for a high-yielding hybrid. In each cross, the transgenic line was used as the female parent with pollen from the elite inbred parent. For the final hybrid seed for grain production, the Stiff Stalk (SP114) line was used as the female parent. All field-based experiments were performed with APHIS permits and inspected facilities.

In high oil germplasm (HO-703) crosses, either the high oil or the transgenic line was used as the female parent in the single cross. Plants from each transgenic line/inbred parent combination were crossed to the HO-703 line. Data are reported for the lines producing the most progeny.

Enzyme activity

Cellulase assays were performed as described earlier (Hood *et al.*, 2007). Corn meal (from coffee grinder) from fifty seeds was ground in the presence of liquid nitrogen, and 0.1 g of sample was extracted in 1 mL of 50 mM sodium acetate buffer, pH 5.0. The homogenate was vortexed and centrifuged at 9.3 g for 10 min or 0.4 g for 15 min at 4 °C. Supernatants from two independent extractions and two or three samples from each extraction were used for enzyme activity assays. Reaction mixtures contained 90 μL of 50 mM sodium acetate buffer, pH 5.0, and 10 μL of extract (approximately 500 ng total protein) with 25 μL of 5 mM methylumbelliferyl β-D-cellobioside (MUC; # M6018 Sigma Chemical Co., St. Louis, MO). A 96-well plate containing the reaction mixtures was incubated at 50 °C. For CBH I and E1 activity assays, incubation time was 120 or 30 min, respectively. Highly purified CBH I and E1 (provided by NREL) were used initially as the standard. As the amount of purified material was limiting and we had a desire to run a standard with every assay (thousands of samples over several years), we calibrated a commercial preparation of cellulase from *T. reesei* (E1 and CBH I mixture; Sigma # C8546) to use as a relative standard for every assay. *Trichoderma reesei*

from Sigma showed 3.6-fold more activity on a weight basis than purified CBH I and 10-fold less activity than purified E1; therefore, these correction values were used when individual assay plates were run for corn extracts. The amount of protein in each sample was estimated using the Bradford method (Bradford, 1976).

Western blot

Polyclonal antibodies were prepared by Cocalico Labs (Reamstown, PA) against purified CBH I and E1 proteins that were individually expressed in the *E. coli* Gateway™ vector system (Invitrogen, Carlsbad, CA). One-half microgram of total protein from a crude protein extract of corn flour was size separated using 12% SDS–PAGE (Invitrogen). Western blots were prepared from gels by transferring proteins to a PVDF membrane (Millipore, Billerica, MA) and blocking with 5% BSA (Fisher Scientific, Atlanta, GA) in 1× Tris-buffered saline buffer (25 mM Tris, 0.8% NaCl, pH 7.4). The blocked membrane was incubated with primary antibody, anti-CBH I or anti-E1 (1 : 2500), for 2 h. After incubation, membrane was washed three times for 5 min each with 1× TBS buffer and then incubated with secondary antibody, goat-anti-rabbit-alkaline phosphatase conjugate (1 : 5000), for 2 h. Colour detection was carried out using the NBT/BCIP reagent per manufacturer's instructions (Promega Corporation, Madison, WI).

Enzyme purification

A 100-g sample of corn meal of each CBH I and E1 transgenic lines was separately soaked and agitated in 500 mL of 50 mM sodium acetate buffer pH 5.0 for 1 h at 4 °C. The suspension was filtered through four layers of cheese cloth and centrifuged at 10 000 rpm (Eppendorf centrifuge 5810R and Eppendorf fixed –angle rotor F-34-6-38) for 10 min at 4 °C. The pellet was resuspended in 500 mL of extraction buffer, thus repeating the enzyme extraction. Supernatants from the two extractions were pooled and used for ammonium sulphate precipitation. For CBH I, a 50%–90% ammonium sulfate pellet contained the protein, whereas a 0%–50% ammonium sulfate pellet contained the E1 protein. Pellets obtained after ammonium sulfate precipitation were dissolved in 15 mL of 20 mM Tris-HCl pH 7.0 buffer containing 150 mM NaCl.

Protein purification was carried out using the Bio-CAD HPLC system (Perceptive Biosystems; Global Medical Instrumentation, Inc., Ramsey, MN). Protein samples were desalted using a HiPrep 26/10 desalting column (GE Healthcare Bio-Sciences, Piscataway, NJ). Desalted proteins were loaded onto an anion exchange, HiTrap Q XL column (GE Healthcare Bio-Sciences). Before the start of 1 M NaCl gradient, the column was run with 20 mM Tris-HCl buffer pH 7.0 alone for 10 min, and then, the 1 M NaCl gradient was started and run for 20 min. Protein fractions were collected from 0 to 30 min with a flow rate of 5 mL/min (2 mLs per fraction). Eluted protein fractions were used for enzyme analysis and fractions from 41 to 60 and 39 to 60, which showed CBH I and E1 activities, respectively, were pooled separately. Pooled protein solutions were concentrated using a Spin-X UF concentrator (Corning, Corning, NY), 2 mL samples were loaded onto a Sephacryl S-200 Hi Prep, 1.6 × 60 column (GE Healthcare Bio-Sciences), and protein fractions were collected at flow rate of 0.5 mL/min using 20 mM Tris-HCl buffer pH 7.0. Eluted protein fractions were used for enzyme analysis, and fractions from 45 to 55 and 75 to 87, which showed

CBH I and E1 activity, respectively, were pooled separately. Pooled protein fractions were concentrated using Spin-X UF concentrators and buffer exchanged with 50 mM sodium acetate pH 5.0. Purified proteins were assessed for relative purity by SDS–PAGE and remaining volumes lyophilized.

Validation of extraction and quantification of enzyme

Enzyme extraction was carried out as described above. Each resultant pellet was used for an additional four extractions. The supernatant from each extraction was collected separately and used for enzyme assay and Western blot analysis.

Cellulose substrates

The source of cellulose was Sigmacell (Sigma Chemical Co.). 4-Methylumbelliferyl β-D-cellobioside (MUC) was obtained from Sigma Chemical Co., St. Louis, MO (M2018).

BacTAlert/Microbial growth

Microbial growth was monitored using Sigmacell as the sole carbon source to indicate the release of fermentable sugars and to ensure that no major interfering by-products were in the reaction (Jimenez-Flores et al., 2010). Control enzymes were from T. reesei (Sigma # C8546, Celluclast) and β-glucosidase (Sigma # C6105). A starter culture of Saccharomyces cerevisiae was allowed to grow in 5 mL of YPD broth on a rotary shaker at 37 °C and 225 r.p.m for 4–6 h; 0.1 mL of the culture (OD_{550} = 0.5 of a 1 : 60 dilution) was used as inoculum for the cellulose fermentation assays. The assay was carried out under sterile conditions in BacT (BioMerieux, St. Louis, MO) bottles containing 6.25 mg/mL cellulose (Sigma # S3504, Sigmacell) suspended in 40 mL of 140 mM citrate/90 mM bicarbonate buffer pH 5.0. Treatments were as follows: no enzyme added, 4 μL Celluclast, 4 μL β-glucosidase, or 4 μL β-glucosidase with 3.75 mg purified plant-derived CBH I and 0.1 mg of purified plant-derived E1. All treatments contained 250 mg Sigmacell, 100 μL of yeast (grown to log phase in YPD (a standard medium for growing yeast comprising 1% yeast extract, 2% peptone and 2% glucose). Glucose is one of the media components used to grow the yeast. After growing the yeast to log phase, cells were harvested by centrifugation and washed with 140 mM citrate/90 mM bicarbonate buffer pH 5.0, to remove the sugar and other components of YPD medium. After washing, cells were resuspended in the citrate buffer, aliquoted into glycerol stocks and stored at −80 °C.

Bottles were incubated at 37 °C and consisted of the following treatments: No enzyme, 1 μL Celluclast, 1 μL β-glucosidase, or 1 μL β-glucosidase with 939 μg purified plant-derived CBH I and 27.5 μg of purified plant-derived E1. All treatments contained 62.5 mg Sigmacell and citrate buffer to a total volume of 10 mL.

Glucose oxidase assay

The GLOX assay was conducted as described by the Worthington Biochemical website (http://www.worthington-biochem. com/gop/assay.html) with the following modifications. Peroxidase and glucose oxidase were resuspended to 1mg/mL, nonoxygenated o-dianisidine (Sigma Chemical Co.) was resuspended in DMSO (Sigma Chemical Co.) to a stock concentration of 2% (v/v) and used in the assay at a concentration of 0.016%, and a 10% D-glucose stock solution was left to mutarotate for a minimum of 1 h prior to use. The total assay volume was 200 μL: 150 μL o-dianisidine solution in 0.1 M potassium phosphate buffer pH 6.0, 10 μL peroxidase, 10 μL glucose oxidase and 30 μL of glucose standard or cellulase reaction sample. Three independent GLOX reactions were performed for each treatment, and mean glucose concentrations are reported. GLOX reactions were conducted at room temperature, and readings were taken at 460 nm every 30 s for 5 min. SoftmaxPro5.4 (Molecular Devices, Sunnyvale, CA) software was used to analyse reaction rates. Samples for both the GLOX assay and HPLC analysis were removed from the same reaction bottles.

HPLC methods

Carbohydrate concentrations were obtained using protocols established by the National Renewable Energy Laboratory Technical Report (NREL/TP-510-42623, available online at: http:// www.nrel.gov/biomass/pdfs/42623.pdf). Analysis was performed on a Shimadzu Prominence Series HPLC with a Bio-Rad Aminex (HPX-87P) column, a Bio-Rad de-ashing pre-column and an Agilent 1200 Series Refractive Index Detector. Results shown are the median of three replicate samples.

Statistical analysis

The analysis was performed in SAS version 9.1 (SAS® business analytics software and services, Cary, NC) using PROC MIXED. Four parallel analyses were performed for BCH0101/SP122, BCH0101/SP114, BCC0206/SP122 and BCC0206/SP114. Each analysis was a one-way ANOVA with E1 (or CBH I) per cent dry weight as the response and generation as the only factor. The LSMEANS command was used to compare the mean response at each generation to every other generation and to the initial T1 value. These comparisons were adjusted using Tukey's method. Each analysis used a 5% significance level.

Acknowledgements

This work was supported by a grant from the US Department of Energy (DE FG36 GO88025) with cost share from the Wal-Mart Foundation, the Walton Family Foundation, and Arkansas State University. We also thank Dr John Walker, Cal Poly State University, for his help in statistical analysis of the data.

References

Armstrong, C., Green, C. and Phillips, R. (1991) Development and availability of germplasm with high type II culture formation response. Maize Genet. Coop. Newslett. 65, 92–93.

Baker, J.O., Ehrman, C.I., Adney, W.S., Thomas, S.R. and Himmel, M.E. (1998) Hydrolysis of cellulose using ternary mixtures of purified celluloses. Appl. Biochem. Biotechnol. 70–72, 395–403.

Beguin, P. and Aubert, J.P. (1994) The biological degradation of cellulose. FEMS Microbiol. Rev. 13, 25–58.

Belanger, F. and Kriz, A. (1991) Molecular basis for allelic polymorphism of the maize Globulin-1 gene. Genetics, 129, 863–872.

Bradford, M. (1976) A rapid and sensitive method for the quantitation of microgram quantities of protein utilizing the principle of protein-dye binding. Anal. Biochem. 72, 248–254.

Clough, R.C., Pappu, K., Thompson, K., Beifuss, K., Lane, J., Delaney, D.E., Harkey, R., Drees, C., Howard, J.A. and Hood, E.E. (2006) Manganese peroxidase from the white-rot fungus Phanerochaete chrysosporium is enzymatically active and accumulates to high levels in transgenic maize seed. Plant Biotechnol. J. 4, 53–62.

Cosgrove, D.J. (1999) Enzymes and other agents that enhance cell wall extensibility. Annu. Rev. Plant Physiol. Plant Mol. Biol. 50, 391–417.

Duensing, W., Roskens, A. and Alexander, R. (2003) Corn dry milling: processes, products, and applications. In *Corn: Chemistry and Technology* (White, P. and Johnson, L. eds), St Paul MN: Amer Assoc of Cereal Chemists.

Hayashi, T., Yoshida, K., Park, Y.W., Konishi, T. and Baba, K. (2005) Cellulose metabolism in plants. *Int. Rev. Cytol—a Surv. Cell Biol.* **247**: 1–34.

Hood, E.E. (2004) Bioindustrial and Biopharmaceutical products from plants. In *New Directions for a Diverse Planet: Proceedings for the 4th International Crop Science Congress; 26 September—1 October 2004*, Brisbane, Australia: The Regional Institute Ltd.

Hood, E., Bailey, M.R., Beifuss, K., Magallanes-Lundback, M., Horn, M., Callaway, E., Drees, C., Delaney, D., Clough, R and Howard, J. (2003) Criteria for high-level expression of a fungal laccase gene in transgenic maize. *Plant Biotechnol J* **1**: 129–140.

Hood, E., Howard, J. and Delaney, D. (2002) Method of Increasing Heterologous Protein Expression in Plants. US patent # 7,541,515.

Hood, E.E., Love, R., Lane, J., Bray, J., Clough, R., Pappu, K., Drees, C., Hood, K.R., Yoon, S., Ahmad, A. and Howard, J.A. (2007) Subcellular targeting is a key condition for high-level accumulation of cellulase protein in transgenic maize seed. *Plant Biotechnol. J.* **5**, 709–719.

Howard, J.A. and Hood, E. (2005) Bioindustrial and biopharmaceutical products produced in plants. *Adv. Agron.* **85**, 91–124.

Howard, J.A. and Hood, E.E. (2007) Methods for growing nonfood products in transgenic plants. *Crop Sci.* **47**, 1255–1262.

Howard, J., Nikolov, Z. and Hood, E. (2011) Enzyme production systems for biomass conversion. In *Plant Biomass Conversion* (Hood, E., Nelson, P. and Powell, R. eds), pp. 227–253. Ames, IA: Wiley-Blackwell.

Jimenez-Flores, R., Fake, G., Carroll, J., Hood, E. and Howard, J. (2010) A novel method for evaluating the release of fermentable sugars from cellulosic biomass. *Enzyme Microb. Technol.* **47**, 206–211.

Jorgensen, H., Kristensen, J.B. and Felby, C. (2007) Enzymatic conversion of lignocellulose into fermentable sugars: challenges and opportunities. *Biofuels Bioprod. Bioref.* **1**, 119–134.

Kabel, M., Bos, G., Zeevalking, J., Voragen, A. and Schols, H. (2007) Effect of pretreatment severity on xylan solubility and enzymatic breakdown of the remaining cellulose from wheat straw. *Bioresour. Technol.* **98**, 2034–2042.

McElroy, D. (2003) Sustaining agbiotechnology through lean times. *Nat. Biotechnol.* **21**, 996–1002.

Meksem, K. and Kahl, G. (2010) *The Handbook of Plant Mutation Screening: Mining of Natural and Induced Alleles.* Hoboken, NJ: Wiley-Blackwell.

Merino, S.T. and Cherry, J. (2007) Progress and challenges in enzyme development for biomass utilization. *Adv. Biochem. Eng. Biotechnol.* **108**, 95–120.

Miles, S., Arellano, S., Krebs, H., Nelso, A., Batie, C. and Betts, S. (2007) Requirement of an ER-directed signal peptide for accumulation of active Trichoderma reesei cellobiohydrolase I in transgenic corn seed. *Plant Biol. Bot.* Abstract # P03010 http://abstracts.aspb.org/pb2007/public/P03/P03010.html.

Perlack, R.D., Wright, L.L., Turhollow, A.F., Graham, R.L., Stokes, B.J. and Erbach, D.C. (2005) *Biomass as Feedstock for a Bioenergy and Bioproducts Industry: The Technical Feasibility of a Billion-Ton Annual Supply.* http://handle.dtic.mil/100.2/ADA436753, Washington DC: Department of Energy USDoA ed.

Saville, B. (2011) Pretreatment Options. In *Plant Biomass Conversion* (Hood, E., Nelson, P. and Powell, R. eds), pp. 199–226. Ames, IA: Wiley-Blackwell.

Sticklen, M.B. (2008) Plant genetic engineering for biofuel production: towards affordable cellulosic ethanol. *Nat. Rev. Genet.* **9**, 433–443.

Streatfield, S.J., Mayor, J.M., Barker, D.K., Brooks, C., Lamphear, B.J., Woodard, S.L., Beifuss, K.K., Vicuna, D.V., Massey, L.A., Horn, M.E., Delaney, D.E., Nikolov, Z.L., Hood, E.E., Jilka, J.M. and Howard, J.A. (2002) Development of an Edible Subunit Vaccine in Corn against Enterotoxigenic Strains of Escherichia coli. *In Vitro Cellular and Developmental Biology – Plant* **38**, 11–17.

Wood, B. and Ingram, L. (1992) Ethanol production from cellobiose, amorphous cellulose, and crystalline cellulose by recombinant Klebsiella oxytoca containing chromosomally integrated Zymomonas mobilis genes for ethanol production and plasmids expressing thermostable cellulase genes from Clostridium thermocellum. *Appl. Environ. Microbiol.* **58**, 2103–2110.

6

Modulation of kernel storage proteins in grain sorghum (*Sorghum bicolor* (L.) Moench)

Tejinder Kumar[1], Ismail Dweikat[1], Shirley Sato[2], Zhengxiang Ge[2], Natalya Nersesian[2], Han Chen[2], Tom Elthon[1,2], Scott Bean[3], Brian P. Ioerger[3], Mike Tilley[3] and Tom Clemente[1,4,*]

[1]Department of Agronomy and Horticulture, University of Nebraska-Lincoln, Lincoln, NE, USA
[2]Center for Biotechnology, University of Nebraska-Lincoln, Lincoln NE, USA
[3]Grain Quality and Structure Research Unit, USDA/ARS, Manhattan, KS, USA
[4]Center for Plant Science Innovation, University of Nebraska-Lincoln, Lincoln, NE, USA

*Correspondence

email tclemente1@unl.edu

Keywords: maize, agrobacterium tumefaciens, sorghum, wheat, kafirin, transgenic, digestibility.

Summary

Sorghum prolamins, termed kafirins, are categorized into subgroups α, β, and γ. The kafirins are co-translationally translocated to the endoplasmic reticulum (ER) where they are assembled into discrete protein bodies that tend to be poorly digestible with low functionality in food and feed applications. As a means to address the issues surrounding functionality and digestibility in sorghum, we employed a biotechnology approach that is designed to alter protein body structure, with the concomitant synthesis of a co-protein in the endosperm fraction of the grain. Wherein perturbation of protein body architecture may provide a route to impact digestibility by reducing disulphide bonds about the periphery of the body, while synthesis of a co-protein, with known functionality attributes, theoretically could impact structure of the protein body through direct association and/or augment end-use applications of sorghum flour by stabilizing ß-sheet formation of the kafirins in sorghum dough preparations. This in turn may improve viscoelasticity of sorghum dough. To this end, we report here on the molecular and phenotypic characterizations of transgenic sorghum events that are down-regulated in γ- and the 29-kDa α-kafirins and the expression of a wheat Dy10/Dx 5 hybrid high-molecular weight glutenin protein. The results demonstrate that down-regulation of γ-kafirin alone does not alter protein body formation or impacts protein digestibility of cooked flour samples. However, reduction in accumulation of a predicted 29-kDa α-kafirin alters the morphology of protein body and enhances protein digestibility in both raw and cooked samples.

Introduction

Grain sorghum (*Sorghum bicolor* (L.) Moench) is a major staple for a large portion of the world. The crop ranks fifth among the cereals world-wide with respect to its importance for food and feed applications. To this end, the grain harvested from sorghum, and the millets provides an important source for dietary calories and protein for approximately one billion people in the semi-arid regions of the world (Belton and Taylor, 2004). However, grain sorghum products are known to have relatively poor digestibility, only approximately 50%–70%, in comparison with other grains, such as wheat and maize, which tend to have digestibility percentages over 80% and 70%, respectively (Aboubacar *et al.*, 2001; MacLean *et al.*, 1981). Protein with high digestibility is by definition nutritionally superior owing to the increased availability of amino acids. Digestibility can be impacted by both protein–protein and/or protein–nonprotein interactions (Duodu *et al.*, 2003; Taylor *et al.*, 2007). However, with respect to grain sorghum, it is thought that the major factor influencing digestibility is the former because of high protein cross-linking around the protein body (Duodu *et al.*, 2003).

Protein content of grain sorghum is approximately 13% (Beta *et al.*, 1995), of which the kafirins comprise over 80% of the protein in the endosperm component of the grain (Hamaker *et al.*, 1995). The kafirins are categorized into α, β, γ groupings, with α kafirins approximately 26–27 kDa, and β and γ having molecular masses of 18.7 and 20 kDa, respectively (Belton *et al.*, 2006). The protein component of the endosperm is estimated to contain between 9%–21% γ-kafirin and 66%–84% α-kafirins (de Mesa-Stonestreet *et al.*, 2010). The kafirins are assembled into discrete protein bodies (PB) in the ER, whereby the α-kafirins compose the core and the β- and γ-kafirins decorate the periphery of the PB. It is thought that the organizational structure of the PB has a major impact on protein digestibility of sorghum food and feed products (Hicks *et al.*, 2001). As mentioned above increased cross-linking through disulphide bonds of the β- and γ-kafirins blocks access to the more digestible α-kafirin core (Hamaker *et al.*, 1994; Oria *et al.*, 1995, 2000), which tends to be further exaggerated upon cooking. This model is supported by the observation that the addition of reducing agents during cooking improves the *in vitro* digestibility of sorghum (Arbab and El Tinay, 1997; Hamaker *et al.*, 1987). Moreover, studies evaluating highly digestible sorghum mutants further support the structural role of the PB on this parameter (Weaver *et al.*, 1998). Furthermore, in the sorghum mutant with a high digestible, high lysine (HDHL) phenotype, the PB are highly folded, with reduction in the γ-kafirin around the periphery (Oria *et al.*, 2000).

In addition to digestibility, overall functionality of sorghum flour is relatively poor. Similar to digestibility, functionality is also influenced by both protein–protein and protein–nonprotein interactions. Therefore, to mirror the viscoelastic properties of wheat dough, it is necessary to functionalize the kafirins in such a fashion to allow them to mimic polymeric structures formed during processing between high- and low-molecular weight glutenins and gliadins. Recently, it has been communicated that blending of co-proteins, such as casein, with prolamins, such as zeins and kafirins, during the baking process can stabilize the β-sheet structures assembled during dough formation, thereby mirroring the functionality of wheat glutens (Hamaker et al., 2009). Hence, it maybe feasible to simultaneously improve upon the functionality and digestibility of grain sorghum by attempting to mirror the in vitro blending approach described by Hamaker et al. (2009) via an in vivo strategy whereby perturbation of the PB would expose the α-kafirins, which in turn would theoretically improve digestibility, while the in vivo production of a co-protein would aid in the functionalization α-kafirins during the baking process.

We describe here the generation and characterization of novel sorghum genotypes in which deliberate reduction in both γ-kafirin and a predicted 29-kDa α-kafirin storage proteins have been achieved. In addition, we report here on the molecular and phenotypic analyses of transgenic sorghum in which high levels of a prototype co-protein is produced in the endosperm, a wheat high-molecular weight glutenin subunit (HMW-GS).

Results

Characterization of transgenic sorghum events

As a means to address end-use quality of grain sorghum, with respect to nutrition, digestibility and functionality, we designed a strategy for the deliberate down-regulation of both γ- and α-kafirins, alone and in combination, and stacking of these phenotypes with the endosperm-specific expression of a HMW-GS. This in turn will create a gene stack that attempts to alter the rigid structure of the PB, with the simultaneous production of a co-protein, HMW-GS, with the latter addressing functionality and the former, enhanced digestibility, thereby creating transgenic gene stacks in sorghum that will mirror the blending strategy communicated by Hamaker et al. (2009). To this end, we assembled a set of binary vectors that harbour genetic elements to specifically down-regulate the γ-kafirin and a 29-kDa α-kafirin. The binary vector designated pPTN915 (Figure S1b) carries the γ-kafirin open reading frame (ORF), under control of its own promoter (Mishra et al., 2008) and terminated by a self-cleaving ribozyme (RZ) derived from a satellite RNA of tobacco ringspot virus. The expression of an ORF terminated with this self-cleaving RZ generates aberrant transcripts, which are retained in the nuclear compartment where they are acted on by an RNA-dependent RNA polymerase, which in turn are processed to siRNAs (Buhr et al., 2002). A total of 14 independent sorghum transformants were derived from transformations conducted with pPTN915. Among the 14 events, two events were significantly reduced in the accumulation of γ-kafirin transcripts. Shown in Figure 1 are the molecular analyses of one pPTN915 event designated 133-3-1-1. A Southern blot analysis on progeny derived from 133-3-1-1 indicates two transgenic loci, at approximately 6.2 kb, are integrated within the genome (Figure 1a). A northern analysis on T_2 seed derived from a single

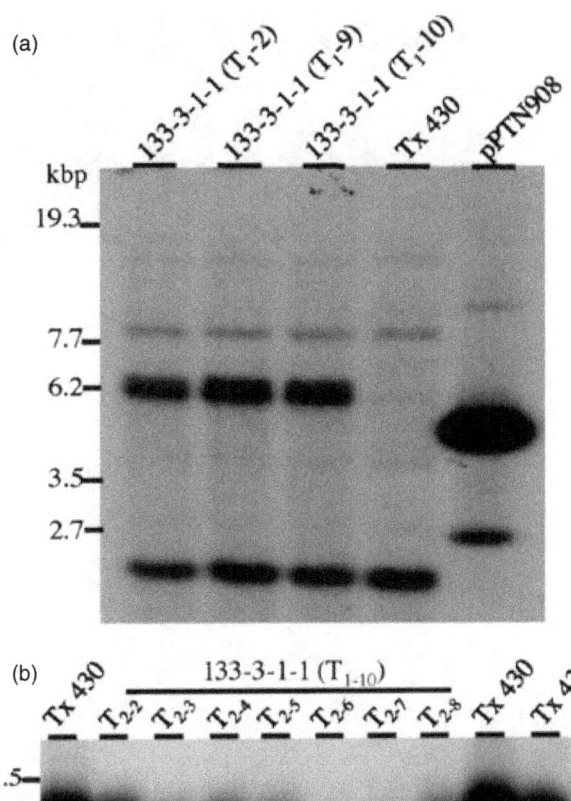

Figure 1 Molecular characterization of down-regulated γ-kafirin event 133-3-1-1. (a) Southern blot analysis on three T_1-derived individuals of event 133-3-1-1. Total genomic DNA (10 μg) was digested with Sst I. Blot was hybridized with the γ-kafirin ORF. Lanes 1–3 T_1 plants derived from event 133-3-1-1. Lane 4 contains control wild-type Tx 430 genomic DNA, and Lane 5 contains 50 pg of plasmid pPTN908, an intermediate vector of pPTN915 that harbours γ-kafirin ORF. (b) Northern blot analysis on T_2-derived seed from individual T_1-10 (a) from event 133-3-1-1. Blot hybridized with 800-bp region of γ-kafirin ORF. Lane 1 control RNA from immature Tx 430 seed. Lanes 2–8 RNA isolated from immature T_2 seed derived from T_1-10 event 133-3-1-1. Lanes 9 and 10 RNA isolated from control Tx 430 immature seed (lower panel ethidium bromide gel image of blot).

T_1 individual shows the significant reduction in γ-kafirin transcript accumulation in the developing seed (Figure 1b).

Seed derived from the 133-3-1-1 lineage displayed an opaque phenotype (Figure 2a). To confirm reduction in transcript accumulation translated to lowering of the γ-kafirin protein, 2-D SDS–PAGE gel analysis was conducted on the transgenic and control seed. Differential spots observed in the overview of the proteome of transgenic versus the control seed revealed the absence of a protein spot confirmed to be γ-kafirin by tandem mass spectrometry (Figure 3). Quantification of the γ-kafirin spot indicated that it was approximately 1.1% of the proteins in the 2-D gel (Figure 3a). This is relatively low given the

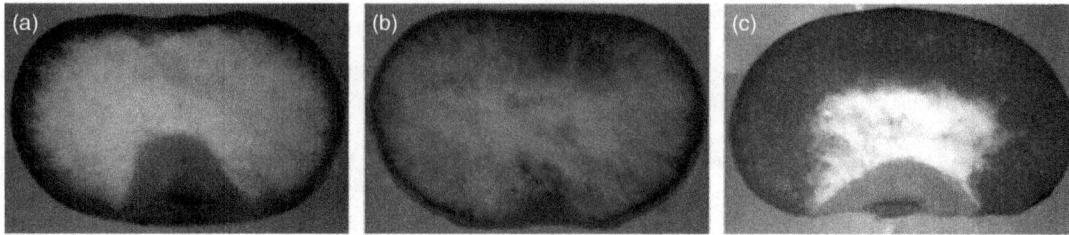

Figure 2 Opaque phenotype observed in down-regulated γ- and 29-kDa α-kafirin seed. (a) Opaque phenotype observed in down-regulated γ-kafirin event 133-3-1-1. (b) Opaque phenotype observed in down-regulated 29-kDa α-kafirin event 285-1-2-1. (c) Opaqueness observed in control Tx 430 seed.

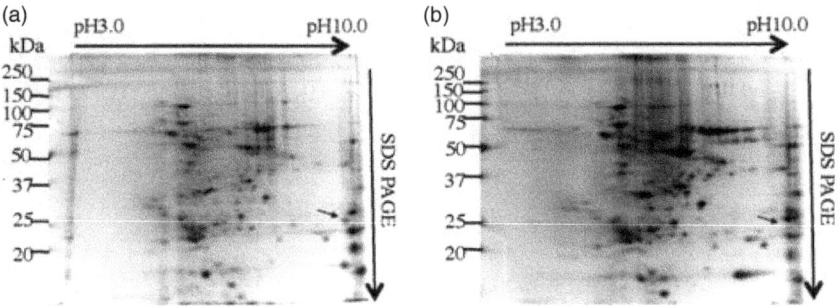

Figure 3 Kernel proteome image of down-regulated γ-kafirin event 133-3-1-1. (a) 2-D gel image of proteome of control Tx 430 seed (arrow indicates confirmed γ-kafirin spot). (b) 2-D gel of proteome of event 133-3-1-1 T_2 seed (arrow indicates location where γ-kafirin spot is expected to reside).

reported levels of γ-kafirin in the literature range from 9% to 21% (de Mesa-Stonestreet *et al.*, 2010). The apparent under-representation of γ-kafirin in the gel likely results from it not entering the IPG strip owing to its aggregation in the IEF sample buffer because of the high cysteine content and its known property of forming intermolecular and intramolecular disulphide bonds.

Down-regulation of a 29-kDa α-kafirin in grain sorghum

As a means to disrupt the central core of the PB, deliberate silencing of a 29-kDa α-kafirin (GenBank accession EU424175) was targeted using an RNAi approach. Here, a binary vector, designated pPTN1017 (Figure S1b), was built that carries a hairpin cassette, with an inverted 500-bp region of the 29-kDa α-kafirin ORF under control of an α-kafirin promoter (DeRose *et al.*, 1989). A total of 12 independent transgenic sorghum events were derived from transformations with pPTN1017. Among the 12 events generated, six displayed down-regulation of the target transcript, and as seen with the down-regulated γ-kafirin events, an opaque phenotype was observed in the seed (Figure 2b). Northern blot analysis on segregating seed derived from two of the events designated 288-1-1-2 and 285-1-2-1 is shown in Figure 4. In these events, the 29-kDa α-kafirin transcript accumulation is clearly significantly reduced. To gain insight on the changes in the proteome at maturity, total protein preparations from null, vitreous seed and down-regulated, opaque seed were separated using 2-D SDS–PAGE (Figure S2). A number of differential spots that were observed in the 2-D SDS–PAGE gel of the vitreous sample and absent in the opaque sample were analysed via tandem mass spectrometry. One of those spots was identified as the 29-kDa α-kafirin with a predicted isoelectric point of 8.97 (Figure S2). The series of spots

above and below the location of the 29-kDa α-kafirin ID spot, residing about the upper pI range all had hits with the maize globulin-1 (pI 9.02). This might reflect a tendency of the sorghum globulin proteins to nonspecifically bind to other seed proteins even under the denaturing conditions of the assay. In addition, a number of differential spots, not observed in the vitreous sample were picked from the separated proteins derived from the opaque sample, down-regulated in the 29-kDa α-kafirin, seeds. The IDs of a subset of these are listed in Table 1 and positions on 2-D gel shown in Figure S2b.

The differential proteins identified from the down-regulated 29-kDa α-kafirin sample include an S-like RNAse, 2-cysteine peroxiredoxin, a desiccation-related protein, an isoflavone reductase-like (IRL) along with a glyoxalase I and xylanase inhibitor peptide (Table 1). The common thread underlying these proteins is that their expression pattern can be tied back to stress biology (Dodds *et al.*, 1996, 2010; Haslekås *et al.*, 2003; Kim *et al.*, 2003). The opaque endosperm phenotype in maize associated with either *floury2* (*fl2*) mutation (Gillikin *et al.*, 1997), *defective endosperm** (*De*-B30*) mutation (Kim *et al.*, 2004) and the *Macronate* (*Mc*) mutation (Kim *et al.*, 2006) trigger up-regulation of proteins linked with endoplasmic reticulum (ER) stress (Schröder, 2006) as a consequence of harbouring mutations in the signal peptides of the 22-kDa α-zein, 19-kDa γ-zein, and the 16-kDa γ-zein in *fl2*, *De*-B30* and *Mc* mutants, respectively. In such situations, the ER stress-induced cellular response is designed to aid in enhancing proper protein folding and hence the observed increase in transcript levels of various molecular chaperones such as BiP, hsp70 and PDI in *fl2* and *Mc* (Hunter *et al.*, 2002) and in BiP peptide accumulation in *De*B30* (Kim *et al.*, 2004).

In contrast to the kernel opaque phenotype in maize manifested by the various signal peptide mutations that reside in *fl2*,

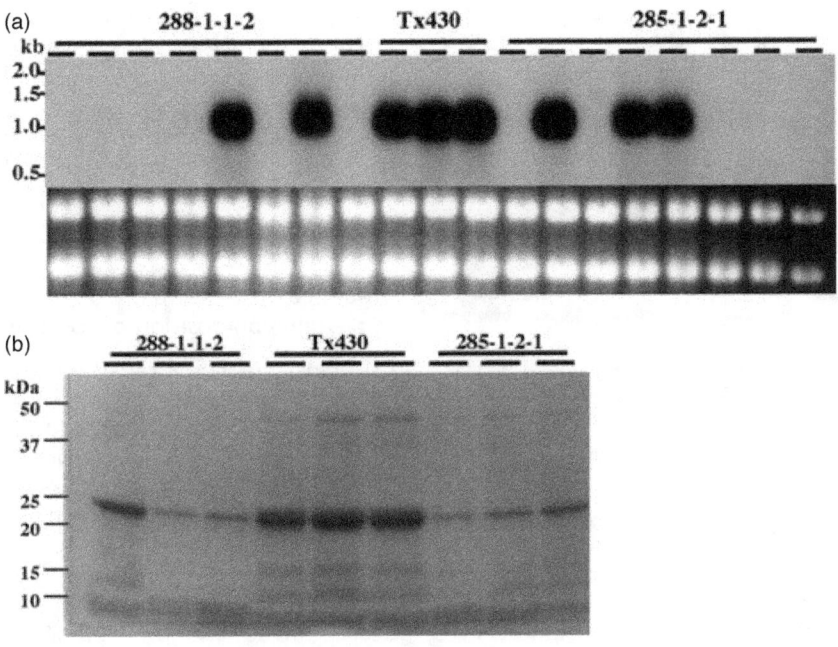

Figure 4 Northern blot and kernel storage protein accumulation visualization in down-regulated 29-kDa α-kafirin events. (a) Northern blot analysis on RNA derived from immature kernels. Lanes 1–8 event 288-1-1-2. Lanes 9–11 Tx 430 control samples. Lanes 12–19 event 285-1-2-1. Lower panel is an image of the corresponding ethidium bromide gel. Probe used in the hybridization is 500-bp region of the 29-kDa α-kafirin ORF. (b) 1-D SDS–PAGE gel image of kernel proteome. Lanes 1–3, a 1-D image of kernel proteome of pPTN1017 event 285-1-2-1. Lanes 4–6, a 1-D image of kernel proteome image of Tx 430 controls and lanes 7–9 a 1-D image of kernel proteome of pPTN1017 event 288-1-1-2.

Table 1 Novel proteins identified in kernel proteome of down-regulated 29-kDa α-kafirin events

Protein	ID	IP	% Coverage	Score
S-like RNAse	gi242081561	5.32	72	722
2-cysteine peroxiredoxins	gi195626524	5.81	27	394
Desiccation PCC 13-62	gi242075650	5.06	37	619
Isoflavone reductase	gi242052385	5.38	66	1197
Glyoxalase I	gi1808684	5.73	24	404
Xylanase inhibitor	gi242047612	6.59	56	785

Protein column refers to name of the identified protein. The ID and IP columns indicate the GenInfo Identifier (gi) number and predicted isoelectric point of the corresponding protein, while the percent coverage and score columns refer to the percent amino acid sequence coverage and probability score from mass spec analysis data.

De*-B30 and *Mc*, the *opaque2* (*o2*) mutant is because of a defect in a basic leucine zipper type transcription factor (Lohmer *et al.*, 1991; Schmidt *et al.*, 1992), which governs expression of a number of maize endosperm transcripts, including that which encodes the 22-kDa α-zein. Hence, changes in the transcriptome and proteome of the kernel triggered by silencing a major storage protein, such as the 29-kDa α-kafirin, described herein or the maize zeins (Huang *et al.*, 2006) would more likely mirror that associated with *o2* than the ER stress triggering opaque mutants. Indeed, transcript profiling of *Mc*, *fl2* and *o2* revealed the two ER stress-inducing mutants display similar transcript accumulation of the molecular chaperonins, BiP, hsp70 and cyclophilin, while *o2* is more like wild type with respect to these chaperonins (Hunter *et al.*, 2002). Moreover, an IRL transcript was found to be significantly up-regulated in the embryo of

α-zeins silenced maize event (Frizzi *et al.*, 2010), one of the differential protein spots identified in the silenced 29-kDa α-kafirin sorghum (Table 1).

Production of wheat HMW-GS

As a prototype genetic approach to introduce a co-protein with potential to impact functionality and/or digestibility of sorghum grain a binary vector, designated pPTN883 (Figure S1c), was assembled that carried a HMW-GS cassette that has the promoter, 5′ UTR and the genetic region to encode for the first 124 amino acids residues of the Dy10 HMW-GS, fused to the C-terminal portion of the Dx5 HMW-GS encompassing amino acid residues 130–848, coupled with the 3′ polyadenylation element of the Dx5 HMW-GS (Blechl and Anderson, 1996). A total of 23 independent transgenic sorghum events were derived from transformations with pPTN883. Relative transcript levels of the HMW-GS gene were monitored via Northern blot analysis. Expression of the gene in a subset of the events is shown in Figure 5. It is noteworthy that the expression of this HMW-GS under control of the Dy10 promoter in sorghum grain is extremely strong, the hybridization signals on the autoradiograph seen in Figure 5 were observed within 4 h of exposure. Moreover, a comparative proteome image between wild-type and transgenic seed reveals major new protein spots in the transgenic sample, verified to be HMW-GS via tandem mass spectrometry (Figure 6).

A sequential extraction scheme was utilized to investigate the level of the HMW-GS cross-linking into high-molecular weight protein structures. Initial extraction of ground kernels under nonreducing conditions removes primarily monomeric kafirins, and any monomeric forms of the HMW-GS, along with lower-molecular weight disulphide bonded polymeric protein structures, which together comprise the 'soluble protein' component

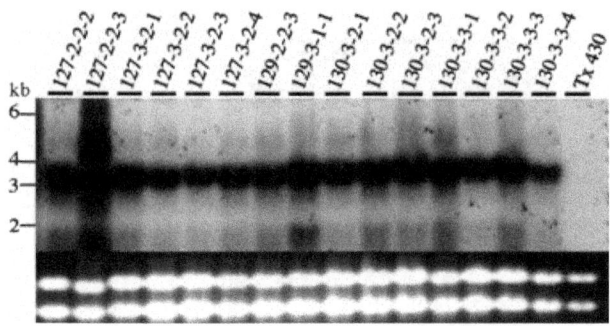

Figure 5 Northern blot on sorghum events carrying the wheat HMW-GS cassette. Lanes 1–15 RNA isolated from immature seed derived from pPTN883 sorghum events. Lane 16 RNA isolated from immature seed derived from control Tx 430 (lower panel ethidium bromide image of blot).

of the kernel (Shimoni et al., 1997). The remaining pellet was subsequently extracted under reducing conditions incorporating ß-mercaptoethanol as the reducing agent. Under these conditions, disulphide bonded larger molecular weight polymeric protein structures are recovered, which are categorized as 'insoluble proteins' (Shimoni et al., 1997). The soluble and insoluble protein components were analysed via SDS–PAGE. As can be seen in Figure 7, the vast majority of the HMW-GS resides in the insoluble protein component, implying that this transgenically expressed protein is either directly disulphide bonded to the kafirin/PB complex or is self-disulphide bonded in extremely large molecular weight molecules.

Imaging of PB and amino acid profiles of transgenic events

Imaging of PB in immature kernels was monitored via transmission electron microscopy (TEM). The morphology of the PB present in immature kernels derived from transgenic events either expressing the HMW-GS (pPTN883) or displaying a down-regulation of γ-kafirin (pPTN915) was not altered in morphology as compared to control, wild-type (nontransgenic) PB imaged at the same time point in development (data not shown). However, visualization of the PB in immature grain in which 29-kDa α-kafirin was down-regulated (pPTN1017) revealed distortion of the PB (Figure 8). The major morphological change observed was deep invaginations to the central core of the PB, reminiscent of PB structure found in the highly digestible sorghum mutant (Oria et al., 2000).

Total amino acid profile of mature grain was determined for the selected pPTN883, pPTN915 and pPTN1017 events. As can be seen in Figure 9, down-regulation of γ-kafirin led to decreases in proline, glutamate and leucine, while down-regulation of 29-kDa α-kafirin enhanced lysine, arginine and aspartate, along with slight reductions in glutamate and leucine. With respect to the pPTN883 events, production of the HMW-GS had minimal impact on the amino acid profile of the grain, with only a slight lowering of leucine observed (Figure 9).

Influence on digestibility by the targeted alterations in protein composition of sorghum grain

In vitro digestibility was carried out on flour derived from a subset of the transgenic events. Grain protein extracts were first characterized via reversed phase HPLC (RP-HPLC; Bean et al., 2011). Traces obtained from events 128-2-1-1 (pPTN883), 133-3-1-1 (pPTN915) and wild type are shown in Figure 10a. A novel peak falls out just under 4 min in the 128-2-1-1 event, and the known γ-kafirin peak is drastically diminished in the 133-3-1-1 event (Figure 10a). RP-HPLC traces on ground seed samples derived from progeny of the down-regulated 29-kDa α-kafirin event 288-1-1-2 are shown in Figure 10b, and corresponding Coomassie gel in Figure 4b. The traces reveal drastic changes between the 8.5 and 12.0 min ranges, which are reflected in the loss of major peaks, which correspond to the reduced α-kafirin band imaged in Figure 4b.

Seed samples, ground to uniformity, were subjected to a pepsin digestion assay (Mertz et al., 1984). In vitro digestibility was measured in both raw flour and cooked samples. Although variation in per cent protein digestibility of control (genotype Tx 430), raw flour samples was observed, when compared to the corresponding transgenics samples, assayed at the same time period, raw flour derived from some of the HMW-GS and down-regulated γ-kafirin events displayed improvement in this parameter over 10% (Table 2). However, upon cooking in vitro digestibility of the HMW-GS and silenced γ-kafirin transgenic events and control samples did not vary significantly, with percentage of digested protein observed ranging from 24% to just under 30% across these samples assayed (Table 2). In regards to the pPTN883 events assayed, correlating HMW-GS production in the seed, based on RP-HPLC peak area (Figure 10a), with raw digestibility of raw flour, a slight but positive relationship was observed ($r^2 = 0.39$). This data, along with the results observed from the sequential extraction procedure is suggestive that indeed the HMW-GS is directly interacting with the PB.

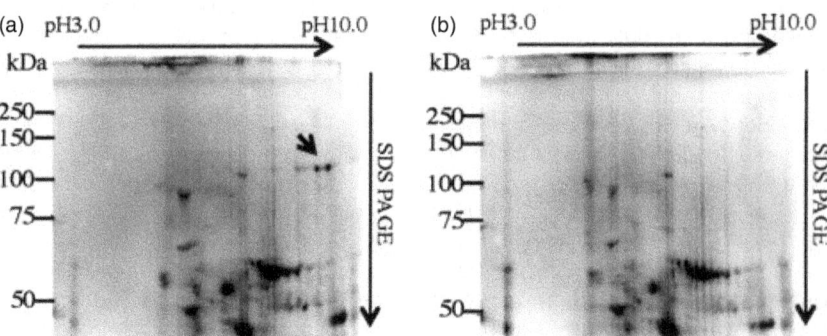

Figure 6 Kernel proteome image of wheat HMW-GS sorghum event. (a) kernel proteome image from pPTN883 sorghum event. Arrow highlights distinct spots identified as the wheat HMW-GS. (b) kernel proteome image of Tx 430 control kernel.

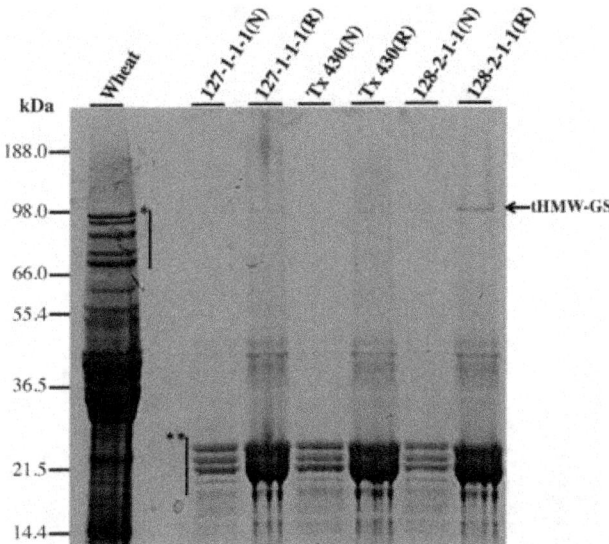

Figure 7 Sequential extraction of HMW-GS from sorghum. Wheat lane refers to extracts from wheat seed under reducing conditions. Lanes 127-1-1-1 (pPTN883 event), Tx 430 (control seed) and 128-2-1-1 (pPTN883 event). N and R indicate nonreducing and reducing conditions, respectively. Highlight the endogenous wheat *HMW-GS and **sorghum kafirins.

In vitro digestibility of raw flour derived from two independent events down-regulated in 29-kDa α-kafirin (pPTN1017) designated 285-1-2-1 and 288-1-1-2 also displayed improvement in this parameter, including over 30% improved digestibility of the cooked samples (Table 2). Moreover, size exclusion chromatography revealed that silencing of 29-kDa α-kafirin led to a significant reduction in percentage of the insoluble protein component of the flour (Table 3). Importantly, owing to the compensation mechanism of seeds, overall total protein levels were not changed (Table 2).

Discussion

A set of transgenic sorghum events has been generated with targeted modulation of kernel storage protein accumulation. The long-term goal is to assemble a series of creative gene stacks from these transgenic events as a means to simultaneously alter PB morphology and express a co-protein, HMW-GS, as a means to address improvements in both digestibility and viscoelasticity of the flour for end-use food and feed applications. Thereby creating a genetic approach to mirror the *in vitro* blending strategy described by Hamaker *et al.,* (2009). To this end, we have used selected events, 285-1-2-1 (pPTN1017), 133-3-1-1 (pPTN915) and 128-2-1-1 (pPTN883) in a series of crossing blocks and have successfully generated double stack combinations along with the triple gene stack. The novel sorghum events described here will aid in dissecting out some of the parameters influencing protein and perhaps indirectly starch (Wong *et al.,* 2009) digestibility, leading to improved functionality of grain sorghum end-use applications for feed and food. For example, a long-standing hypothesis has been that disulphide cross-linking of the storage proteins plays a negative role in digestibility of both uncooked and cooked sorghum (Duodu *et al.,* 2003; El Nour *et al.,* 1998; Winn *et al.,* 2009). In regards to cooked sorghum samples, it has been shown that disulphide-linked polymeric kafirins are formed which are more resistant to digestion (Emmambux and Taylor, 2009). In a study that partitioned the level of polymeric proteins in vitreous and opaque components of sorghum kernels, it was shown that the level of γ-kafirin in the vitreous portion of the endosperm was negatively correlated with amount of soluble proteins and γ-kafirin levels appear to be a strong indicator of polymeric proteins in the vitreous portion of the kernel (Ioerger *et al.,* 2007). In a related study employing both statistical and biochemical approaches as a means to identify genetic targets for the improvement of end-use quality of grain sorghum, it was concluded that γ-kafirin is the most resistant to digestion and the β-kafirin, the other storage protein about the periphery of the PB, is readily digested, and therefore, based on the analyses, the former was a logical target for down-regulation as a means to improve digestibility (Wong *et al.,* 2010). However, the event 133-3-1-1 serves as the direct test of this hypothesis, and while a modest increase (approximately 11%) in protein digestibility was observed in raw flour preparations, no change was seen upon cooking (Table 2). Moreover, no change in PB morphology was observed in the 133-3-1-1 event, indicating reduction in γ-kafirin alone is not sufficient to impact PB shape or digestibility. In maize, reduction in both peripheral prolamins around the PB, β- and γ-zeins, is required to alter PB formation

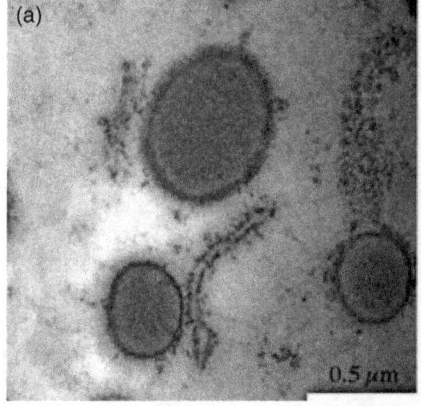
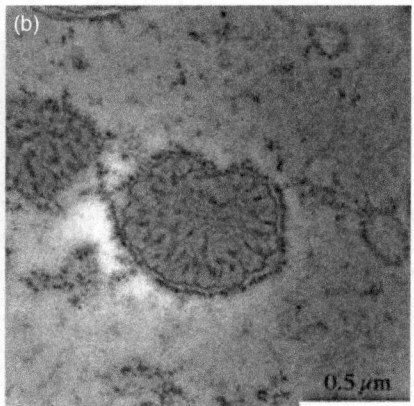

Figure 8 Transmission electron microscopy image of sorghum protein bodies. (a) TEM image (30×) of PB in endosperm of Tx 430 kernel 12 days postanthesis. (b) TEM image (30×) of PB in endosperm of down-regulated 29-kDa α-kafirin event (pPTN1017) kernel 12 days postanthesis.

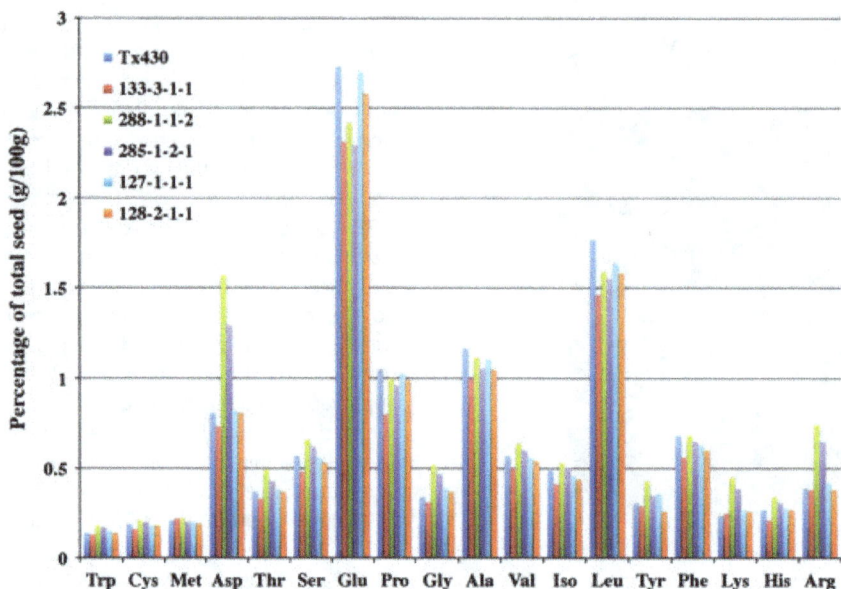

Figure 9 Amino acid profiles obtained from whole kernels. Numbers corresponding to the respective amino acids are percentages (g/100 g) of total kernel. Amino acid profiles were ascertained from ground kernel samples derived from 133-3-1-1 (pPTN915), 288-1-1-2 (pPTN1017), 285-1-2-1 (pPTN1017), 127-1-1-1 (pPTN883) and 128-2-1-1 (pPTN883).

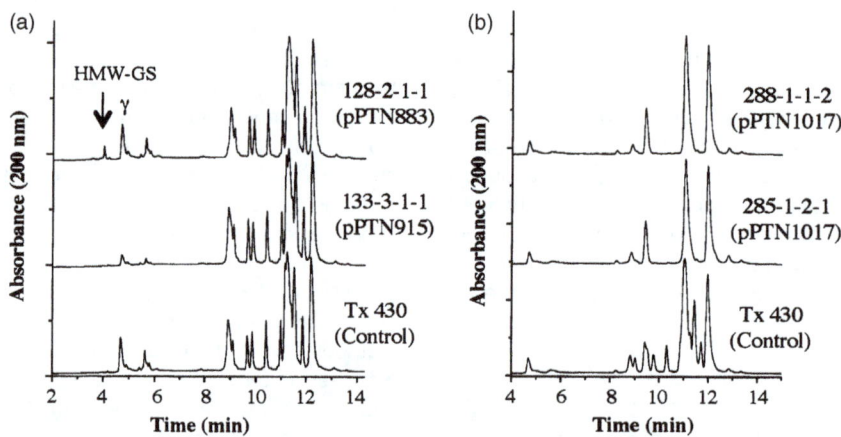

Figure 10 RP-HPLC chromatograms derived from sorghum kernel samples. (a) Displays trace obtained from sorghum event 128-2-1-1 accumulating wheat HMW-GS (arrow), middle trace obtained from sorghum event 133-3-1-1 (pPTN915) showing distinct reduction in γ-kafirin peak and bottom trace corresponding to a Tx 430 control sample. (b) Displays two RP-HPLC traces obtained from the two pPTN1017 events and lower trace corresponds to control Tx 430.

(Wu and Messing, 2010). Hence, the next logical approach in sorghum would be to explore the impact on PB morphology and digestibility in events silenced in β-kafirin level alone and stacked with down-regulated γ-kafirin or exploit the genetic variation present in the sorghum germplasm for β-kafirin (Laidlaw *et al.*, 2010) to pyramid with the respective silenced kafirin events described here.

The opaque phenotype observed here in the down-regulated γ- and 29-kDa α-kafirin events (Figure 2) is also seen when the corresponding maize homologues are silenced (Wu and Messing, 2010); however, the penetrance of the phenotype is greater in the γ-kafirin silenced events, as compared to 29-kDa α-kafirin silenced events (Figure 2), while the opposite effect is noticed in maize. However, the observation that reduction in the major kernel storage protein, 29-kDa α-kafirin, leads to

accumulation of nonkafirin proteins, which in turn translates to amino acid changes (Figure 9), without compromising total protein levels (Table 2), is in agreement with the findings in other grains such as maize (Frizzi *et al.*, 2010) and rice (Kawakatsu *et al.*, 2010) and is the underlying biology that accounts for the seminal report communicated by Mertz *et al.* (1964).

The ability of plants to compensate for the loss of a major seed storage protein by up-regulating nonstorage protein genes is a phenomenon that has been observed in both dicotyledonous (Schmidt and Herman, 2008) and monocotyledonous (Hunter *et al.*, 2002; Kawakatsu *et al.*, 2010) species. We undertook a proteomics approach to get a glimpse of the nonkafirin proteins that accumulate in mature sorghum grain upon down-regulation of the 29-kDa α-kafirin. All the proteins identified are linked to a generalized stress response (Table 1). Changes in

Table 2 *In vitro* digestibility of flour samples derived from sorghum kernels

Event	Construct	% Protein (Pre-R)	% Digestibility (Raw)	% Digestibility (Cooked)
125-4-3-3	pPTN883	12.08 ± 0.09	51.33 ± 1.51	25.77 ± 4.39
126-4-1-3	pPTN883	13.80 ± 0.14	35.67 ± 0.56	26.99 ± 0.88
127-1-1-1	pPTN883	14.48 ± 0.09	43.57 ± 1.41	24.59 ± 7.07
128-2-1-1	pPTN883	12.64 ± 0.13	50.14 ± 2.10	29.36 ± 1.83
128-2-1-1	pPTN883	14.97 ± 0.06	66.57 ± 0.23	28.98 ± 0.90
133-3-1-1	pPTN915	14.55 ± 0.03	59.76 ± 4.19	28.46 ± 4.74
288-1-1-2	pPTN1017	15.76 ± 0.11	67.33 ± 2.98	35.27 ± 2.09
285-1-2-1	pPTN1017	18.52 ± 0.19	63.04 ± 0.61	39.20 ± 1.91
Tx 430	—	13.62 ± 0.75	44.87 ± 5.60	24.48 ± 1.99

Event column indicates from which sorghum event flour sample was derived. Construct column refers to the respective plasmid used for transformation. Percent protein (Pre-R) refers to mean protein percentage ± standard deviation in pre *in vitro* digested raw flour. Percent digestibility (Raw) and (Cooked) columns indicate the mean percentage of digested protein ± standard deviation, following *in vitro* digestion assay.

Table 3 Percentage of soluble proteins present in derived sorghum flour

Event	Construct	%SP	%IP	%RP
127-1-1-1	pPTN883	40.8 ± 0.25	45.5 ± 3.93	13.7 ± 3.67
128-2-1-1	pPTN883	40.8 ± 0.10	43.9 ± 3.45	15.3 ± 3.35
133-3-1-1	pPTN915	49.0 ± 0.60	42.7 ± 3.08	8.2 ± 2.47
288-1-1-2	pPTN1017	44.6 ± 5.30	17.4 ± 4.32	38.0 ± 4.56
285-1-2-1	pPTN1017	40.1 ± 2.27	20.6 ± 3.66	39.3 ± 2.25
Tx 430	WT	47.0 ± 0.53	41.4 ± 4.63	11.59 ± 4.34

Event column indicates the origin of flour samples analysed, while construct column reflects corresponding plasmid. Per cent SP, IP and RP refer to mean (±standard deviation) percentage of soluble, insoluble and residual proteins, respectively.

carbon flux, at the initiation of synthesis of storage reserves, typically 14–18 days after pollination in maize, include up-regulation of a number of stress related transcripts in the various maize mutants with an opaque phenotype (Hunter *et al.*, 2002) and silenced α-zeins through an RNAi approach (Frizzi *et al.*, 2010). However, the global changes, in immature kernels of maize, in both transcript and protein levels that occur upon reduction in one or more major kernel storage proteins covers a much broader array of function (Frizzi *et al.*, 2010; Hartings *et al.*, 2011; Hunter *et al.*, 2002).

Elucidation of the mechanism controlling these global changes in carbon flux will aid in our understanding of the compensation mechanism of seeds (Kawakatsu *et al.*, 2010; Schmidt and Herman, 2008), which in turn will translate to designing of optimal genetic strategies for targeted improvements in seed quality, nutritional content and end-use functionality. Having a snapshot of the proteome changes at various stages of seed development will provide insight into the regulation underpinning this phenomenon. With respect to the differential up-expressed spots identified in mature sorghum kernels in which the 29-kDa α-kafirin is down–regulated the identified proteins fall under a single functionality class, stress response. For example, a S-like RNAse protein was one of the identified up-regulated spots observed. Homologs of S-like RNAse have been communicated to be elevated in expression upon leaf senescence and phosphate starvation (Liang *et al.*, 2002) and pathogen ingress (Dodds *et al.*, 1996). The other novel spots identified in the 29-kDa α-kafirin down-regulated event include 2-cysteine peroxiredoxin whose activity is associated with maintenance of proper redox potential of a cell (Dietz *et al.*, 2002), protein PCC 13-62 putatively linked to desiccation (Bartels *et al.*, 1990), isoflavone reductase-like genes implicated in both biotic and abiotic stress responses (Kim *et al.*, 2003, 2010), glyoxalase 1 which has been shown to combat salt stress (Singla-Pareek *et al.*, 2003) and xylanase inhibitors which play a role in abating fungal pathogenesis (Dornez *et al.*, 2010).

The wheat HMW-GS (Blechl and Anderson, 1996) was selected as co-protein for downstream investigations into potential impacts on improving leavening properties of sorghum flour. This co-protein was chosen because of the known endogenous influence that HMW-GS have on functionality of wheat flour (Shewry, 2009). Grain sorghum is often a sought after ingredient in formulating diets for individuals who suffer from coeliac disease (de Mesa-Stonestreet *et al.*, 2010). The prevalence of coeliac disease throughout the world is generally under 2%, but has been reported to be up to 5.6% (Tack *et al.*, 2010). One of the proteins thought to play a role in the triggering of coeliac disease are the wheat HMW-GS (Sollid, 2002). Hence, if this co-protein proves to be successful in aiding in the leaven properties of baked goods with sorghum products, then proper identity preservation from conventional sorghum grain may need to be implemented. Moreover, products with wheat ingredients fall under the Food Allergen Labeling Consumer Act (FALCPA), therefore, within the US food labels would be required to list a sorghum flour with HMW-GS as wheat-based ingredient. On the flip side, utilizing such a sorghum biological as the pPTN883 events described herein, can serve as a useful tool for elucidating the dietary components that trigger coeliac disease by isolating a putative coeliac disease trigger away from its endogenous seed matrix, which is thought to influence other, more serious, food allergens (van Wiljk *et al.*, 2005).

When over expressed in wheat, the hybrid HMW-GS (Blechl and Anderson, 1996) was shown to have the capacity to form intramolecular disulphide bonds, because of the unique position of cysteine residues at the N- and C-terminal regions of the peptide; however, a portion of the transgenically expressed peptide was still present within glutenin polymers of the grain (Shimoni *et al.*, 1997). Employing a sequential extraction scheme, we have shown that the vast majority of the hybrid HMW-GS is present in reduced extracts (Figure 7) of sorghum, suggestive of being incorporated into large polymeric structures, most likely the PB. However, we cannot rule out the possibility that the HMW-GS is competent to form large self-bonded polymeric structures in sorghum kernels. Either way, pyramiding of this trait with the altered PB phenotype observed in the pPTN1017 events (Figure 8) will in essence assemble the *in vivo* blending of a co-protein in a sorghum kernel matrix with improved digestibility (Table 2), thereby creating a novel sorghum flour composite with potential improvements in end-use baking applications.

In both maize and sorghum, the ratio of vitreous and opaqueness of the kernel will influence texture of the endosperm. A hard endosperm typically will possess a larger vitreous area about an opaque core of the kernel, while a softer endosperm will have a significantly larger area of opaqueness. Endosperm hardness is a desired trait for selection in cereal breeding programs because of its association with improved functionality and positive correlation with yield and grain resistance to fungal ingress (Tesso et al., 2006; Wu et al., 2010). On the other hand, soft endosperm has been linked with improved digestibility and nutritional quality in sorghum (Oria et al., 2000) and maize (Sofi et al., 2009; Wu et al., 2010). The maize opaque2 (o2) mutant is a lesion in a transcription factor that among its targets is the 22-kDa α-zein (Habben et al., 1993). Maize germplasm carrying o2 possess enhanced nutritional quality, but lack desired grain hardness. The stacking of o2 with various o2 modifiers has resulted in selections of vitreous kernels that maintain the enhanced nutritional quality of such o2 combinations, without compromising agronomic performance of the crop, this germplasm is referred to as quality protein maize (QPM; Crow and Kermicle, 2002). QTLs linked to the o2 modifiers have been mapped to chromosomes, 1, 7 and 9 (Holding et al., 2008), and interestingly, the o2 modifying phenotype is associated with increased levels of γ-zeins (Wu et al., 2010). Importantly, the genetic dissection of the o2 modifying phenotype has permitted the development of molecular markers that have greatly aided maize breeding programs in introgression of the QTLs associated with o2 modifiers into elite genotypes (Gupta et al., 2009; Sofi et al., 2009).

The high digestible/high lysine sorghum germplasm is currently following an analogous path as that of the o2/QPM story but some hurdles still remain (Tesso et al., 2006, 2008). Nonetheless, the genetic loci associated with the sorghum endosperm modifiers (Tesso et al., 2008) may complement the biotechnology strategy outlined herein thereby laying the foundation towards the development of quality protein sorghum coupled with enhanced digestibility and improved functionality.

Experimental procedures

Assembly of binary vectors

The 29-kDa α-kafirin gene was isolated via RT-PCR (genotype Tx 430) from mRNA derived from immature seed. Primers used in the RT-PCR reaction were Kaf-5: ATGGCTACCAAGATATTTG-TCCTCCTTGCG and Kaf-3: AATCTAGAAGATGGCACTTCCAAC-GATGGG, based on GenBank accession number EU424175. A 500-bp element derived from the PCR product was subsequently re-amplified to incorporate convenient cloning sites at the 5' and 3' ends, Hpa I and Xba I, along with Sst I and Xho I, respectively. The 29-kDa α-kaf 500-bp elements, with incorporated restriction sites were cloned as inverted repeats into pUC18-RNAi (gift H. Cerutti U. of Nebraska), a plasmid which harbours the second intron of the Arabidopsis small nuclear ribonuclear protein D1 (locus At 4g02840). The intron is delineated by a set of restriction sites that facilitates assembly of inverted repeat elements. The derived 29-kDa α-kafirin RNAi element was then subcloned downstream of the 832-bp α-kafirin promoter (GenBank accession X16104), which was cloned out by PCR using primer set p-alphakaf-1: AGA-CCTCCCAACCCATGCTCGCCACGTTTG and p-alphakaf-2: TTG-GAAGGACGTTGCTAGTTCGTTTCAC. The RNAi cassette was terminated by the CaMV 35S polyadenylation signal. The resultant RNAi cassette was sublconed into the binary vector

pPZP212 (Hajdukiewicz et al., 1994) to make the final vector designated pPTN1017 (Figure S1a).

The γ-kafirin promoter was isolated by PCR using primer set gKaf-2-5:CCGTGTACAACGAAGTGGTGAGTCATGAG and gKaf-2-3:GGTGTCGAGTTCTTGTCTGCTCTG based on GenBank accession number X62480, which amplified a 493-bp region upstream of the translational start site of the gene. The γ-kafirin ORF was cloned using RT-PCR derived from mRNA from immature seed (Tx 430) using primer set, based on GenBank accession number X62480, gKaf5: ATGAAGGTGTTGCTC-GTTGCCCTCGCTC and gKaf3: TTAATAGTGGACACCACCGG-CAAAAGG. The γ-kafirin promoter and ORF were assembled into a cassette terminated by a self-cleaving ribozyme derived from the satellite RNA of tobacco ringspot virus (Buhr et al., 2002). The resultant element was subcloned into the binary vector pPZP212 (Hajdukiewicz et al., 1994) and the final vector designated pPTN915 (Figure S1b).

A hybrid HMW-GS cassette consisting of the 5' UTR and first 124 amino acid residues of the Dy 10 HMW-GS, fused to the C-terminal portion of the Dx5 HMW-GS encompassing amino acids residues 130–848, coupled with the 3' polyadenylation element of the Dx5 HMW-GS (Blechl and Anderson, 1996) was subcloned as an Eco RI fragment into the binary vector pPZP211 (Hajdukiewicz et al., 1994) and the resultant plasmid referred to as pPTN883 (Figure S1c).

The three final binary vectors, pPTN883, pPTN915 and pPTN1017 were mobilized in to A. tumefaciens strain NTL$_4$/pKPSF2 (Palanichelvam et al., 2000) and the transconjugants used for sorghum transformation, genotype Tx 430, as previously described (Howe et al., 2006).

Molecular characterization of transgenic sorghum events

Total genomic DNA was isolated from sorghum leaves 133-3-1-1 event (pPTN915) following a modification of the protocol described by Dellaporta et al. (1983). Genomic DNA was restriction digested with Sst I, which has a single recognition site within the T-DNA of pPTN915. The membrane was hybridized with dCT^{32}P labelled γ-kafirin ORF (Prime-It II kit Agilent Technologies Cat # 300385, La Jolla, CA).

Northern blot analyses were conducted on total RNA isolated from individual seeds harvested 20 days postanthesis. RNA sample isolation, separation and hybridizations were carried out as previously described (Buhr et al., 2002). Membranes were hybridized with dCT^{32}P labelled HMW-GS, 2.2 kb region of the ORF (pPTN883) and 500-bp region of 29-kDa α-kafirin ORF (pPTN1017) or 820-bp region of γ-kafirin ORF (pPTN915).

Seed protein extraction

Individual seeds were ground in liquid nitrogen. Protein extraction buffer was added to the ground seed and thoroughly mixed. The extraction buffer was composed of 0.1 M Trizma base (pH 8.0), 10 mM EDTA, 0.9 M sucrose and 0.4% (v/v) ß-mercaptoethanol. The mixture was subsequently extracted for 30 min at 4 °C, with slight agitation, with an equal volume of Tris-saturated phenol. Following the extraction step, the suspension was centrifuged at 4350 g in a Beckman JA-20 rotor for 10 min. The upper phenol phase was collected and precipitated overnight at −20 °C by the addition of 5× volume of 0.1 M ammonium acetate in 100% methanol. Precipitated proteins were pelleted at 17,400 g in a Beckman JA-20 rotor for 10 min. The protein pellet was subsequently washed with 0.1 M ammonium acetate in 100% methanol, followed by washings in 80% (v/v) acetone and 70% (v/v) ethanol. The washed protein pellets were suspended in 8 M urea, 2 M thiourea, 2% (w/v) CHAPS and 2% (v/v) Triton X-100. Protein concentrations were quantified using the Bradford assay (Bio-Rad Cat# 500-0116, Hercules, CA).

Two-dimensional gel electrophoresis

A total of 300 μg of protein was loaded onto each 7 cm pH 3.0–10.0 Ready Strip IPG strip (Bio-Rad Cat# 163-2000). Following focusing in the first dimension, the strips were run in the second dimension using 14% (w/v) SDS–PAGE and the gel was subsequently stained with Coomassie G-250. Differential spots selected were analysed using tandem mass spectrometry.

Imaging of protein bodies

Immature seeds harvested 12 days after anthesis were transversely sliced into 1–2 mm pieces and fixed in 4% (v/v) paraformaldehyde and 1% (v/v) glutaraldehyde in 50 mM potassium phosphate buffer (pH 6.8) at 4 °C for 16 h. The samples were stained with 2% (w/v) osmium tetroxide in 50 mM potassium phosphate buffer (pH 6.8), followed by three washes in buffer. Following the last wash, the samples were dehydrated in a graded ethanol series (v/v) for 15 min at each step from 10%, 30%, 50%, 70%, 90%, 95% and 100% ethanol. Following the dehydration step, samples were infiltrated for 2 h at each gradient in 20%, 40%, 60% and 80% LR White resin in (w/v) ethanol, followed by an over night infiltration in 100% LR white resin. The 100% LR white resin infiltration was continued for 48 h with 12 h changes in resin. Following the infiltration procedure, the samples were placed in plastic moulds and polymerized for 2 days at 55 °C. The prepared specimens were sectioned with a diamond knife using an ultramicrotome (LKB Ultratome III, Stockholm, Sweden). The sections were overlaid to copper grids coated with formvar membrane and carbon. Sections were poststained for 5 min with 2.5% (w/v) uranyl acetate, followed by 3 min step in 0.1% (w/v) lead citrate. The prepared samples were imaged with a Hitachi H7500 transmission electron microscope (Hitachi, Tokyo, Japan) at 80 kV at the University of Nebraska's Microscopy Core Research Facility.

In vitro digestibility assay, RP-HPLC and SDS–PAGE analyses of storage protein extracts

Kafirin fractions were isolated from 100 mg seed samples that were uniformly ground using a Udy mill (Udy Corp., Fort Collins, CO). Protein extractions and kafirin selective isolation procedures were conducted as previously described (Bean *et al.*, 2011), with HPLC separations carried out on a Agilent 1100 HPLC, and reverse-phase separations using a Poroshell C18 column following the procedure outlined by Bean *et al.* (2000, 2011).

In vitro pepsin digestibility assay was conducted with duplicated samples. The samples were reground with a UDY mill coupled with a 0.5-m screen. Pepsin digestion assay was conducted as outlined by Mertz *et al.* (1984). Digestibility was measured in raw flour and in some cases cooked as porridge. Amino acid profile of the ground grain was ascertained through a commercial source (Eurofins, Des Moines, IA).

To determine whether the HMW-GS resides primarily as a monomer or polymeric protein structures within the cell a sequential extraction scheme was used based on that reported on by Shimoni *et al.* (1997). Albumin and globulin proteins were pre-extracted from the samples as previously described (Bean *et al.*, 2011). The residual remaining following the removal of albumin and globulin fractions was subsequently extracted with 1 mL of 60% (w/v) t-butanol containing 0.5% (w/v) sodium acetate for 5 min with continuous vortexing. This extraction step was repeated and supernatants from each extraction were pooled in a 1 : 1 ratio. Samples were then analysed by SDS–PAGE using 4–12% NuPAGE® gel (InVitrogen, Grand Island, NY) with MOPS runner buffer. Prior to SDS–PAGE analysis, samples were mixed with Novex sample buffer (InVitrogen) in a 1 : 4 ratio containing 2% (v/v) ß-mercaptoethanol and boiled for 5 min. Wheat proteins from the cultivar Karl 92 were extracted with 50% (v/v) 1-propanol plus 2% (v/v) ß-mercaptoethanol and mixed with Novex sample buffer. Ten microlitres from each sample was loaded per lane and gels run at 200 V until the dye front entered the buffer. Gels were stained using Imperial™ protein stain (Thermo Scientific, Rockford, IL) following the manufacturer's directions.

Acknowledgements

This work was supported by the Center for Biotechnology and Center for Plant Science Innovation through funds provided by the Nebraska Sorghum Board and the Nebraska Research Initiative. TK was supported through a USDA-NRI graduate training grant award number USDA 2007-55100-17788. The authors wish to thank Amy Hilske for the greenhouse care of plants and Ron Cerny from the Nebraska Center for Mass Spectrometry.

Disclaimer

Mention of trade names or commercial products in this publication is solely for the purpose of providing specific information and does not imply recommendation or endorsement by the U.S. Department of Agriculture. USDA is an equal opportunity provider and employer.

References

Aboubacar, A., Axtell, J.D., Huang, C.P. and Hamaker, B.R. (2001) A rapid digestibility assay for identifying highly digestible sorghum lines. *Cereal Chem.* **78**, 160–165.

Arbab, M.E. and El Tinay, A.H. (1997) Effect of cooking and treatment with sodium bissulpite or ascorbic acid on the *in vitro* protein digestibility of two sorghum cultivars. *Food Chem.* **59**, 339–343.

Bartels, D., Schneider, K., Terstappen, G., Piatkowski, D. and Salamini, F. (1990) Molecular cloning of abscisic acid-modulated genes which are induced during desiccation of the resurrection plant *Craterostigma plantagineum. Planta*, **181**, 27–34.

Bean, S.R., Lookhart, G.L. and Bietz, J.A. (2000) Acetonitrile as a buffer additive for free zone capillary electrophoresis separation and characterization of maize (*Zea mays* L.) and sorghum (*Sorghum bicolor* Moench.) storage proteins. *J. Agric. Food. Chem.* **48**, 318–327.

Bean, S.R., Ioerger, B.P. and Blackwell, D.L. (2011) Separation of kafirins on surface porous reversed phase-high performance liquid chromatography columns. *J. Agric. Food. Chem.* **59**, 85–91.

Belton, P.S. and Taylor, J.R.N. (2004) Sorghum and millets: protein sources for Africa. *Trends Food Sci. Technol.* **15**, 94–98.

Belton, P.S., Delgadillo, I., Halford, N.G. and Shewry, P.R. (2006) Kafirin structure and functionality. *J. Cereal Sci.* **44**, 272–286.

Beta, T., Rooney, L.W. and Waniska, R.D. (1995) Malting characteristics of sorghum cultivars. *Cereal Chem.* **72**, 533–538.

Blechl, A.E. and Anderson, O.D. (1996) Expression of a novel high-molecular-weight glutenin subunit in transgenic wheat. *Nat. Biotechnol.* **14**, 875–879.

Buhr, T., Sato, S., Ebrahim, F., Xing, A., Zhou, Y., Mathiesen, M., Schweiger, B., Kinney, A.J., Staswick, P. and Clemente, T. (2002) Ribozyme termination of RNA transcripts down-regulate seed fatty acid genes in transgenic soybean. *Plant J.* **30**, 155–163.

Crow, J.F. and Kermicle, J. (2002) Oliver Nelson and quality protein maize. *Genetics*, **160**, 819–821.

Dellaporta, S.L., Wood, J. and Hicks, J.B. (1983) A plant DNA minipreparation: version II. *Plant Mol. Biol. Rep.* **1**, 19–21.

DeRose, R.T., Ma, D.-P., Kwon, I.-S., Hasain, S.E., Klassy, R.C. and Hall, T.C. (1989) Characterization of the kafirin gene family from sorghum reveals extensive homology with zein from maize. *Plant Mol. Biol.* **12**, 245–256.

Dietz, K.J., Horling, F., König, J. and Baier, M. (2002) The function of the chloroplast 2-cysteine peroxiredoxin in peroxide detoxification and its regulation. *J. Exp. Bot.* **53**, 1321–1329.

Dodds, P.N., Clarke, A.E. and Newbigin, E. (1996) Molecular characterisation of an S-like RNase of *Nicotiana alata* that is induced by phosphate starvation. *Plant Mol. Biol.* **31**, 227–238.

Dornez, E., Croes, E., Gebruers, K., Carpentier, S., Swennen, R., Laukens, K., Witters, E., Urban, M., Delcour, J.A. and Courtin, C.M. (2010) 2-D DIGE reveals changes in wheat xylanase inhibitor protein families due to Fusarium graminearum *ΔTri5* infection and grain development. *Proteomics*, **10**, 2303–2319.

Duodu, K.G., Taylor, J.R.N., Belton, P.S. and Hamaker, B.R. (2003) Factors affecting sorghum protein digestibility. *J. Cereal Sci.* **38**, 117–131.

El Nour, I.N.A., Peruffo, A.D.B. and Curioni, A. (1998) Characterization of sorghum kafirins in relation to their cross-linking behaviour. *J. Cereal Sci.* **28**, 197–208.

Emmambux, M.N. and Taylor, J.R.N. (2009) Properties of heat-treated sorghum and maize meal and their prolamin proteins. *J. Agric. Food. Chem.* **57**, 1045–1050.

Frizzi, A., Caldo, R., Morrell, J.A., Wang, M., Lutfiyya, L.L., Brown, W.E., Malvar, T.M. and Huang, S. (2010) Compositional and transcriptional analyses of reduced zein kernels derived from the *opaque2* mutation and RNAi suppression. *Plant Mol. Biol.* **73**, 569–585.

Gillikin, J.W., Zhang, F., Coleman, C.E., Bass, H.W., Larkins, B.A. and Boston, R.S. (1997) A defective signal peptide tethers the *floury-2* zein to the endoplasmic reticulum membrane. *Plant Physiol.* **114**, 345–352.

Gupta, H.S., Agrawal, P.K., Mahajan, V., Bisht, G.S., Kumar, A., Verma, P., Srivastava, A., Saha, S., Babu, R., Pant, M.C. and Mani, V.P. (2009) Quality protein maize for nutritional security: rapid development of short duration hybrids through molecular marker assisted breeding. *Curr. Sci.* **96**, 230–237.

Habben, J.E., Kirleis, A.W. and Larkins, B.A. (1993) The origin of lysine-containing proteins in opaque-2 maize endosperm. *Plant Mol. Biol.* **23**, 825–838.

Hajdukiewicz, P., Svab, Z. and Maliga, P. (1994) The small, versatile *pPZP* family of *Agrobacterium* binary vectors for plant transformation. *Plant Mol. Biol.* **25**, 989–994.

Hamaker, B.R., Kirleis, A.W., Butler, L.G., Axtell, J.D. and Mertz, E.T. (1987) Improving the *in vitro* protein digestibility of sorghum with reducing agents. *Proc. Natl Acad. Sci. USA*, **84**, 626–628.

Hamaker, B.R., Mertz, E.T. and Axtell, J.D. (1994) Effect of extrusion on sorghum kafirin solubility. *Cereal Chem.* **71**, 515–517.

Hamaker, B.R., Mohamed, A.A., Habben, J.E., Huang, C.P. and Larkins, B.A. (1995) Efficient procedure for extracting maize and sorghum kernal proteins reveals higher prolamin contents than conventional method. *Cereal Chem.* **72**, 583–588.

Hamaker, B.R., Mejia, C.D., Mauer, L.J. and Campanella, O.H. (2009) Leavened products made from non-wheat cereal proteins. United States Patent Application 20090304861. 10 December 2009.

Hartings, H., Lauria, M., Lazzaroni, N., Pirona, R. and Motto, M. (2011) The *Zea mays* mutants opaque-2 and opaque-7 disclose extensive changes in endosperm metabolism as revealed by protein, amino acid, and transcriptome-wide analyses. *BMC Genomics*, **12**, 41.

Haslekås, C., Viken, M.K., Grini, P.E., Nygaard, V., Nordgard, S.H., Meza, T.J. and Aalen, R.B. (2003) Seed 1-cysteine peroxiredoxin antioxidants are not involved in dormancy, but contribute to inhibition of germination during stress. *Plant Physiol.* **133**, 1148–1157.

Hicks, C., Bean, S.R., Lookhart, G.L., Pedersen, J.F., Kofoid, K.D. and Tuinstra, M.R. (2001) Genetic analysis of kafirins and their phenotypic correlations with feed quality traits, *in vitro* digestibility and seed weight in grain sorghum. *Cereal Chem.* **78**, 412–416.

Holding, D.R., Hunter, B.G., Chung, T., Gibbon, B.C., Ford, C.F., Bharti, A.K., Messing, J., Hamaker, B.R. and Larkins, B.A. (2008) Genetic analysis of opaque2 modifier loci in quality protein maize. *Theor. Appl. Genet.* **117**, 157–170.

Howe, A., Sato, S., Dweikat, I., Fromm, M. and Clemente, T. (2006) Rapid and reproducible *Agrobacterium*-mediated transformation of sorghum. *Plant Cell Rep.* **25**, 784–791.

Huang, S., Frizzi, A., Florida, C.A., Kruger, D.E. and Luethy, M.H. (2006) High lysine and high tryptophan transgenic maize resulting from the reduction of both 19- and 22-kD alpha-zeins. *Plant Mol. Biol.* **61**, 525–535.

Hunter, B.G., Beatty, M.K., Singletary, G.W., Hamaker, B.R., Dilkes, B.P., Larkins, B.A. and Jung, R. (2002) Maize opaque endosperm mutations create extensive changes in patterns of gene expression. *Plant Cell*, **14**, 2591–2612.

Ioerger, B., Bean, S.R., Tuinstra, M.R., Pedersen, J.F., Erpelding, J., Lee, K.H. and Herrman, T.J. (2007) Characterization of polymeric proteins from vitreous and floury sorghum endosperm. *J. Agric. Food Chem.* **55**, 10232–10239.

Kawakatsu, T., Hirose, S., Yasuda, H. and Takaiwa, F. (2010) Reducing rice seed stoarge protein accumulation leads to changes in nutrient quality and storage organelle formation. *Plant Physiol.* **154**, 1842–1854.

Kim, S.-T., Cho, K.-S., Kim, S.-G., Kang, S.-K. and Kang, K.-Y. (2003) A rice isoflavone reductase-like gene, *OsIRL*, is induced by rice blast fungal elicitor. *Mol. Cells*, **16**, 224–231.

Kim, C.S., Hunter, B.G., Kraft, J., Boston, R.S., Yans, S., Jung, R. and Larkins, B.A. (2004) A defective signal peptide in a 19-kD alpha-zein protein causes the unfolded protein response and an opaque endosperm phenotype in the maize *De-B30* mutant. *Plant Physiol.* **134**, 380–387.

Kim, C.-S., Gibbon, B.C., Gillikin, J.W., Larkins, B.A., Boston, R.S. and Jung, R. (2006) The maize *Mucronate* mutation is a deletion i the 16-kDa γ-zein gene that induces the unfolded protein response. *Plant J.* **48**, 440–451.

Kim, S.G., Kim, S.T., Wang, Y., Kim, S.-K., Lee, C.H., Kim, K.-K., Kim, J.-K., Lee, S.Y. and Kang, K.Y. (2010) Overexpression of rice isoflavone reductase-like gene (OsIRL) confers tolerance to reactive oxygen species. *Physiol. Plant.* **138**, 1–9.

Laidlaw, H.K.C., Mace, E.S., Williams, S.B., Sakrewski, K., Mudge, A.M., Prentis, P.J., Jordan, D.R. and Godwin, I.D. (2010) Allelic variation of the β-γ- and δ-kafirin genes in diverse *Sorghum* genotypes. *Theor. Appl. Genet.* **121**, 1227–1237.

Liang, L., Lai, Z., Ma, W., Zhang, Y. and Xue, Y. (2002) AhSL28, a senescence-and phosphate starvation-induced S-like RNase gene in *Antirrhinum. Biochim. Biophys. Acta*, **1579**, 64–71.

Lohmer, S., Maddaloni, M., Motto, M., Di Fonzo, N., Hartings, H., Salamini, F. and Thompson, R.D. (1991) The maize regulatory locus *Opaque-2* encodes a DNA-binding protein which activates the transcription of the b-32 gene. *EMBO J.* **10**, 617–624.

MacLean, W.C., Lopez de Romana, G. and Placko, R.P. (1981) Protein quality and digestibility of sorghum in preschool children: balances studies and plasma free amino acids. *J. Nutr.* **111**, 1928–1936.

Mertz, E.T., Nelson, O.E. and Bates, L.S. (1964) Mutant gene that changes protein composition and increases lysine content of maize endosperm. *Science*, **145**, 279–280.

Mertz, E.T., Hassen, M.M., Cairns-Whittern, C., Kirleis, A.W., Tu, L. and Axtell, J.D. (1984) Pepsin digestibility of poteins in sorghum and other major cereals. *Proc. Natl Acad. Sci. USA*, **81**, 1–2.

de Mesa-Stonestreet, N.J., Alavi, S. and Bean, S.R. (2010) Sorghum proteins: the concentration, isolation, modification, and food applications of kafirins. *J. Food Sci.* **75**, 90–104.

Mishra, A., Tomar, A., Bansal, S., Khanna, V.K. and Garg, G.K. (2008) Temporal and spatial expression analysis of gamma kafirin promoter from

Sorghum (*Sorghum bicolor* L. Moench) var. M 35-1. *Mol. Biol. Rep.* **35**, 81–88.

Oria, M.P., Hamaker, B.R. and Shull, J.M. (1995) Resistance of sorghum α-β- and γ-kafirins to pepsin digestion. *J. Agric. Food Chem.* **43**, 2148–2153.

Oria, M.P., Hamaker, B.R., Axtell, J.D. and Huang, C.-P. (2000) A highly digestible sorghum mutant cultivar exhibits a unique folded structure of endosperm protein bodies. *Proc. Natl Acad. Sci. USA*, **97**, 5065–5070.

Palanichelvam, K., Oger, P., Clough, S.J., Cha, C., Bent, A.F. and Farrand, S.K. (2000) A second T-region of the soybean-supervirulent chrysopine-type Ti plasmid pTiChry5, and construction of a fully disarmed *vir* helper plasmid. *Mol. Plant Microbe Interact.* **13**, 1081–1091.

Schmidt, M.A. and Herman, E.M. (2008) Proteome rebalancing in soybean seeds can be exploited to enhance foreign protein accumulation. *Plant Biotechnol. J.* **6**, 832–842.

Schmidt, R.J., Ketudat, M., Aukerman, M.J. and Hoschek, G. (1992) Opaque-2 is a transcriptional activator that recognizes a specific target site in 22-kD zein genes. *Plant Cell*, **4**, 689–700.

Schröder, M. (2006) The unfolded protein response. *Mol. Biotechnol.* **34**, 279–290.

Shewry, P.R. (2009) Wheat. *J. Exp. Bot.* **60**, 1537–1553.

Shimoni, Y., Blechl, A.E., Anderson, O.D. and Galili, G. (1997) A recombinant protein of two high molecular weight glutenins alters gluten polymer formation in transgenic wheat. *J. Biol. Chem.* **272**, 15488–15495.

Singla-Pareek, S.L., Reddy, M.K. and Sopory, S.K. (2003) Genetic engineering of the glyoxalase pathway in tobacco leads to enhanced salinity tolerance. *Proc. Natl Acad. Sci. USA*, **100**, 14672–14677.

Sofi, P.A., Wani, S.A., Rather, A.G. and Wani, S.H. (2009) Quality protein maize (QPM): genetic manipulation for the nutrituional fortification of maize. *J. Plant Breed. Crop Sci.* **1**, 244–253.

Sollid, L.M. (2002) Coeliac disease: dissecting a complex inflammatory disorder. *Nat. Rev. Immunol.* **2**, 647–655.

Tack, G.J., Verbeek, W.H.M., Schreurs, M.W.J. and Mulder, C.J.J. (2010) The spectrum of celiac disease: epidemiology, clinical aspects and treatment. *Nat. Rev. Gastroenterol. Hepatol.* **7**, 204–213.

Taylor, J., Bean, S.R., Ioerger, B.P. and Taylor, J.R.N. (2007) Preferential binding of sorghum tannins with gamma-kafirin and the influnece of tannin binding on kafirin digestibility and biodegradation. *J. Cereal Sci.* **46**, 22–31.

Tesso, T., Ejeta, G., Chandrashekar, A., Huang, C.-P., Tandjung, A., Lewamy, M., Axtell, J.D. and Hamaker, B.R. (2006) A novel modified endosperm texture in a mutant high-protein digestibility/high-lysine grain sorghum (*Sorghum bicolor* (L.) Moench). *Cereal Chem.* **83**, 194–201.

Tesso, T., Hamaker, B.R. and Ejeta, G. (2008) Sorghum protein digestibility is affected by dosage of mutant alleles in endosperm cells. *Plant Breed.* **127**, 579–586.

Weaver, C.A., Hamaker, B.R. and Axtell, J.D. (1998) Discovery of grain sorghum germplasm with high uncooked and cooked in vitro digestibility. *Cereal Chem.* **75**, 665–670.

van Wiljk, F., Nierkens, S., Hassing, I., Feijen, M., Koppelman, S.J., de Jong, G.A.H., Pieters, R. and Knippels, L.M.J. (2005) The effect of the food matrix on In vivo immune responses to purified peanut allergens. *Toxicol. Sci.* **86**, 333–341.

Winn, J.A., Mason, R.E., Robbins, A.L., Rooney, W.L. and Hays, D.B. (2009) QTL mapping of a high protein digestibility trait in *Sorghum bicolor*. *Int. J. Plant Genomics*, **2009**, 6.

Wong, J.H., Lau, T., Cai, N., Singh, J., Pedersen, J.F., Vensel, W., Hurkman, W.J., Wilson, J.D., Lemaux, P.G. and Buchanan, B.B. (2009) Digestibility of protein and starch from sorghum (*Sorghum bicolor*) is linked to biochemical and structural features of grain endosperm. *J. Cereal Sci.* **49**, 73–82.

Wong, J.H., Marx, D.B., Wilson, J.D., Buchanan, B.B., Lemaux, P.G. and Pedersen, J.F. (2010) Principal component analysis and biochemical characterization of protein and starch reveal primary targets for improving sorghum grain. *Plant Sci.* **179**, 598–611.

Wu, Y. and Messing, J. (2010) RNA interference-mediated change in protein body morphology and seed opacity through loss of different zein proteins. *Plant Physiol.* **153**, 337–347.

Wu, Y., Holding, D.R. and Messing, J. (2010) γ-Zeins are essential for endosperm modification in quality protein maize. *Proc. Natl Acad. Sci. USA* **107**, 12810–12815.

7

The *Brassica napus* receptor-like protein RLM2 is encoded by a second allele of the *LepR3/Rlm2* blackleg resistance locus

Nicholas J. Larkan, Lisong Ma and Mohammad Hossein Borhan*

Saskatoon Research Centre, Agriculture and Agri-Food Canada, Saskatoon, SK, Canada

Correspondence

email hossein.borhan@agr.gc.ca
GenBank accession numbers: KM097068,
KM097069, KM097070, KM097071,
KM097072, KM097073, KM097074,
KM097075, KM097076, KM097077,
KM097078, KM097079, KM097080,
KM097081.

Keywords: blackleg, *Brassica napus*,
disease resistance, *Leptosphaeria
maculans*, receptor-like protein,
SOBIR1.

Summary

Leucine-rich repeat receptor-like proteins (LRR-RLPs) are highly adaptable parts of the signalling apparatus for extracellular detection of plant pathogens. Resistance to blackleg disease of *Brassica* spp. caused by *Leptosphaeria maculans* is largely governed by host race-specific *R*-genes, including the LRR-RLP gene *LepR3*. The blackleg resistance gene *Rlm2* was previously mapped to the same genetic interval as *LepR3*. In this study, the *LepR3* locus of the *Rlm2 Brassica napus* line 'Glacier DH24287' was cloned, and *B. napus* transformants were analysed for recovery of the *Rlm2* phenotype. Multiple *B. napus*, *B. rapa* and *B. juncea* lines were assessed for sequence variation at the locus. *Rlm2* was found to be an allelic variant of the *LepR3* LRR-RLP locus, conveying race-specific resistance to *L. maculans* isolates harbouring *AvrLm2*. Several defence-related LRR-RLPs have previously been shown to associate with the RLK SOBIR1 to facilitate defence signalling. Bimolecular fluorescence complementation (BiFC) and co-immunoprecipitation of RLM2-SOBIR1 studies revealed that RLM2 interacts with SOBIR1 of *Arabidopsis thaliana* when co-expressed in *Nicotiana benthamiana*. The interaction of RLM2 with *At*SOBIR1 is suggestive of a conserved defence signalling pathway between *B. napus* and its close relative *A. thaliana*.

Introduction

Blackleg disease, caused by the hemibiotrophic fungal pathogen *Leptosphaeria maculans* (anamorph *Phoma lingam*) (Howlett *et al.*, 2001), impacts production of canola/oilseed rape (*Brassica napus* and *B. rapa*) in most growing regions of the world (Fitt *et al.*, 2006). The prevention of catastrophic crop loss is achieved primarily through rotation strategies and the incorporation of genetic resistance into canola varieties, primarily race-specific resistance (*R*) genes. Plant *R* proteins convey recognition, either directly or through intermediary protein complexes, of specific pathogen avirulence (Avr) factors, often small secreted proteins termed 'effectors' which interfere with host cell targets (Bent and Mackey, 2008; Dudler, 2013; Jones and Dangl, 2006; Katagiri and Tsuda, 2010; Oliva *et al.*, 2010). Many *Brassica R*-genes responding in a race-specific manner to *L. maculans* isolates have been genetically defined, most residing in the A-genome of *B. napus* (AACC) and *B. rapa* (AA) (Delourme *et al.*, 2004; Larkan *et al.*, 2013; Leflon *et al.*, 2007; Raman *et al.*, 2013; Yu *et al.*, 2005). While several of the corresponding *L. maculans* effectors have been characterized (Balesdent *et al.*, 2013; Fudal *et al.*, 2007; Gout *et al.*, 2006; Parlange *et al.*, 2009; Van de Wouw *et al.*, 2014), to date, only one *Brassica* blackleg *R*-gene, *LepR3*, has been cloned (Larkan *et al.*, 2013).

The best described plant *R*-genes are the *Arabidopsis* NBS-LRR class of genes, encoding intracellular proteins that respond to effectors produced by many bacterial, fungal and oomycete pathogens after translocation or delivery into the host cell's cytoplasm (Ali and Bakkeren, 2011; Belkhadir *et al.*, 2004; Grant *et al.*, 2006). During the initial infection of *Brassica* leaves, the hyphae of the invading *L. maculans* colonize the intercellular spaces between mesophyll cells (Howlett *et al.*, 2001). While the intimate interactions of the fungus and host cells have not been extensively studied in the *Brassica–Leptosphaeria* pathosystem, it is believed that effector proteins are secreted by *L. maculans* into the apoplastic fluid, where they may interact with extracellular host targets or be translocated into the host cells (Kale *et al.*, 2010; Oliva *et al.*, 2010). Regardless of their final destination, detecting the AVR proteins of invading *L. maculans* as soon as they are released into the host apoplast would be advantageous to the plant in terms of mounting a co-ordinated host defence response, particularly when the effectors are targeted to disrupt host cell defensive signalling pathways, providing impetus for the evolution of extracellular detection components.

Extracellular detection of pathogen elicitors is often achieved through deployment of host proteins featuring extracellular leucine-rich repeat (eLRR) motifs capable of facilitating protein–protein interactions, most notably the cell membrane-bound receptor-like proteins (LRR-RLPs) and receptor-like kinases (LRR-RLKs) (Kruijt *et al.*, 2005; Stotz *et al.*, 2014; Yang *et al.*, 2012). Plant LRR-RLPs have a basic primary structure consisting of seven domains (A through G). The eLRR region domain C is further defined into three subdomains C1–C3, with C1 and C3 containing strings of LRRs while C2 is a short 'loop out' break in the LRR consensus (Jones *et al.*, 1994; Zhang and Thomma, 2013). Plant eLRR motifs are typically 24 residues in length and characterized by the consensus amino acid sequence **LxxLxLxxNxL**(t/s) **GxIPxxLGxLxx**. The first 11 residues contain the β-strand portion of the repeat, while the more variable tail end of the motif forms an α-helix (Kajava, 1998; Matsushima *et al.*, 2010). The β-strands come together in parallel to form concave β-sheet structure, with ligand specificity being determined by the solvent-exposed residues (the x's in xxLxLxx) at the β-sheet surface (Van der Hoorn *et al.*, 2001; Wulff *et al.*, 2001, 2009b).

In many cases, it has been shown that LRR-RLPs, lacking any cytoplasmic signalling domains, require the formation of protein complexes with a LRR-RLK partner for stable accumulation in the host cell membrane and for the transduction of extracellular signals. The most notable example is the LRR-RLK SUPPRESSOR OF BIR1-1 (SOBIR1), which is widely conserved throughout the plant kingdom and has a role in a broad range of plant developmental and defence responses (Gao *et al.*, 2009; Liebrand *et al.*, 2014). SOBIR1 has been shown to be active in response to several fungal plant pathogens, being required for effector-triggered immunity mediated by the LRR-RLPs Cf-4 and Ve1 of tomato (Jeong *et al.*, 1999; Liebrand *et al.*, 2014), in response to infection by the fungal pathogens *Cladosporium fulvum* (Thomas *et al.*, 1997) and *Verticillium dahliae* (Kawchuk *et al.*, 2001), respectively. A *Sobir1* orthologue of the Brassicaceae *Sinapis alba* (white mustard) was also found to be marginally up-regulated during infection with the fungal pathogen *Alternaria brassicicola* (Ghose *et al.*, 2008). SOBIR1 itself can also potentially be the target of pathogen effectors, as evidenced by its interaction with a protein from *Pseudomonas* spp. (Liebrand *et al.*, 2014; Rioja *et al.*, 2013).

We have previously shown that the *LepR3* gene of *B. napus* encodes a LRR-RLP which conveys race-specific resistance to isolates of *L. maculans* carrying the effector *AvrLm1* (Larkan *et al.*, 2013). A second blackleg resistance gene, *Rlm2*, is co-localized within the same genetic interval of *B. napus* chromosome A10, suggesting *Rlm2* and *LepR3* may be allelic variants of the same LRR-RLP locus (Larkan *et al.*, 2014). In this study, we report the cloning of *Rlm2*, detail the allelic variation of the *LepR3/Rlm2* LRR-RLP locus in several *Brassica* accessions and also demonstrate the interaction of RLM2 with *At*SOBIR1 during transient co-expression in *Nicotiana benthamiana* leaves.

Results

Cloning and transformation of *Rlm2*

As the *LepR3* RLP locus was the most promising resistance gene candidate previously identified within the *Rlm2* map interval (Larkan *et al.*, 2014), we decided to clone the corresponding allele from the *Rlm2 B. napus* line 'Glacier DH24287'. PCR amplification of the candidate LRR-RLP locus using the same primers originally used to clone *LepR3* produced a single amplicon of approximately 4.5 kb from each of the *Rlm2* lines 'Glacier DH24287', 'Tapidor DH' and 'Bristol', suggesting all three *Rlm2* lines contained a similar allele at the target locus that was distinct from the *LepR3* or *lepR3* alleles previously described (Larkan *et al.*, 2013). Transformation of the susceptible *B. napus* line 'Topas DH16516' was performed after transferring the 'Glacier DH24287' allele to the plant transformation vector pMDC123 (Curtis and Grossniklaus, 2003). Fifteen putative T_0 transformants survived selection and were allowed to set selfed seed (T_1). Initial phenotypic screening (cotyledon test; eight plants per T_1 seed lot) using the *L. maculans* isolate '00-100' (*AvrLm2*) revealed that four of the 15 T_1 seed lots were segregating for resistance to the pathogen, while the remaining 11 were completely susceptible (cotyledon test scores of 7–9). Droplet digital PCR analysis of T_0 and T_1 individuals confirmed the transgenic status of the four

resistant lines. No target signal was detected for the 'Topas DH16516' and 'Glacier DH24287' negative controls, while a single homozygous insertion was confirmed for the *LepR3* transgenic line 'NLA8-2' used as a positive control. One *Rlm2* T_0 individual (NLA51) was revealed to harbour a single heterozygous insertion event, and homozygous T_1 individuals were retained to produce stable T_2 lines (Figure 1). Segregation analysis of 43 NLA51 T_1 individuals detected 36 resistant and seven susceptible plants, conforming to a 3:1 ratio ($\chi^2 = 1.74$, $P = 0.187$) and confirming the presence of a single transgenic event.

Confirmation of *Rlm2* phenotype

To verify that the 'Glacier DH28247' candidate LRR-RLP allele is indeed *Rlm2*, a set of *L. maculans* isolates varying in their pathogenicity towards *LepR3*, *Rlm2* and *Rlm3* was employed (Table 1). We used *B. napus* lines 'Topas DH16516' as susceptible and 'NLA8-2' (*LepR3* transgenic, Larkan *et al.*, 2013) and Quantum (*Rlm3*) as resistance controls and proof of *Rlm2* specificity. All isolates were highly virulent (mean cotyledon test score > 8.0) on the susceptible control line 'Topas DH16516'. The isolate '00-100' (*avrLm1*, *AvrLm2*, *AvrLm3*), originally used to screen the transgenic plants, was virulent on the *LepR3*-tansgenic line 'NLA8-2' and avirulent on both 'Glacier DH24287' (*Rlm2*, *Rlm3*) and 'Quantum' (*Rlm3*), and confirmed the resistant phenotype of the T_1 and T_2 lines. The reference isolate 'v23.1.3' as expected was virulent towards both 'Glacier DH24287' and 'Quantum', and avirulent on 'NLA8-2', confirming its previously described '*AvrLm1*, *avrLm2*, *avrLm3*' pathotype (Balesdent *et al.*, 2002). 'v23.1.3' was also completely virulent on all T_1 and T_2 plants. Three additional *L. maculans* isolates, '98-15', 'Lifolle 6' and 'B13-17', with the genotype *avrLm1*, *AvrLm2*, *avrLm3* were all virulent on 'Quantum' and 'NLA8-2' yet avirulent on 'Glacier DH24287'. These three isolates all allowed for the detection of the segregating resistance phenotype in the T_1 generation and homozygous resistance in the T_2 line. Finally, the '*avrLm1*, *avrLm2*, *AvrLm3*' isolate 'WA30' was avirulent on both *Rlm3* lines ('Glacier DH24287' and 'Quantum'), virulent on the *LepR3* line 'NLA8-2' and completely virulent on all T_1 and T_2 plants tested. Together, these results demonstrate that only

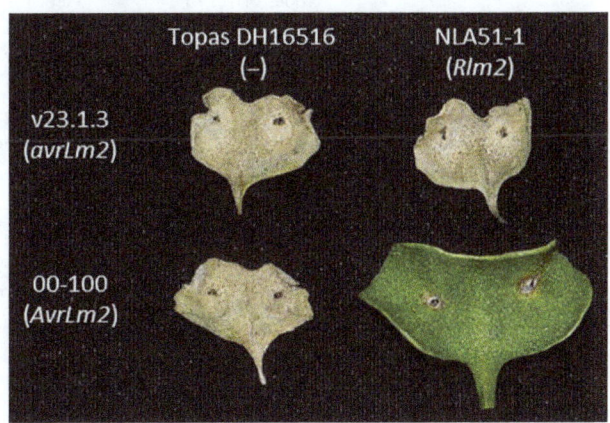

Figure 1 Phenotypic interaction of *avrLm2* and *AvrLm2 L. maculans* isolates with *Rlm2*-transgenic *B. napus*. Cotyledons of susceptible *B. napus* 'Topas DH16516' and Topas:*Rlm2* transgenic T_2 line 'NLA51-1' 14 days post-inoculation with *avrLm2 L. maculans* isolate v23.1.3 and *AvrLm2* isolate 00-100.

Table 1 Phenotypic verification of *Rlm2* transformant. Reaction of control *B. napus* lines and *Rlm2*-transgenic T_1 (segregating) and T_2 (homozygous) generations to cotyledon infection by *L. maculans* isolates of various *Avr* profiles. S = susceptible (shaded red), R = resistant (shaded green), average test score (0–9 scale) 14 days postinoculation given in parentheses

L. maculans Isolate		*B. napus* control				*Rlm2* transgenic	
		Topas DH16516	Glacier DH24287	Quantum	NLA8-2	NLA51 T_1	NLA51-1 T_2
		–	*Rlm2, Rlm3*	*Rlm3*	*LepR3*	*Rlm2*	*Rlm2*
00-100	*avrLm1, AvrLm2, AvrLm3*	S (9.0)	R (2.5)	R (3.0)	S (9.0)	36R:7S	R (2.9)
v23.1.3	*AvrLm1, avrLm2, avrLm3*	S (9.0)	S (8.6)	S (9.0)	R (2.2)	0R:8S	S (9.0)
98-15	*avrLm1, AvrLm2, avrLm3*	S (9.0)	R (3.1)	S (8.5)	S (9.0)	4R:4S	R (2.7)
Lifolle 6	*avrLm1, AvrLm2, avrLm3*	S (8.5)	R (3.0)	S (8.4)	S (8.8)	5R:2S	R (3.0)
B13-17	*avrLm1, AvrLm2, avrLm3*	S (9.0)	R (3.2)	S (9.0)	S (9.0)	5R:3S	R (3.5)
WA30	*avrLm1, avrLm2, AvrLm3*	S (8.4)	R (3.0)	R (3.0)	S (9.0)	0R:8S	S (9.0)

AvrLm2, and not *AvrLm1* or *AvrLm3*, induces the resistance phenotype imparted by the 'Glacier DH24287' *LepR3* locus allele; thus, we have cloned *Rlm2*.

Allelic variation of the *LepR3/Rlm2* locus

To gain insight into allelic variation of the *LepR3/Rlm2* locus, alleles from 16 *B. napus*, *B. juncea* and *B. rapa* lines were analysed (Table 2). Predicted coding sequences (CDS) and proteins revealed five additional *lepR3/rlm2 B. napus* lines ('Westar N-o-1', 'Polo', 'Columbus', 'Quantum' and 'Ning You') plus the *B. napus* var. *napobrassica* landrace 'Hvammsrofa' contained alleles at the target locus identical to that of the 'Topas DH16516' susceptible type, encoding a single-exon CDS of 2853 bp and a predicted protein of 950 aa (hereafter referred to as 'BnlepR3'). All three *Rlm2* lines ('Glacier DH24287', 'Tapidor' and 'Bristol') carried identical alleles (single-exon 2778-bp CDS, 925 aa protein). The *LepR3* line 'Hyola 60' contained an identical allele (single-exon

2556-bp CDS, 851 aa predicted protein) to that of 'Surpass 400'. The *B. napus* line 'Marnoo' and the *B. juncea* line 'AC Vulcan' (both *rlm2/lepR3*) harbour alleles encoding highly similar two-exon coding sequences (2219 and 2218 bp; 97.52% identity) and proteins (719 aa and 720 aa; 90.66% identity) that are distinct from any other alleles defined in this study. *B. rapa* subsp. *sylvestris* '006-1-1' also produced a unique allele (single-exon 2916-bp CDS, 971 aa protein) distinct from the 'Surpass 400' *LepR3* allele (77.7% identity between predicted proteins). Finally, the oilseed *B. rapa* variety 'Torch' also produced a distinct *lepR3/rlm2* allele (2772-bp CDS, 923 aa protein) that most closely matched the *Rlm2 B. napus* lines (90.38% protein identity). A full comparison of amino acid identities, including related proteins from *A. thaliana* and *Capsella rubella*, and the *S. lycopersicum* LRR-RLP Cf-9, is presented in Figure S1. All *LepR3/Rlm2* locus sequences were deposited to the GenBank database (accession numbers KM097068–KM097081).

Table 2 Allelic variation of the *LepR3/Rlm2* locus. Summary of the predicted coding sequence (CDS) and protein sizes of the sequenced allelic variants of *LepR3/Rlm2* locus from multiple *Brassica* species and cultivars

Species	Cultivar	GenBank no.	Genotype	Allele (CDS)	Protein (Size)
B. napus	Topas DH16516	JX880109	*lepR3/rlm2*	*lepR3* (2853 bp)	BnlepR3 (950 aa)
B. napus	Westar N-o-1	KM097073			
B. napus	Polo	KM097074			
B. napus	Columbus	KM097072			
B. napus	Quantum	KM097081			
B. napus	Ning You	KM097075			
B. napus var. *napobrassica*	Hvammsrofa	KM097076			
B. napus	Glacier DH24297	KM097068	*Rlm2*	*Rlm2* (2778 bp)	RLM2 (925 aa)
B. napus	Tapidor DH	KM097069			
B. napus	Bristol	KM097070			
B. napus	Surpass 400	JX880110	*LepR3*	*LepR3* (2556 bp)	LEPR3 (851 aa)
B. napus	Hyola 60	KM097071			
B. rapa	Torch	KM097077	*lepR3/rlm2*	*lepR3-2* (2772 bp)	BrlepR3 (923 aa)
B. rapa subsp. *sylvestris*	006-1-1	KM097078	*lepR3/rlm2*	*lepR3-3* (2916 bp)	BrlepR3-2 (971 aa)
B. napus	Marnoo	KM097079	*lepR3/rlm2*	*lepR3-4* (2219 bp)	BjlepR3 (719 aa)
B. juncea	AC Vulcan	KM097080	*lepR3/rlm2*	*lepR3-5* (2218 bp)	BjlepR3-2 (720 aa)

Identification of putative protein motifs

All of the predicted proteins encoded by the various *LepR3/Rlm2* locus alleles feature the N-terminus signal peptide, N-terminal LRR region and multiple LRR motifs of typical RLPs. However, only the *B. juncea* 'AC Vulcan' and the *B. napus* 'Marnoo' variants lacked the TM domain and a short cytoplasmic C-terminus region featured in the other variants. Phylogenetic and sequence alignment comparisons of the predicted proteins encoded by each unique *Brassica* spp. allele, including Bra008930.1 from the *B. rapa* 'Chiifu' reference genome, revealed RLM2 to be the most similar to the susceptible BnlepR3 protein, both of which showed a higher level of relatedness to lepR3 from the *B. rapa* oilseed variety 'Torch' than to LEPR3 or the protein predicted from the *B. rapa* subsp. *sylvestris* allele. The *B. juncea* 'AC Vulcan' and *B. juncea*-like protein from *B. napus* 'Marnoo' displayed a high degree of similarity to each other, forming a clade removed from the other proteins. All of the alleles defined in this study only showed a distant relationship to Bra008930.1 due to the apparent truncated nature of the Bra008930 allele (Figure S2).

We further analysed the predicted primary structure of the predicted protein variants LEPR3, RLM2 and BnlepR3. The primary structure of all three protein variants closely matched the A to G domains described for the tomato RLP Cf-9: Domain A, signal peptide; B, N-terminal LRR domain; C, LRR domain; D, connecting domain; E, acidic domain; F, transmembrane (TM) domain; and G, cytoplasmic domain. Domain C is further defined into three subdomains C1–C3, with C1 and C3 containing strings of LRRs while C2 is a short 'loop out' break in the LRR consensus (Jones *et al.*, 1994; Zhang and Thomma, 2013). Through a combination of computational analysis and manual inspection of protein alignments, we defined 30 putative LRR motifs within the BnlepR3 protein, most of which fit the plant-specific PGIP-type LRR consensus sequence LxxLxLxxNxL(t/s)GxIPxxLGxLxx, where bold uppercase letters have a higher than 70% occurrence at a given position, normal uppercase letters 40% and lowercase letters 30%; x = any residue, L = Leu, Ile, Val or Phe, N = Asn, Thr, Ser or Cys (Figure 2a; Table S1). Twenty-nine of the 30 LRRs are present in RLM2 and 26 in LEPR3. A span of 44–45 residues (domain C2) interrupts the LRR region of each protein, dividing 22–26 amino terminal LRRs (C1) from the four carboxyl terminal LRRs of domain C3. The major variation in LRR content between the proteins occurs at LRRs 4-8; LRR4, occurring only in BnlepR3, is highly similar to LRR3 and may have resulted from a duplication event. An apparent deletion event in *LepR3* has resulted in the removal of LRRs 6 and 7, along with the last eight residues of LRR5 and the first 16 residues of LRR8, creating a chimeric LRR in LEPR3 which we denote as LRR5* (Figure 2b).

LEPR3, RLM2 and BnlepR3 all contain identical N-terminal LLxxK (ILELK), CxWxGVxC (CSWDGIRC) and potential truncated LssW motif (VSW), located in the B domain (Table S1). Another feature of LRR-RLPs is a large number of *N*-linked glycosylation site (PGS) motifs (NxS/T) which can be crucial to RLP activity particularly when contained within the α-helices of the LRRs (Van Der Hoorn *et al.*, 2005). A search revealed a high number of PGS motifs in each protein variant, approximately one for every 35 residues, with 22 of the 30 putative LRRs containing at least one PGS. Polymorphism between BnlepR3 and RLM2 affecting the PGS content of LRRs was detected in the domain C1, within LRRs 1,3,5 and 7 (Figure 2a, Table S1).

RLM2 and BnlepR3 are highly conserved within the LRR region, sharing identical amino acid sequence for 22 of the 30 LRRs, including an interval of 21 completely conserved LRR motifs (LRRs 8-28) with most of the variation occurring at the ends of the LRR region (LRRs 1-5, 7, 29-30; Table S1). In particular, amino acid substitutions between BnlepR3 and RLM2 occur within the β-strand solvent-exposed residues of LRRs 1, 3, 5 and 7. Given the amount of potentially critical variation observed between the alleles within the first eight LRRs, we have designated this region

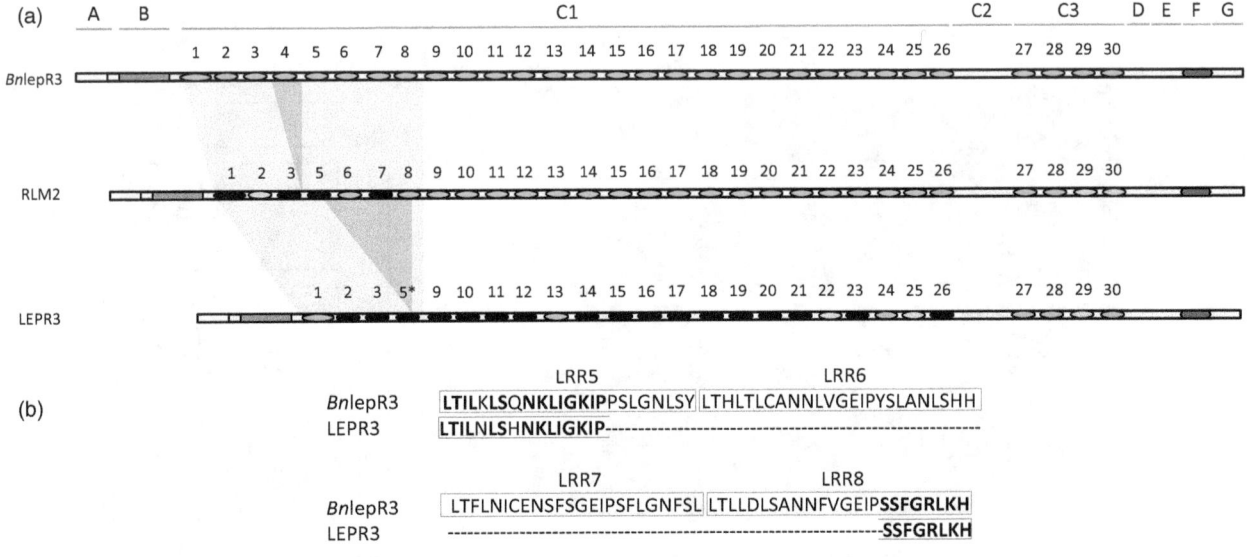

Figure 2 Putative primary structure of LEPR3/RLM2 proteins. (a) Schematic representation of the predicted protein motifs of BnlepR3, RLM2 and LEPR3. Bars at top denote LRR-RLP domains A through G. Yellow box = signal peptide; blue box = N-terminal LRR; ovals = LRR, where green = identical, orange = residue substitution(s) and red = residue substitution(s) affecting solvent-exposed residues or PGS; purple box = transmembrane domain. Blue shaded area denotes 'variable region' of C1, and red shading denotes gaps in protein alignments. (b) Alignments of BnlepR3 LRR5-8 with chimeric LEPR3 LRR5*. Blue boxes define BnlepR3 LRR motifs, and open red boxes define portions of LRR5 and LRR8 combined to form LRR5*.

of the C1 LRR domain as the 'variable LRR' region for this locus (Figure 2a).

Membrane localization and physical interaction between RLM2 and AtSOBIR1

Recently, the *Arabidopsis thaliana* LRR-receptor-like kinase (LRR-RLK) suppressor of Bir1-1 (*At*SOBIR1) and its homologues from tomato were shown to interact specially with a number of LRR-receptor-like proteins (LRR-RLPs), such as tomato Cf-4 and Ve1, to play a role in LRR-RLPs-mediated resistance against the corresponding fungal pathogen (Liebrand *et al.*, 2013, 2014). To examine whether *At*Sobir1 interacts with RLM2, we first confirmed the presence of RLM2 at the cell membrane by observing the co-localization of a RLM2-GFP fusion protein with the cell membrane marker mCherry-HVR (Pietraszewska-Bogiel *et al.*, 2013) during transient co-expression in *N. benthamiana* leaves (Figure S3). Bimolecular fluorescence complementation (BiFC) assay was then performed by transiently co-expressing *Rlm2-VYCE* and *AtSobir1-VYNE* or empty vector (GW-VYCE) and *AtSobir1-VYNE* in *N. benthamiana* leaves. Figure 3a shows that YFP fluorescent signal was only observed in cells co-expressing *Rlm2-VYCE* and *AtSobir1-VYNE*, but not in the negative control cells. To further prove this association *in vivo*, the co-immunoprecipitation (co-IP) assay was performed upon transient expression in *N. benthamiana*. GFP-tagged RLM2 was capable to co-immunoprecipitate Myc-tagged *At*SOBIR1 (Figure 3b).

Discussion

We report here the cloning of *Rlm2*, the second *B. napus* LRR-RLP conferring resistance to *L. maculans* to be characterized, adding to the growing arsenal of molecular tools available for the dissection of the *Brassica–Leptosphaeria* pathosystem. Although RLM2 and LEPR3 share only 30.7 and 32.25% identity with the tomato RLP Cf-9, respectively (Figure S1), they share a similar arrangement of predicted motifs. Cf-9 encodes 27 LRRs arranged in groups of 23 (C1 domain) and 4 (C3 domain) repeats separated

by a short C2 'loop out' domain (Jones *et al.*, 1994). The model structure predicts a heavily glycosylated extended spring-like form with the specificity-determining surface on the inner concave side and the α-helices providing the outer surface (Van Der Hoorn *et al.*, 2005), with the potential for intramolecular interaction between distal regions of the C1 LRR domain creating the inactive state of the protein (Wulff *et al.*, 2009a). The eLRR regions of LEPR3 and RLM2 are also arranged in similar fashion to Cf-9 with C1 domains containing 22 and 25 LRRs, respectively, C2 domains of 45 and 44 residues, and C3 domains containing four LRRs. Although we have yet to perform any protein modelling studies ourselves, given the similarity in primary protein structures we would hypothesize that LEPR3 and RLM2 form similar spring-like structures as predicted for Cf-9.

N-terminal LRR (B domain) regions can determine specificity of plant RLPs (Van der Hoorn *et al.*, 2001). LEPR3, RLM2 and *Bn*lepR3 all contain identical N-terminal LLxxK (ILELK), CxWxGVxC (CSWDGIRC) and potential truncated LssW motif (VSW), located in the B domain (Table S1). In the tomato RLP Cf-9, substitution of the conserved Trp (W) in either the LssW or CxWxGVxC motifs completely abolishes Cf-9 activity, while substitution of either Cys (C) in the CxWxGVxC causes near-inactivation of the protein (Van Der Hoorn *et al.*, 2005). While the requirement of these motifs in the RLM2 and LEPR3 proteins has yet to be determined, their conserved sequence in both susceptible and resistant variants suggests they are not responsible for the differences in RLP function at the *LepR3/Rlm2* locus. However, variation occurs in the putative C-terminal GNxGLCGxPLxxxC motif (DNPGLxGPSLxxxC) in domain D (Table S1), which may have an effect on the specificity.

The LRR motifs of RLM2 and *Bn*lepR3 are highly conserved over most of the C1 domain, with the major variation between the two occurring within the first seven LRRs where a LRR duplication event, substitutions in solvent-exposed β-sheet residues and differences in PGS may all affect ligand specificity. This makes RLM2 and *Bn*lepR3 ideal candidates for domain-swapping studies, in which we hope to define the specific region(s) of the

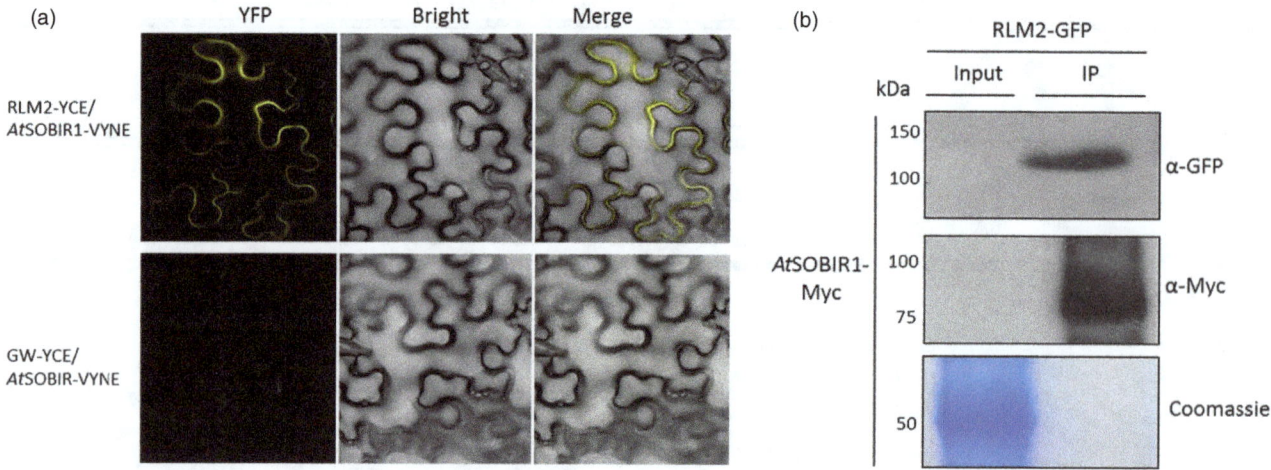

Figure 3 RLM2 interacts with *At*SOBIR1. (a) VYNE-fused *At*SOBIR1 and VYCE-fused RLM2 were transiently co-expressed in *N. benthamiana*. Co-infiltration of empty vector (GW-VYCE) and VYNE-fused *At*SOBIR1 was used as control. YFP fluorescent signal was visualized by Zeiss AxioImager.Z1 (ApoTome) microscopy. (b) GFP-tagged RLM2 and Myc-tagged *At*SOBIR1 were co-expressed in *N. benthamiana*. Proteins were extracted after 48 h and subjected to immunoprecipitation by GFP-Trap_M beads. Total proteins (input) and immunopurified proteins (IP) were subjected to SDS-PAGE, followed by blotting using anti-GFP antibody to detect the immunopurified GFP fusion proteins and anti-Myc antibody to detect co-immuopurifying Myc-*At*SOBIR1 proteins, respectively. Coomassie blue stained protein gel showing the 50-kDa Rubisco protein band in the input sample for the loading control.

proteins which determine specificity towards AVRLM2. In contrast, there were variations detected between *Bnl*epR3 and LEPR3 in all but three of the C1 LRR motifs (Figure 2a), suggesting a high degree of evolutionary divergence at the *LepR3/Rlm2* locus, fitting with the reported introgression of *LepR3* into *B. napus* germplasm from *B. rapa* subsp. *sylvestris* (Buzza and Easton, 2002; Crouch et al., 1994). The recognition specificity of RLM2 may also be imparted by variation in amino acid residues outside of the highly variable region of the C1 domain. Recent domain-swapping experiments with the tomato LRR-RLP Ve1 and its nonfunctional paralog Ve2 have shown that LRRs 1-30 of the two proteins are interchangeable, yet a functional Ve1-mediated resistance response requires both the LRRs 30-35 of Ve1, contained mostly within the C3 domain of the protein, and the cytoplasmic C-terminus of Ve1. Ve2 was shown to also interact with SOBIR1, suggesting impediment of association with the RLK partner was not the basis for loss of function (Fradin et al., 2014).

Variation of individual amino acid residues within RLP homologues often conveys no effect on the specificity of the proteins, as shown in both and Ve1 and the tomato RLP Cf-4, indicating that RLPs are not fully adapted for recognition of particular avirulence triggers, but rather form the basis of a versatile and adaptive signalling complex (Fradin et al., 2014; Van der Hoorn et al., 2001). The *LepR3/Rlm2* locus provides an excellent example of that adaptability, with allelic variants producing alternate forms of the same protein, each conferring specific recognition of separate *L. maculans* avirulence factors. Generating alternate primary structures within RLP alleles, such as the 'recycling' of LRRs observed with the chimeric LRR5* in LEPR3, or the duplication of whole LRR motifs as observed in *Bnl*epR3, may be mechanisms for altering RLP specificity.

The identification of *Rlm2* and *LepR3* as alleles of the same locus would prevent the pyramiding of both genes in a single homozygous *B. napus* cultivar, although this restriction could easily be overcome in hybrid crop development systems. As identical alleles were sequenced from 'Glacier DH24287', 'Tapidor DH' and 'Bristol', the presence of *Rlm2* in those varieties has been confirmed. The fact that there was no sequence variation detected between the three lines over the entire locus is not surprising, given that all three lines are derived from French winter-type rapeseed varieties which may have shared pedigrees. *Rlm2* has also previously been detected in *B. rapa* (Leflon et al., 2007), which could possibly provide a different allelic variant and some insight into the evolution of the locus. We also confirmed the presence of *LepR3* in the hybrid variety 'Hyola 60' although we were unable to show that *LepR3* is derived from *B. rapa* subsp. *sylvestris* as previously reported (Buzza and Easton, 2002), as our accession of this species ('006-1-1') produced a unique allele sequence distinct from that of *LepR3*. Previous introgression of resistance from this material into *B. napus* captured the blackleg resistance genes *LepR1*, *LepR2* and *LepR4* (Yu et al., 2012), yet *LepR3* was never detected in these experiments. Given the low self-compatibility of *B. rapa* subsp. *sylvestris*, we find it likely that the original seed lot was heterozygous for *LepR3*, and while the gene was captured in the original hybridization experiments of Crouch et al. (1994), it is not present in the accession held by our laboratory. The discovery of a *B. juncea*-type allele in the Australian *B. napus* variety 'Marnoo' is also noteworthy. While not specifically listed in the pedigree of 'Marnoo', there was a concerted effort to introduce the highly effective *B. juncea*

(genome AABB) blackleg resistance traits into Australian material after the disease nearly wiped out Australian canola production in the 1970s. While none of the dominant B-genome resistance genes were stably introduced, *B. juncea* appears in the pedigree of many Australian cultivars (Cowling, 2007). We have demonstrated here that, at least for one variety, stable introgression of *B. juncea* A-genome material can be detected in the *B. napus* A-genome.

A rapidly growing number of LRR-RLPs that play a role in plant immunity have been documented (Stotz et al., 2014; Yang et al., 2012). However, to form a signalling-competent receptor complex, LRR-RLPs, lacking an intracellular kinase domain, are required to interact with LRR-RLKs. This hypothesis has been evidenced for some LRR-RLPs involved in plant immunity (Böhm et al., 2014; Liebrand et al., 2014; Zhang et al., 2013, 2014). An example of such a RLK is SOBIR1, the suppressor of BIR1 (BAK1-interacting receptor-like kinase 1). In a recent publication, Liebrand et al. (2013) demonstrated that the LRR-RLP Cf-4 or Ve1 interacts with tomato homologues of LRR-RLK *At*SOBIR1 to trigger plant defence against the infection of *C. fulvum* secreting Avr4 or *V. dahliae* secreting Ave1. Furthermore, *At*SOBIR1 was found to be required for *Arabidopsis* RLP30, perceiving the SsE1 elicitor (*Sclerotinia sclerotiorum* elicitor-1)-dependent resistance against *Sclerotinia sclerotiorum* (Zhang et al., 2013). Another *Arabidopsis* LRR-RLP, RBPG1 (responsiveness to Botrytis polygalacturonase-1), that specifically recognizes fungal endo-polygalacturonases (PGs) was found to interact with *At*SOBIR1, and this is required for PG-induced necrosis and PG-induced resistance to *Hyaloperonospora arabidopsidis* (Zhang et al., 2014). Similarly, the results that we presented here provide the evidence for physical interaction between *At*SOBIR1 and the *B. napus* LRR-RLP RLM2, showing that when co-expressed in *N. benthamiana*, RLM2 is able to heterodimerize with *At*SOBIR1 at the cell membrane. These data support an essential and common role for SOBIR1 as a partner in LRR-RLP-containing receptor complexes. In *B. napus* lines carrying *Rlm2*, it is expected that RLM2 and a *B. napus* SOBIR1 homologue (*Bn*SOBIR1) form a signalling-competent complex to initiate downstream signalling after recognition, either directly or through other protein intermediates, of the *L. maculans* effector AVRLM2. Three *AtSobir1* homologues are present in the A-genome of *B. rapa* (Wang et al., 2011), Bra022870 on chromosome A3, and a pair of homologues on A4 (Bra021758 & Bra021759), fitting with the ancient triplication of the ancestral *Brassica* genome followed by genome shrinkage (Mun et al., 2009; Nelson and Lydiate, 2006; Parkin et al., 2005), and likely meaning there are six *BnSobir1* homologues in the full *B. napus* genome (AACC). Validation of a functional interaction between RLM2 and *Bn*SOBIR1 and identification of potential redundancy between *BnSobir1* homologues will be the focus of further research.

We have provided here the characterization of the second cloned *B. napus* blackleg resistance gene, *Rlm2*; an allelic variant of the *LepR3/Rlm2* LRR-RLP locus, confirming the importance of RLPs in the race-specific responses to *L. maculans* and demonstrating the adaptability of LRR-RLP gene specificity in defence signalling. By showing RLM2 interacts with the RLK *At*SOBIR1, we hope to promote further dissection of *Brassica* defence signalling pathways. Future functional analysis of RLM2 and LEPR3 and characterization of their interaction with their corresponding avirulence factors will provide a greater understanding of the *Leptosphaeria*-*Brassica* pathosystem.

Experimental procedures

L. maculans isolates

Single-spore cultures of the *L. maculans* isolates '00-100', '98-15', 'WA30' and 'Lifolle 6' were sourced from the Rimmer Collection, AAFC Saskatoon. Isolate 'B13-17' was isolated from infected canola field stubble collected near North Battleford, Saskatchewan, in 2013. The reference isolate 'v23.1.3' was sourced from the INRA-Bioger collection, France, and kindly provided by Dr. T. Rouxel.

Plant material

Control lines used for transformation and phenotypic analysis were as follows (blackleg *R*-gene content of each line provided in parentheses): the susceptible doubled-haploid (DH) line 'Topas DH16516' (-), the DH line 'Glacier DH24287' (*Rlm2, Rlm3*), 'Quantum' (*Rlm3*) and the 'Topas DH16516:LepR3' transgenic line 'NLA8-2' (*LepR3*).

Additional *B. napus* variety-derived control lines were used for analysis of variation at the *LepR3* locus along with the above-mentioned lines; 'Surpass 400' (*LepR3, RlmS*), 'Columbus' (*Rlm1, Rlm3*), 'Tapidor DH' (*Rlm2*), 'Bristol' (*Rlm2, Rlm9*), DH line 'Westar N-o-1' (-), 'Polo' (-), 'Marnoo' (-), the hybrid variety 'Hyola 60' (*LepR3*) and 'Ning You' (-). Also included were the *B. rapa* subsp. 'sylvestris' line '006-1-1' (*LepR1, LepR2, LepR3*(?), *LepR4*), *B. rapa* oilseed variety 'Torch' (-), Icelandic *B. napus* var. *napobrassica* (swede) landrace 'Hvammsrofa' (*Rlm1*) and the *B. juncea* line 'AC Vulcan' (*LmJR1, LmJR2*). Seed for all lines except the hybrid variety 'Hyola 60' and *B. rapa* 'Torch' were previously produced at AAFC Saskatoon from at least one single-plant decent with confirmation of homozygous resistance phenotype via cotyledon test (data not shown). Original variety seed for 'Hvammsrofa' was obtained from the Nordic Gene Bank (NordGen accession # NGB9911).

Transgenic analysis of Rlm2 candidate

For the transgenic analysis of the candidate *Rlm2* gene, the same genomic amplicon produced from 'Glacier DH24287' for the allelic variant analysis, including 1179 bp upstream and 531 bp downstream of the predicted coding region, was used for construct production and transformation of susceptible *B. napus* line 'Topas DH16516' as described in Larkan et al. (2013). T_0 putative transgenic individuals were allowed to set selfed seed before the *Rlm2* resistance phenotype was assessed in the T_1 generation using *L. maculans* isolate '00-100'. Further confirmation of phenotype was performed using additional isolates 'v23.1.3', '98-15', 'Lifolle 6', 'B13-17' and 'WA30'.

All T_0 and T_1 putative *B. napus* transformants were further analysed by droplet digital PCR (ddPCR) to assess insertion copy number. A *BnActin2* probe set was used for reference (kindly provided by Dr. Dwayne Hegadus, AAFC Saskatoon), and a target probe set was designed to the Basta^R gene contained within the *Rlm2* construct (Table S2), with each probe containing a 3' Iowa Black quencher (Integrated DNA Technologies, Coralville, IA). 'Topas DH16516' and 'Glacier DH24287' were included as negative controls, while the *LepR3* transgenic line 'NLA8-2' was included as a positive control.

Analysis of LepR3 allelic variants

DNA extraction, PCR amplification and cloning of amplicons were all performed as described previously (Larkan et al., 2013), using

the *attb*-tagged primer pair GW-BnRLP F and GW-BnRLP Rb which amplify a single, A-genome-specific locus. Cloning and sequencing primers are listed in Table S2. Gene and protein translation predictions were performed using FGENESH with *B. rapa* genome-specific parameters (Solovyev et al., 2006). InterProScan 5 (Jones et al., 2014) and LRRfinder 2.0 (Offord and Werling, 2013) were used to detect functional protein domains. Multiple sequence alignments and tree construction (UPGMA, 1000 bootstraps) were performed in CLC Workbench 6.7.1 software (Red eQuus Corporation, Toronto, Canada).

Vectors for Agrobacterium-mediated transient transformation of Nicotiana benthamiana

To generate the bimolecular fluorescence complementation (BiFC) constructs, the *Rlm2* ORF was amplified using primers *Rlm2*-FB and *Rlm2*-RB-S. The obtained PCR product was introduced into entry clone pDONR/Zeo (Invitrogen, Burlington, Canada) using gateway protocol described by the manufacturer. The *Arabidopsis thaliana* (At)-*Sobir1* ORF was cloned in pDONR/Zeo using template pGWB20-*At-Sobir1*-Myc with primer pair *AtSobir1*-FB and *AtSobir1*-RB-S (Liebrand et al., 2013). The hence obtained pDONR/Zeo-*Rlm2* and pDONR/Zeo-*Sobir1* were recombined into the binary vectors pDEST-GW-VYCE and pDEST-GW-VYNE, respectively (Gehl et al., 2009). The resulting plasmids pDEST-*Rlm2*-VYCE and pDEST-*Sobir1*-VYNE were used for *Agrobacterium* transformation.

For co-immunoprecipitation and confocal microscopy, the *Rlm2* coding sequence was cloned into pDONR/Zeo gateway entry vector. After confirmation of the insert by sequencing, the pDONR/Zeo-*Rlm2* was recombined with the binary vector pGWB451 containing C-terminal GFP (Nakagawa et al., 2007), which resulted in pGWB451-*Rlm2*. The pGWB20-*At-Sobir1*-Myc with c-terminally tagged Myc epitope was previously described (Liebrand et al., 2013). All resulting binary plasmids were transformed to *A. tumefaciens* strain GV3101 (pMP90). *Agrobacterium*-mediated transient transformation of *N. benthamiana* was performed according to methods described previously (Ma et al., 2012). To enhance transient expression, the RNA silencing suppressor *p19* from *Tomato bushy stunt virus* (Voinnet et al., 2003) was co-infiltrated into the *N. benthamiana* leaves.

Confocal microscopy

To monitor the localization of RLM2, mCherry-HVR (Pietraszewska-Bogiel et al., 2013) was co-infiltrated with the RLM2-GFP fusion into 4-week-old *N. benthamiana* leaves. The fluorescence was visualized 36 h after infiltration with confocal microscopy LSM710 (Carl Zeiss Canada Ltd., Toronto, Canada). mCherry was excited at 587 nm and GFP at 488 nm. GFP emission was captured with a 505- to 530-nm filter and mCherry with a 590- to 620-nm filter. Images were scanned eight times.

Bimolecular fluorescence complementation

BiFC has been widely employed to test the protein–protein interaction *in vivo* (Gehl et al., 2009). For BiFC assays, infiltration was performed in leaves of 4- to 5-week-old *N. benthamiana* plants with Agrobacteria containing the corresponding constructs OD600 of 0.5. Leaf discs were imaged 48 h after infiltration using a Zeiss AxioImager.Z1 (ApoTome) microscope with an EC Plan-Neofluar 40X/0.75 M27 objective (Carl Zeiss Canada Ltd., Toronto, Canada) as described previously (Kagale et al., 2012).

Co-immunoprecipitation and Western blotting

Co-immunoprecipitation was performed as previously described (Liebrand *et al.*, 2013). Briefly, the membrane fractions were extracted from *N. benthamiana* leaves 48 h after infiltrating with mixture of *A. tumefaciens* GV3101 containing either pGWB451-*Rlm2*, pGW20-*Sobir1* or *p19* in buffer (150 mm NaCl, 1.0% IGEPAL CA-630 [NP-40], 0.5% sodium deoxycholate, 0.1% SDS, 50 mm Tris, pH 8.0, 1x Roche complete protease inhibitor cocktail). Extracts were centrifuged at 14 000 rpm, 4 °C for 15 min, and 2 mL supernatant was added to 50 µL of GFP-Trap_M beads (ChromoTek, Martinsried, Germany), which was incubated for 2 h at 4 °C in a rotator. After washing the beads four times with extraction buffer, immunoprecipitated proteins were separated by 8% SDS-PAGE gels and blotted onto PVDF membrane (Bio-Rad, Mississauga, Canada) using wet blotting for overnight. Skimmed milk powder (5%) was used as blocking agent. A 1 : 1000 dilution of anti-GFP antibody (Sigma, Oakville, Canada) or 1:2000 diluted anti-Myc (cMyc 9E10, sc-40-HRP; Santa Cruz, Dallas, TX) was used. The secondary antibody goat anti-mouse (Pierce, Rockford, IL) was used as a 1 : 15 000 dilution. The luminescent signal was visualized by Immobilon Western Chemiluminescent HRP Substrate using BioMax MR film (Kodak, Burnaby, Canada).

Acknowledgements

The authors would like to thank Delwin Epp for performing *B. napus* transformations; Catherine Guenther for technical assistance; Dr. Gordon Gropp, Dr. Dwayne Hegadus and Dianna Bekkaoui for assistance with the droplet digital PCR; Merek Wigness for assistance with the confocal microscopy; and Clinton Jurke, Canola Council of Canada, for providing the infected field stubble from which 'B13-17' was isolated. Funding for this work was provided by the Western Grains Research Foundation, SaskCanola and the Government of Saskatchewan Agricultural Development Fund. Authors declare no conflict of interest in regard to this manuscript.

References

Ali, S. and Bakkeren, G. (2011) Fungal and oomycete effectors – strategies to subdue a host. *Can. J. Plant Pathol.* **33**, 425–446.

Balesdent, M.H., Attard, A., Kuhn, M.L. and Rouxel, T. (2002) New avirulence genes in the phytopathogenic fungus *Leptosphaeria maculans*. *Phytopathology*, **92**, 1122–1133.

Balesdent, M.H., Fudal, I., Ollivier, B., Bally, P., Grandaubert, J., Eber, F., Chèvre, A.M., Leflon, M. and Rouxel, T. (2013) The dispensable chromosome of *Leptosphaeria maculans* shelters an effector gene conferring avirulence towards *Brassica rapa*. *New Phytol.* **198**, 887–898.

Belkhadir, Y., Subramaniam, R. and Dangl, J.L. (2004) Plant disease resistance protein signaling: NBS-LRR proteins and their partners. *Curr. Opin. Plant Biol.* **7**, 391–399.

Bent, A.F. and Mackey, D. (2008) Elicitors, effectors, and R genes: the new paradigm and a lifetime supply of questions. *Annual Review of Phytopathology*, **45**, 399–436.

Böhm, H., Albert, I., Fan, L., Reinhard, A. and Nürnberger, T. (2014) Immune receptor complexes at the plant cell surface. *Curr. Opin. Plant Biol.* **20**, 47–54.

Buzza, G. and Easton, A. (2002) *A new source of blackleg resistance from Brassica sylvestris*. In: GCIRC Technical Meeting p. Bulletin No.18. Poznan, Poland.

Cowling, W.A. (2007) Genetic diversity in Australian canola and implications for crop breeding for changing future environments. *Field. Crop. Res.* **104**, 103–111.

Crouch, J.H., Lewis, B.G. and Mithen, R.F. (1994) The effect of A genome substitution on the resistance of *Brassica napus* to infection by *Leptosphaeria maculans*. *Plant Breed.* **112**, 265–278.

Curtis, M.D. and Grossniklaus, U. (2003) A gateway cloning vector set for high-throughput functional analysis of genes *in planta*. *Plant Physiol.* **133**, 462–469.

Delourme, R., Pilet-Nayel, M.L., Archipiano, M., Horvais, R., Tanguy, X., Rouxel, T., Brun, H., Renard, M. and Balesdent, M.H. (2004) A cluster of major specific resistance genes to *Leptosphaeria maculans* in *Brassica napus*. *Phytopathology*, **94**, 578–583.

Dudler, R. (2013) Manipulation of host proteasomes as a virulence mechanism of plant pathogens. *Annual Review of Phytopathology*, **51**, 521–542.

Fitt, B.D.L., Brun, H., Barbetti, M.J. and Rimmer, S.R. (2006) World-wide importance of phoma stem canker (*Leptosphaeria maculans* and *L. biglobosa*) on oilseed rape (*Brassica napus*). *Eur. J. Plant Pathol.* **114**, 3–15.

Fradin, E.F., Zhang, Z., Rovenich, H., Song, Y., Liebrand, T.W.H., Masini, L., Van Den Berg, G.C.M., Joosten, M.H.A.J. and Thomma, B.P.H.J. (2014) Functional analysis of the tomato immune receptor Ve1 through domain swaps with its non-functional homolog Ve2. *PLoS ONE*, **9**, e88208.

Fudal, I., Ross, S., Gout, L., Blaise, F., Kuhn, M.L., Eckert, M.R., Cattolico, L., Bernard-Samain, S., Balesdent, M.H. and Rouxel, T. (2007) Heterochromatin-like regions as ecological niches for avirulence genes in the *Leptosphaeria maculans* genome: map-based cloning of *AvrLm6*. *Mol. Plant Microbe Interact.* **20**, 459–470.

Gao, M., Wang, X., Wang, D., Xu, F., Ding, X., Zhang, Z., Bi, D., Cheng, Y.T., Chen, S., Li, X. and Zhang, Y. (2009) Regulation of cell death and innate immunity by two receptor-like kinases in Arabidopsis. *Cell Host Microbe*, **6**, 34–44.

Gehl, C., Waadt, R., Kudla, J., Mendel, R.R. and Hänsch, R. (2009) New GATEWAY vectors for high throughput analyses of protein-protein interactions by bimolecular fluorescence complementation. *Molecular Plant*, **2**, 1051–1058.

Ghose, K., Dey, S., Barton, H., Loake, G.J. and Basu, D. (2008) Differential profiling of selected defence-related genes induced on challenge with *Alternaria brassicicola* in resistant white mustard and their comparative expression pattern in susceptible India mustard. *Mol. Plant Pathol.* **9**, 763–775.

Gout, L., Fudal, I., Kuhn, M.-L., Blaise, F., Eckert, M., Cattolico, L., Balesdent, M.-H. and Rouxel, T. (2006) Lost in the middle of nowhere: the *AvrLm1* avirulence gene of the Dothideomycete *Leptosphaeria maculans*. *Mol. Microbiol.* **60**, 67–80.

Grant, S.R., Fisher, E.J., Chang, J.H., Mole, B.M. and Dangl, J.L. (2006) Subterfuge and manipulation: type III effector proteins of phytopathogenic bacteria. *Annu. Rev. Microbiol.* **60**, 425–449.

Howlett, B.J., Idnurm, A. and Pedras, M.S.C. (2001) *Leptosphaeria maculans*, the causal agent of blackleg disease of Brassicas. *Fungal Genet. Biol.* **33**, 1–14.

Jeong, S., Trotochaud, A.E. and Clark, S.E. (1999) The Arabidopsis *CLAVATA2* gene encodes a receptor-like protein required for the stability of the CLAVATA1 receptor-like kinase. *Plant Cell*, **11**, 1925–1933.

Jones, J.D.G. and Dangl, J.L. (2006) The plant immune system. *Nature*, **444**, 323–329.

Jones, D.A., Thomas, C.M., Hammond-Kosack, K.E., Balint-Kurti, P.J. and Jones, J.D.G. (1994) Isolation of the tomato *Cf-9* gene for resistance to *Cladosporium fulvum* by transposon tagging. *Science*, **266**, 789–793.

Jones, P., Binns, D., Chang, H.Y., Fraser, M., Li, W., McAnulla, C., McWilliam, H., Maslen, J., Mitchell, A., Nuka, G., Pesseat, S., Quinn, A.F., Sangrador-Vegas, A., Scheremetjew, M., Yong, S.Y., Lopez, R. and Hunter, S. (2014) InterProScan 5: genome-scale protein function classification. *Bioinformatics*, **30**, 1236–1240.

Kagale, S., Uzuhashi, S., Wigness, M., Bender, T., Yang, W., Hossein Borhan, M. and Rozwadowski, K. (2012) TMV-Gate vectors: gateway compatible tobacco mosaic virus based expression vectors for functional analysis of proteins. *Sci. Rep.* **2**, 874.

Kajava, A.V. (1998) Structural diversity of leucine-rich repeat proteins. *J. Mol. Biol.* **277**, 519–527.

Kale, S.D., Gu, B., Capelluto, D.G.S., Dou, D., Feldman, E., Rumore, A., Arredondo, F.D., Hanlon, R., Fudal, I., Rouxel, T., Lawrence, C.B., Shan, W.

and Tyler, B.M. (2010) External lipid PI3P mediates entry of eukaryotic pathogen effectors into plant and animal host cells. *Cell*, **142**, 284–295.

Katagiri, F. and Tsuda, K. (2010) Understanding the plant immune system. *Mol. Plant Microbe Interact.* **23**, 1531–1536.

Kawchuk, L.M., Hachey, J., Lynch, D.R., Kulcsar, F., Van Rooijen, G., Waterer, D.R., Robertson, A., Kokko, E., Byers, R., Howard, R.J., Fischer, R. and Prüfer, D. (2001) Tomato *Ve* disease resistance genes encode cell surface-like receptors. *Proc. Natl Acad. Sci. USA*, **98**, 6511–6515.

Kruijt, M., De Kock, M.J.D. and De Wit, P.J.G.M. (2005) Receptor-like proteins involved in plant disease resistance. *Mol. Plant Pathol.* **6**, 85–97.

Larkan, N.J., Lydiate, D.J., Parkin, I.A.P., Nelson, M.N., Epp, D.J., Cowling, W.A., Rimmer, S.R. and Borhan, M.H. (2013) The *Brassica napus* blackleg resistance gene *LepR3* encodes a receptor-like protein triggered by the *Leptosphaeria maculans* effector AVRLM1. *New Phytol.* **197**, 595–605.

Larkan, N.J., Lydiate, D.J., Yu, F., Rimmer, S.R. and Borhan, M.H. (2014) Co-localisation of the blackleg resistance genes *Rlm2* and *LepR3* on *Brassica napus* chromosome A10. *BMC Plant Biol.* **14**, 1595.

Leflon, M., Brun, H., Eber, F., Delourme, R., Lucas, M.O., Vallée, P., Ermel, M., Balesdent, M.H. and Chèvre, A.M. (2007) Detection, introgression and localization of genes conferring specific resistance to *Leptosphaeria maculans* from *Brassica rapa* into *B. napus*. *Theor. Appl. Genet.* **115**, 897–906.

Liebrand, T.W.H., Van Den Berg, G.C.M., Zhang, Z., Smit, P., Cordewener, J.H.G., America, A.H.P., Sklenar, J., Jones, A.M.E., Tameling, W.I.L., Robatzek, S., Thomma, B.P.H.J. and Joosten, M.H.A.J. (2013) Receptor-like kinase SOBIR1/EVR interacts with receptor-like proteins in plant immunity against fungal infection. *Proc. Natl Acad. Sci. USA*, **110**, 10010–10015.

Liebrand, T.W.H., van den Burg, H.A. and Joosten, M.H.A.J. (2014) Two for all: Receptor-associated kinases SOBIR1 and BAK1. *Trends Plant Sci.* **19**, 123–132.

Ma, L., Lukasik, E., Gawehns, F. and Takken, F.L.W. (2012) The use of agroinfiltration for transient expression of plant resistance and fungal effector proteins in *Nicotiana benthamiana* leaves. *Methods in Molecular Biology*, **835**, 61–74.

Matsushima, N., Miyashita, H., Mikami, T. and Kuroki, Y. (2010) A nested leucine rich repeat (LRR) domain: the precursor of LRRs is a ten or eleven residue motif. *BMC Microbiol.* **10**, 235.

Mun, J.H., Kwon, S.J., Yang, T.J., Seol, Y.J., Jin, M., Kim, J.A., Lim, M.H., Kim, J.S., Baek, S., Choi, B.S., Yu, H.J., Kim, D.S., Kim, N., Lim, K.B., Lee, S.I., Hahn, J.H., Lim, Y.P., Bancroft, I. and Park, B.S. (2009) Genome-wide comparative analysis of the *Brassica rapa* gene space reveals genome shrinkage and differential loss of duplicated genes after whole genome triplication. *Genome Biol.* **10**, R111.

Nakagawa, T., Suzuki, T., Murata, S., Nakamura, S., Hino, T., Maeo, K., Tabata, R., Kawai, T., Tanaka, K., Niwa, Y., Watanabe, Y., Nakamura, K., Kimura, T. and Ishiguro, S. (2007) Improved gateway binary vectors: high-performance vectors for creation of fusion constructs in transgenic analysis of plants. *Biosci. Biotechnol. Biochem.* **71**, 2095–2100.

Nelson, M.N. and Lydiate, D.J. (2006) New evidence from *Sinapis alba* L. for ancestral triplication in a crucifer genome. *Genome*, **49**, 230–238.

Offord, V. and Werling, D. (2013) LRRfinder2.0: a webserver for the prediction of leucine-rich repeats. *Innate Immun.* **19**, 398–402.

Oliva, R., Win, J., Raffaele, S., Boutemy, L., Bozkurt, T.O., Chaparro-Garcia, A., Segretin, M.E., Stam, R., Schornack, S., Cano, L.M., van Damme, M., Huitema, E., Thines, M., Banfield, M.J. and Kamoun, S. (2010) Recent developments in effector biology of filamentous plant pathogens. *Cell. Microbiol.* **12**, 705–715.

Parkin, I.A.P., Gulden, S.M., Sharpe, A.G., Lukens, L., Trick, M., Osborn, T.C. and Lydiate, D.J. (2005) Segmental structure of the *Brassica napus* genome based on comparative analysis with *Arabidopsis thaliana*. *Genetics*, **171**, 765–781.

Parlange, F., Daverdin, G., Fudal, I., Kuhn, M.-L., Balesdent, M.-H., Blaise, F., Grezes-Besset, B. and Rouxel, T. (2009) *Leptosphaeria maculans* avirulence gene *AvrLm4-7* confers a dual recognition specificity by the *Rlm4* and *Rlm7* resistance genes of oilseed rape, and circumvents *Rlm4*-mediated recognition through a single amino acid change. *Mol. Microbiol.* **71**, 851–863.

Pietraszewska-Bogiel, A., Lefebvre, B., Koini, M.A., Klaus-Heisen, D., Takken, F.L.W., Geurts, R., Cullimore, J.V. and Gadella, T.W.J. (2013) Interaction of *Medicago truncatula* Lysin motif receptor-like kinases, NFP and LYK3,

produced in *Nicotiana benthamiana* induces defence-like responses. *PLoS ONE*, **8**, e65055.

Raman, H., Raman, R. and Larkan, N. (2013) Genetic dissection of blackleg resistance loci in rapeseed (*Brassica napus* L.). In *Plant Breeding from Laboratories to Fields* (Andersen, S.B., ed.), pp. 85–120. Rijeka, Croatia: InTech.

Rioja, C., Van Wees, S.C., Charlton, K.A., Pieterse, C.M.J., Lorenzo, O. and García-Sánchez, S. (2013) Wide screening of phage-displayed libraries identifies immune targets in planta. *PLoS ONE*, **8**, e54654.

Solovyev, V., Kosarev, P., Seledsov, I. and Vorobyev, D. (2006) Automatic annotation of eukaryotic genes, pseudogenes and promoters. *Genome Biol.* **7**(Suppl 1), S10.11–S10.12.

Stotz, H.U., Mitrousia, G.K., de Wit, P.J.G.M. and Fitt, B.D.L. (2014) Effector-triggered defence against apoplastic fungal pathogens. *Trends Plant Sci.* **19**, 491–500.

Thomas, C.M., Jones, D.A., Parniske, M., Harrison, K., Balint-Kurti, P.J., Hatzixanthis, K. and Jones, J.D.G. (1997) Characterization of the tomato *Cf-4* gene for resistance to *Cladosporium fulvum* identifies sequences that determine recognitional specificity in Cf-4 and Cf-9. *Plant Cell*, **9**, 2209–2224.

Van de Wouw, A.P., Lowe, R.G.T., Elliott, C.E., Dubois, D.J. and Howlett, B.J. (2014) An avirulence gene, *AvrLmJ1*, from the blackleg fungus, *Leptosphaeria maculans*, confers avirulence to *Brassica juncea* cultivars. *Mol. Plant Pathol.* **15**, 523–530.

Van der Hoorn, R.A.L., Roth, R. and De Wit, P.J.G.M. (2001) Identification of distinct specificity determinants in resistance protein Cf-4 allows construction of a Cf-9 mutant that confers recognition of avirulence protein AVR4. *Plant Cell*, **13**, 273–285.

Van Der Hoorn, R.A.L., Wulff, B.B.H., Rivas, S., Durrant, M.C., Van Der Ploeg, A., De Wit, P.J.G.M. and Jones, J.D.G. (2005) Structure-function analysis of Cf-9, a receptor-like protein with extracytoplasmic leucine-rich repeats. *Plant Cell*, **17**, 1000–1015.

Voinnet, O., Rivas, S., Mestre, P. and Baulcombe, D. (2003) An enhanced transient expression system in plants based on suppression of gene silencing by the p19 protein of tomato bushy stunt virus. *Plant J.* **33**, 949–956.

Wang, X., Wang, H., Wang, J., Sun, R., Wu, J., Liu, S., Bai, Y., Mun, J., Bancroft, I., Cheng, F., Huang, S., Li, X., Hua, W., Wang, J., Wang, X., Freeling, M., Pires, J.C., Paterson, A.H., Chalhoub, B., Wang, B., Hayward, A., Sharpe, A.G., Park, B., Weisshaar, B., Liu, B. and Li, B. (2011) The genome of the mesopolyploid crop species *Brassica rapa*. *Nat. Genet.* **43**, 1035–1039.

Wulff, B.B.H., Thomas, C.M., Smoker, M., Grant, M. and Jones, J.D.J. (2001) Domain swapping and gene shuffling identify sequences required for induction of an Avr-dependent hypersensitive response by the tomato Cf-4 and Cf-9 proteins. *Plant Cell*, **13**, 255–272.

Wulff, B.B.H., Chakrabarti, A. and Jones, D.A. (2009a) Recognitional specificity and evolution in the tomato-*Cladosporium fulvum* pathosystem. *Mol. Plant Microbe Interact.* **22**, 1191–1202.

Wulff, B.B.H., Hesse, A., Tomlinson-Buhot, L., Jones, D.A., De La Peña, M. and Jones, J.D.G. (2009b) The major specificity-determining amino acids of the tomato Cf-9 disease resistance protein are at hypervariable solvent-exposed positions in the central leucine-rich repeats. *Mol. Plant Microbe Interact.* **22**, 1203–1213.

Yang, X., Deng, F. and Ramonell, K.M. (2012) Receptor-like kinases and receptor-like proteins: keys to pathogen recognition and defense signaling in plant innate immunity. *Front. Biol.* **7**, 155–166.

Yu, F., Lydiate, D.J. and Rimmer, S.R. (2005) Identification of two novel genes for blackleg resistance in *Brassica napus*. *Theor. Appl. Genet.* **110**, 969–979.

Yu, F., Lydiate, D.J., Gugel, R.K., Sharpe, A.G. and Rimmer, S.R. (2012) Introgression of *Brassica rapa* subsp. *sylvestris* blackleg resistance into *B. napus*. *Mol. Breed.* **30**, 1495–1506.

Zhang, Z. and Thomma, B.P.H.J. (2013) Structure-function aspects of extracellular leucine-rich repeat-containing cell surface receptors in plants. *J. Integr. Plant Biol.* **55**, 1212–1223.

8

Production of non-glycosylated recombinant proteins in *Nicotiana benthamiana* plants by co-expressing bacterial PNGase F

Tarlan Mamedov*, Ananya Ghosh, R. Mark Jones, Vadim Mett, Christine E. Farrance, Konstantin Musiychuk, April Horsey and Vidadi Yusibov

Fraunhofer USA Center for Molecular Biotechnology, Newark, DE, USA

*Correspondence

email tmammedov@fraunhofer-cmb.org

Keywords: *N*-linked glycosylation, PNGase F, deglycosylation, plant transient expression, recombinant proteins, malaria vaccine candidate Pfs48/45.

Summary

Application of tools of molecular biology and genomics is increasingly leading towards the development of recombinant protein-based biologics. As such, it is leading to an increased diversity of targets that have important health applications and require more flexible approaches for expression because of complex post-translational modifications. For example, *Plasmodium* parasites may have complex post-translationally modified proteins such as Pfs48/45 that do not carry *N*-linked glycans (*Exp. Parasitol.* 1998; **90**, 165.) but contain potential *N*-linked glycosylation sites that can be aberrantly glycosylated during expression in mammalian and plant systems. Therefore, it is important to develop strategies for producing non-glycosylated forms of these targets to preserve biological activity and native conformation. In this study, we are describing *in vivo* deglycosylation of recombinant *N*-glycosylated proteins as a result of their transient co-expression with bacterial PNGase F (Peptide: *N*-glycosidase F). In addition, we show that the recognition of an *in vivo* deglycosylated plant-produced malaria vaccine candidate, Pfs48F1, by monoclonal antibodies I, III and V raised against various epitopes (I, III and V) of native Pfs48/45 of *Plasmodium falciparum*, was significantly stronger compared to that of the glycosylated form of plant-produced Pfs48F1. To our knowledge, neither *in vivo* enzymatic protein deglycosylation has been previously achieved in any eukaryotic system, including plants, nor has bacterial PNGase F been expressed in the plant system. Thus, here, we report for the first time the expression in plants of an active bacterial enzyme PNGase F and the production of recombinant proteins of interest in a non-glycosylated form.

Introduction

N-linked glycosylation is a post-translational modification which is critical for correct folding, stability and biological activity of many proteins including recombinant subunit vaccines and therapeutic proteins produced in heterologous expression systems (Gomord *et al.*, 2010; Wujek *et al.*, 2004). Some eukaryotic (as well as bacterial) proteins may not contain *N*-glycans in the native host, but their proteins may contain multiple potential glycosylation sites that are aberrantly glycosylated when these proteins are expressed in heterologous eukaryotic expression systems, potentially leading to impaired functional activity. Indeed, the attachment of carbohydrates strongly affects the physico-chemical properties of a protein, therefore can alter its essential biological properties such as specific activity, ligand–receptor interactions and immunogenicity and may pose a safety risk when used *in vivo*. For example, aberrant *N*-linked glycosylation of cell surface receptors such as integrins and cadherins has been shown to stimulate carcinoma progression and metastazing (Guo *et al.*, 2002; Partridge *et al.*, 2004), indicating significant changes in these proteins' behaviour. When a malaria vaccine candidate, Pfs48/45, containing seven putative *N*-glycosylation sites, was expressed in a yeast system (*Pichia pastoris*), it did not have a transmission-blocking (TB) activity

and did not induce TB antibodies in mice (Milek *et al.*, 2000). Pfs48/45 was also expressed in plants (*Nicotiana benthamiana*) at Fraunhofer USA Center for Molecular Biotechnology (FhCMB). FhCMB has engineered Pfs48/45 as a poly-histidine (6xHis)-tagged protein, which was purified using immobilized metal ion affinity chromatography (IMAC), but the TB activity of this plant-derived vaccine candidate was low (data not shown). The low TB activity of Pfs48/45 may be associated with an incorrect/altered folding or masking of important epitopes of the protein because of glycosylation. Although according to a recent report (Bushkin *et al.*, 2010) *Plasmodium* has the potential to transmit GlcNAc proteins, it was shown that the native Pfs48/45 protein does not contain *N*-linked glycans (Milek *et al.*, 1998). Because the Pfs48/45 protein has seven putative *N*-linked glycosylation sites, these sites can be aberrantly glycosylated when the protein is expressed in any of the available eukaryotic hosts. Indeed, the C-terminal fragment (containing 10 cysteine residues) of the Pfs48/45 protein of *Plasmodium falciparum* expressed in *Escherichia coli* as a maltose-binding protein fusion has been shown to be correctly folded (when co-expressed with four *E. coli* chaperones) and elicits functional TB antibodies in mice (Outchkourov *et al.*, 2008). However, the microbial expression system has drawbacks: (i) the full-length form of Pfs48/45 (M-Pfs16C, 16 cysteine residues and seven putative glycosylation sites) did not properly fold and had poor solubility and much weaker epitope recognition; (ii) the Pfs48/45 fragments expressed in *E. coli* are rapidly and significantly degraded by host cell proteases in the *E. coli* periplasm; (iii) the expression level is low; and (iv) purified vaccine candidates may not be free from bacterial endotoxin contamination.

Plants, on the other hand, offer several advantages compared to other recombinant protein expression systems; these include the possession of eukaryotic post-translational modification machinery, simple low-cost scale up for manufacturing, and safety of use of plant-derived products in humans or animals because of the lack of any harboured microbial or mammalian pathogens (Mett *et al.*, 2008; Yusibov and Mamedov, 2010). Plant viral vector-based transient expression in *N. benthamiana* plants has been successfully used for a time- and cost-efficient production of recombinant vaccine candidates against various infectious diseases (Farrance *et al.*, 2011; Mett *et al.*, 2011). The biological activity of many therapeutic proteins is dependent on their glycosylation status, and mammalian glycoproteins are efficiently glycosylated when they are expressed in plants. However, plants can also efficiently produce *N*-linked glycosylated recombinant proteins that do not require *N*-linked glycosylation. Thus, for improved functional activity and immunogenicity of certain vaccine candidates and therapeutic proteins including antibodies, it is important to develop a strategy to produce non-glycosylated forms of proteins using transient expression in *N. benthamiana* plants.

Here, we describe a method for producing deglycosylated proteins in plant cells, by transiently expressing bacterial PNGase F (Peptide: *N*-glycosidase F) in combination with a target protein of interest. PNGase F is a 34.8-kDa enzyme secreted by a Gram-negative bacterium *Flavobacterium meningosepticum* (Plummer *et al.*, 1984; Tarentino *et al.*, 1990). It cleaves a bond between the innermost GlcNAc and asparagine residues of high-mannose, hybrid and complex oligosaccharides in *N*-linked glycoproteins, except when the α (1–3) core is fucosylated. In this study, we transiently co-expressed several recombinant proteins including the malaria vaccine candidate Pfs48F1, protective antigen (PA) of *Bacillus anthracis* and an antibody against PA of *B. anthracis* with bacterial PNGase F in *N. benthamiana* plants. We have demonstrated that bacterial PNGase F is fully active *in vivo* and successfully cleaves *N*-linked oligosaccharides in target glycoproteins, resulting in the uniformity of the expressed proteins.

Results

Expression, purification and characterization of bacterial PNGase F produced in *N. benthamiana* plants

The bacterial PNGase F gene sequence encompassing 314 amino acids (the full length of the catalytically active protein without a signal sequence) was optimized for the expression in *N. benthamiana* plants, cloned into the pGRD4 vector (Shoji *et al.*, 2009) and expressed in *N. benthamiana* plants as described in Experimental procedures. All constructs generated and used in this study are schematically shown in Figure 1. The average expression level of PNGase F was approximately 150 mg/kg of fresh leaf biomass. The expression of PNGase F was confirmed by the Western blot analysis using an anti-FLAG monoclonal antibody (mAb) (Figure 2a).

The recombinant enzyme was purified from *N. benthamiana* leaves using an anti-FLAG agarose column as described in Experimental procedures. As shown by SDS–PAGE and Coomassie staining, the purified PNGase F protein was highly homogeneous (Figure 2b).

To evaluate the deglycosylating activity of recombinant PNGase F *in vitro*, the purified plant-produced enzyme was incubated with a glycosylated malaria antigen, the purified Pfs48F1 protein of *P. falciparum*, which was also expressed in *N. benthamiana* plants. The results of the Western blot analysis indicated that purified plant-derived PNGase F was able to deglycosylate Pfs48F1 *in vitro* and that the degree of deglycosylation was proportional to the amount of purified PNGase F used in the incubation reaction (Figure 2c). Similar results were obtained when Pfs48F1 was incubated with a crude extract from PNGase F-expressing plants or commercial PNGase F (New England Biolabs, Ipswich, MA) (Figure 2c).

In addition, *N. benthamiana* plants expressing PNGase F remained healthy at 7, 8 and 9 days post-infiltration (dpi) (data not shown) with no visible symptom development or change in growth when co-expression of target reached the highest level.

In vivo deglycosylation of recombinant Pfs48F, *B. anthracis* PA and an anti-PA antibody in *N. benthamiana* plants by co-expressing PNGase F

To evaluate *in vivo* cleavage of *N*-linked oligosaccharides decorating Pfs48F1, bacterial PNGase F and Pfs48F1 were transiently co-expressed in *N. benthamiana* plants via co-agroinfiltration

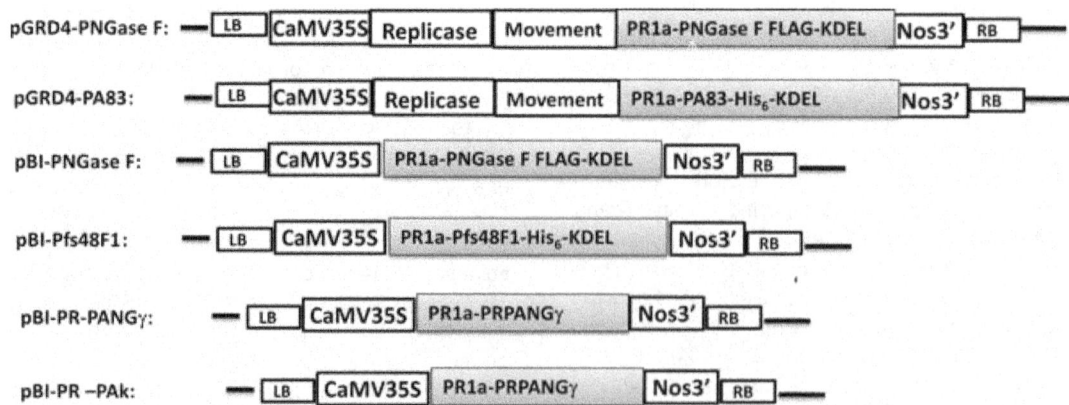

Figure 1 Schematic representation of different plant expression cassettes based on the pGRD4 and binary vectors generated in this study. CaMV35S, Cauliflower mosaic virus promoter; LB, left border; RB, right border; nos, nopaline synthase gene terminator.

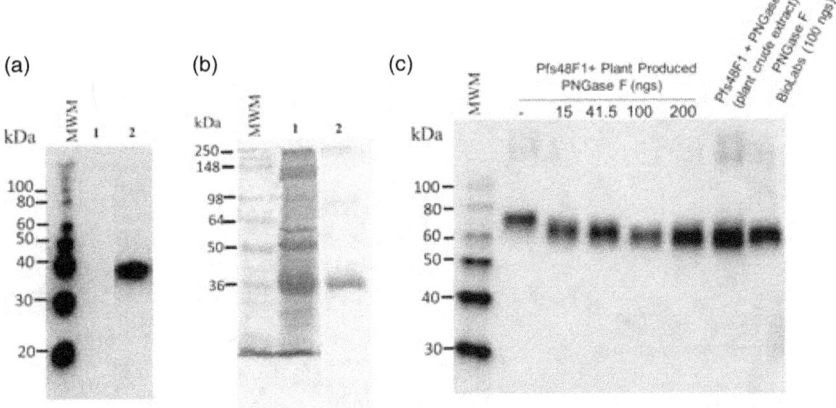

Figure 2 (a) Western blot analysis of bacterial PNGase F produced in *Nicotiana benthamiana* plants. Samples were prepared as described in Experimental procedures. Briefly, leaf samples were taken at 7 dpi and homogenized in three volumes of extraction buffer. After centrifugation at 13 000 **g** for 20 min, 10 μL of 10-fold diluted samples were run on SDS–PAGE prior to Western blotting. PNGase F was detected using the anti-FLAG polyclonal antibody (Sigma). Lanes: 1—crude extract prepared from control plant; 2—crude extract prepared from plant infiltrated with bacterial PNGase F (pGRD4-PNGase F). (b) SDS–PAGE of purified PNGase F. PNGase F was purified from *N. benthamiana* plants using an anti-FLAG agarose column as described in Experimental procedures, and the purified protein was separated by SDS–PAGE and stained with Coomassie. Lanes: 1—TSP; 2—purified PNGase F. (c) Western blot analysis of *in vitro* and *in vivo* Pfs48F1 deglycosylation by plant-produced PNGase F. Purified Pfs48F1 was incubated with either PNGase F expressed in and purified from plants, or with the commercial enzyme (New England Biolabs). The 0.75 μg amounts of the deglycosylated and intact Pfs48F1 protein preparations were run on SDS–PAGE followed by Western blotting. Pfs48F1 was detected using the anti-4xHis tag mAb. MWM refers to molecular weight markers: (a, c) SeeBlue® Plus2 Pre-Stained Standard, Cat. No. LC5925, Invitrogen, Carlsbad, CA; (b) Magic-Mark™ XP Western Protein Standard, Cat. No. LC5602, Invitrogen.

with both pGRD4-PNGase F and pBI121-Pfs48F1 constructs. As shown by SDS–PAGE and the Western blot analysis, co-expression with PNGase F led to the accumulation of Pfs48F1 that was similar in size to that of the *in vitro* deglycosylated molecule (Figure 3a), suggesting that Pfs48F1 was enzymatically deglycosylated (Figure 3a). The expression level of deglycosylated Pfs48F1 co-expressed with PNGase F was about 50 mg/kg of fresh leaf biomass, and solubility was about 95%. This degly-

cosylated form of Pfs48F1 had significantly enhanced rates of recognition by mAbs raised against various epitopes (I, IIb, III and V) of the Pfs48/45 protein (Outchkourov *et al.*, 2007) when measured using the enzyme-linked immunosorbent assay (ELISA, Figure 4). As shown in Figure 4, epitope-specific mAbs I, III and V recognize the deglycosylated form of Pfs48F1 2–6-fold better than the glycosylated form of the same protein. It should also be noted that mAbs I and III are conformation-specific. This

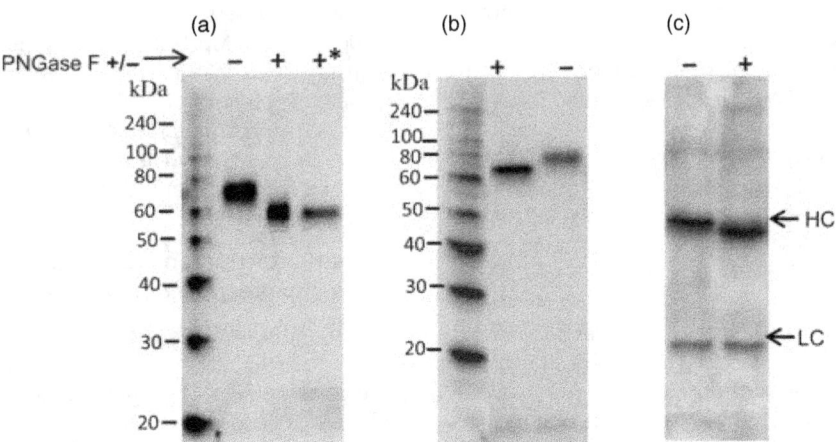

Figure 3 Western blot analysis of co-expression of Pfs48F1 (a), *Bacillus anthracis* PA (b) or an antibody against *B. anthracis* PA (c) with bacterial PNGase F in *N. benthamiana* plants. *N. benthamiana* plants were infiltrated with combinations of the pGRD4-PNGase F/pBI-Pfs48F1, pGRD4-PA83-1/pBI-PNGase F or pBI-PA/pGRD4-PNGase F constructs, for the production of Pfs48F1, PA and the anti-PA antibody, respectively. Leaf samples were taken at 7 (for Pfs48F1), 5 (for PA) and 6 (for the anti-PA antibody) dpi and were homogenized in three volumes of extraction buffer. After centrifugation at 13 000 **g**, samples were diluted 10-fold in the SDS sample buffer. Ten microlitre of samples were run on SDS–PAGE followed by Western blotting. Pfs48F1 bands were probed using the anti-4xHis tag mAb (a). PA bands were probed with the anti-4xHis tag mAb (b), while the anti-PA antibody bands were probed with a rabbit anti-human IgG polyclonal antibody (c). Pfs48F1, PA and the anti-PA antibody were expressed alone (–) and with PNGase F (+). Purified Pfs48F1 was incubated with the commercial PNGase F enzyme (+*, Cat. No. P0704S, New England BioLabs).

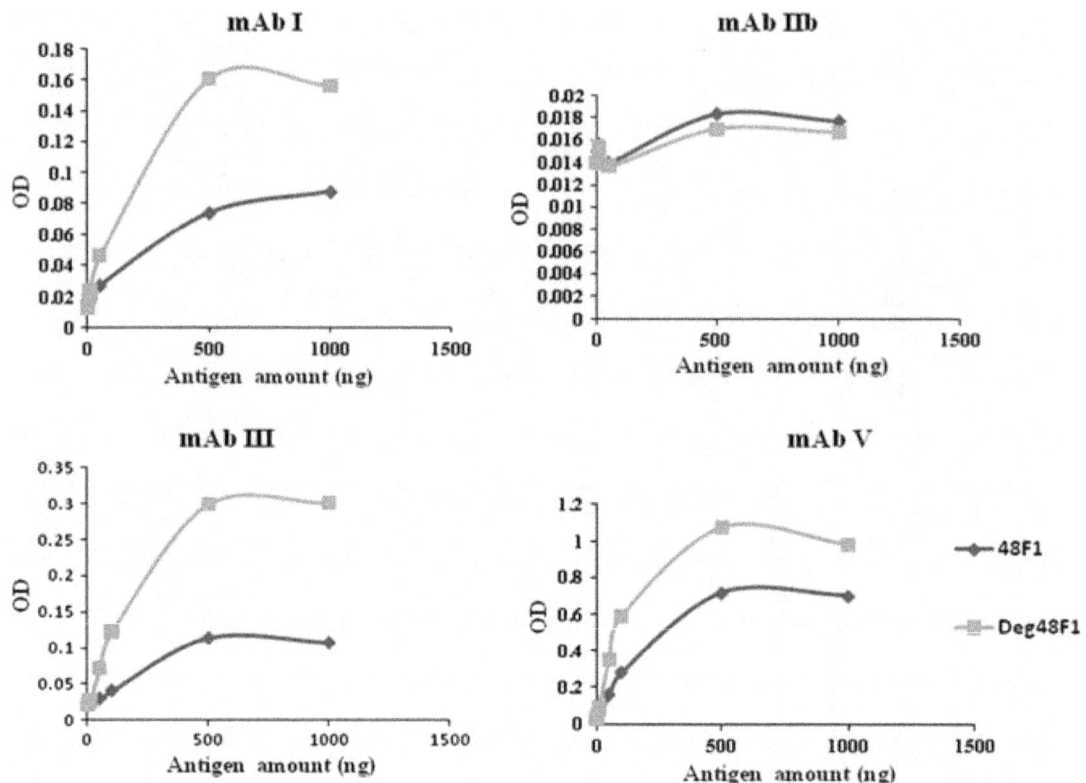

Figure 4 Comparative ELISA analysis of the glycosylated (Pfs48F1, ◇) and deglycosylated forms of Pfs48F1 (■). Recognition of the glycosylated and deglycosylated forms of Pfs48F1 by various rat mAbs that detect epitopes I, IIb, III and V of Pfs48/45 was measured by ELISA as described in Experimental procedures. ELISA plates were coated overnight with the anti-4xHis tag mAb followed by incubation with 1–1000 ng of the glycosylated or deglycosylated forms of the Pfs48F1 antigen for 2 h. Bound antibodies were detected by a HRP-conjugated goat anti-rat polyclonal antibody and visualized using OPD as a substrate. Values shown represent the optical density at the wavelength of 490 nm.

indirectly suggests that, perhaps, deglycosylated Pfs48F1 has native-like folding.

This approach was further tested by co-expressing the enzyme with PA of *B. anthracis* or with an antibody against PA. It should be noted that PA of *B. anthracis* is not a glycoprotein, but it is glycosylated when expressed in *N. benthamiana* plants. The Western blot analysis performed at 5 dpi demonstrated a shift in the mobility of PA co-expressed with PNGase F (Figure 3b), indicating protein deglycosylation. When PNGase F was co-expressed with an anti-PA antibody, the protein mobility shift on SDS–PAGE was observed with the heavy chain (HC) that has one glycosylation site, but not with the light chain (LC) that lacks glycosylation sites (Figure 3c). These results, along with the observation of *in vivo* deglycosylation of Pfs48F1 following co-expression with PNGase F, demonstrate that PNGase F successfully cleaved *N*-linked glycans from all tested glycoproteins and suggest that the PNGase F co-expression strategy can be used to produce therapeutic proteins in a non-glycosylated form in the *N. benthamiana*-based transient expression system.

Mass spectrometry/peptide mapping analysis of glycosylated and deglycosylated Pfs48F1 variants

To further demonstrate *in vivo* deglycosylation of Pfs48F1 co-expressed with PNGase F, we performed the mass spectrometry/peptide mapping analysis of glycosylated and deglycosylated Pfs48F1 variants as described in Experimental procedures. Pfs48F1 contains seven predicted *N*-linked glycosylation sites: N23, N104, N163, N177, N227, N272 and N276.

Mass spectrometry data confirmed that all sites in Pfs48F1, except for N104, have some level of glycosylation. The site N104 showed no increase in detection either before or after treatment with PNGase F. The mass spectrometry data for Pfs48F1 co-expressed with PNGase F *in vivo* showed a high level of deamidated Asp (change of Asn to Asp during deglycosylation) similar to that in Pfs48F1 treated *in vitro* with PNGase F. *In vitro* treatment of Pfs48F1 co-expressed with PNGase F (*in vivo* deglycosylated Pfs48F1) showed no enhancement of Asp residues which were deamidated, implying that co-expression with PNGase F offers the same level of deglycosylation as *in vitro* treatment.

Glycan detection analysis of glycosylated and deglycosylated Pfs48F1 variants

To demonstrate the *in vivo* deglycosylation of Pfs48F1, we also performed the glycan detection analysis of glycosylated and deglycosylated Pfs48F1 variants as described in Experimental procedures. As shown in Figure 5a, the glycan was only detected in plant-produced glycosylated Pfs48F1 (lane 1, Figure 5a). Figure 5b shows the Western blot analysis of the same samples (1/10 diluted) probed with an anti-4xHis tag mAb.

Affinity analysis of mAb V binding to glycosylated and PNGase F-deglycosylated Pfs48F1

Solution binding affinity measurements were performed using mAb V and plant-produced glycosylated, *in vitro* deglycosylated (treated with PNGase F) and *in vivo* deglycosylated

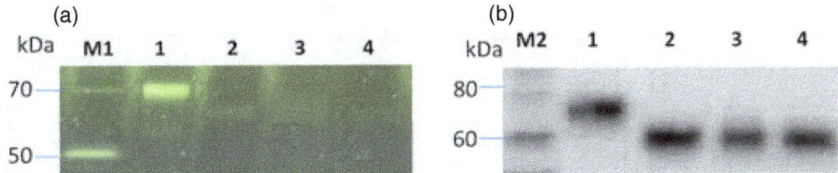

Figure 5 Glycan detection and Western blot analysis of glycosylated and deglycosylated Pfs48F1 variants. (a) 0.25 μg of the plant-produced Pfs48F1 protein from each sample was run on 10% SDS–PAGE followed by in-gel glycan detection using the Pro-Q Emerald 300 glycoprotein stain kit (Cat. No. P33378, Molecular Probes, Grand Island, NY). Stained proteins were visualized using UV illumination. Lanes: 1—glycosylated Pfs48F1, 2—glycosylated Pfs48F1 treated with commercial PNGase F, 3—*in vivo* deglycosylated Pfs48F1, 4—*in vivo* deglycosylated Pfs48F1 treated with commercial PNGase F *in vitro*. (b) Western blot analysis of the same sample using the anti-4xHis tag antibody. PNGase F treatment and Western blot analysis were performed as described above. M1, CandyCane™ glycoprotein molecular weight standards (Molecular Probes), 250 ng of each protein per lane; M2, MagicMark™ XP Western Protein Standard (Cat. No. LC5602; Invitrogen).

(co-expressed with PNGase F) Pfs48F1 as described in Experimental procedures. mAb V was equilibrated at three different concentrations of each antigen. The resulting affinities for glycosylated, and *in vitro* and *in vivo* deglycosylated Pfs48F1 were 9.7 nM (17.5–4.5 nM), 3.8 nM (11.7–1.1 nM) and 2.6 nM (6.3–1.0 nM), respectively (Figure 6). These findings indicate that the affinity of mAb V binding to the deglycosylated form of Pfs48F1 is higher than to the glycosylated form of Pfs48F1, but only a minimal difference exists between the forms of Pfs48F1 deglycosylated *in vitro* and *in vivo*.

The active concentration of mAb V was determined in experiments with glycosylated Pfs48F1 (Figure 6a). From this activity, a ligand concentration multiplier (LCM) was assigned to the two deglycosylated Pfs48F1 versions (Figure 6b,c). This concentration assignment was made because of an unknown nature of the *in vivo* deglycosylated Pfs48F1 concentration. It also normalizes the affinity values based on the activity of the antibody. Without this, any errors in the antigen concentrations would directly impact the calculated dissociation constant (K_d). Because the antigen concentrations are determined based on the antibody activity, exact concentrations of the antigens are not required for analysis. The LCM for *in vitro* deglycosylated Pfs48F1 was determined to be 0.97 or 97% the nominal value. The LCM for *in vivo* deglycosylated Pfs48F1 was 58% the nominal value, which is not unexpected based on the method of the concentration estimation.

Inhibition of mAb III with glycosylated and PNGase F-deglycosylated Pfs48F1

Experiments to compare mAb III binding to each of the Pfs48F1 variants were also performed using the Kinetic Exclusion Assay (KinExA) 3200 instrument (Sapidyne, Boise, ID). Initial tests of the mAb III signal indicated that the binding of mAb III to Pfs48F1 was much weaker compared with mAb V (data not shown). The limited availability of reagents reduced the extent of mAb III titration that could be performed; therefore, we used a single-point inhibition assay to assess differences in the affinity of mAb III to the three forms of Pfs48F1 (glycosylated, *in vivo* deglycosylated and *in vitro* deglycosylated with PNGase F). Solutions containing 140 nM of mAb III (nominal binding site concentration) with 250 nM of one of the three Pfs48F1 forms or without Pfs48F1 were tested using the KinExA 3200 instrument on a single-point inhibition screen. Concentration of Pfs48F1 was made based on the LCM values determined in the experiments with mAb V. Figure 7 shows the results of the mAb III signal inhibition observed when mAb III was incubated

with each of the three Pfs48F1 forms. The strongest binding to and the maximum mAb III signal inhibition were observed with *in vivo* PNGase F-deglycosylated Pfs48F1, while the *in vitro* deglycosylated and glycosylated forms of Pfs48F1 were equivalent in their ability to inhibit the mAb III signal. Although these results are qualitative in nature, they suggest that mAb III has the higher affinity to *in vivo* deglycosylated Pfs48F1 compared to both the *in vitro* deglycosylated and glycosylated forms of Pfs48F1.

Discussion

Recently, an increasing interest in using plants for recombinant protein production, including vaccine antigens, antibodies and therapeutic proteins, has been observed. Plants offer several advantages as a production platform, including eukaryotic post-translational modifications that are required for the biological activity of recombinant therapeutic proteins. Plants also represent an excellent production system because of safety, reduced costs and the eukaryotic-type expression of pharmaceuticals that do not require *N*-glycosylation. von Horsten *et al.* (2010) have reported the production of non-fucosylated antibodies by co-expression of the prokaryotic enzyme GDP-6-deoxy-D-lyxo-4-hexulose reductase. Some eukaryotic proteins (as well as bacterial proteins, for example, PA of *B. anthracis*) contain no *N*-glycans in the native host, but because of the presence of multiple sites for potential glycosylation, these proteins may become aberrantly glycosylated when expressed in plants. Therefore, it is very important to develop a strategy to express these types of recombinant proteins in plants in the non-glycosylated form. It has been previously demonstrated that blocking of *N*-glycosylation by tunicamycin, a specific inhibitor of the enzyme that transfers acetylglucosaminephosphate (GlcNAc-1-P) onto dolichol phosphate (Dol-P), results in a non-uniform expression of proteins in plants (Frank *et al.*, 2008; Hori and Elbein, 1981). However, this strategy may not be useful for the production of recombinant proteins in the non-glycosylated form for a number of reasons: (i) tunicamycin is very toxic; (ii) even short-term treatment of plants with tunicamycin significantly affects protein folding (D'Amico *et al.*, 1992; Sparvoli *et al.*, 2000) and (iii) inhibits extracellular secretion of proteins (Faye and Chrispeels, 1989); and (iv) moreover, a long-term treatment with tunicamycin has a lethal effect on plants (Gomord *et al.*, 2010).

In this study, we have demonstrated for the first time the expression of a deglycosylating bacterial enzyme PNGase F in plants using FhCMB's plant virus-based transient expression

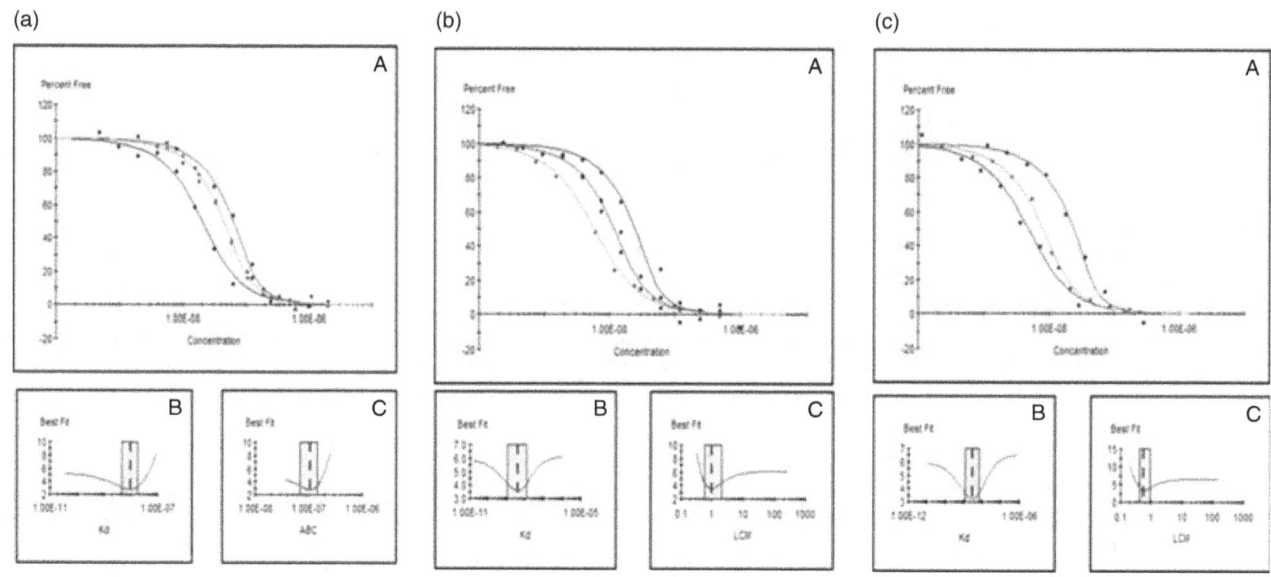

Figure 6 Affinity of mAb V binding to plant-produced Pfs48F1. (a) Affinity of mAb V binding to plant-produced glycosylated Pfs48F1. A. Three-curve binding titration fit for mAb V and glycosylated Pfs48F1. mAb V binding site concentrations were held constant at 107 nM, 69 nM and 22 nM in titrations of Pfs48F1. B. K_d determined for mAb V and glycosylated Pfs48F1 was 9.7 nM (17.5–4.5 nM). C. The active binding site concentration, reported in A, was determined from the three-curve fit to be 107 nM (144–69 nM) for the initial curve. (b) Affinity of mAb V binding to *in vitro* deglycosylated Pfs48F1. A. Three-curve binding titration fit for mAb V and *in vitro* deglycosylated Pfs48F1. mAb V binding site concentrations were held constant at 42.4 nm, 16.2 nM and 4.2 nM in titrations of Pfs48F1. B. K_d determined for mAb V and *in vitro* deglycosylated Pfs48F1 was 3.8 nM (11.7–1.1 nM). C. LCM is 97%, consistent with the nominal concentration after PNGase F treatment. (c) Affinity of mAb V binding to *in vivo* deglycosylated Pfs48F1. A. Three-curve binding titration fit for mAb V and *in vivo* deglycosylated Pfs48F1. mAb V binding site concentrations were held constant at 42.4 nM, 10.6 nM and 4.2 nM in titrations of Pfs48F1. B. K_d determined for mAb V and *in vivo* deglycosylated Pfs48F1 was 2.6 nM (6.3–1.0 nM). C. LCM indicates that the concentration is 54% of that expected from the Western blot analysis.

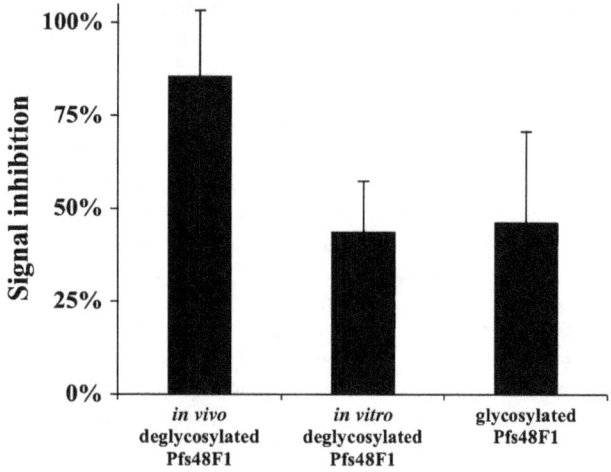

Figure 7 Inhibition of the mAb III signal with glycosylated and PNGase F-deglycosylated Pfs48F1. Pfs48F1 was measured using a single-point inhibition test as described in Experimental procedures. The error bars indicate the range of values from duplicate samples.

spectrometry analyses. The mass spectrometry analysis of Pfs48F1 co-expressed with PNGase F *in vivo* showed a high level of Asn deamidation (Asn23, 163, 177, 227, 272 and 276 of *in vivo* deglycosylated Pfs48F1 were deamidated into Asp) similarly to that in Pfs48F1 treated *in vitro* with PNGase F, indicating the same activity of PNGase F *in vivo* and *in vitro*. In addition, when the deglycosylated and glycosylated forms of Pfs48F1 were compared using ELISA, the deglycosylated form was recognized by mAbs I, III and V at a greater level than the glycosylated form. The binding affinity of mAb V for glycosylated Pfs48F1 and *in vitro* and *in vivo* deglycosylated Pfs48F1 determined using the KinExA 3200 instrument was 9.7 nM, 3.8 nM and 2.7 nM, respectively, suggesting that aberrant glycosylation might have led to masking of important epitopes or caused incorrect/altered folding of the Pfs48F1 protein. In addition, qualitative results of the signal inhibition analysis suggest that mAb III had the highest affinity to *in vivo* deglycosylated Pfs48F1 compared to both *in vitro* deglycosylated and glycosylated Pfs48F1. Based on all these results, we anticipate that the PNGase F deglycosylation strategy would allow for production of a malaria vaccine candidate Pfs48F1 which can provide a high TB activity. In an ongoing animal study, mice were immunized with either a glycosylated or a deglycosylated form of Pfs48F1, and TB activities of their sera are being evaluated.

In plants, most proteins of the extracellular compartment and the endomembrane system are glycosylated, and N-linked glycosylation of proteins has a great impact on their biological functions (Rayon *et al.*, 1998). Therefore, we have tested the deglycosylating effect of plant-expressed PNGase F on endogenous total soluble proteins (TSP) of *N. benthamiana* plants during the short transient expression period. No significant

system and utilized the strategy of *in vivo* enzymatic deglycosylation to produce glycan-free target proteins. We found that the recombinant plant-produced PNGase F is fully active *in vivo* and successfully cleaved N-linked glycans from all tested glycoproteins, including the malaria vaccine candidate Pfs48F1, PA of *B. anthracis* and an antibody against PA of *B. anthracis*.

The efficiency of *in vivo* deglycosylation of the Pfs48F1 protein was demonstrated by the glycan detection and mass difference in SDS–PAGE mobility patterns was observed for TSP from control (non-transformed) plants and plants transformed with bacterial PNGase F (Figure 8). Importantly, *N. benthamiana* plants transformed with PNGase F remained healthy when the target expression reached the maximum level. Taken together, these data suggest that because of the transient nature of expression and brief time span, the effect of PNGase F on the endogenous protein folding and extracellular secretion is not significant. Therefore, the enzymatic *in vivo* deglycosylation strategy can be used to produce therapeutic proteins and antibodies in the non-glycosylated form in *N. benthamiana* plants using the transient expression system. As mentioned above, this is especially important for those proteins that are not glycosylated in their native hosts but become aberrantly glycosylated when expressed in other eukaryotic hosts, potentially leading to reduced functionality and immunogenicity because of incorrect/altered folding and/or masking of epitopes. Our findings can be also important for producing fucose- and xylose-free recombinant proteins, which would reduce a potential risk of immunologic and allergic reactions to these epitopes in humans. We suggest that co-expression with a deglycosylating enzyme such as PNGase F can be applied to various targets including antibodies and vaccine candidates that can be post-translationally modified in the endoplasmic reticulum (ER) of *N. benthamiana* plants. Finally, because the recombinant enzyme PNGase F has not been previously expressed in plants, our results on PNGase F expression support the utility of plants as an expression system for the production of an active, endotoxin-free PNGase F at reduced costs.

Experimental procedures

Cloning, expression and purification of recombinant PNGase F produced in *N. benthamiana* plants, and evaluation of its deglycosylating activity *in vitro* and *in vivo*

The PNGase F gene was optimized for expression in *N. benthamiana* plants (for codon optimization, mRNA stability, etc.) and

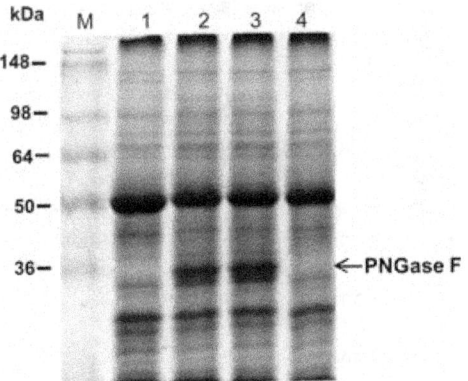

Figure 8 SDS–PAGE of soluble proteins from control (non-transformed) *Nicotiana benthamiana* plants and *N. benthamiana* plants transformed with bacterial PNGase F. Samples were prepared as described in Experimental procedures. Briefly, leaf samples from either control (non-transformed, lane 1) *N. benthamiana* plants or plants transformed with pGRD4-PNGase F (lane 2), pGRD4-PNGase F plus p19 (lane 3) or p19 (lane 3) were taken at 7 dpi and ground in three volumes of extraction buffer. Ten microlitre of the 10-fold diluted samples were run on SDS–PAGE followed by staining with Coomassie.

synthesized by GENEART AG (Regensburg, Germany) with flanking *PacI* (5′ terminus) and *XhoI* (3′ terminus, after stop codon) restriction enzyme sites. To transiently express PNGase F in *N. benthamiana* plants, the signal peptide (amino acids 1–40) was removed from the PNGase F sequence, and *Nicotiana tabacum* PR-1a signal peptide (MGFVLFSQLPSFLLVSTLLLFLVISHSCRA) was added to the N-terminus. In addition, the KDEL sequence (the ER retention signal) and the FLAG epitope (the affinity purification tag) were added to the C-terminus. The resulting sequence was inserted into the launch vector pGRD4 (Shoji *et al.*, 2009) or the binary expression vector pBI121 (Chen *et al.*, 2003) using *PacI/XhoI* sites to obtain pGRD4-PNGase F and pBI-PNGase F, respectively. pGRD4-PNGase F and pBI-PNGase F (together with pSoup for pGRD4-PNGase F, which provides a replication function *in trans* [Hellens *et al.*, 2000]) were then introduced into the *Agrobacterium tumefaciens* strain GV3101. The resulting bacterial strain was grown in BBL medium (10 g/L soy hydrolysate, 5 g/L yeast extract, 5 g/L NaCl, 50 mg/L kanamycin) overnight at 28 °C. The bacteria were introduced by manual infiltration into 6-week-old *N. benthamiana* plants grown in soil. At 5, 6 and 7 dpi, leaf tissues were harvested and homogenized using a bullet blender (Zymo Research, Irvine, CA). Extracts were clarified by centrifugation (13 000 *g* for 30 min) and analysed using Western blotting.

PNGase F was purified from *N. benthamiana* leaves using an anti-FLAG antibody column chromatography. For purification of PNGase F, 8 g of frozen leaves was ground in 24 mL of TBS buffer (50 mM Tris–HCl, pH 7.4, 150 mM NaCl) using a mortar and a pestle. Plant debris was removed by filtration through Miracloth (Calbiochem, San Diego, CA) followed by centrifugation at 13 000 *g* for 10 min and then filtered through a 0.22-μ syringe filter (ACROS Organics, Morris Plains, NJ). An anti-FLAG M2 affinity gel column (Anti-FLAG M2 Affinity gel, Cat. No. A2220; Sigma, St. Louis, MO) was prepared according to the manufacturer's instructions (Sigma). Twenty four millilitre of a clear supernatant were mixed with the gel and rotated at 4 °C for 1 h, after which the whole mixture was returned to the column and the column was washed with 20 volumes of TBS buffer. Bound proteins were eluted using 0.1 M glycine-HCl buffer, pH 3.5 into tubes containing 1 M Tris solution to neutralize glycine buffer.

To assess the deglycosylating activity of the recombinant plant-produced PNGase F *in vitro*, different amounts (10–200 ng) of the purified enzyme in phosphate buffered saline (PBS) were incubated at 37 °C for 1 h with Pfs48F1 expressed in *N. benthamiana* plants and purified by IMAC, and 0.75 μg amounts of glycosylated and *in vitro* PNGase F-treated Pfs48F1 were subjected to SDS–PAGE and the Western blot analysis. Similarly, *in vivo* deglycosylated Pfs48F1 (co-expressed with PNGase F in *N. benthamiana*) was analysed by SDS–PAGE and Western blotting.

Potential deglycosylation of *N. benthamiana* endogenous proteins by plant-produced PNGase F was evaluated by comparing TSP from control (non-transformed) *N. benthamiana* plants and *N. benthamiana* plants expressing bacterial PNGase F, PNGase F plus p19 (plant suppressor of RNA silencing [Voinnet *et al.*, 2003]) or p19 alone on SDS–PAGE. Leaf punches from each sample were homogenized using a bullet blender (Zymo Research) at 4 °C for 2 min, homogenate was centrifuged twice at 13 000 *g* for 10 min, samples were boiled with 1 × SDS sample buffer and 10 μL of each sample was loaded on SDS–PAGE.

Co-expression of PNGase F with Pfs48F1, *B. anthracis* PA and an anti-PA antibody

To co-express PNGase F with Pfs48F1, *B. anthracis* PA or an antibody against PA, pBI-Pfs48F1/pGRD4-PNGase F, pGRD4-PA/pBI-PNGase F or pBI-PA/pGRD4-PNGase F plasmids were constructed and used for infiltration of *N. benthamiana* plants. The sequences of Pfs48F1 (amino acids 28–401, GenBank accession number EU366251), *B. anthracis* PA (amino acids 30–764, GenBank accession number AAA22637) and HC and LC of the anti-PA antibody (Mett *et al.*, 2011) were inserted into the launch vector pGRD4 or the binary expression vector pBI using *PacI/XhoI* sites. All genes described above were codon-optimized for expression in *N. benthamiana* plants. The Pfs48F1 gene optimized for expression in *N. benthamiana* plants encoded the wild-type Pfs48/45 protein (Kocken *et al.*, 1993). All targets were expressed with a signal sequence of *N. tabacum* PR-1a protein at their N-termini. PA and Pfs48F1 were expressed with the 6xHis tag followed by the ER retention signal KDEL at their C-termini. *Agrobacterium* transformation, plant infiltration and leaf protein extraction were performed as described above.

Purification of Pfs48F1 from *N. benthamiana* plants

For purification of the Pfs48F1 recombinant protein from *N. benthamiana* plants, 750 g of plant material was homogenized in 2.25 L of extraction buffer (50 mM sodium phosphate buffer, pH 8.0, 0.5 M NaCl, 20 mM Imidazole, 20% glycerol, 1 mM Dieca) and incubated with 0.5% Triton (final concentration) for 20 min at 4 °C with stirring. After incubation, the lysate was centrifuged at 48 000 *g* for 40 min, and crude extract was filtered through Miracloth and loaded onto column with 70 mL Chelating Sepharose Big Beads charged with Ni. Column was washed with 15 volumes of 50 mM sodium phosphate buffer, pH 8.0, 0.5 M NaCl, 20 mM Imidazole, 20% glycerol. Proteins were eluted with 50 mM sodium phosphate buffer, pH 8.0, 0.5 M NaCl, 300 mM Imidazole, 20% glycerol. Eluted protein fraction was dialysed first against 10 mM sodium phosphate buffer, pH 6.5, 50 mM NaCl, 10 mM EDTA, 10% glycerol and then into 10 mM sodium phosphate buffer, pH 6.5, 10% glycerol. After spin down, dialysed sample was loaded onto a 5-mL Capto Q column equilibrated with 10 mM sodium phosphate buffer, pH 6.5, 10% glycerol. Proteins were eluted with 10 mM sodium phosphate buffer, pH 6.5, 10% glycerol, 600 mM NaCl. Eluted fraction was concentrated up to 2.6 mg/mL and dialysed against PBS, pH 7.5. The purity of Pfs48F1 estimated by Coomassie-stained SDS–PAGE was 72%.

Purification of *in vivo* deglycosylated Pfs48F1 from *N. benthamiana* plants

For purification of *in vivo* deglycosylated recombinant Pfs48F1 protein from *N. benthamiana* plants, 50 g of plant material infiltrated with pBI-Pfs48F1 and pGRD4-PNGase F was homogenized in 150 mL of extraction buffer (50 mM sodium phosphate buffer, pH 8.0, 0.5 M NaCl, 20 mM Imidazole, 1 mM Dieca) and incubated with 0.5% Triton (final concentration) for 20 min at 4 °C with stirring. After incubation, the lysate was centrifuged at 48 000 *g* for 40 min, and crude extract was filtered through Miracloth and loaded onto a 5-mL HisTrap FF column (Cat. No. 17-5255-01; GE Healthcare, Waukesha, WI) followed by washing with 15 volumes of 50 mM sodium phosphate buffer, pH 8.0, 0.5 M NaCl, 20 mM Imidazole. Proteins were eluted with

50 mM sodium phosphate buffer, pH 8.0, 0.5 M NaCl, 100 mM Imidazole. Eluted fraction was concentrated and dialysed against PBS, pH 7.5.

Mass spectrometry/peptide mapping analysis

Plant-produced, partially purified Pfs48F1 samples (Pfs48F1 glycosylated, Pfs48F1 glycosylated treated with commercial PNGase F, *in vivo* deglycosylated Pfs48F1 and *in vivo* deglycosylated Pfs48F1 treated with commercial PNGase F) were electrophoresised on a 10% SDS–PAGE, and the target bands were excised and sent for in-gel enzymatic protein digestion followed by the liquid chromatography—tandem mass spectrometry analysis of the resulting peptide map (MS Bioworks, Ann Arbor, MI). Peptide mapping utilized three different peptidases, trypsin, chymotrypsin and elastase, in an attempt to maximize sequence coverage. The peptide maps were assessed for the presence of *N*-glycan-containing peptides by the shift of deamidated Asn residues to Asp. The gel slices that were digested with trypsin were washed with 25 mM ammonium bicarbonate followed by an acetonitrile wash. Samples were then reduced with 10 mM dithiothreitol at 60 °C followed by alkylation with 50 mM iodoacetamide at room temperature. Trypsin was added, and digestion was performed for 4 h at 37 °C. The reaction was quenched with formic acid, and the supernatant was analysed directly without further processing. Chymotrypsin and elastase digestion was performed in a similar manner, except digestion was performed at 37 °C overnight. Quenched supernatants were analysed using a Waters nanoACQUITY high-performance liquid chromatography system (Waters Corporation, Milford, MA) with peaks detected with a Thermo Scientific Orbitrap Velos Pro (Thermo Fisher Scientific, Waltham, MA). Resulting data were searched using a local copy of Mascot (Matrix Science, Boston, MA) in the UniProt *Nicotiana* and *A. tumefaciens* databases against the Pfs48F1 sequence. The Mascot Database (DAT) files were parsed into the Scaffold software (Proteome Software, Portland, OR) for validation, filtering and creation of a non-redundant list per sample.

Glycan detection analysis

The presence of glycans in plant-produced, partially purified Pfs48F1 samples (Pfs48F1 glycosylated, Pfs48F1 glycosylated treated with commercial PNGase F, *in vivo* deglycosylated Pfs48F1, and *in vivo* deglycosylated Pfs48F1 treated with commercial PNGase F) was detected by Pro-Q Emerald 300 glycoprotein staining. For this, 0.25 µg of the plant-produced Pfs48F1 protein from each sample was run on 10% SDS–PAGE followed by the detection of glycans in the gel using the Pro-Q Emerald Glycoprotein Stain Kit according to the manufacturer's protocol (Molecular Probes). Stained proteins were visualized using UV illumination.

Western blot analysis

Samples from infiltrated *N. benthamiana* leaves were separated on 10% SDS–PAGE, transferred onto a polyvinylidene fluoride membrane (Millipore, Billerica, MA) and blocked with 0.5% I-block (Applied Biosystems, Carlsbad, CA). The membrane was then incubated with a primary anti-4xHis tag mAb (for His-tagged proteins; Roche Applied Science, Indianapolis, IN) followed by a secondary horseradish peroxidase (HRP)-conjugated goat anti-mouse polyclonal antibody. For detection of PNGase F, a rabbit anti-FLAG mAb (Cat. No. F2555; Sigma) was used as a primary antibody and a HRP-conjugated anti-rabbit IgG was used as a secondary antibody. Proteins reacting with the anti-4xHis tag and the anti-FLAG mAbs were visualized using Super-Signal West Pico Chemiluminescent Substrate (Pierce, Rockford, IL). The images were taken using the GeneSnap software on a GeneGnome and quantified using the Gene Tools software (Syngene Bioimaging, Frederick, MD).

Comparative ELISA analysis

Recognition of the deglycosylated and glycosylated forms of Pfs48F1 by rat mAbs raised against various epitopes (I, IIb, III and V) of the *P. falciparum* surface protein Pfs48/45 (Out-chkourov *et al.*, 2007; Roeffen *et al.*, 2001) was assessed using ELISA. ELISA plates (96-well MaxiSorp plates [NUNC, Rochester, NY]) were coated with the anti-4xHis tag mAb (Cat. No. 34670; Qiagen, Valencia, CA) in PBS at 50 μL/well (5 μg/mL) overnight at 4 °C. After blocking with 0.5% I-block in PBS, desired amounts (1–1000 ng) of the deglycosylated and glycosylated forms of Pfs48F1 were added and incubated for 2 h. After washing plates, 50 μL (2 μg/mL in I-block) of various mAbs raised against Pfs48/45 was added and incubated for 2 h. Bound antibodies were detected using a HRP-conjugated goat anti-rat polyclonal antibody (1 : 25 000 in I-block) and visual-ized using o-phenylenediamine (OPD) as a substrate, at the wavelength of 490 nm.

Affinity analysis of mAb V binding to glycosylated and deglycosylated Pfs48F1 variants

The K_d of mAb V binding to glycosylated and deglycosylated (*in vitro* and *in vivo*) variants of Pfs48F1 were assessed using the KinExA 3200 instrument (Sapidyne, Boise, ID; Blake *et al.*, 1999) by determining the amount of free antibody remaining in the solution after equilibration with the respective Pfs48F1 bind-ing partner was reached. mAb V concentrations were held con-stant while the selected glycosylated or deglycosylated Pfs48F1 protein was serially diluted. mAb V was mixed with glycosylated Pfs48F1 at 107 nM, 69 nM and 22 nM; with *in vitro* deglycosylat-ed Pfs48F1 at 42.4 nM, 16.2 nM and 4 nM; and with *in vivo* de-glycosylated Pfs48F1 at 42.4 nM, 10.6 nM and 4 nM. Running buffer was 1x PBS, pH 7.5 with 0.02% sodium azide as a pre-servative. The samples and the secondary antibody were diluted using running buffer augmented with 1 mg/mL bovine serum albumin (BSA). The secondary antibody was Dylight 649-conju-gated goat anti-rat (Jackson ImmunoResearch, West Grove, PA) used at 0.5 μg/mL. PNGase-treated Pfs48F1 was used as the coating reagent for the flow cell bead pack (PMMA beads; Sapidyne) in all experiments. The flow rate for all samples and for the labelling antibody was 0.25 mL/min. Sample volume ranged from 0.35 to 3.0 mL.

Titration data resulting from three different antibody concen-trations equilibrated with each Pfs48F1 were fit to a global 1 : 1 binding model included in the KinExA software (version 3.1.2; Sapidyne, Boise, ID). Data generated for mAb V and Pfs48F1 were fit to the standard 1 : 1 binding model where K_d, active binding site concentration (ABC), and signal maximums and minimums were determined. In this case, the concentration of Pfs48F1 (glycosylated) was determined by densitometry vs. BSA using Coomassie-stained SDS–PAGE. Data generated for mAb V with either deglycosylated version of Pfs48F1 were fit to the same 1 : 1 binding model, but using an unknown antigen concentration model (Xie *et al.*, 2005). In this model, the K_d

and signal parameters were determined from the fit as well as the LCM. The LCM relates an unknown concentration of the ligand to the determined ABC (from the experiment with mAb V and Pfs48F1). This allows for directly comparing binding data when the concentration of antigen is less well characterized or unknown. In this case, the concentration of *in vivo* deglycosylat-ed Pfs48F1 was estimated using Western blotting, so the bind-ing analysis was standardized to the concentration of glycosylated Pfs48F1.

Qualitative analysis of mAb III inhibition by glycosylated and deglycosylated Pfs48F1 variants

Qualitative analysis of mAb III inhibition by the Pfs48F1 variants (glycosylated and PNGase F-deglycosylated *in vitro* and *in vivo*) was performed using the KinExA 3200 instrument. Solutions containing mAb III at 140 nM binding site and 70 nM nominal protein concentrations and one of the Pfs48 variants (glycosylat-ed or PNGase F-deglycosylated) at a final concentration of 250 nM were prepared and incubated, and the amount of free mAb III in each reaction mixture was determined using the KinExA instrument. The initial concentration of each Pfs48F1 variant in the reaction mixture was based on the LCM value determined in the mAb V experiments. The signal generated by mAb III in the solution without Pfs48F1 was considered as 100%.

Acknowledgements

The authors are grateful to Dr. Stephen J. Streatfield for useful discussion and Dr. Natasha Kushnir for editorial assistance. This study was supported by Fraunhofer USA Center for Molecular Biotechnology. The authors have no conflict of interest to declare.

References

Blake, R.C. 2nd, Pavlov, A.R. and Blake, D.A. (1999) Automated kinetic exclusion assays to quantify protein binding interactions in homogeneous solution. *Anal. Biochem.* **272**, 123–134.

Bushkin, G.G., Ratner, D.M., Cui, J., Banerjee, S., Duraisingh, M.T., Jennings, C.V., Dvorin, J.D., Gubbels, M.J., Robertson, S.D., Steffen, M., O'Keefe, B.R., Robbins, P.W. and Samuelson, J. (2010) Suggestive evidence for Darwinian Selection against asparagine-linked glycans of *Plasmodium falciparum* and *Toxoplasma gondii*. *Eukaryot. Cell*, **9**, 228–241.

Chen, P., Wang, C., Soong, S. and To, K. (2003) Complete sequence of the binary vector pBI121 and its application in cloning T-DNA insertion from transgenic plants. *Mol. Breed.* **11**, 287–293.

D'Amico, L., Valsasina, B., Daminati, M.G., Fabbrini, M.S., Nitti, G., Bollini, R., Ceriotti, A. and Vitale, A. (1992) Bean homologs of the mammalian glucose-regulated proteins: induction by tunicamycin and interaction with newly synthesized seed storage proteins in the endoplasmic reticulum. *Plant J.* **2**, 443–455.

Farrance, C.E., Chichester, J.A., Musiychuk, K., Shamloul, M., Rhee, A., Manceva, S.D., Jones, R.M., Mamedov, T., Sharma, S., Mett, V., Streatfield, S.J., Roeffen, W., van de Vegte-Bolmer, M., Sauerwein, R.W., Wu, Y., Muratova, O., Miller, L., Duffy, P., Sinden, R. and Yusibov, V. (2011) Antibodies to plant-produced *Plasmodium falciparum* sexual stage protein Pfs25 exhibit transmission blocking activity. *Hum. Vaccin.* **7**, 191–198.

Faye, L. and Chrispeels, M.J. (1989) Apparent Inhibition of beta-Fructosidase Secretion by Tunicamycin May Be Explained by Breakdown of the Unglycosylated Protein during Secretion. *Plant Physiol.* **89**, 845–851.

Frank, J., Kaulfürst-Soboll, H., Rips, S., Koiwa, H. and von Schaewen, A. (2008) Comparative analyses of Arabidopsis complex glycan1 mutants and genetic interaction with staurosporin and temperature sensitive 3a. *Plant Physiol.* **148**, 1354–1367.

Gomord, V., Fitchette, A.C., Menu-Bouaouiche, L., Saint-Jore-Dupas, C., Plasson, C., Michaud, D. and Faye, L. (2010) Plant-specific glycosylation patterns in the context of therapeutic protein production. *Plant Biotechnol. J.* **8**, 564–587.

Guo, H.B., Lee, I., Kamar, M., Akiyama, S.K. and Pierce, M. (2002) Aberrant *N*-glycosylation of beta1 integrin causes reduced alpha5beta1 integrin clustering and stimulates cell migration. *Cancer Res.* **62**, 6837–6845.

Hellens, R.P., Edwards, E.A., Leyland, N.R., Bean, S. and Mullineaux, P.M. (2000) pGreen: a versatile and flexible binary Ti vector for Agrobacterium-mediated plant transformation. *Plant Mol. Biol.* **42**, 819–832.

Hori, H. and Elbein, A. (1981) Tunicamycin Inhibits Protein Glycosylation in Suspension Cultured Soybean Cells. *Plant Physiol.* **67**, 882–886.

von Horsten, H.H., Ogorek, C., Blanchard, V., Demmler, C., Giese, C., Winkler, K., Kaup, M., Berger, M., Jordan, I. and Sandig, V. (2010) Production of non-fucosylated antibodies by co-expression of heterologous GDP-6-deoxy-D-lyxo-4-hexulose reductase. *Glycobiology*, **20**, 1607–1618.

Kocken, C.H., Jansen, J., Kaan, A.M, Beckers, P.J., Ponnudurai, T., Kaslow, D.C., Konings, R.N. and Schoenmakers, J.G. (1993) Cloning and expression of the gene coding for the transmission blocking target antigen Pfs48/45 of *Plasmodium falciparum*. *Mol. Biochem. Parasitol.* **61**, 59–68.

Mett, V., Farrance, C.E., Green, B.J. and Yusibov, V. (2008) Plants as biofactories. *Biologicals*, **36**, 354–358.

Mett, V., Chichester, J.A., Stewart, M.L., Musiychuk, K., Bi, H., Reifsnyder, C.J., Hull, A.K., Albrecht, M.T., Goldman, S., Baillie, L.W. and Yusibov, V. (2011) A non-glycosylated, plant-produced human monoclonal antibody against anthrax protective antigen protects mice and non-human primates from *B. anthracis* spore challenge. *Hum. Vaccin.* **7**, 183–190.

Milek, R.L, DeVries, A.A., Roeffen, W.F., Stunnenberg, H., Rottier, P.J. and Konings, R.N. (1998) *Plasmodium falciparum*: heterologous synthesis of the transmission-blocking vaccine candidate Pfs48/45 in recombinant vaccinia virus-infected cells. *Exp. Parasitol.* **90**, 165–174.

Milek, R.L., Stunnenberg, H.G. and Konings, R.N. (2000) Assembly and expression of a synthetic gene encoding the antigen Pfs48/45 of the human malaria parasite *Plasmodium falciparum* in yeast. *Vaccine*, **18**, 1402–1411.

Outchkourov, N., Vermunt, A., Jansen, J., Kaan, A., Roeffen, W., Teelen, K., Lasonder, E., Braks, A., van de Vegte-Bolmer, M., Qiu, L.Y., Sauerwein, R. and Stunnenberg, H.G. (2007) Epitope analysis of the malaria surface antigen pfs48/45 identifies a subdomain that elicits transmission blocking antibodies. *J. Biol. Chem.* **282**, 17148–17156.

Outchkourov, N.S., Roeffen, W., Kaan, A., Jansen, J., Luty, A., Schuiffel, D., van Gemert, G.J., van de Vegte-Bolmer, M., Sauerwein, R.W. and Stunnenberg, H.G. (2008) Correctly folded Pfs48/45 protein of *Plasmodium falciparum* elicits malaria transmission-blocking immunity in mice. *Proc. Natl. Acad. Sci. USA*, **105**, 4301–4305.

Partridge, E.A., Le Roy, C., Di Guglielmo, G.M., Pawling, J., Cheung, P., Granovsky, M., Nabi, I.R., Wrana, J.L. and Dennis, J.W. (2004) Regulation of cytokine receptors by Golgi *N*-glycan processing and endocytosis. *Science*, **306**, 120–124.

Plummer Jr, T.H.., Elder, J.H., Alexander, S., Phelan, A.W. and Tarentino, A.L. (1984) Demonstration of peptide:*N*-glycosidase F activity in endo-beta-*N*-acetylglucosaminidase F preparations. *J. Biol. Chem.* **259**, 10700–10704.

Rayon, C., Lerouge, P. and Faye, L. (1998) The protein *N*-glycosylation in plants. *J. Exp. Bot.* **49**, 1463–1472.

Roeffen, W., Teelen, K., van As, J., vd Vegte-Bolmer, M., Eliing, W. and Sauerwein, R. (2001) *Plasmodium falciparum*: production and characterization of rat monoclonal antibodies specific for the sexual-stage Pfs48/45 antigen. *Exp. Parasitol.* **97**, 45–49.

Shoji, Y., Bi, H., Musiychuk, K., Rhee, A., Horsey, A., Roy, G., Green, B., Shamloul, M., Farrance, C.E., Taggart, B., Mytle, N., Ugulava, N, Rabindran, S., Mett, V., Chichester, J.A. and Yusibov, V. (2009) Plant-derived hemagglutinin protects ferrets against challenge infection with the A/Indonesia/05/05 strain of avian influenza. *Vaccine*, **27**, 1087–1092.

Sparvoli, F., Faoro, F., Daminati, M.G., Ceriotti, A. and Bollini, R. (2000) Misfolding and aggregation of vacuolar glycoproteins in plant cells. *Plant J.* **24**, 825–836.

Tarentino, A.L., Quinones, G., Trumble, A., Changchien, L.M., Duceman, B., Maley, F. and Plummer, T.H. Jr. (1990) Molecular cloning and amino acid sequence of peptide-N4-(*N*-acetyl-beta-D glucosaminyl)asparagine amidase from flavobacterium meningosepticum. *J. Biol. Chem.* **265**, 6961–6966.

Voinnet, O., Rivas, S., Mestre, P. and Baulcombe, D. (2003) An enhanced transient expression system in plants based on suppression of gene silencing by the p19 protein of tomato bushy stunt virus. *Plant J.* **33**, 949–956.

Wujek, P., Kida, E., Walus, M., Wisniewski, K.E. and Golabek, A.A. (2004) *N*-glycosylation is crucial for folding, trafficking, and stability of human tripeptidyl-peptidase I. *J. Biol. Chem.* **279**, 12827–12839.

Xie, L., Jones, R.M., Glass, T.R., Navoa, R., Wang, Y. and Grace, M.J. (2005) Measurement of the functional affinity constant of a monoclonal antibody for cell surface receptors using kinetic exclusion fluorescence immunoassay. *J. Immunol. Meth.* **304**, 1–14.

Yusibov, V.M. and Mamedov, T.G. (2010) Plants as an alternative system for expression of vaccine antigens. *Proc. ANAS (Biol. Sci.).* **65**, 195–200.

9

The expression of a recombinant glycolate dehydrogenase polyprotein in potato (*Solanum tuberosum*) plastids strongly enhances photosynthesis and tuber yield

Greta Nölke[1], Marcel Houdelet[1], Fritz Kreuzaler[1], Christoph Peterhänsel[2] and Stefan Schillberg[1,3,*]

[1]*Fraunhofer Institute for Molecular Biology and Applied Ecology IME, Aachen, Germany*
[2]*Institute of Botany, Leibniz-University Hannover, Hannover, Germany*
[3]*Phytopathology Department, Institute for Phytopathology and Applied Zoology, Justus-Liebig University Giessen, Giessen, Germany*

*Correspondence

email stefan.schillberg@ime.fraunhofer.de

Keywords: biomass, carbon metabolism, photorespiration, transgenic plants, tuber size.

Summary

We have increased the productivity and yield of potato (*Solanum tuberosum*) by developing a novel method to enhance photosynthetic carbon fixation based on expression of a polyprotein (DEFp) comprising all three subunits (D, E and F) of *Escherichia coli* glycolate dehydrogenase (GlcDH). The engineered polyprotein retained the functionality of the native GlcDH complex when expressed in *E. coli* and was able to complement mutants deficient for the D, E and F subunits. Transgenic plants accumulated DEFp in the plastids, and the recombinant protein was active *in planta*, reducing photorespiration and improving CO_2 uptake with a significant impact on carbon metabolism. Transgenic lines with the highest DEFp levels and GlcDH activity produced significantly higher levels of glucose (5.8-fold), fructose (3.8-fold), sucrose (1.6-fold) and transitory starch (threefold), resulting in a substantial increase in shoot and leaf biomass. The higher carbohydrate levels produced in potato leaves were utilized by the sink capacity of the tubers, increasing the tuber yield by 2.3-fold. This novel approach therefore has the potential to increase the biomass and yield of diverse crops.

Introduction

Potato is an important nongrain food commodity, with global production exceeding 373 million tonnes in 2011 (FAOSTAT, 2012). It is the primary staple food for more than one billion people worldwide, reflecting its high nutritional value, adaptability to diverse environments and potential yield (Wang, 2008). The tuber starch content also makes potato a versatile and sustainable raw material for the starch and biofuel industries. The additive effects of population growth, the emerging bioenergy economy and the loss of agricultural land to urbanization and land degradation means that greater agricultural productivity is required per hectare of land to meet the demands for food, feed, biomaterials and biofuel (FAO, 2009; OECD/FAO, 2011).

One way to increase the productivity of potato crops is to enhance their photosynthetic efficiency, because carbon metabolism in higher plants drives growth and helps to determine the yield (Long *et al.*, 2006). Previous studies have shown that carbon metabolism can be optimized by genetic engineering (Peterhansel *et al.*, 2008). Enhanced photosynthesis and biomass production has been achieved in tobacco by overexpressing the Calvin cycle enzyme SBPase/FBPase (Lefebvre *et al.*, 2005; Miyagawa *et al.*, 2001). Therefore, enhancing the efficiency of photosynthesis, and thus the amount of fixed carbon could be a straightforward approach to boost the productivity of important agronomic crops.

In C3 plants, photorespiration metabolizes the products of the ribulose-1,5-biphosphate (RuBisCO) oxygenation reaction, and thus reduces the efficiency of photosynthesis by removing carbon from the Calvin cycle, nitrogen and consuming reducing power (reviewed by Maurino and Peterhansel, 2010; Peterhansel and Maurino, 2011). Although the RuBisCO carboxylation reaction produces two molecules of phosphoglycerate that enter the Calvin cycle (ultimately to produce glucose, sucrose and starch and to regenerate ribulose-1,5-biphosphate), the oxygenation reaction produces single molecules of phosphoglycerate and phosphoglycolate, the latter being converted to phosphoglycerate by photorespiration (Leegood *et al.*, 1995; Tolbert, 1997). The balance between these two activities depends mainly on the CO_2/O_2 ratio in the leaves (Laing *et al.*, 1974). Abiotic stress, such as drought and heat, increase the rate of photorespiration. Although photorespiration is necessary for plant growth and survival (Peterhansel and Maurino, 2011), it is an inefficient and wasteful process that lowers the overall efficiency of photosynthesis by removing 25% of the fixed carbon and reduced nitrogen (Wingler *et al.*, 2000). Therefore, reducing metabolic flux through the photorespiration pathway should increase resource-use efficiency, promote growth and increase yields.

The first attempts to increase yields by inhibiting photorespiration were unsuccessful because the photorespiration mutant phenotype was lethal under ambient CO_2 conditions (Somerville, 1984; Somerville, 2001). This is because photorespiration is necessary to protect plants from the accumulation of toxic photosynthesis inhibitors (Campbell and Ogren, 1990; Givan and Kleczkowski, 1992), the effects of excess light (Kozani and Takeba, 1996; Wingler *et al.*, 2000) and to provide reducing power for nitrate assimilation (Bloom *et al.*, 2010; Rachmilevitch *et al.*, 2004).

Photorespiration is an inevitable consequence of the RuBisCO catalytic mechanism, as O_2 and CO_2 compete for the same active site on the RuBisCO enzyme, but its impact on the plant can be reduced by engineering a photorespiratory bypass in the chloroplast (Kebeish *et al.*, 2007; reviewed by Peterhansel *et al.*, 2012), or by the complete oxidation of glycolate to CO_2 in the chloroplast (Fahnenstich *et al.*, 2008; Maier *et al.*, 2012). Stepwise transformation has been used to introduce the bacterial glycolate catabolic pathway into *Arabidopsis thaliana* chloroplasts to convert glycolate into glycerate (Kebeish *et al.*, 2007). The plants were sequentially transformed with genes encoding the three subunits of glycolate dehydrogenase (D, E and F), glyoxylate carboxyligase (G) and tartronic semialdehyde reductase (T). The resulting transgenic plants showed higher levels of net photosynthesis and produced more shoot and root biomass. Similar results were achieved when only the genes encoding glycolate dehydrogenase (DEF), the first enzyme in the pathway, were introduced.

Here, we tested the ability of *Escherichia coli* glycolate dehydrogenase to increase photosynthetic efficiency and biomass accumulation in potato, which is an ideal model because of its high source-to-sink capacity. The three subunits of *E. coli* glycolate dehydrogenase are expressed at different levels (Pellicer *et al.*, 1996), and they assemble to form a functional GlcDH enzyme. The introduction of GlcDH into plants therefore typically requires multiple gene transfer. Data from the transgenic *Arabidopsis thaliana* plants mentioned above showed that the relative scarcity of the F subunit could limit the assembly of the complete DEF complex (Kebeish *et al.*, 2007). The three corresponding bacterial genes (*glcD*, *glcE* and *glcF*) were therefore fused to create a polyprotein construct with intervening flexible linkers. The activity of the engineered GlcDH polyprotein (DEFp) was initially verified in *E. coli*, and then by expression in the plastids of potato plants to test the impact of DEFp on carbon metabolism, photosynthetic efficiency and biomass accumulation.

Results and Discussion

DEFp displays glycolate dehydrogenase activity in *E. coli*

The D, E and F subunits of *E. coli* glycolate dehydrogenase have previously been introduced step-wise into *Arabidopsis thaliana* plants to reduce the loss of fixed carbon and nitrogen during photorespiration (Kebeish *et al.*, 2007). However, to avoid the time-consuming and cumbersome process of multigene transformation, we designed a recombinant glycolate dehydrogenase polyprotein (DEFp) by fusing the corresponding *E. coli glcD* (1493 bps), *glcE* (1046 bps) and *glcF* (1220 bps) cDNAs with intervening flexible (Gly₄Ser)₃ linkers (Figure 1a). This strategy ensured that the three subunits were expressed in stoichiometric amounts, which is not always possible when separate transgenes are expressed (Kebeish *et al.*, 2007). Prior to plant transformation, the engineered DEFp construct was overexpressed in *E. coli* allowing us to carry out rapid glycolate dehydrogenase activity assays and complementation analysis. The 3926 bp multisubunit fusion gene encoding the ~140 kDa DEFp polyprotein was therefore transferred to the bacterial expression vector pET22 downstream of the *pelB* leader peptide using the EcoRI and NotI restriction sites to generate pET-DEFp.

Crude extracts of *E. coli* strain ER2566 overexpressing the DEFp polyprotein were used for enzyme activity assays, with the intact

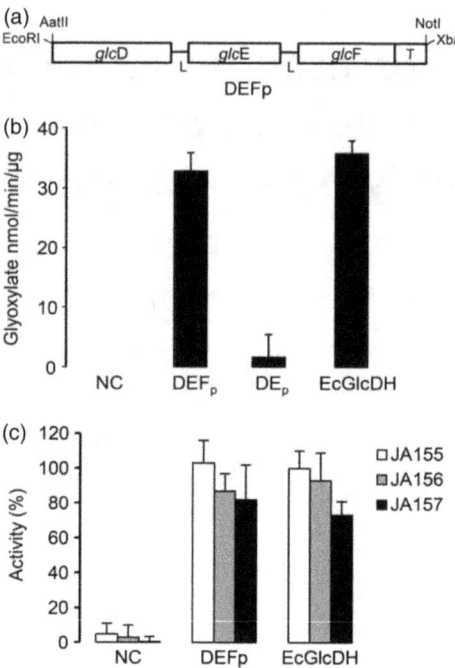

Figure 1 DEFp is functional in *Escherichia coli*. (a) Schematic presentation of the synthetic DEFp multisubunit fusion cassette. Abbreviations: *glcD*, *glcE* and *glcF*: coding sequences for bacterial glycolate dehydrogenase subunits D, E and F; L: (Gly₄Ser)₃ linker; T: His₆ polyhistidine tag. Restriction sites used for cloning of DEFp into bacterial and plant expression vectors are indicated. (b) DEFp shows glycolate dehydrogenase activity *in vitro*. (c) DEFp can complement the glycolate dehydrogenase activity of three *E. coli* glycolate oxidase mutants. Protein was extracted from (b) *E. coli* strain ER2566 and (c) the *E. coli* glycolate oxidase mutants JA155 (*glcD*), JA156 (*glcE*) and JA157 (*glcF*), and was tested for glycolate dehydrogenase activity by detecting glyoxylate formation directly. NC: overexpression of the empty pET (b) or pTrc (c) vectors. DEFp: overexpression of the glcD-glcE-glcF polyprotein. DEp: overexpression of the glcD-glcE polyprotein, EcGlcDH: overexpression of *E. coli* glycolate dehydrogenase subunits glcD-F. In (c), GlcDH activities were expressed as a percentage of the *E. coli* wild-type enzyme, which was set to 100%. The data represent the mean ± standard deviation of three biological replicates.

E. coli glycolate dehydrogenase (*Ec*GlcDH) and DEp polyprotein lacking the F subunit used as controls. As shown in Figure 1b, DEFp showed significant glycolate dehydrogenase activity comparable to that of the intact *E. coli* enzyme. The absence of the F subunit (DEp) completely abolished enzyme activity, confirming the importance of this subunit for the maintenance of GlcDH activity.

We next investigated whether DEFp was able to complement *E. coli* glycolate oxidase mutants carrying transposon insertions in the *glcD*, *glcE* and *glcF* subunits of the *glc* operon. These mutants cannot grow on glycolate as a sole carbon source and do not produce detectable glycolate dehydrogenase activity (Pellicer *et al.*, 1996). The overexpression of DEFp in all three mutant strains restored the ability of the bacteria to grow on glycolate-containing media (data not shown). Furthermore, all three mutants yielded detectable levels of glycolate dehydrogenase activity when transformed with DEFp (Figure 1c). Taken together, these results confirmed that the engineered polyprotein retained the functionality of the native *E. coli* glycolate dehydrogenase complex.

Generation and characterization of DEFp transgenic potato plants

To study the effects of DEFp accumulation *in planta*, the pTRA-35S-rbcs-cTP:DEFp construct (Figure 2a) was introduced into potato plants by *Agrobacterium*-mediated transformation. Transformed plants were regenerated under kanamycin selection and screened for the presence of the DEFp transcript and recombinant protein. Sixty-four transgenic lines (named DEFp-1 to DEFp-64) contained the transgene and produced DEFp with the anticipated molecular size of 142 kDa (Figure 2b). DEFp accumulated to levels between 0.05 and 0.15% of total soluble protein (TSP). Transgenic lines accumulating low (~0.05% TSP, DEFp-27), moderate (~0.09% TSP, DEFp-12) and high (~0.14% TSP, DEFp-21) levels of recombinant DEFp were selected for vegetative propagation and were analyzed for GlcDH activity and photosynthetic performance. The recombinant protein accumulated in the vegetatively propagated plants to similar levels as the parental lines (~0.04% TSP, DEFp-27; ~0.1% TSP, DEFp-12; ~0.15% TSP, DEFp-21).

To investigate whether the DEFp was active *in planta*, GlcDH activity was measured in chloroplast extracts of 6-week-old transgenic plants accumulating different amounts of the polyprotein (lines 12, 21 and 27). Wild-type potato plants assayed under the same conditions showed significant background activity as previously described in *Arabidopsis* (Kebeish *et al.*, 2007).

Chloroplast preparations from the DEFp plants produced significantly more glyoxylate from glycolate than the wild-type plants (Figure 2c), indicating that the engineered DEFp was correctly localized in the plastids and possessed GlcDH activity. The addition of cyanide inhibited GlcDH activity, confirming that the observed activity was due to the recombinant DEFp and not endogenous peroxisomal glycolate oxidase (Nelson and Tolbert, 1970). The DEFp activity correlated with the level of recombinant protein accumulation, i.e. highest in line 21, lower in line 12, and lowest in line 27 (correlation coefficient R^2 = 0.8935, data not shown). Therefore, we concluded that the novel DEFp polyprotein was targeted to the chloroplast and was functional *in planta*.

Photosynthetic activity of DEFp transgenic potato plants

The impact of recombinant DEFp on the photosynthetic performance of the transgenic potato plants was determined by monitoring gas-exchange parameters in the youngest fully expanded leaf. The apparent photosynthetic rate (A) in DEFp lines 12 and 21 (with the highest GlcDH activity) increased by 18% and 34%, respectively, under ambient conditions (400 ppm CO_2, 21% O_2, 24 °C) (Table 1). The maximum CO_2 fixation rate did not change significantly in any of the lines at saturating CO_2 concentration, indicating that the photosynthetic rates under ambient conditions is dependent on the competition of oxygen and CO_2 for binding to RuBisCO, suggesting that the increased photosynthetic rate in the transgenic lines reflects the greater availability of CO_2 in the vicinity of RiBisCO. To estimate the amount of CO_2 competing with O_2 for RuBisCO binding in the plastids, the O_2 inhibition of photosynthesis was calculated in the control plants and DEFp lines 12 and 21 (Table 1). Oxygen inhibition was significantly lower in the transgenic plants, falling 12% in line 12 and 19% in line 21 ($P < 0.05$ and $P < 0.005$, respectively) compared to the control, providing evidence for a

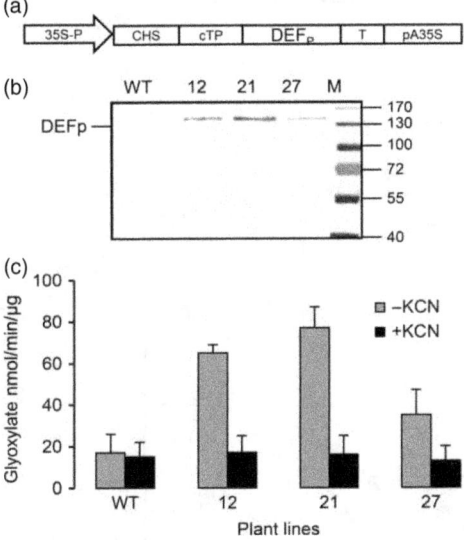

Figure 2 Generation of DEFp transgenic potato plants. (a) Plant expression cassette targeting DEFp to the chloroplast. Abbreviations: 35S-P, double-enhanced CaMV 35S promoter; CHS, 5′-untranslated region of the chalcone synthase gene; cTP: potato *rbc*S1 gene coding for the chloroplast targeting peptide; T: His6 polyhistidine tag; pA35S: CaMV terminator sequence. (b) Immunoblot analysis of 10 μg crude total soluble protein (TSP) extracted from the leaves of DEFp transgenic potato plants. The anti-His6 antibody was used for detection of the 140 kDa DEFp. M: prestained protein marker; 12, 21, and 27: TSP extracts from representative transgenic lines; WT: TSP extract from wild-type potato leaves. (c) DEFp is active in potato chloroplasts. The chloroplast extract was assayed for glycolate dehydrogenase activity by direct measurement of glyoxylate formation in the presence and absence of KCN. WT: wild-type potato plant (*n* = 3); 12, 21, 27: Transgenic potato lines producing DEFp (*n* = 3). The data represent means ± standard deviation.

Table 1 Photosynthetic performance of wild-type and DEFp transgenic potato plants

Photosynthesis parameter	Potato lines		
	WT	DEFp12	DEFp21
A_{max} (μmol/m²/s¹) (at C_a = 400 p.p.m)	16.5 ± 0.9	19.8 ± 0.5*	22.7 ± 1.3**
A_{max} (μmol/m²/s) (at C_a = 2000 p.p.m)	40.6 ± 2.7	41.5 ± 2.4	40.2 ± 1.6
O_2 inhibition (%)	34.9 ± 4.2	30.75 ± 2.1**	28.25 ± 1.3**
Γ (p.p.m CO_2)	49.6 ± 6.3	45.7 ± 1.52**	41.15 ± 3.9**
q_P	0.54 ± 0.07	0.55 ± 0.06	0.53 ± 0.09
q_N	0.77 ± 0.01	0.76 ± 0.02	0.76 ± 0.02
Fv/Fm	0.80 ± 0.01	0.79 ± 0.01	0.79 ± 0.01
ΦPSII	0.25 ± 0.04	0.25 ± 0.05	0.26 ± 0.06

A: apparent CO_2 assimilation; C_a: CO_2 concentration in the measuring cuvette; Γ, apparent CO_2 compensation point; F_v/F_m: maximum quantum efficiency of photosystem II; ΦPSII: efficiency of PSII photochemistry; q_P: photochemical quenching; q_N: nonphotochemical quenching. Vegetatively propagated plants cultivated in the phytotron or greenhouse (six plants per line) were analyzed for each measurement in two independent experiments. Values represent means ± SD for wild-type (WT) control and transgenic lines (*n* = 12). No significant difference in photosynthetic performance was observed between plants grown in the phytotron or greenhouse.

*$P < 0.05$; **$P < 0.005$; ±, standard deviation.

higher CO_2/O_2 ratio in the vicinity of RuBisCO. Furthermore, there were no differences in the maximum quantum efficiency of photosystem II, photochemical and nonphotochemical quenching, and the Fv/Fm ratio. Plastid electron transport, membrane energization and energy dissipation therefore remained unchanged in the transgenic lines. Lines 21 and 12 were characterized by a significant ($P < 0.01$) reduction in the apparent CO_2 compensation point (18% and 9.4%, respectively), indicating a higher rate of photosynthesis at low c_i (supplementary Figure S1). These results provided additional evidence to confirm the functionality of the engineered DEFp enzyme, which was able to catalyze the conversion of glycolate to glyoxylate. The reduction in CO_2 compensation point and O_2 inhibition indicated that the resulting glyoxylate was oxidized further to CO_2 within the plastids. Chloroplasts may oxidize glycolate completely to CO_2, producing glyoxylate as an intermediate step (Frederick et al., 1973; Goyal and Tolbert, 1996; Kebeish et al., 2007). Recently, Blume et al. (2013) reported that the tobacco chloroplast pyruvate dehydrogenase complex (PDC) can decarboxylate glyoxylate and may therefore participate in the conversion of glycolate to CO_2 in the plastid. The CO_2 released due to the activity of the PDC complex could be directly re-fixed by RuBisCO,

generating higher levels of photoassimilates. This hypothesis is supported by the higher carbohydrate content (Figure 3) and biomass production (see below) in DEFp lines. Furthermore, less ammonia would need to be re-assimilated, thus potentially improving nitrogen use efficiency. Preliminary results from our laboratory indicate that DEFp expression in chloroplast of tobacco plants improves their ability to thrive under conditions of nitrogen limitation (unpublished results).

Leaf carbohydrates and other metabolites in DEFp transgenic plants

To determine the impact of DEFp on primary carbon metabolism, we evaluated the ability of transgenic and wild-type plants to accumulate photosynthesis end products using GC-MS to measure the corresponding metabolite pool. We focused on 31 metabolites, but only those selected for physiological importance and/or significant modulation (t-test $P < 0.05$) are shown in Figure 3a,b. Metabolite levels relative to the standard ribitol, are presented in the supplementary Table S1. The DEFp transgenic lines showed a significant ($P < 0.05$) increase in the rapidly metabolized monosaccharaides glucose (2.1/5.8-fold in lines 12/21) and fructose (1.5/3.8-fold in lines 12/21), and the major

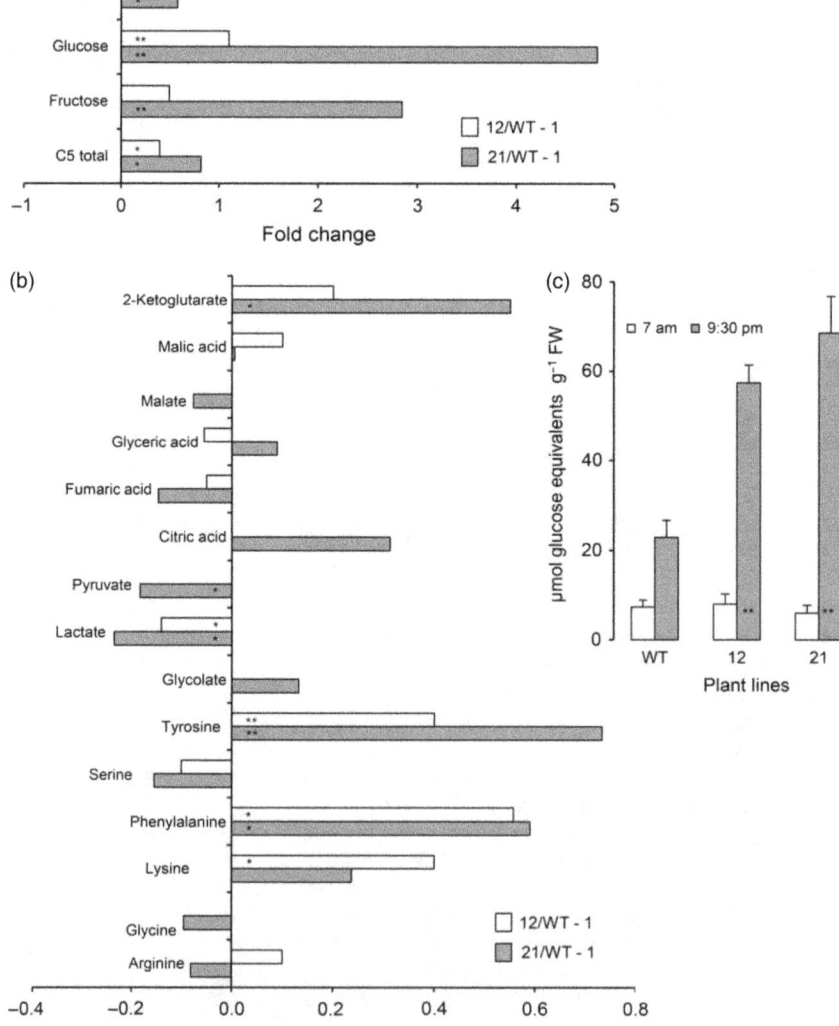

Figure 3 Metabolite levels in the leaves of transgenic and wild-type potato plants. (a) Comparative carbohydrate analysis. (b) Comparative amino and organic acid analysis. Samples were collected 5 h after illumination from five well-expanded leaves representing six plants per transgenic line (and four wild-type controls). The samples from each plant were pooled and analyzed by GC-MS. The ratio between mean relative response from the transgenic lines (12 = white bars, 21 = grey bars) ($n = 6$), and wild-type plants ($n = 4$) are plotted. Values <0 indicate lower metabolite levels in the transgenic lines compared to the control, whereas values >0 represent an increase. The significance of the changes was evaluated using Student's t-test. (c) Starch content of leaves of DEFp and wild-type plants. Starch content was analyzed at the beginning (7 : 00 am) and end (21 : 30 pm) of the illumination period. Data represent means ± standard deviation from three independent pools extracted from five well-expanded leaves of each line ($n = 3$). Statistical differences are indicated: * = $P < 0.05$, ** = $P < 0.005$.

exported disaccharide sucrose (1.3/1.6-fold in lines 12/21). The most dramatic increase was observed in line 21, with the highest DEFp levels and highest GlcDH activity. In addition, the amounts of C5 carbohydrates in these lines were also significantly higher (1.4-fold and 1.8-fold, respectively).

Figure 3b shows the relative levels of additional metabolites. Lactate levels declined in both transgenic lines, whereas pyruvate and maleate levels declined only in line 21. Both DEFp lines accumulated significantly higher amounts of 2-ketoglutarate, tyrosine and lysine than wild-type plants, but there were no significant differences in the levels of glycolate, glyceric and citric acids. The glycine to serine ratio also declined in the transgenic lines compared to controls, suggesting a reduction in the level of photorespiration metabolites by diversion to the peroxisomes and mitochondria, although the tendency was not statistically significant (data not shown).

In many plant species the immediate products of photosynthetic carbon assimilation in the light are divided between sucrose (immediately available for growth) and starch, which accumulates in the leaf through the day and is degraded to produce sucrose at night (Geiger et al., 2000; Gibon et al., 2004; Lu et al., 2005). We analyzed the content of starch (the major storage compound) in the leaves at two different time points: the beginning and end of the illumination period (Figure 3c). The leaves of all lines contained low levels of starch early in the morning but significant increases (2.5-fold/$P < 0.005$ in line 12 and threefold/$P < 0.005$ in line 21) were observed by the end of the day in transgenic lines compared to wild-type control. The greater increase in the transgenic lines is reflected the more efficient assimilation of CO_2. Ferreira et al. (2010) showed that only 70% of the starch synthesized in potato leaves during the day is broken down during the dark phase to provide nocturnal sucrose to sink tissues, whereas 100% is broken down in Arabidopsis (Gibon et al., 2004; Schneider et al., 2002). However, the results presented in Figure 3c show almost complete starch degradation during the night in the best-performing DEFp lines, indicating that the

transitory starch is either converted to sucrose and exported from the leaf to the sink organs, or used for dark respiration and biomass accumulation in the leaf (Graf and Smith, 2011).

The increase in starch content did not occur at the expense of other carbohydrates, because DEFp lines 12 and 21 (with the highest GlcDH activity and photosynthesis rates) also produced higher levels of glucose, fructose and sucrose (Figure 3a). Photosynthetic carbon assimilation was sufficient to support not only growth, but also the accumulation of storage compounds in the leaf, which are then mobilized to provide carbon for growth during the night.

Phenotypic effects of DEFp expression

We also investigated whether the enhanced photosynthetic performance of the transgenic plants resulted in phenotypic differences and higher biomass accumulation. Indeed, tubers harvested from lines 21 and 12 with the highest levels of GlcDH activity, and the most significant reduction in CO_2 compensation point and O_2 inhibition, showed accelerated sprouting (1 week earlier than normal) and developed 25% more ($P < 0.05$; $n = 10$) shoots per plant than control lines. The transgenic lines also produced more (9 ± 2; $P < 0.05$; $n = 10$) leaves and broader stems than the control plants (Figure 4a). Studies on the manipulation of photosynthetic carbon fixation suggest that plants evolved mechanisms to ensure co-ordination of leaf development and metabolism (reviewed by Raines and Paul, 2006). Our results support the idea of Raines and Paul (2006) that changes in photosynthetic carbon assimilation, resulting in increased availability of carbohydrates not only have an impact on plant yield, but also play a role in modulating the developmental programme of the plant.

Although the mechanisms governing release from dormancy and the initiation of sprouting are not completely understood (Suttle, 1996), starch degradation and sucrose biosynthesis are tightly linked to tuber sprouting (Farré et al., 2001). Furthermore, sucrose levels are responsible for the regulation of metabolic

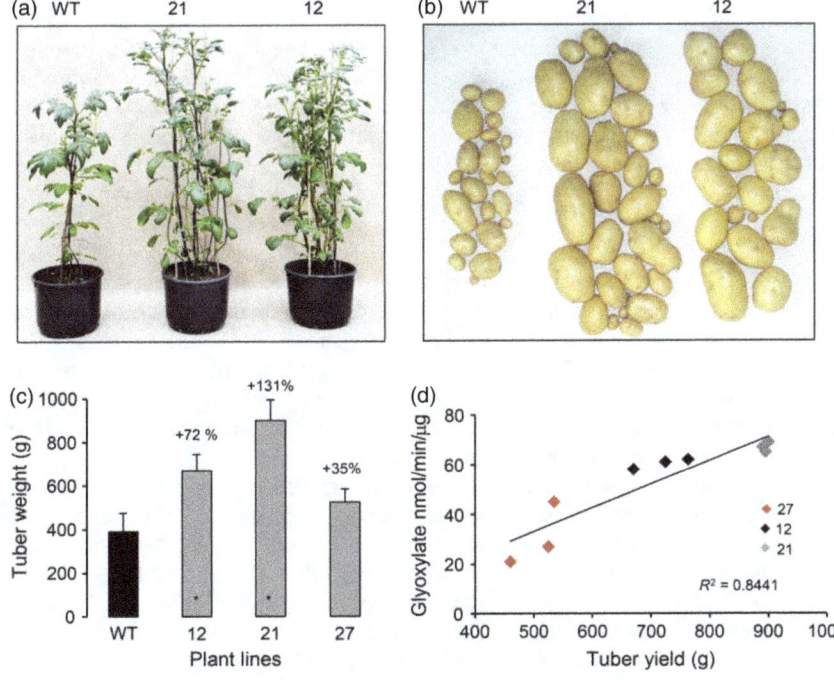

Figure 4 Impact of DEFp expression on potato phenotype and tuber yield. (a) Phenotypic appearance of 8-week-old representative wild-type plant and DEFp lines 21 and 12, planted in soil 6 months after harvest. (b) The phenotype of tubers from one representative wild-type plant, and transgenic lines 12 and 21, 2 days after harvest. (c) Tuber yield per plant in wild-type and DEFp transgenic lines 12, 21 and 27. Data represent means ± standard deviations ($n = 15$). *$P < 0.005$. The increase in tuber yield of the DEFp lines is indicated as a percentage compared to wild-type tubers. (d) Relationship between GlcDH activity (in nmol/min/µg) and tuber yield of transgenic lines 12, 21 and 27. R^2: correlation coefficient. The GlcDH activity of the wild-type control was subtracted from the activity measured in the DEFp lines.

processes during the sink to source transition in potato tubers (Hajirezaei et al., 2003). The metabolic analysis of DEFp-21 tubers showed they contained 1.2-fold higher ($P < 0.05$) levels of sucrose than wild-type tubers, which would affect their sprouting behaviour.

The constitutive production of DEFp led to a significant increase in tuber yield in the transgenic lines (Figure 4b,c), 2.3-fold/ $P < 0.005$ in line 21, followed by 1.7-fold/$P < 0.05$ in line 12 and 1.3-fold (not statistically significant but nevertheless a trend in line 27 (p < 0.1). The tuber yield in the transgenic lines correlated strongly with GlcDH activity ($R^2 = 0.8441$; Figure 4d). These results were consistent in three different experiments carried out over a period of 2 years.

Similar phenotypic effects were observed in transgenic Arabidopsis plants expressing all three subunits of the bacterial glycolate dehydrogenase (Kebeish et al., 2007). The transgenic plants grew faster, produced more shoot and root biomass and contained more soluble sugars, reflecting the increased CO_2 concentration in the vicinity of RuBisCO. Furthermore, the introduction of a complete glycolate catabolic cycle into Arabidopsis chloroplasts reduced the photorespiratory flux, increased the photosynthesis rates and improved plant growth (Maier et al., 2012).

The strong correlation between GlcDH activity, photosynthesis rates and tuber yield suggest that the significant increase in photosynthesis rates and the accumulation of end-product metabolites was able to enhance the sink capacity of DEFp transgenic plants. The higher tuber yield did not occur at the expense of leaf starch and primary metabolites (Figure 3). Sucrose is the major form in which fixed carbon is transported to the sink organs. Thus, the rate of sucrose synthesis is likely to be important for sink development and final crop yield, demonstrating the central role of sugars in the coordination of carbon supply and plant growth as previously described by Smith and Stitt (2007). There was no impact on photosynthetic parameters and tuber yield when photosynthetic sucrose biosynthesis was reduced by suppressing cytosolic fructose-1,6-bisphosphate in potato plants (Zrenner et al., 1996). However, our study demonstrated that a 1.7-fold increase in GlcDH activity led to a significant increase in photosynthesis, boosting the accumulation of assimilates, and ultimately biomass and tuber yield. The yield differences reported here are much higher than expected, but transgenic potatoes modified to simultaneously enhance sugar export form the leaf and sugar uptake in the tuber also had a higher starch content and tuber yield (Jonik et al., 2012). This indicates that substantial changes in yield can be achieved by the targeted manipulation of metabolic pathways.

Conclusion

We have established a powerful approach to increase the biomass of potato plants by improving photosynthetic carbon fixation using a glycolate dehydrogenase polyprotein. The constitutive expression of DEFp in potato chloroplasts boosted photosynthetic efficiency and carbohydrate metabolism. Changes in the photosynthetic capacity of the plants were directly reflected in the phenotype, i.e. more leaves, a thicker stem and a 2.3-fold increase in tuber yield. Molecular and biochemical analysis revealed a strong correlation between GlcDH activity, the photosynthetic performance and overall yield. This is the first study describing such a substantial photorespiratory bypass effect in a crop species. As previously suggested, it may be possible to

improve the yield further by boosting the sink strength (Jonik et al., 2012; Zhang et al., 2008). Further work is required to understand the mode of action of DEFp, the extent of the metabolic changes, and the feasibility of this strategy to improve yields in other crop species.

Experimental procedures

Engineering the recombinant GlcDH polyprotein

The multi-subunit fusion cassette DEFp contained the E. coli genes glcD, glcE and glcF in tandem (E. coli K12 genome sequence: gi49175990) and was codon optimized for Brassica napus and synthesized by Entelechon GmbH (Bad Abbach, Germany). A genetic algorithm was also used to optimize the synthetic genes simultaneously for a large set of competing parameters, such as mRNA secondary structure, cryptic splice sites, codon and motif repeats, and homogenous GC content. The synthesized DEFp-cDNA was transferred to the EcoRI and XbaI restriction sites of pUC18 to generate plasmid pUC-DEFp.

Plasmid DNA, bacteria and plants

The plasmids pET22b(+) (Novagen, Darmstadt, Germany) and pTrc99a (Pharmacia, Freiburg, Germany) were used for protein expression in bacteria, and pTRA was used in plants (Sack et al., 2007). Escherichia coli strain DH5α was used for general cloning and ER2566 (New England Biolabs, Frankfurt, Germany) was used for protein expression. Strains JA155 (araD Δlac rpsL flbB deoC ptsF rbsR glcD::cat), JA156 (araD Δlac rpsL flbB deoC ptsF rbsR glcE::cat) and JA157 (araD Δlac rpsL flbB deoC ptsF rbsR glcF::cat) were used for complementation experiments (Pellicer et al., 1996). Agrobacterium tumefaciens strain GV3101 (pMP90RK, Gm^R, Km^R, Rif^R) was used for plant transformation (Konz and Schell, 1986). Leaf discs from 4–5-week-old wild-type potato plants (Solanum tuberosum cv. Bintje) were transformed by infection with Agrobacterium tumefaciens carrying binary vector pTRA-DEFp (Dietze et al., 1995). Potato plants were cultivated in the phytotron and greenhouse in DE73 standard soil in 13-L pots with a 16-h natural daylight photoperiod and 21/18 °C day/night temperature. In phytotrons, metal-halide lamps at light intensity of 100–110 µE were used. Unless otherwise specified, the data presented here are average of greenhouse conditions. Tubers were harvested when the plants entered senescence. Plants were propagated by cuttings and vegetative multiplication of tubers. Tubers of the same weight (~30 g) were chosen for plant propagation.

Construction of the bacterial and plant expression cassettes

DEF$_p$ was subcloned from pUC18-DEFp into the bacterial expression vector pET22b(+) downstream of the pelB leader peptide using the EcoRI and NotI restriction sites, generating vector pET-DEFp. This was facilitated by shuttling the DEFp cassette into pTRA upstream of the Cauliflower mosaic virus (CaMV) 35S terminator sequence at the EcoRI/XbaI restriction sites. The constitutive CaMV 35S promoter (Kay et al., 1987), the 5′-untranslated region of the chalcone synthase gene and the chloroplast targeting peptide from the potato rbcS1 gene (gi21562) were amplified by PCR using vector pTRAkc-rbcs1-cTP as the template (Kebeish et al., 2007). The amplified 35S-rbcs-cTP PCR product was subcloned into the shuttle vector using AscI/AatII restriction sites introduced by PCR, yielding the final plant expression vector pTRA-35S-rbcs-cTP: DEFp.

Protein expression and complementation analysis

Escherichia coli ER2566 cells (carrying pET-DEFp, pER-DEp, pET-EcGlcDH or empty pET) were cultivated in 200 mL LB medium until the OD_{600} reached 0.5. The pET-EcGlcDH vector contained part of the *E. coli glc* operon encoding the separate glycolate dehydrogenase D, E and F subunits (Bari *et al.*, 2004). Expression was induced by adding β-D-isopropyl-thiogalactopyranoside (IPTG) to a final concentration of 1 mM and incubating for 2 h at 37 °C. The cells were then washed in 10 mM potassium phosphate (pH 8.0) and resuspended in 1 mL of the same buffer. The cells were lysed by sonication on ice (4 cycles, 4 × 30 s; Sonicator Bandelin Sonopulus GM 70, Berlin Germany) and centrifuged at 30 000 ***g*** for 25 min at 4 °C.

Escherichia coli mutants JA155, JA156 and JA157 deficient for *glc*D, *glc*E and *glc*F subunits of endogenous glycolate dehydrogenase, respectively (Pellicer *et al.*, 1996), were transformed with the vectors pTrc99a-DEFp and pTrc-EcGlcDH (Bari *et al.*, 2004) plasmids and cultivated for 2 days in minimal medium (Miller, 1972) using glycolate as the sole carbon source, supplemented with appropriate antibiotics (25 µg/mL chloramphenicol and 100 µg/mL ampicillin). The cells were diluted into 3 L of fresh medium and grown to an OD_{600} of 0.7–0.9. Protein expression was induced by adding 0.5 mM IPTG and the cells were incubated for a further 3 h at 37 °C followed by the extraction of soluble proteins for the glycolate dehydrogenase assay as described below. The concentration of extracted soluble protein was determined in triplicate by the Bradford assay against bovine serum albumin (BSA) standards.

Total protein extraction and immunoblot analysis

The upper fully expanded leaves from 6-week-old potato plants were ground to a fine powder under liquid nitrogen, and total soluble protein (TSP) was extracted as described by Nölke *et al.* (2008) with two volumes of extraction buffer [50 mM Tris-HCl, pH 8, 100 mM NaCl, 10 mM dithiothreitol (DTT), 5 mM ethylenediaminetetraacetic acid (EDTA) and 0.1% (v/v) Tween-20]. The extracts were centrifuged at 8500 ***g*** for 20 min at 4 °C and used for immunoblot analysis. DEFp was detected with a rabbit anti-His_6 monoclonal antibody (RAb-His; 200 ng/mL) and a horseradish peroxidase-conjugated goat-anti-rabbit secondary antibody (GAR^{AP}; 120 ng/mL) (Jackson ImmunoResearch Laboratories, Suffolk, UK). DEFp band intensities were quantified using Aida software (Raytest, Straubenhardt, Germany) against known concentrations of bacterial affinity-purified DEFp as a standard.

Isolation of chloroplasts from potato leaves

Intact chloroplasts were isolated from 6-week-old potato plants as described by Goyal *et al.* (1988). Leaf material (5 g) was ground in 100 mL grinding buffer (50 mM HEPES-KOH pH 7.5, 1 mM $MgCl_2$, 1 mM EDTA, 1 g/L BSA, 0.2 g/L sodium ascorbate, 0.3 M mannitol, 5 g/L polyvinylpyrrolidone) and all subsequent steps were performed in the dark at 0 °C. Crude protein extract was filtrated through three layers of Miracloth and the solution was centrifuged at 1000 ***g*** for 10 min at 0 °C. The pellets were resuspended in 1 mL SH-buffer (50 mM HEPES-KOH pH 7.5, 0.33 M sorbitol) and 1 mL of the solution was loaded onto 8-mL 35% (v/v) Percoll gradient (35% Percoll, 65% SH-buffer). The gradient was centrifuged at 500 ***g*** for 5 min at 0°C. The chloroplast pellet was washed in 1 mL SH-buffer and chloroplast proteins were extracted in 500 µL extraction buffer (50 mM HEPES-NaOH pH 7.5, 2 mM EDTA, 5 mM $MgCl_2$, 0.1% (v/v) Triton

X-100, 20% (v/v) glycerol). The purity (> 95%) of the chloroplast fraction was confirmed by catalase and fumarase activity assays (Figure S2) as previously described (Ferri *et al.*, 1978). The chloroplast proteins were directly used in the glycolate dehydrogenase activity assay.

Glycolate dehydrogenase assay

Glycolate dehydrogenase activity was determined as described by Lord (1972) using 40 µg of bacterial crude protein extract or chloroplast proteins added to 150 µL of buffer (10 mM potassium phosphate, pH 8.0, 0.025 mM phenazine methosulfate, 10 mM potassium glycolate). At fixed time points (1, 2, 3, 4, 5 and 10 min) individual assays were terminated by adding 30 µL 12 M HCl. After incubation for 10 min, 70 µL 0.1 M phenylhydrazine was added and the mixture was again incubated at room temperature for 10 min. The extinction due to the formation of glyoxylate phenylhydrazone was measured at 324 nm in a quartz 96-well microtiter plate using a Synergy HT-I multiplate reader (BioTec, Bad Friedrichshall, Germany).

Gas-exchange measurements

Fully expanded upper leaves from 7-week-old potato plants were used for gas-exchange measurements in LI-6400 system (Li-Cor), as previously described (Kebeish *et al.*, 2007). The following parameters were used: photon flux density 1 000 mmol/m²/s, chamber temperature 26 °C, flow rate 150 mmol/s, relative humidity 60–70%. The oxygen inhibition of carbon assimilation (A) was calculated from A at $C_a = 400$ ppm, and atmospheric oxygen concentrations of 21% and 2%, using the following equation:

$$\text{oxygen-inhibition} (\%) = (1 - A_{21}/A_2) \times 100.$$

The CO_2 compensation point (Γ) was determined by measuring the photosynthesis rates at 400, 300, 200, 100, 80, 60 and 40 ppm CO_2. The apparent CO_2 compensation point (Γ) was deduced from A/Ci curves by regression analysis in the linear range of the curve. Measurements were taken from the same plants after 4 h light on two different days. After these measurements, leaf samples were used to determine glycolate dehydrogenase activity and carbohydrate levels.

Sugar and starch quantities in potato leaves

For sugar analysis, 40 mg of leaf material harvested from five well-expanded leaves from 7-week-old plants after 4 h of illumination was flash frozen in liquid nitrogen and ground in 1 mL prechilled chloroform/methanol/water (2.5/1/1 v/v/v). The extract was incubated at 4 °C for 10 min with moderate shaking and centrifuged at 16 000 ***g*** for 2 min at 4 °C before 500 µL of the supernatant was mixed with 250 µL of water. The samples were centrifuged as above, the top layer was collected and dried in a speed vacuum concentrator and the glucose, fructose and sucrose concentrations were determined enzymatically (Stitt *et al.*, 1989). For starch measurements, 50 mg leaf material was collected in two different time points, at the beginning (7:00 am, after 1.5 h illumination) and at the end of the illumination period (21:30). The frozen leaf material was ground in liquid nitrogen and resuspended in 80% (v/v) ethanol. The extract was mixed for 10 min at 80 °C and centrifuged for 20 min. The pellet was resuspended in 80% (v/v) and 50% (v/v) ethanol, respectively, followed by mixing at 80 °C and centrifugation as above. The resulting pellet was washed with 90% (v/v) ethanol, resuspended in 400 µL 0.2 KOH and incubated at 95 °C for 1 h. Finally, samples were mixed with 70 µL 1 M acetic acid and the starch content was measured enzymatically.

Analysis of metabolites from potato leaves

For metabolite analysis, samples were collected 5 h after illumination from five well-expanded leaves representing six plants per transgenic line and four wild-type controls. The samples from each plant were pooled and the leaf material was homogenized in liquid nitrogen. Metabolites were quantified in 20 mg of homogenized material by GC-MS as described by Lisec et al. (2006). Chromatograms were analyzed using CHROMA TOF software (Leco Corporation, Mönchengladbach, Germany) and TAGFINDER (Luedemann et al., 2008).

Statistical analysis

Significance was determined according to Student's t-test using Excel software (Microsoft). Two-sided tests were performed for homoscedastic matrices.

Acknowledgements

The authors gratefully acknowledge Dr. Flora Schuster for producing the transgenic potato plants, Holger Spiegel for helpful discussion for construct design, Birgit Lippmann for help with GC/MS analysis, and Dr. Richard M Twyman for critical reading of the manuscript. This work was supported by the BMBF—GABI Improve FKZ 0315038C.

References

Bari, R., Kebeish, R., Kalamajka, R., Rademacher, T. and Peterhänsel, C. (2004) A glycolate dehydrogenase in the mitochondria of Arabidopsis thaliana. J. Exp. Bot. **55**, 623–630.

Bloom, A.J., Burger, M., Asensio, J.S.R. and Cousins, A.B. (2010) Carbon dioxide enrichment inhibits nitrate assimilation in wheat and Arabidopsis. Science, **328**, 899–903.

Blume, C., Behrens, C., Eubel, H., Braun, H.-P. and Peterhansel, C. (2013) A possible role for the chloroplast pyruvate dehydrogenase complex in plant glycolate and glyoxylate metabolism. Phytochemistry, **95**, 168–176.

Campbell, W.J. and Ogren, W.L. (1990) Glyoxylate inhibition of ribulosebiphosphate carboxylase-oxygenase: activation in intact, lysed and reconstituted chloroplasts. Photosynth. Res. **23**, 257–268.

Dietze, J., Blau, A. and Willmitzer, L. (1995) Agrobacterium-mediated transformation of potato (Solanum tuberosum). In: Gene Transfer to Plants (Potrykus, I. and Spangenberg, G., eds), pp. 24–29. Berlin: Springer-Verlag.

Fahnenstich, H., Scarpeci, T.E., Valle, E.M., Flügge, U.-I. and Maurino, V.G. (2008) Generation of hydrogen peroxide in chloroplasts of Arabidopsis overexpressing glycolate oxidase as an inducible system to study oxidative stress. Plant Physiol. **148**, 719–729.

FAO (2009) Coping with a changing climate: considerations for adaptation and mitigation in agriculture. In Environment and Natural Resources Service Series No. 15 (Glantz, M.H., Gommes, R., Ramasamy, S. and FAO, eds), pp. 1–20. Rome: FAO.

FAOSTAT (2012) Crop processed. Potato production in metric tonnes. GeoHive, Available at: http://www.geohive.com/charts/ag_potato.aspx

Farré, E.M., Bachmann, A., Willmitzer, L. and Trethewey, R.N. (2001) Acceleration of potato tuber sprouting by the expression of bacterial pyrophosphatase. Nat. Biotechnol. **19**, 268–272.

Ferreira, S., Senning, M., Sonnewald, S., Keßling, P.-M., Goldstein, R. and Sonnewald, U. (2010) Comparative transcriptome analysis coupled to X-ray CT reveals sucrose supply and growth velocity as major determinants of potato tuber starch synthesis. BMC Genomics, **11**, 1471–2164.

Ferri, G., Comerio, G., Ladarola, P., Zaponi, M.C. and Speranza, M.L. (1978) Subunit structure and activity of glyceraldehyde-3-phosphate dehydrogenase from spinach chloroplast. Biochim. Biophys. Acta, **522**, 19–31.

Frederick, S.E., Gruber, P.J. and Tolbert, N.E. (1973) The occurrence of glycolate dehydrogenase and glycolate oxidase in green parts: an evolutionary survey. Plant Physiol. **52**, 318–323.

Geiger, D.R., Geigenberger, P. and Stitt, M. (2000) Role of starch in carbon translocation and partitioning at the plant level. Aust. J. Plant Physiol. **27**, 571–582.

Gibon, Y., Bläsing, O.E., Palacios-Rojas, N., Pankovic, D., Hendriks, J.H., Fisahn, J., Höhne, M., Günther, M. and Stitt, M. (2004) Adjustment of diurnal starch turnover to short days: depletion of sugar during the night may leads to a temporary inhibition of carbohydrate utilization, accumulation of sugar and post-tranlational activation of ADP-glucose pyrophosphorylase in the following light period. Plant J. **39**, 847–862.

Givan, C.V. and Kleczkowski, L.A. (1992) The enzymic reduction of glyoxylate and hydroxypyruvate in leaves of higher plants. Plant Physiol. **100**, 552–556.

Goyal, A. and Tolbert, N.E. (1996) Association of glycolate oxidation with photosynthetic electron transport in plant and algal chloroplasts. Proc. Natl Acad. Sci. USA, **93**, 3319–3324.

Goyal, A., Betsche, T. and Tolbert, N.E. (1988) Isolation of intact chloroplasts from Dunaliella tertiolecta. Plant Physiol. **88**, 543–546.

Graf, A. and Smith, A.M. (2011) Starch and the clock: the dark side of plant productivity. Trends Plant Sci. **16**, 169–175.

Hajirezaei, M.R., Börnke, F., Peisker, M., Takahata, Y., Lerchl, J., Kirakosyan, A. and Sonnewald, U. (2003) Decreased sucrose content triggers starch breakdown and respiration in stored potato tubers (Solanum tuberosum). J. Exp. Bot. **54**, 477–488.

Jonik, C., Sonnewald, U., Hajirezaei, M., Flüge, U.-I. and Ludewig, F. (2012) Simultenous boosting of source and sink capacities doubles tuber starch yield of potato plants. Plant Biotechnol. J., **10**, 1088–1098.

Kay, R., Chan, A. and McPherson, J. (1987) Duplication of CaMV 35S promoter sequences creates a strong enhancer for plant genes. Science, **236**, 1299–1302.

Kebeish, R., Niessen, M., Thiruveedhi, K., Bari, R., Hirsch, H.-J., Rosenkranz, R., Stäbler, N., Schönfeld, B., Kreuzaler, F. and Peterhansel, C. (2007) Chloroplast photorespiratory bypass increases photosynthesis and biomass production in Arabidopsis thaliana. Nat. Biotechnol. **25**, 593–599.

Konz, C.B. and Schell, J. (1986) The promoter of TL-DNA gene 5 controls the tissue-specific expression of chimaeric genes carried by a novel type of Agrobacterium binary vector. Mol. Gen. Genet. **204**, 382–396.

Kozani, A. and Takeba, G. (1996) Photorespiration protects C3 plants from photooxidation. Nature, **384**, 557–560.

Laing, W.A., Ogren, W.L. and Hageman, R.H. (1974) Regulation of soybean net photosynthetic CO_2 fixation by the interaction of CO_2, O_2 and ribulose 1,5-diphosphate carboxylase. Plant Physiol. **54**, 678–685.

Leegood, R.C., Lea, P.J., Adcock, M.D. and Häusler, R.E. (1995) The regulation and control of photorespiration. J. Exp. Bot. **46**, 1397–1414.

Lefebvre, S., Lawson, T., Zakhleniuk, O.V., Lloyd, J.C., Raines, C.A. and Fryer, M. (2005) Increased sedoheptulose-1,7-bisphosphatase activity in transgenic tobacco plants stimulates photosynthesis and growth from an early stage in development. Plant Physiol. **138**, 451–460.

Lisec, J., Schauer, N., Kopka, J., Willmitzer, L. and Fernie, A.R. (2006) Gas chromatography mass spectrometry-based metabolite profiling in plants. Nat. Protoc. **1**, 387–396.

Long, S.P., Zhu, X.-G., Naidu, S.L. and Ort, D.R. (2006) Can improvement in photosynthesis increase crop yields? Plant, Cell Environ. **29**, 315–330.

Lord, J.M. (1972) Glycolate oxidoreductase in Escherichia coli. Biochim. Biophys. Acta, **267**, 227–237.

Lu, Y., Gehan, J.P. and Sharkey, T.D. (2005) Daylength and circadian effects on starch degradation and maltose metabolism. Plant Physiol. **138**, 2280–2291.

Luedemann, A., Strassburg, K., Erban, A. and Kopka, J. (2008) TagFinder for the quantitative analysis of gas chromatography and mass spectrometry (GC-MS)-based metabolite profiling experiments. Bioinformatics, **24**, 732–737.

Maier, A., Fahnenstich, H., Von Caemmerer, S., Engqvist, M.K., Weber, A.P.M., Flugge, U.-I. and Maurino, V.G. (2012) Glycolate oxidation in A. thaliana chloroplasts improves biomass production. Front. Plant Sci. **3**, 38.

Maurino, V.G. and Peterhansel, C. (2010) Photorespiration: current status and approaches for metabolic engineering. Curr. Opin. Plant Biol. **13**, 249–256.

Miller, J.H. (1972) *Experiments in Molecular Genetics*. New York: Cold Spring Harbour Laboratory Press.

Miyagawa, Y., Tamoi, M. and Shigeoka, S. (2001) Overexpression of cyanobacterial fructose-1,6-/sedoheptulose-1,7-bisphosphatase in tobacco enhances photosynthesis and growth. *Nat. Biotechnol.* **19**, 965–969.

Nelson, E.B. and Tolbert, N.E. (1970) Glycolate dehydrogenase in green algae. *Arch. Biochem. Biophys.* **141**, 102–110.

Nölke, G., Cobanov, P., Uhde-Holzem, K., Reustle, G., Fischer, R. and Schillberg, S. (2008) Grapevine fanleaf virus (GFLV)-specific andibodies confer GFLV and Arabis mosaic virus (ArMV) resistance in *Nicotiana benthamiana*. *Mol. Plant Pathol.* **9**, 41–49.

OECD/FAO (2011) *OECD-FAO Agricultural Outlook 2011–2020*. OECD Publishing and FAO, Available at: http://dx.doi.org/10.1787//agr_outlook-2011-en

Pellicer, M.T., Badia, J., Aguilar, J. and Baldoma, L. (1996) glc locus of *Escherichia coli*: characterization of genes encoding the subunits of glycolate oxidase and the glc regulator protein. *J. Bacteriol.* **178**, 2051–2059.

Peterhansel, C. and Maurino, V.G. (2011) Photorespiration redesigned. *Plant Physiol.* **155**, 49–55.

Peterhansel, C., Niessen, M. and Kebeish, R.M. (2008) Metabolic engineering towards the enhancement of photosynthesis. *Photochem. Photobiol.* **84**, 1317–1323.

Peterhansel, C., Blume, C. and Offermann, S. (2012) Photorespiratory bypasses: how can they work? *J. Exp. Bot.* **64**, 709–715.

Rachmilevitch, S., Cousins, A.B. and Bloom, A.J. (2004) Nitrate assimilation in plant shoots depends on photorespiration. *Proc. Natl Acad. Sci. USA*, **101**, 11506–11510.

Raines, C.A. and Paul, M.J. (2006) Products of leaf primary carbon metabolism modulate the developmental programme determining plant morphology. *J. Exp. Bot.* **57**, 1857–1862.

Sack, M., Paetz, A., Kunert, R., Bomble, M., Hesse, F., Stiegler, G., Fischer, R., Katinger, H., Stoeger, E. and Rademacher, T. (2007) Functional analysis of the broadly neutralizing human anti-HIV-1 antibody 2F5 produced in transgenic BY-2 suspension cultures. *FASEB J.* **21**, 1655–1664.

Schneider, A., Häusler, R.E., Kolukisaoglu, U., Kunze, R., van der Graaff, E., Schwacke, R., Catoni, E., Desimone, M. and Flügge, U.I. (2002) An *Arabidopsis thaliana* knock-out mutant of the chloroplast triose phosphate/phosphate translocator is severely compromised only when starch synthesis, but not starch mobilization is abolished. *Plant J.* **32**, 685–699.

Smith, A.M. and Stitt, M. (2007) Coordination of carbon supply and plant growth. *Plant, Cell Environ.* **30**, 1126–1149.

Somerville, C.R. (1984) The analysis of photosynthetic carbon dioxide fixation and photorespiration by mutant selection. *Oxford Surveys Plant Mol. Cell Biol.* **1**, 103–131.

Somerville, C.R. (2001) An early *Arabidopsis* demonstration. Resolving a few issues concerning photorespiration. *Plant Physiol.* **125**, 20–24.

Stitt, M., Lilley, R., Gerhard, R. and Heldt, H. (1989) Determination of metabolite levels in specific cells and subcellular compartments of plant leaves. *Methods Enzymol.* **174**, 518–552.

Suttle, J.C. (1996) Dormancy in tuberous organs: problems and perspectives. In *Plant Dormancy, Physiology, Biochemistry and Molecular Biology*. (Lang, G.A., ed.), pp. 133–143. Oxon, UK: Lab International.

Tolbert, N.E. (1997) The C2 oxidative photosynthetic carbon cycle. *Annu. Rev. Plant Physiol. Plant Mol. Biol.* **48**, 1–25.

Wang, F. (2008) *The importance of quality potato seed in increasing potato production in Asia and the Pacific region*. Workshop to commemorate the international year of the potato. FAO.

Wingler, A., Lea, P.J., Quick, W.P. and Leegood, R.C. (2000) Photorespiration: metabolic pathways and their role in stress protection. *Philos. Trans. R. Soc. Lond. B Biol. Sci.* **355**, 1517–1529.

Zhang, L., Häusler, R.E., Greiten, C., Hajirezaei, M.R., Haferkamp, I., Neuhaus, H.E., Flügge, U.I. and Ludewig, F. (2008) Overriding the co-limiting import of carbon and energy into tuber amyloplasts increases starch content and yield of transgenic potato plants. *Plant Biotechnol. J.* **6**, 453–464.

Zrenner, R., Krause, K.P., Apel, P. and Sonnerwald, U. (1996) Reduction of the cytosolic fructose-1,6-bisphosphatase in transgenic potato plants limits photosynthetic sucrose biosynthesis with no impact on plant growth and tuber yield. *Plant J.* **9**, 671–681.

10

Tissue-specific and pathogen-inducible expression of a fusion protein containing a *Fusarium*-specific antibody and a fungal chitinase protects wheat against *Fusarium* pathogens and mycotoxins

Wei Cheng[1,2], He-Ping Li[1,3], Jing-Bo Zhang[1,2], Hong-Jie Du[1,3], Qi-Yong Wei[1,3], Tao Huang[1,3], Peng Yang[1,2], Xian-Wei Kong[1,2] and Yu-Cai Liao[1,2,4,*]

[1]*Molecular Biotechnology Laboratory of Triticeae Crops, Huazhong Agricultural University, Wuhan, China*
[2]*College of Plant Science and Technology, Huazhong Agricultural University, Wuhan, China*
[3]*College of Life Science and Technology, Huazhong Agricultural University, Wuhan, China*
[4]*National Center of Plant Gene Research (Wuhan), Huazhong Agricultural University, Wuhan, China*

*Correspondence

email yucailiao@mail.hzau.edu.cn

Keywords: antibody fusion, chitinase, Fusarium head blight, mycotoxins, pathogen-inducible expression, tissue-specific promoter, transgenic wheat.

Summary

Fusarium head blight (FHB) in wheat and other small grain cereals is a globally devastating disease caused by toxigenic *Fusarium* pathogens. Controlling FHB is a challenge because germplasm that is naturally resistant against these pathogens is inadequate. Current control measures rely on fungicides. Here, an antibody fusion comprised of the *Fusarium* spp.-specific recombinant antibody gene *CWP2* derived from chicken, and the endochitinase gene *Ech42* from the biocontrol fungus *Trichoderma atroviride* was introduced into the elite wheat cultivar Zhengmai9023 by particle bombardment. Expression of this fusion gene was regulated by the lemma/palea-specific promoter *Lem2* derived from barley; its expression was confirmed as lemma/palea-specific in transgenic wheat. Single-floret inoculation of independent transgenic wheat lines of the T_3 to T_6 generations revealed significant resistance (type II) to fungal spreading, and natural infection assays in the field showed significant resistance (type I) to initial infection. Gas chromatography–mass spectrometry analysis revealed marked reduction of mycotoxins in the grains of the transgenic wheat lines. Progenies of crosses between the transgenic lines and the FHB-susceptible cultivar Huamai13 also showed significantly enhanced FHB resistance. Quantitative real-time PCR analysis revealed that the tissue-specific expression of the antibody fusion was induced by salicylic acid drenching and induced to a greater extent by *F. graminearum* infection. Histochemical analysis showed substantial restriction of mycelial growth in the lemma tissues of the transgenic plants. Thus, the combined tissue-specific and pathogen-inducible expression of this *Fusarium*-specific antibody fusion can effectively protect wheat against *Fusarium* pathogens and reduce mycotoxin content in grain.

Introduction

Fusarium head blight (FHB) or head scab is a devastating disease of wheat and other small grain cereals worldwide, that is caused by *Fusarium* species (Bai and Shaner, 2004; Kazan *et al.*, 2012; Xu and Nicholson, 2009; Zhang *et al.*, 2013). Since the mid-1990s, FHB has re-emerged as a serious problem to agriculture in North America and Europe (Kazan *et al.*, 2012; Nganje *et al.*, 2002). In the U.S. between 1998 and 2000, FHB caused estimated losses of about 3 billion U.S. dollars (Nganje *et al.*, 2002). Global climate change during recent years has been implicated in the aggravation of the spread and severity of FHB to ever wider regions; FHB is thus now considered to be one of the most deleterious factors in global cereal production (Goswami and Kistler, 2004). In China, FHB epidemics that cause huge economic losses occur frequently in the middle and lower regions of the Yangtze River and have recently extended to other regions. Furthermore, beyond reductions in yields, *Fusarium* pathogens produce various trichothecene mycotoxins in grains that are

highly toxic to both humans and domestic animals (Pestka and Smolinski, 2005). Mycotoxicosis caused by the consumption of FHB-affected wheat flour has been reported in China (Chen *et al.*, 2003) and continues to pose a serious threat to human health.

Typically with fungal diseases, the best control strategy is to prevent infection in fields by growing cultivars that are resistant to fungal pathogens. However, FHB-resistant cultivars are not yet available because naturally *Fusarium*-resistant germplasm is inadequate (Liu, 2001; Xue *et al.*, 2009). Current protective measures rely heavily on the application of chemical fungicides. Fungicide application only provides a 50%-60% reduction in FHB incidence under optimal conditions (Leonard and Bushnell, 2003), and such applications have resulted in undesirable environmental and ecological consequences (Zhang *et al.*, 2009). Worryingly, the incidence of fungicide-resistant *Fusarium* pathogens in wheat fields has increased dramatically in many regions of China since the mid-1990s (Yuan and Zhou, 2005). Moreover, the application of fungicides increases mycotoxin production in wheat grains

(D'Mello et al., 2000; Zhang et al., 2009). The use of biological control measures has often proved inconsistent or is simply ineffective under field conditions (Xu and Nicholson, 2009). Therefore, the introduction of alien resistance genes into the wheat genome via transgenic approaches has been proposed as an important strategy to protect plants against Fusarium pathogens and to reduce mycotoxin production. Different antifungal peptide-encoding genes from both plants and microbes have been genetically transformed into wheat to generate wheat plants under regulation of constitutive promoters that showed improved FHB resistance; most such studies have been conducted with the wheat model cultivar Bobwhite (Anand et al., 2003; Chen et al., 1999; Han et al., 2012; Lakshman et al., 2013; Li et al., 2008; Makandar et al., 2006), that is less recalcitrant for stable transformation than elite wheat varieties.

Antibodies produced by all vertebrates recognize and bind pathogen-specific antigens. It has been demonstrated that various antibodies including monoclonal antibodies and single-chain antibodies can be functionally expressed in plants for various applications. Such antibodies have been used for the targeted protection of plants against a range of agronomically important pathogens (Safarnejad et al., 2011). In the first application of this technology, a single-chain variable fragment antibody (scFv) specific to artichoke mottled crinkle virus was used to confer specific resistance to this virus in transgenic tobacco plants (Tavladoraki et al., 1993). Since then, many different virus-specific scFvs and monoclonal antibodies have been used to improve the resistance of plants to particular viruses (Cervera et al., 2010; Nölke et al., 2009; Voss et al., 1995; Xiao et al., 2000). The first example of antibody-mediated fungal resistance was demonstrated in Arabidopsis thaliana: CWP2, a chicken-derived Fusarium spp.-specific scFv, recognized a surface antigen of F. graminearum and conferred resistance to Fusarium pathogens (Peschen et al., 2004). More importantly, when this antibody was fused to one of three antifungal peptides and transformed into Arabidopsis thaliana and wheat, transgenic plants showed high levels of resistance to Fusarium pathogens (Li et al., 2008; Peschen et al., 2004). Immunofluorescence localization and biological assays demonstrated that these antibody fusions retained their two functions: the binding of the antibody to the antigen and the antifungal action of the peptide (Peschen et al., 2004). Subsequently, fungus-specific scFv antibodies have been applied for the protection of canola against Sclerotinia sclerotiorum (Yajima et al., 2010) and the protection of soybean against Fusarium virguliforme (Brar and Bhattacharyya, 2012), both cases in which there were no naturally resistant sources of germplasm available. Thus, pathogen-specific antibodies with defined specificity can be used to improve disease resistance in plants; this is particularly vital in situations where there is a lack of naturally resistant germplasm to use in conventional breeding efforts.

Wheat spikes are the organs infected by Fusarium pathogens in FHB. In resistance assays, disease symptoms of the blighted spikes of a given wheat genotype are visually assessed by counting the diseased spikelets after some period of fungal infection (Bai and Shaner, 2004). Two types of FHB resistance have been proposed: type I refers to resistance to initial infection, and type II describes resistance to the spread of colonizing fungus within a spike (Bai and Shaner, 2004; Schroeder and Christensen, 1963). Different inoculation methods are often used in experiments to differentiate between these two types of resistance. Type I resistance is

typically evaluated in the field; wheat grains that have been inoculated with the fungus are scattered over the soil surface prior to anthesis, or spore suspensions are sprayed over flowering spikes. Type II resistance is assessed by delivering conidial spores onto a single floret. A third type of resistance (type III) has been proposed to determine the amounts of mycotoxins in harvested grains. The characterization of type III resistance was based on the observation that low deoxynivalenol (DON) content in wheat grains result from either fewer infected kernels or high levels of DON in spike tissues other than kernels (Miller et al., 1986). In all three types of resistance, the lemma/palea is among the first barriers to confront invading Fusarium fungi (Bai and Shaner, 2004; Xu and Nicholson, 2009) and/or the translocation of mycotoxins into kernels (Miller et al., 1986; Snijders and Krechting, 1992). Thus, the lemma/palea-specific expression of an introduced resistance gene in transgenic wheat could efficiently restrict the initial infection and spread of Fusarium pathogens on spikes and reduce the translocation of mycotoxins from lemma/palea into kernels. Moreover, it is possible to combine a pathogen-inducible component to the transgene alongside with the localized lemma/palea expression component, thereby adding specificity and effectiveness to transgenic plants to combat FHB pathogens.

In this study, a fusion gene comprised of the CWP2 antibody and the fungal Ech42 chitinase gene, under the control of the barley lemma/palea-specific Lem2 promoter (Abebe et al., 2005, 2006), was constructed and introduced into the elite wheat cultivar Zhengmai9023. The transgene was preferentially expressed in lemma/palea organs, and expression of the gene was quickly activated upon Fusarium infection. Inoculation of plants of the T_3 to T_6 generations of different transgenic lines showed significantly enhanced Type I, II and III resistance to Fusarium pathogens. Thus, the lemma/palea-specific and Fusarium-inducible regulation of transgenes in wheat by the Lem2 promoter may provide a promising approach for the efficient control of FHB and associated mycotoxins in cereals.

Results

Transformation and selection of transgenic wheat plants

Immature embryos from an elite wheat cv. Zhengmai9023 (Z9023) were cultured to induce calli that were used for bombardment with an expression cassette containing an antibody fusion gene Ech42-CWP2 and a PMI gene as a selection marker (Figure 1a). CPR-positive plantlets were identified by PCR with three pairs of primers annealing to different regions of the expression cassette (Figure 1a). Five independent T_0 transgenic plants were obtained to generate T_1 transgenic lines. To select genetically stable and homozygous transgenic plants for subsequent analyses, individual plants of the T_1, T_2 and T_3 generations were analysed by PCR. Two transgenic lines, Z1 and Z4, showed no segregation at the T_3 generation. PCR analyses of T_3 transgenic plants of these two lines with the Lem2F/Ech42R primer pair showed a DNA fragment of 394 bp, as expected (Figure 1b). Southern blot analyses revealed that the Z1 transgenic line contained four copies of the transgene whilst one copy was integrated into the Z4 transgenic line (Figure 1c). These two lines were regenerated to produce transgenic lines of the T_4 to T_6 generations. PCR was used to identify individual transgenic wheat plants in each generation.

Figure 1 Structure of the pUL-PMI-Lem2-Ech42-CWP2 construct and molecular identification of the presence of the construct in transgenic wheat plants. (a) Structure of a minimal cassette containing *Ech42-CWP2* gene for wheat transformation. Ubi-P, maize ubiquitin1 promoter; *pmi*, phosphomannose isomerase gene; Lem2-P, barely *Lem2* promoter; *Ech42*, an endochitinase gene from *Trichoderma atroviride*; *CWP2*, a *Fusarium*-specific antibody gene derived from chicken; $(G_4S)_3$ Linker, a 15-amino acid glycine–serine linker; Nos-T, *Nos* terminator. The solid line indicates the DNA fragment amplified by PCR for confirmation of the construct in transgenic wheat; this is also the sequence used as a probe in the Southern blotting experiment; dotted lines indicate fragments amplified by PCR for the identification of plants with the transgene. (b) PCR products amplified from transgenic wheat. DNA isolated from leaves of T_3 transgenic wheat lines Z1, Z4 and nontransgenic Z9023 used as the template for PCR amplification using primer pair Lem2F/Ech42R. (c) Southern blot analysis of transgenic wheat. DNA from T_3 transgenic wheat lines Z1, Z4 and nontransgenic Z9023 was digested with *Hind*III, resolved on an agarose gel, transferred to a Hybond-N$^+$ nylon membrane and hybridized with an α-[^{32}P]-dCTP-labelled DNA fragment that was amplified from a junction region (the solid line in Figure 1a) of the *Lem2* promoter and the *Ech42* gene with primer pair Lem2F/Ech42R. (d) Wheat tissues used for RNA isolation. En: endosperm; Fl: flag leaf; L/P: lemma/palea; Pe: pericarp. (e) Reversely transcribed PCR products were amplified from RNA extracted from wheat tissues indicated above the panel.

Expression pattern of the transgene in transgenic wheat plants

As the *Lem2* promoter from barley has been reported to regulate the tissue-specific expression in lemma/palea in barley, RNA from flag leaves, lemma/palea, pericarp and endosperm (Figure 1d) of the two homozygote T_3 transgenic wheat lines and nontransgenic Z9023 was used for RT-PCR analysis. The results showed that the antibody fusion gene *Ech42-CWP2* was indeed expressed at the highest level in the lemma/palea, with only a trace level present in flag leaves, pericarp and endosperm, in both transgenic lines (Figure 1e). No amplification of the transgene was detected in the nontransgenic control. These results indicated that the barely *Lem2* promoter regulated the specific expression of the *Ech42-CWP2* fusion gene at the outer floret organ of transgenic wheat, a pattern similar to that described for

Lem2 promoter constructs evaluated in barley (Abebe *et al.*, 2005, 2006).

FHB resistance in T_3 transgenic wheat in the greenhouse after single-floret injection

To investigate the response of the transgenic plants to FHB pathogens, the two transgenic Z1 and Z4 lines of T_3 generation were assayed by single-floret injection in a greenhouse, and their percentages of infected spikelets were scored at 21 dpi. Nontransgenic cv. Z9023 and FHB-resistant cv. Sumai3 were similarly inoculated and used as controls. The results showed significant differences in FHB resistance between the transgenic wheat and nontransgenic controls (Table 1). The Z1 and Z4 plants had 7.13% and 8.78% infected spikelets at 21 dpi, with significant disease reductions ($P < 0.01$) of 83% and 79%, respectively, compared with the nontransgenic Z9023 (42.09%). Thus, the tissue-specific expression in lemma/palea of the *Ech42-CWP2* gene efficiently restricted the spread of *Fusarium* pathogens on wheat spikes. Encouraged by these results, the two transgenic wheat lines were further characterized for their response to FHB under field conditions.

FHB resistance in field-grown T_4 to T_6 transgenic wheat following single-floret injection

T_4 to T_6 generation plants of the Z1 and Z4 lines and the control plants grown in fields were inoculated by single-floret injection and their percentages of infected spikelets were scored at 21 dpi (Table 1). The transgenic wheat line Z1 had 12.31%, 10.69% and 12.14% infected spikelets, respectively, in T_4, T_5 and T_6 generations, with significant disease reductions ($P < 0.01$) of 76%, 67%, and 72% compared with the nontransgenic Z9023 (50.29%, 32.89%, and 43.79%). The transgenic wheat line Z4 showed a similar level of FHB resistance to that of the Z1 line. Compared with the control Z9023, significant disease reductions of 69%-75% were observed for this line in the T_4 to T_6 generations. These results indicated that the *Ech42-CWP2* gene conferred durable FHB resistance (i.e. a type II resistance) to fungal spread on wheat spikes in the field in different years. Representative resistant and susceptible wheat spikes from single-floret injection at 21 dpi are presented in Figure 2a.

FHB resistance in field-grown T_4 to T_6 transgenic wheat under natural infection conditions

To ascertain whether the two transgenic wheat lines displayed type I resistance in field conditions, Z1 and Z4 plants of the T_4 to T_6 generations were assayed in fields by natural infection; there was a rich variety of different *Fusarium* pathogens present in the test fields. The percentages of infected spikelets were scored at 30 days postanthesis (dpa) (Table 1). The transgenic wheat line Z1 had percentages of infected spikelets ranging from 3.2% to 6.74% in T_4, T_5 and T_6 generations, with significant reductions by 70%-84%, compared with the nontransgenic Z9023 (19.88%-22.88%). Transgenic line Z4 also showed enhanced FHB resistance; the proportion of infected spikelets ranged from 3.35% to 9.37% in the three generations, with a significant reduction of 54%-83% relative to that of the control Z9023. Therefore, the lemma/palea-specific expression of the *Ech42-*

CWP2 gene conferred a durable FHB resistance to initial infection by *Fusarium* pathogens, that is a type I resistance, in fields under natural conditions. Distinct disease symptoms of wheat spikes from transgenic and nontransgenic controls by natural infection at 30 dpa were shown in Figure 2b.

FHB resistance in F_4 progenies from crosses between transgenic wheat and an FHB-susceptible cultivar in field conditions

To reveal the functionality of the *Ech42-CWP2* gene in an additional genetic background of wheat, T_3 plants of the homozygous Z1 and Z4 lines were crossed with the FHB-susceptible cultivar Huamai13 (H13). The resulting F_1 to F_4 progenies derived from the H13 × Z1 and H13 × Z4 crosses were genotyped by PCR. Plants in the F_4 progenies were inoculated by single-floret injection and natural infection in fields, as described above. Two F_4 progenies had infected spikelets of 21.7% from a cross H13 × Z1 and 16.8% from a cross H13 × Z4 at 21 dpi, respectively, with significant reductions ($P < 0.01$) of 62%-71% and 50%-62% compared with the nontransgenic parental cultivars H13 (57.1%) and Z9023 (43.8%) (Figure 3). Further natural infection assays showed similar FHB resistance for the progenies: the two progenies had significant disease reduction ($P < 0.01$) of 62%-67% and 41%-49%, respectively, relative to the two nontransgenic parents at 30 dpa. These results indicated that the *Ech42-CWP2* gene regulated by the *Lem2* promoter functioned well in more than one wheat genotype.

Mycotoxins in the grains of T_6 transgenic wheat plants

To determine whether the transgenic wheat plants displayed type III resistance (i.e. to reduce *Fusarium* mycotoxin contents in grains), the amounts of trichothecene mycotoxins from grains harvested after maturation in the field were determined by GC/MS analysis. T_6 transgenic wheat plants of the Z1 and Z4 lines and control Z9023 plants were used in this analysis. In single-floret injection experiment, grains of the Z1 and Z4 contained 8.26 μg/g and 8.81 μg/g of DON, with significant reductions of 48% and 44%, respectively, compared with the nontransgenic control Z9023 grains that contained 15.83 μg/g (Table 2). In natural infection, the two transgenic lines contained only 4.12 μg/g and

Table 1 Fusarium head blight resistance of transgenic and nontransgenic wheat

Genotype*	Single-floret injection Infected spikelets (%)[†]				Natural infection Infected spikelets (%)[†]		
	Greenhouse	Field			Field		
	T_3 October 2011	T_4 April 2012	T_5 April 2013	T_6 April 2014	T_4 April 2012	T_5 April 2013	T_6 April 2014
Z1	7.13 ± 0.78 a	12.31 ± 3.42 a	10.69 ± 1.24 a	12.14 ± 1.04 a	3.20 ± 1.70 a	6.19 ± 1.73 a	6.74 ± 0.77 a
Z4	8.78 ± 0.60 a	15.77 ± 2.94 a	8.36 ± 0.81 a	12.43 ± 1.33 a	3.35 ± 0.77 a	9.37 ± 2.02 a	5.21 ± 0.67 a
Z9023	42.09 ± 6.03 b	50.29 ± 12.27 b	32.89 ± 6.60 b	43.79 ± 5.21 b	19.88 ± 2.66 b	20.58 ± 3.27 b	22.88 ± 1.83 b
Su3	9.57 ± 1.72	8.94 ± 1.87	8.77 ± 1.53	13.22 ± 2.34	4.38 ± 3.71	8.75 ± 1.61	7.85 ± 1.08
Hua13	–	63.90 ± 4.53	55.96 ± 6.86	57.08 ± 5.78	29.44 ± 2.69	35.34 ± 4.50	36.45 ± 3.83

*Z1, Z4: the two independent transgenic lines Z1 and Z4; Z9023: nontransgenic control cv. Zhengmai9023; Su3: FHB-resistant control cv. Sumai3; Hua13: FHB-susceptible control cv. Huamai13.

[†]Percentages of infected spikelets are the mean ± standard error from 40 spikes of 40 plants per genotype. Values followed by different letters within one generation indicate statistical significance ($P < 0.01$) assessed by Student's *t*-tests. –: not determined.

Sumai3   Hua13   Z9023   Z1   Z4

(a)

(b)

Figure 2 Phenotype of representative spikes from transgenic and nontransgenic wheat inoculated with *Fusarium* pathogens in the field. (a) Spikes of single-floret injection (inoculated floret indicated by an arrow) at 21 days postinoculation with *F. graminearum* conidia. (b) Spikes naturally infected by *Fusarium* spp. at 30 days postanthesis.

4.02 µg/g of DON, with significant reductions of 63% and 64% relative to the control Z9023 (11.25 µg/g). Thus, the expression of the *Ech42-CWP2* gene in wheat lemma/palea reduces mycotoxin accumulation in grains after infection by *Fusarium* pathogens.

Expression of the Ech42-CWP2 fusion in response to SA and Fusarium pathogens

It is known that the *Lem2* promoter contains putative *cis*-elements that are responsive to the defence signalling molecule SA in barley (Abebe *et al.*, 2005). Therefore, we investigated the expression level of the *Ech42-CWP2* fusion in transgenic wheat with qRT-PCR analysis following SA treatment and *F. graminearum* challenge. To differentiate the induction effects of SA and *F. graminearum*, T_6 generation of Z1 and Z4 plants was irrigated with SA or water for 24 h and then inoculated with either *F. graminearum* or water. At 12, 24, 48 and 72 h postinoculation (hpi), RNA from the lemma/palea of the wheat plants was isolated for qRT-PCR analysis of *Ech42-CWP2* expression. To compara-

tively analyse the expression, the transcript levels from plants that were both irrigated and inoculated with water (H_2O+H_2O) were used as controls. We calculated the fold changes of the various samples relative to the controls. The results of these experiments are illustrated in Figure 4. Compared to the control (H_2O+H_2O), a clear induction of the transgene was seen after SA treatment and *Fusarium* inoculation at 12 hpi, and this induction became more pronounced throughout the time courses extended. For instance, at 24 hpi, SA irrigation ($SA+H_2O$) induced a 1.7-fold increase of *Ech42-CWP2* expression in the transgenic plants; *Fusarium* inoculation (H_2O+Fg) activated a higher level of *Ech42-CWP2* expression, with 3- and 3.6-fold increase in the two lines. Furthermore, SA treatment together with *Fusarium* inoculation ($SA+Fg$) induced the highest level of *Ech42-CWP2* expression, up to 3.4- to 4.8-fold. At 48 and 72 hpi, similar patterns with higher *Ech42-CWP2* transcript levels were observed after both SA irrigation ($SA+H_2O$) and *Fusarium* inoculation (H_2O+Fg).

More interestingly, *F. graminearum* activated higher *Ech42-CWP2* expression than did SA, particularly at the later time points

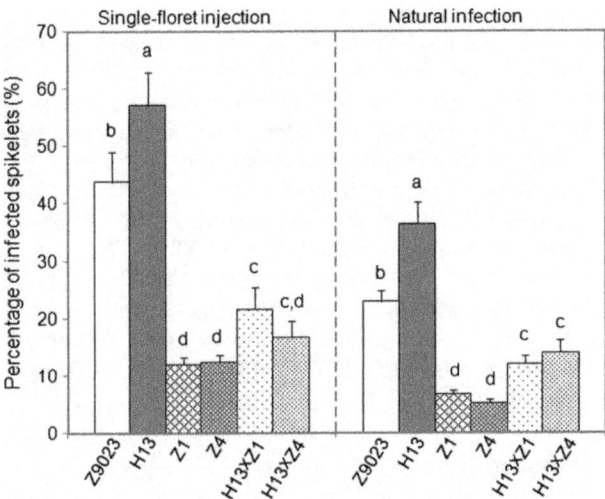

Figure 3 Percentages of infected spikelets of F_4 progenies derived from two crosses (H13 × Z1; H13 × Z4) and their parent genotypes. Crosses were made between an FHB-susceptible wheat cv. Huamai13 (H13) and T_3 generation plants of the two transgenic lines Z1 and Z4. Individual plants from the F_1 to F_4 progenies were genotyped by PCR. F_4 progenies, the two transgenic wheat lines (T_6 generation), and nontransgenic wheat cultivars Huamai13 (H13) and Zhengmai9023 (Z9023) were inoculated by single-floret injection with *F. graminearum* conidia, or naturally infected with a mix of four *Fusarium* pathogens in the field. Percentages of infected spikelets were scored from 40 spikes (one spike per plant) per genotype at 21 days postinoculation after single-floret injection or 30 days postanthesis for the natural infection experiments. Different letters indicate statistical significance ($P < 0.05$) assessed by Student's *t*-tests.

Table 2 Mycotoxin content in grains of T_6 transgenic wheat lines and nontransgenic wheat cultivars

Genotype*	Single-floret injection DON (μg/g)[†]	Natural infection DON (μg/g)[†]
Z1	8.26 ± 0.07 a	4.12 ± 0.16 a
Z4	8.81 ± 0.35 a	4.02 ± 0.10 a
Z9023	15.83 ± 1.42 b	11.25 ± 0.42 b
Su3	6.73 ± 0.08	1.83 ± 0.14
Hua13	22.52 ± 0.89	17.56 ± 0.74

*Z1, Z4: the two independent transgenic lines Z1 and Z4; Z9023: nontransgenic control cv. Zhengmai9023; Su3: FHB-resistant control cv. Sumai3; Hua13: FHB-susceptible control cv. Huamai13; DON = deoxynivalenol.

[†]Mycotoxin values are given as the mean ± standard error from three measurements as determined by gas chromatography–mass spectrometry from the grain following harvest from the field. For single-floret injection, plants were inoculated with *F. graminearum* strain 5035, which is known to produce DON. For natural infection, plants were assayed under natural field conditions where there were a variety of different *Fusarium* strains, all of which were DON producers. Values followed by different letters within one column indicate statistical significance ($P < 0.01$) assessed by Student's *t*-tests.

(Figure 4, bottom panel). At 24 hpi, *Fusarium* infection induced 3- to 3.6-fold (H_2O+Fg vs. H_2O+H_2O) and 2- to 2.7-fold (SA+Fg vs. SA+H_2O) increases in *Ech42-CWP2* expression in the transgenic plants, whereas SA induced only 1.7-fold (SA+H_2O vs. H_2O+H_2O) and 1.1- to 1.3-fold (SA+Fg vs. H_2O+Fg) increases.

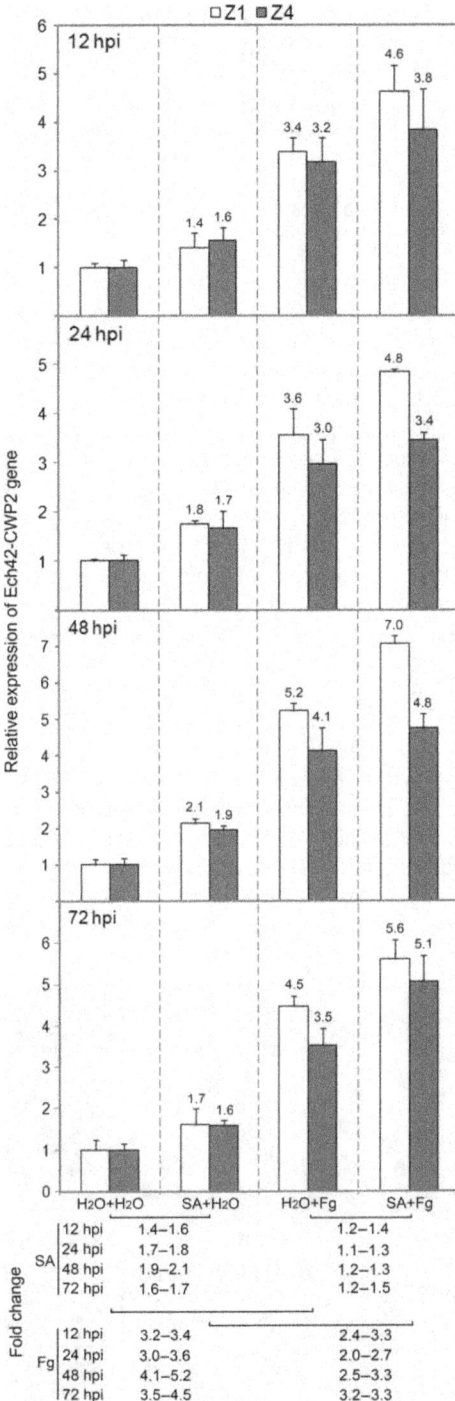

Figure 4 Quantitative real-time PCR (qRT-PCR) analyses of the *Ech42-CWP2* gene in lemma/palea tissues of the transgenic lines Z1 and Z4 at 12, 24, 48 and 72 h postinoculation (hpi) with *Fusarium graminearum* (Fg) or water (H_2O). Wheat plants were irrigated with either SA or water (H_2O) for 24 h prior to single-floret injection of *F. graminearum* or water. Five lemmas/paleas per treatment were used for RNA isolation, and the experiment was performed in triplicate. Each sample was replicated in triplicate (technical replicates) in the qRT-PCR analysis, and data were normalized to the wheat β-*actin* gene. Relative expression of the transgene was calculated using water irrigation and inoculation (H_2O+H_2O) as a control (the transcript level was set as 1). The respective fold changes relative to the control are indicated at the top of the column. Fold changes between the SA and water treatments, as well as the Fg and water treatments are presented at the bottom of the panel.

Similarly, increased fold changes of the transcripts were seen at 48 and 72 hpi. These consistent results congruently indicated that both SA and *F. graminearum* induced the expression of the *Ech42-CWP2* gene regulated by *Lem2* promoter in transgenic wheat and that *F. graminearum* was more potent inducer than was SA.

Histochemical analyses of wheat lemmas inoculated with Fusarium pathogens

To visually investigate the role of SA and the *Ech42-CWP2* gene under the regulation of *Lem2* promoter, lemma tissues from T$_6$ generation of transgenic wheat plants and control Z9023 plants inoculated with *F. graminearum* for 3 days were stained with lactophenol blue. Prior to inoculation, the wheat plants were irrigated with SA or water for 24 h, as described above. In lactophenol blue staining, infected mycelia and necrotic wheat cells are stained deep blue, whilst most of living plant tissues remain unchanged. The results are shown in Figure 5. In the nontransgenic Z9023 lacking SA treatment, deep blue colour was distributed in a large area of lemma surface, indicating the presence of enormous mycelia and damaged wheat cells (Figure 5a). In contrast, after irrigation with SA, the stained blue area was substantially reduced; a small area was still deep blue (Figure 5b). In the transgenic wheat plants lacking SA treatment, lemma surfaces contained discontinuous light blue colour (Figures 5c,e); moreover, lemmas from transgenic plants irrigated with SA contained the lowest intensity of blue and the fewest mycelia. These results indicated a severe restriction of mycelial growth and a substantial decrease of necrotic cells in lemma tissues (Figures 5d,f). These results therefore indicated that the lemma/palea-specific expression of the *Ech42-CWP2* gene

strongly inhibited fungal growth and that treatment with SA further strengthened this inhibitory activity, resulting in yet greater restriction of fungal growth.

The FHB symptoms of Z1 and Z4 plants and nontransgenic Z9023 plants, with or without SA treatment, were then evaluated at 21 dpi after single-floret injection. For the nontransgenic control plants, irrigation with SA significantly ($P < 0.05$) reduced fungal spread by 20% as compared with the control lacking SA irrigation (Figure 6). As for the transgenic Z1 and Z4 plants, SA irrigation slightly reduced the proportion of infected spikelets, but the reductions were not significant compared with the water control. These results suggested that SA treatment increased FHB resistance in wheat genotypes that had a rather low level of basal FHB resistance.

Discussion

Controlling toxigenic *Fusarium* pathogens in cereal crops is challenging due to several factors beyond just the lack of naturally resistant germplasm. The expression levels of alien genes in transgenic wheat are typically low, and it is difficult to specifically target particular pathogens. Additional obstacles include no efficient means of reducing mycotoxin accumulation in grain. In this study, an *Ech42-CWP2* fusion comprising a *Fusarium*-specific antibody and a chitinase was expressed preferentially in the lemma/palea of wheat. Further, the expression of this fusion gene was found to be further activated following fungal infection. These combined strategies enabled the defined expression of a transgene that significantly enhanced three types of resistance to FHB. These results indicated that the tissue-specific and pathogen-inducible expression of a resistance gene

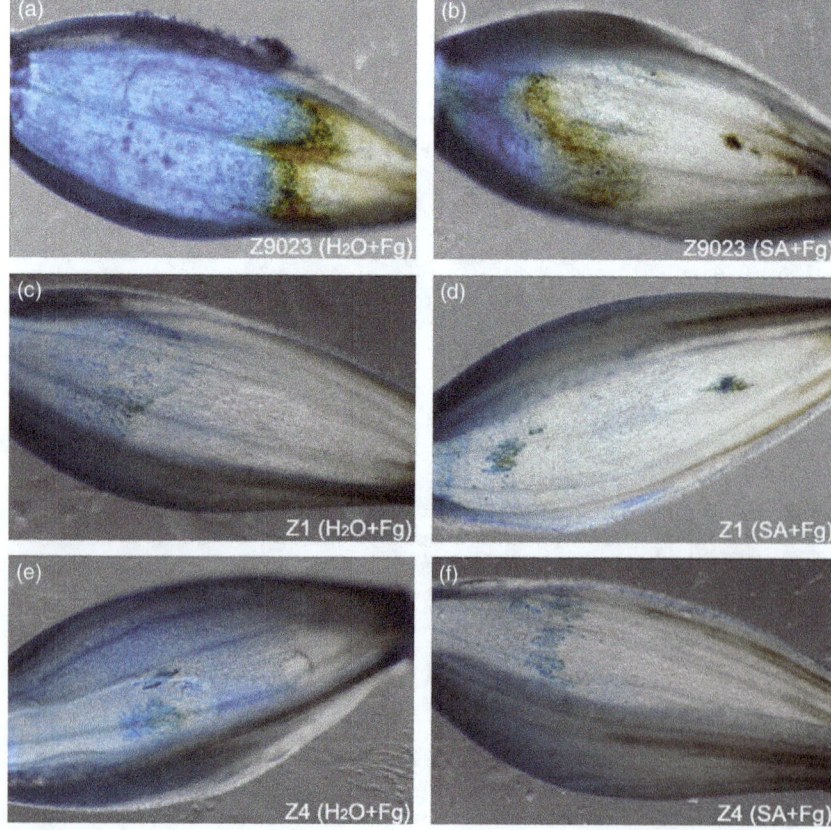

Figure 5 Difference in the quantity of *Fusarium* mycelia observed on lemma surfaces at 3 days postinoculation by single-floret injection of *F. graminearum* (Fg). Transgenic wheat lines Z1, Z4 and nontransgenic wheat cv. Zhengmai9023 (Z9023) were irrigated with either SA or water (H$_2$O) for 24 h prior to inoculation. The lemmas ($n = 5$) were stained with lactophenol blue.

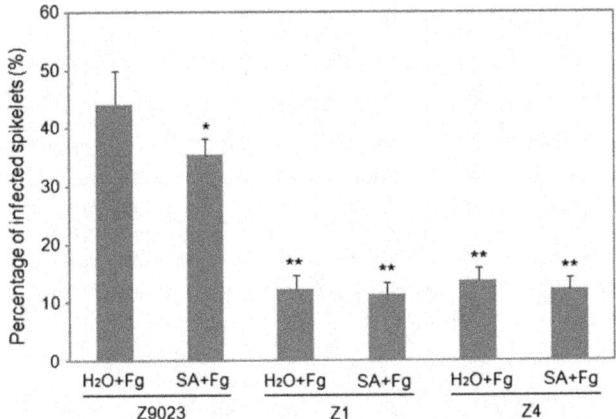

Figure 6 Percentages of infected spikelets of transgenic lines Z1, Z4 and nontransgenic Zhengmai9023 (Z9023) at 21 days postinoculation by single-floret injection of *Fusarium graminearum* (Fg). Wheat plants grown in a greenhouse were continuously irrigated with either SA or water (H_2O) until harvest. Percentages of infected spikelets were scored from a total of 40 spikes per genotype for each treatment. Asterisks indicate statistically significant variation (*$P < 0.05$, **$P < 0.01$) assessed by Student's t-tests.

effectively protected crops against *Fusarium* pathogens and mycotoxins in field conditions.

A predefined pattern for the heterologous expression of a transgene is often achieved through the use of regulatory elements such as promoters. In nature, defence-related genes in plants often have rather low basal expression in conditions without pathogen challenge, but then display rapidly local and/or systemic activation upon pathogen attack (van Loon et al., 2006). In efforts to engineer pathogen resistance in plants, it is more desirable to drive transgene expression in a tissue-specific and/or pathogen-inducible manner to avoid unexpected toxic and/or allergic responses against transgene products and to hypothetically increase plant productivity because transgene products would be produced only in specific tissues and/or accumulated quickly following pathogen attack (Hensel et al., 2011). Because FHB and mycotoxins in grains result from the infection of floret organs by *Fusarium* fungi and lemma/palea tissues are the exterior protection structures of florets that are directly exposed to fungal spores at the onset of colonization (Bai and Shaner, 2004), these tissues are ideal targets for the specific and inducible expression of a resistance gene to control FHB.

Qualitative and quantitative RT-PCR analyses together with histochemical analyses showed a lemma/palea-specific expression of the transgene and its functional effect in wheat (Figs. 1E, 4 and 5). This expression pattern was similar to that observed in barley (Abebe et al., 2005). These results indicated the efficacy of the *Lem2* promoter in regulating lemma/palea-specific gene expression in a heterologous system.

Fusarium infection quickly activated the expression of *Ech42-CWP2* gene, suggesting the presence of *cis*-elements within *Lem2* promoter sequence that are responsive to defence signalling molecules during fungal colonization. The *Lem2* promoter contains two SA-responsive *cis*-elements and its regulatory roles can be enhanced by both SA and the SA functional analogue 2, 6-dichloroisonicotinic acid (Abebe et al., 2005). In this study, SA irrigation induced the expression of *Ech42-CWP2* implying an active response of the *Lem2* promoter resulting from interaction between SA and its *cis*-elements (Figure 4). These results

suggested that the *cis*-elements within the *Lem2* promoter may be functional in wheat. More interestingly, *Fusarium* infection activated a more profound induction than did the exogenous SA drenching treatment. The differential responses to SA drenching and *F. graminearum* infection by the *Lem2* promoter may reflect different modes of action in response to the two stimuli. Initial fungal infection and subsequent invasion took place in lemma/palea, in which the regulatory *cis*-elements in the tissue-specific *Lem2* promoter can quickly and actively respond to the fungal stimulus. Previous reports showed that within 6- 12 h of surface colonization of wheat floret organs, conidial spores of *F. graminearum* germinated and spread over surface (Boenisch and Schäfer, 2011). Penetration into the floret tissues ('true infection') takes place 24-48 hpi (Boddu et al., 2006). The rapid activation patterns of the *Ech42-CWP2* gene at 12 hpi and the succeeding time points of the postinoculation time series in the transgenic lines (Figure 4) are in accordance with the observed infection processes of the fungus. Infection of wheat by *Fusarium* pathogens has been shown to increase SA biosynthesis in spikes (Makandar et al., 2012). These results imply that *Fusarium* pathogens may generate signals for SA biosynthesis, directly and/or indirectly, as soon as floret organs are colonized, and thus actively induce local accumulation of defence molecules including SA, and thereby activate *Ech42-CWP2* expression. In the SA irrigation treatment, plants took up SA from soil and transferred it into different tissues (Makandar et al., 2012); the *Lem2* promoter may perceive a limited amount of SA during this translocation from roots if no challenge from *Fusarium* is available. It is therefore conceivable that the local activation by *Fusarium* infection was apparently more efficient and active in increasing *Ech42-CWP2* expression than exogenous SA treatment alone.

Fusarium-specific CWP2 antibody was generated against *F. graminearum* 5035 (formally named Wuchang 1; Qu et al., 2008). *F. graminearum* 5035 is a representative strain of the predominant *Fusarium* pathogens in Wuhan, China, a region with frequent FHB epidemics in both wheat and barley. CWP2 fused to any of three antifungal peptides has been demonstrated to possess two functions: (i) the binding of the antibody to the surface of *Fusarium* fungi and (ii) antifungal activity damaging the fungi by the peptide (Peschen et al., 2004). Among the fusions, the CWP2 antibody fused to a wheat chitinase was the most active inhibitor, completely destroying *Fusarium* hyphae *in vitro* (Peschen et al., 2004). The *Ech42* gene in the fusion gene used in this study encodes an endochitinase from the biocontrol fungus *T. atroviride* (Hayes et al., 1994). This endochitinase is the main chitin degrading enzyme responsible for inhibiting a wide range of fungal phytopathogens, including some *Fusarium* spp. (Benítez et al., 2004; Gruber and Seidl-Seiboth, 2012; Howell, 2003; Li, 2006). Chitin is a major component of fungal cell walls and is not present in plants or mammals. Reduction of chitin biosynthesis in *F. graminearum* through disruption of a chitin synthase gene has been demonstrated to attenuate fungal virulence and development (Kim et al., 2009; Xu et al., 2010). Therefore, chitin has been considered as an ideal target for controlling FHB pathogens (Xu et al., 2010). The expression of the bi-functional *Ech42-CWP2* molecule in lemma/palea was quickly activated upon *Fusarium* infection. This local activation of the transgene apparently favoured the rapid accumulation of the antibody fusion in the lemma/palea, the organs most liable to infection by *Fusarium* pathogens. The plant-expressed antibody fusion could bind to the surface antigen of the invading *Fusarium* to interfere with its function, whilst at the same time degrading chitin with the chitinase moiety, resulting in efficient restriction of fungal infection and thus FHB resistance.

Resistance assays revealed that transgenic wheat lines expressing an *Ech42-CWP2* gene displayed significantly enhanced resistance to both initial infection (Type I) and fungal spread (Type II) (Table 1). A similar resistance pattern was also observed in two F_4 progenies containing the transgene derived from crosses between an FHB-susceptible variety Huamai13 and transgenic lines Z1 and Z4 (Figure 3). These results indicated that the *Ech42-CWP2* gene is genetically stable and the antibody fusion encoded by the transgene under regulation of the *Lem2* promoter is functionally active in different wheat genetic backgrounds and different environments. Thus, *Ech42-CWP2* fusion-mediated resistance can be transferred into other elite wheat varieties through conventional breeding. Moreover, mycotoxins in the grains of transgenic wheat plants were significantly reduced in both the single-floret injection and the natural field infection experiments, indicating that *Ech42-CWP2* gene confers type III resistance. Previous studies reported that restriction of fungal colonization in chaff (glumes, lemma and palea) and kernels and the inhibition of DON translocation from chaff to kernels can be considered to contribute to a reduction in the accumulation of mycotoxins in wheat grains (Snijders and Krechting, 1992). In this study, the *Lem2* promoter coordinated a tissue-specific and pathogen-inducible expression of the transgene, thus targeting the antibody fusion to lemma/palea. Apparently, the restriction of fungal colonization in these organs plays an important role for reduction of mycotoxins in wheat grains.

In this study, we evaluated the tissue-specific expression and pathogen-inducible activation of a *Fusarium*-specific antibody-based fusion gene in the lemma/palea organs of wheat florets, the organs where initial infection and further spread of FHB pathogens occur. Histochemical analyses provided visual differentiation of mycelial growth on lemma tissues of transgenic wheat and nontransgenic controls. This targeted expression regulated by a lemma/palea-specific promoter conferred three types of FHB resistance, and the transgene was shown to be genetically stable and functionally active in different genetic backgrounds and environments. This study demonstrated the use of the targeting a pathogen-specific antibody fusion to the organs being colonized by fungus to protect against a devastating fungal pathogen; such a strategy is a promising approach for generating 'user-friendly' transgenic crops for pest management in sustainable agriculture.

Experimental procedures

Gene constructs

The coding sequence of endochitinase gene *Ech42* (accession no. L14614; Hayes *et al.*, 1994) from a biocontrol fungus *Trichoderma atroviride* was joined to a *Fusarium*-specific single-chain antibody gene *CWP2* (accession no. AJ517190; Peschen *et al.*, 2004) to construct the *Ech42-CWP2* fusion gene. The fusion sequence was ligated to a barley lemma/palea-specific *Lem2* promoter (Abebe *et al.*, 2005), generating the final pUL-PMI-Lem2-Ech42-CWP2 construct (Figure 1a). The entire expression cassette *PMI-Ech42-CWP2* containing the *Ech42-CWP2* fusion gene driven by *Lem2* promoter and a *PMI* (phosphomannose isomerase) gene as a nonantibiotic selectable marker under the control of a maize ubiquitin promoter (Figure 1a) was released by double digestion with *KpnI/SpeI* (Appendix S1).

Plant materials and transformation

Triticum aestivum L. cv. Zhengmai9023 (Z9023), an elite hexaploid wheat cultivar that is widely cultivated in the middle and lower regions of the Yangtze River valley of China was grown in an experimental field in Wuhan, China. Spikes were collected at 13–14 days postanthesis (dpa). Immature embryos were isolated and placed on callus induction medium for 6 days. Embryonic calli were then bombarded with the minimal cassette *PMI-Ech42-CWP2* using a PDS-1000 He biolistic gun (BioRad, Hercules, CA) at the pressure of 1100 psi. The callus induction and selection of plantlets were performed as previously described (Wright *et al.*, 2001).

PCR, RT-PCR and Southern blot analysis

Genomic DNA was extracted from young leaves of transgenic wheat from different generations and nontransgenic wheat controls using the CTAB method (Nicholson and Parry, 1996). Three sets of primers (PmiF/NosR, Lem2F/Ech42R and scFvP1/scFvP2; Table S1) were used for PCR to confirm the presence of the *PMI-Ech42-CWP2* cassette in transgenic plants.

Total RNA from flag leaves, lemma/palea (at anthesis), pericarp and endosperm (14 dpa) in transgenic and nontransgenic wheat plants was extracted with TRIzol reagent (Invitrogen, Carlsbad, CA) according to manufacturer's instructions. Aliquots of 5 μg of total RNA treated by RNase-free DNase I (Takara, Dalian, China) was reversely transcribed into cDNA. Gene-specific primers (E1/E2; Table S1) were used for qualitative RT-PCR analysis using the PCR cycling programme described above. The wheat β-actin gene was co-amplified and used as an internal control.

For Southern blot analysis, a total of 15 μg DNA from T_3 transgenic and nontransgenic Z9023 plants was digested with 60 U restriction enzyme *Hind*III (Takara) overnight and separated via electrophoresis on 0.8% agarose gels. DNA fragments were transferred onto a nylon membrane (Hybond-N$^+$; Amersham, Buckinghamshire, UK) and hybridized with an α-[^{32}P]-dCTP-labelled DNA fragment, that was prepared from a junction region of the *Lem2* promoter and the antibody fusion gene amplified with the Lem2F/Ech42R primer pair (Figure 1a; Table S1). Autoradiography was conducted using a Fujifilm imaging plate and cassette and analysed with a Fuji BAS1800-II system (Fujifilm, Tokyo, Japan).

FHB resistance assay

For single-floret injection, wheat plants were grown in a greenhouse (T_3 generation) or in experimental fields (T_4, T_5 and T_6 generations) at Huazhong Agricultural University, Wuhan, China. A total of 40 plants from each transgenic line or control were inoculated via single-floret injection at anthesis (Li *et al.*, 2008). One head per plant was inoculated with a 10-μL droplet of a macroconidium suspension of *F. graminearum* isolate 5035 (5×10^5 spores ml^{-1}) that was injected by a pipette tip to the central floret of one middle spikelet. The fungus-inoculated plants were kept in high humidity conditions for 5 days. The FHB-resistant cv. Sumai3, the FHB-susceptible cv. Huamai13, and the nontransgenic cv. Z9023 served as controls. Disease symptoms were scored by counting the number of visually infected spikelets at 21 days postinoculation (dpi), and by calculating infected spikelets relative to the total number of spikelets of the respective head, resulting in the percentage of infected spikelets.

Transgenic wheat plants from the T_4 to T_6 generations were evaluated in the natural infection assays. To create an environment in the experimental fields with sufficient *Fusarium* inoculum, wheat grains (cv. Huamai13) were inoculated with a mixture of four *Fusarium* strains (5031, 5035, 5037, and 5063; Qu *et al.*, 2008) and incubated at 25 °C in darkness for 21 days. The infected wheat grains were then distributed evenly over the soil surface in fields at the booting stage of the wheat plants in the contained experimental fields, as previously described (Anand *et al.*, 2003). Infected spikelets of 40 randomly selected spikes per genotype were scored 30 dpa, and the percentages of infected spikelets relative to the total numbers of spikelets were calculated as for the single-floret injection described above.

Mycotoxin profiling

Mycotoxin contents from the grains of T_6 transgenic plants and nontransgenic controls inoculated with *Fusarium* pathogens were profiled using gas chromatography–mass spectrometry (GC-MS) as previously described (Zhang *et al.*, 2013).

Salicylate treatment and qRT-PCR analysis

Pretreatment of wheat plants with 200 μM of sodium salicylate (SA) (Sigma-Aldrich, St. Louis) in a greenhouse was carried out by irrigating the potted plants 24 h prior to single-floret injection with *F. graminearum* or water (Makandar *et al.*, 2012), and consecutively irrigated with the same concentration of SA until harvest. Wheat plants irrigated with water before and after inoculation were used in parallel in the same trials as controls.

For quantitative real-time PCR (qRT-PCR) analysis, total RNA was extracted from lemma/palea of wheat plants at 12, 24, 48 and 72 h postinoculation (hpi) with *F. graminearum* or water, before which the wheat plants had been irrigated with SA or water for 24 h. For each sample, lemma/palea from 5 spikelets was randomly selected for RNA isolation, and the experiment was performed in triplicate. Reverse transcription into cDNA was carried out as described above and used for qRT-PCR (Appendix S1).

Histochemical analysis

Lemma tissues (*n* = 5) from spikelets 3 days postinoculation with *F. graminearum* of T_6 transgenic wheat lines and nontransgenic Z9023 plants were stained in lactophenol blue solution (Lewandowski *et al.*, 2006; Tekle *et al.*, 2012). The samples were observed under an Olympus SZX16 stereomicroscope (Olympus, Tokyo, Japan).

Statistical analysis

All data were analysed using SAS release 6.12 (SAS Institute, Cary, NC, USA), using significance levels of 0.01 or 0.05.

Acknowledgements

This research was supported by the National Basic Research Program of China (2013CB127801), the National Natural Science Foundation of China (31272004, 31271718), and the Ministry of Agriculture of China (2013ZX08002001-003, 2014ZX0800202B-001).

References

Abebe, T., Skadsen, R.W. and Kaeppler, H.F. (2005) A proximal upstream sequence controls tissue-specific expression of *Lem2*, a salicylate-inducible barley lectin-like gene. *Planta*, **221**, 170–183.

Abebe, T., Skadsen, R., Patel, M. and Kaeppler, H. (2006) The *Lem2* gene promoter of barley directs cell- and development-specific expression of *gfp* in transgenic plants. *Plant Biotechnol. J.* **4**, 35–44.

Anand, A., Zhou, T., Trick, H.N., Gill, B.S., Bockus, W.W. and Muthukrishnan, S. (2003) Greenhouse and field testing of transgenic wheat plants stably expressing genes for thaumatin-like protein, chitinase and glucanase against *Fusarium graminearum. J. Exp. Bot.* **54**, 1101–1111.

Bai, G. and Shaner, G. (2004) Management and resistance in wheat and barley to Fusarium head blight. *Annu. Rev. Phytopathol.* **42**, 135–161.

Benítez, T., Rincon, A.M., Limon, M.C. and Codon, A.C. (2004) Biocontrol mechanisms of *Trichoderma* strains. *Int. Microbiol.* **7**, 249–260.

Boddu, J., Cho, S., Kruger, W.M. and Muehlbauer, G.J. (2006) Transcriptome analysis of the barley-*Fusarium graminearum* interaction. *Mol. Plant Microbe Interact.* **19**, 407–417.

Boenisch, M.J. and Schäfer, W. (2011) *Fusarium graminearum* forms mycotoxin producing infection structures on wheat. *BMC Plant Biol.* **11**, 110.

Brar, H.K. and Bhattacharyya, M.K. (2012) Expression of a single-chain variable-fragment antibody against a *Fusarium virguliforme* toxin peptide enhances tolerance to sudden death syndrome in transgenic soybean plants. *Mol. Plant Microbe Interact.* **25**, 817–824.

Cervera, M., Esteban, O., Gil, M., Gorris, M.T., Martínez, M.C., Peña, L. and Cambra, M. (2010) Transgenic expression in citrus of single-chain antibody fragments specific to *Citrus tristeza virus* confers virus resistance. *Transgenic Res.* **19**, 1001–1015.

Chen, W.P., Chen, P.D., Liu, D.J., Kynast, R., Friebe, B., Velazhahan, R., Muthukrishnan, S. and Gill, B.S. (1999) Development of wheat scab symptoms is delayed in transgenic wheat plants that constitutively express a rice thaumatin-like protein gene. *Theor. Appl. Genet.* **99**, 755–760.

Chen, G.S., Wu, K.H. and Chang, J.L. (2003) A survey report of food poisons caused by *Fusarium* head blight in wheat. *Henan J. Prev. Med.* **14**, 366.

D'Mello, J.P., Macdonald, A.M. and Briere, L. (2000) Mycotoxin production in a carbendazim-resistant strain of *Fusarium sporotrichioides. Mycotoxin Res.* **16**, 101–111.

Goswami, R.S. and Kistler, H.C. (2004) Heading for a disaster: *Fusarium graminearum* on cereal crops. *Mol. Plant Pathol.* **5**, 515–525.

Gruber, S. and Seidl-Seiboth, V. (2012) Self versus non-self: fungal cell wall degradation in *Trichoderma. Microbiology*, **158**, 26–34.

Han, J., Lakshman, D.K., Galvez, L.C., Mitra, S., Baenziger, P.S. and Mitra, A. (2012) Transgenic expression of lactoferrin imparts enhanced resistance to head blight of wheat caused by *Fusarium graminearum. BMC Plant Biol.* **12**, 33.

Hayes, C.K., Klemsdal, S., Lorito, M., Di Pietro, A. and Peterbauer, C. (1994) Isolation and sequence of an endochitinase-encoding gene from a cDNA library of *Trichoderma harzianum. Gene*, **138**, 143–148.

Hensel, G., Himmelbach, A., Chen, W., Douchkov, D.K. and Kumlehn, J. (2011) Transgene expression systems in the Triticeae cereals. *J. Plant Physiol.* **168**, 30–44.

Howell, C.R. (2003) Mechanisms employed by *Trichoderma* species in the biological control of plant diseases: the history and evolution of current concepts. *Plant Dis.* **87**, 4–10.

Kazan, K., Gardiner, D.M. and Manners, J.M. (2012) On the trail of a cereal killer: recent advances in *Fusarium graminearum* pathogenomics and host resistance. *Mol. Plant Pathol.* **13**, 399–413.

Kim, J.E., Lee, H.J., Lee, J., Kim, K.W., Yun, S.H., Shim, W.B. and Lee, Y.W. (2009) *Gibberella zeae* chitin synthase genes, *GzCHS5* and *GzCHS7*, are required for hyphal growth, perithecia formation, and pathogenicity. *Curr. Genet.* **55**, 449–459.

Lakshman, D.K., Natarajan, S., Mandal, S. and Mitra, A. (2013) Lactoferrin-derived resistance against plant pathogens in transgenic plants. *J. Agric. Food Chem.* **61**, 11730–11735.

Leonard, K.J. and Bushnell, W.R. (2003) *Fusarium head blight of wheat and barley*. St. Paul, MN, USA: APS Press.

Lewandowski, S.M., Bushnell, W.R. and Evans, C.K. (2006) Distribution of mycelial colonies and lesions in field-grown barley inoculated with *Fusarium graminearum. Phytopathology*, **96**, 567–581.

Li, D.C. (2006) Review of fungal chitinases. *Mycopathologia*, **161**, 345–360.

Li, H.P., Zhang, J.B., Shi, R.P., Huang, T., Fischer, R. and Liao, Y.C. (2008) Engineering Fusarium head blight resistance in wheat by expression of a fusion protein containing a *Fusarium*-specific antibody and an antifungal peptide. *Mol. Plant Microbe Interact.* **21**, 1242–1248.

Liu, D.J. (2001) Breeding wheat for scab resistance – a worldwide hard nut to crack. Proceedings of International Conference on Wheat Genetics and Breeding – Perspectives of the 21st Century for Wheat Genetics and Breeding. Beijing: Agriculture Publisher, 4–7.

van Loon, L.C., Rep, M. and Pieterse, C.M. (2006) Significance of Inducible Defense-related Proteins in Infected Plants. *Annu. Rev. Phytopathol.* **44**, 135–162.

Makandar, R., Essig, J.S., Schapaugh, M.A., Trick, H.N. and Shah, J. (2006) Genetically engineered resistance to Fusarium head blight in wheat by expression of *Arabidopsis NPR1*. *Mol. Plant Microbe Interact.* **19**, 123–129.

Makandar, R., Nalam, V.J., Lee, H., Trick, H.N. and Dong, Y.H. (2012) Salicylic acid regulates basal resistance to Fusarium head blight in wheat. *Mol. Plant Microbe Interact.* **25**, 431–439.

Miller, J.D., Young, J.C. and Arnison, P.G. (1986) Degradation of deoxynivalenol by suspension cultures of a Fusarium head blight resistant wheat cultivar. *Can. J. Plant Pathol.* **8**, 147–150.

Nganje, W.E., Bangsund, D.A., Leistritx, F.L., Wilson, W.W. and Tiapo, N.M. (2002) Estimating the economic impact of a crop disease: the case of Fusarium head blight in U.S. wheat and barley. In *2002 National Fusarium Head Blight Forum Proceedings* (Canty, S.M., Lewis, J., Siler, L. and Ward, R.W., eds), pp. 275–281. East Lansing: Michigan State University.

Nicholson, P. and Parry, D.W. (1996) Development and use of a PCR assay to detect *Rhizoctonia cerealis*, the cause of sharp eyespot in wheat. *Plant. Pathol.* **45**, 872–883.

Nölke, G., Cobanov, P., Uhde-Holzem, K., Reustle, G., Fischer, R. and Schillberg, S. (2009) Grapevine fanleaf virus (GFLV)-specific antibodies confer GFLV and *Arabis mosaic virus* (ArMV) resistance in *Nicotiana benthamiana*. *Mol. Plant Pathol.* **10**, 41–49.

Peschen, D., Li, H.P., Fischer, R., Kreuzaler, F. and Liao, Y.C. (2004) Fusion proteins comprising a *Fusarium*-specific antibody linked to antifungal peptides protect plants against a fungal pathogen. *Nat. Biotechnol.* **22**, 732–738.

Pestka, J.J. and Smolinski, A.T. (2005) Deoxynivalenol: toxicology and potential effects on humans. *J. Toxicol. Environ. Health B Crit. Rev.* **8**, 39–69.

Qu, B., Li, H.P., Zhang, J.B., Xu, Y.B., Huang, T., Wu, A.B., Zhao, C.S., Carter, J., Nicholson, P. and Liao, Y.C. (2008) Geographic distribution and genetic diversity of *Fusarium graminearum* and *F. asiaticum* on wheat spikes throughout China. *Plant. Pathol.* **57**, 15–24.

Safarnejad, M.R., Jouzani, G.S., Tabatabaie, M., Twyman, R.M. and Schillberg, S. (2011) Antibody-mediated resistance against plant pathogens. *Biotechnol. Adv.* **29**, 961–971.

Schroeder, H.W. and Christensen, J.J. (1963) Factors affecting resistance of wheat to scab caused by *Gibberella zeae*. *Phytopathology*, **53**, 831–838.

Snijders, C.H.A. and Krechting, C.F. (1992) Inhibition of deoxynivalenol translocation and fungal colonization in Fusarium head blight resistant wheat. *Can. J. Bot.* **70**, 1570–1576.

Tavladoraki, P., Benvenuto, E., Trinca, S., De Martinis, D., Cattaneo, A. and Galeffi, P. (1993) Transgenic plants expressing a functional single-chain Fv antibody are specifically protected from virus attack. *Nature*, **366**, 469–472.

Tekle, S., Dill-Macky, R., Skinnes, H., Tronsmo, A.M. and Bjornstad, A. (2012) Infection process of *Fusarium graminearum* in oats (*Avena sativa* L.). *Eur. J. Plant Pathol.* **132**, 431–442.

Voss, A., Niersbach, M., Hain, R., Hirsch, H.J., Liao, Y.C., Kreuzaler, F. and Fischer, R. (1995) Reduced virus infectivity in *N. tabacum* secreting a TMV-specific full-size antibody. *Mol. Breeding* **1**, 39–50.

Wright, M., Dawson, J., Dunder, E., Suttie, J. and Reed, J. (2001) Efficient biolistic transformation of maize (*Zea mays* L.) and wheat (*Triticum aestivum* L.) using the phosphomannose isomerase gene, *pmi*, as the selectable marker. *Plant Cell Rep.* **20**, 429–436.

Xiao, X.W., Chu, P.W.G., Frenkel, M.J., Tabe, L.M., Shukla, D.D., Hanna, P.J., Higgins, T.J.V., Muller, W.J. and Ward, C.W. (2000) Antibody-mediated improved resistance to CIYVV and PVY infections in transgenic tobacco plants expressing a single-chain variable region antibody. *Mol. Breeding* **6**, 421–431.

Xu, X.M. and Nicholson, P. (2009) Community ecology of fungal pathogens causing wheat head blight. *Annu. Rev. Phytopathol.* **47**, 83–103.

Xu, Y.B., Li, H.P., Zhang, J.B., Song, B., Chen, F.F., Duan, X.J., Xu, H.Q. and Liao, Y.C. (2010) Disruption of the chitin synthase gene *CHS1* from *Fusarium asiaticum* results in an altered structure of cell walls and reduced virulence. *Fungal Genet. Biol.* **47**, 205–215.

Xue, A.G., Voldeng, H.D., Savard, M.E., Fedak, G., Tian, X. and Hsiang, T. (2009) Biological control of Fusarium head blight of wheat with *Clonostachys rosea* strain ACM941. *Can. J. Plant Pathol.* **31**, 169–179.

Yajima, W., Verma, S.S., Shah, S., Rahman, M.H., Liang, Y. and Kav, N.N. (2010) Expression of anti-sclerotinia scFv in transgenic *Brassica napus* enhances tolerance against stem rot. *New Biotechnol.* **27**, 816–821.

Yuan, S.K. and Zhou, M.G. (2005) A major gene for resistance to carbendazim, in field isolates of *Gibberella zeae*. *Can. J. Plant Pathol.* **27**, 58–63.

Zhang, Y.J., Yu, J.J., Zhang, Y.N., Zhang, X., Cheng, C.J., Wang, J.X., Hollomon, D.W., Fan, P.S. and Zhou, M.G. (2009) Effect of carbendazim resistance on trichothecene production and aggressiveness of *Fusarium graminearum*. *Mol. Plant Microbe Interact.* **22**, 1143–1150.

Zhang, J.B., Wang, J.H., Gong, A.D., Chen, F.F., Song, B., Li, X., Li, H.P., Peng, C.H. and Liao, Y.C. (2013) Natural occurrence of Fusarium head blight, mycotoxins and mycotoxin-producing isolates of *Fusarium* in commercial fields of wheat in Hubei. *Plant. Pathol.* **62**, 92–102.

11

Scale-up of hydrophobin-assisted recombinant protein production in tobacco BY-2 suspension cells

Lauri J. Reuter, Michael J. Bailey, Jussi J. Joensuu and Anneli Ritala*

VTT Technical Research Centre of Finland, Espoo, Finland

*Correspondence

email anneli.ritala@vtt.fi

Summary

Plant suspension cell cultures are emerging as an alternative to mammalian cells for production of complex recombinant proteins. Plant cell cultures provide low production cost, intrinsic safety and adherence to current regulations, but low yields and costly purification technology hinder their commercialization. Fungal hydrophobins have been utilized as fusion tags to improve yields and facilitate efficient low-cost purification by surfactant-based aqueous two-phase separation (ATPS) in plant, fungal and insect cells. In this work, we report the utilization of hydrophobin fusion technology in tobacco bright yellow 2 (BY-2) suspension cell platform and the establishment of pilot-scale propagation and downstream processing including first-step purification by ATPS. Green fluorescent protein-hydrophobin fusion (GFP-HFBI) induced the formation of protein bodies in tobacco suspension cells, thus encapsulating the fusion protein into discrete compartments. Cultivation of the BY-2 suspension cells was scaled up in standard stirred tank bioreactors up to 600 L production volume, with no apparent change in growth kinetics. Subsequently, ATPS was applied to selectively capture the GFP-HFBI product from crude cell lysate, resulting in threefold concentration, good purity and up to 60% recovery. The ATPS was scaled up to 20 L volume, without loss off efficiency. This study provides the first proof of concept for large-scale hydrophobin-assisted production of recombinant proteins in tobacco BY-2 cell suspensions.

Keywords: aqueous two-phase separation, bioreactor, hydrophobin fusion, scale-up, suspension cells, tobacco bright yellow 2.

Introduction

Plants and suspension cultures of dedifferentiated plant cells are emerging as an alternative to mammalian cell cultures as eukaryotic production platforms for complex recombinant proteins. Plant cell suspensions combine some of the benefits of molecular farming in whole plants and cultivation of mammalian cells (Doran, 2013; Hellwig et al., 2004; Xu et al., 2011). In comparison with field- or greenhouse-grown plants, cell suspensions enable propagation in standardized bioreactors, thereby offering control over the culture conditions and higher batch to batch consistency. Full containment also enables adherence to current good manufacturing practise throughout the production chain (Fischer et al., 2012) and avoids the concern of gene flow to the environment (Doran, 2013). Furthermore, downstream processing of cell suspensions is significantly facilitated due to the lack of fibres, waxes, many secondary metabolites and possible residues of agrochemicals. In comparison with mammalian cells, plant suspension cells grow rapidly in very simple and inexpensive, chemically defined media and most importantly are devoid of any known human pathogens (Doran, 2013). Plant cells can also be used to produce proteins that require plant-like post-transcriptional modifications or proteins that would be harmful or toxic for mammalian host cells.

Tobacco bright yellow 2 (BY-2) cell line has been referred to as the 'HeLa-cells in the biology of higher plants' (Nagata et al., 1992) due to its various applications in fundamental research. However, it would be equally justified to refer to BY-2 as the 'CHO-cells of molecular farming'. The cell line has been utilized as an expression host for numerous recombinant proteins (Bortesi et al., 2012; Kaldis et al., 2013; Kirchhoff et al., 2012; Sack et al., 2007; Schiermeyer et al., 2005; Schinkel et al., 2005; Sun et al.,

2011) and exhibits an exceptional growth rate, multiplying 80- to 100-fold over 1 week in optimal conditions (Nagata et al., 1992). Moreover, the propagation has been established in simple bioreactor systems (Holland et al., 2010; Schmale et al., 2006; Xu et al., 2011). However, the yields of recombinant proteins produced in the BY-2 cell line are still generally low, only occasionally reaching levels of 0.1 to 0.5 g/l (Hellwig et al., 2004; Kaldis et al., 2013; Xu et al., 2011), whereas yields of 5 g/l are common in mammalian cell cultures (Walsh, 2010). Thus, further development is needed in order to meet the general industrial demand of grams per litre product titre.

Recently, several fusion tags, including elastin-like polypeptides (ELP; Conley et al., 2009; Kaldis et al., 2013), zein-derived peptides (Joseph et al., 2012; Torrent et al., 2009) and hydrophobins (HFB; Joensuu et al., 2010), have been introduced as alternative strategies to increase yields of recombinant proteins in plants by stabilizing the fusion partner and directing accumulation of the fusion protein in discrete storage structures (Conley et al., 2011; Khan et al., 2012). HFBs, ubiquitously produced by filamentous fungi, are small (7–15 kD) globular proteins with amphiphilic properties (Hakanpää et al., 2006). Interestingly, when expressed as fusion protein, the hydrophobin 1 (HFBI) of *Trichoderma reesei* has been shown to induce the formation of protein bodies in plant leaves (Gutiérrez et al., 2013; Joensuu et al., 2010) and in filamentous fungi (Mustalahti et al., 2013). Protein bodies are structures typically present in developing seeds, but the detailed mechanism of protein body formation in vegetative tissues induced by foreign proteins remains to be clarified. Nevertheless, the phenomenon appears to be comparable to the formation of similar structures by ELP (Conley et al., 2009; Kaldis et al., 2013) and zein-derived peptides (Torrent et al., 2009). Previously, transient expression of green fluorescent protein-hydrophobin fusion (GFP-HFBI) in *Nicotiana benthamiana* leaves (Joensuu *et al.*, 2010) as well as stable expression in tobacco plants (Gutiérrez *et al.*, 2013) has resulted in twofold yields in comparison with free GFP.

In addition to the potential increase in yield, the HFB-fusion enables a simple and efficient non-chromatographic method for recovering recombinant protein products by aqueous two-phase separation (ATPS; Linder *et al.*, 2004; Penttilä *et al.*, 2008). In this process, HFB-fusion proteins are captured in micellar structures and concentrated in a surfactant phase, while most of the native proteins remain in the aqueous phase. Subsequently, the HFB-fusion is recovered by removing the surfactant with isobutanol back extraction. ATPS has previously been reported as a potential method for first-step purification and concentration of recombinant proteins from *T. reesei* cultures (Linder *et al.*, 2004; Mustalahti *et al.*, 2013), insect cells (Lahtinen *et al.*, 2008) and *N. benthamiana* leaf material (Joensuu *et al.*, 2010).

Here, we report incorporation of HFB-fusion technology in large-scale tobacco BY-2 suspension cell culture, formation of protein bodies and efficient purification of GFP-HFBI fusion by ATPS.

Results

Generation of transgenic callus lines

To investigate the function of the *T. reesei* HFBI fusion tag, two expression vectors (Joensuu *et al.*, 2010), carrying expression constructs for endoplasmic reticulum (ER-) targeted free GFP and ER-targeted HFBI-fused GFP, respectively, were introduced to tobacco BY-2 cells through *Agrobacterium tumefaciens*-mediated transformation. A total of 29 transgenic callus lines were recovered: 10 lines expressing GFP and 19 lines expressing GFP-

HFBI. Many of the callus lines exhibited heterogenic GFP expression, observed as sectorial or mosaic patterns of visible fluorescence under UV-light. To obtain high and consistent expression levels, the calli were subcultured by visually selecting fragments with the most intensive fluorescence, but after multiple rounds of fluorescence-based selection, the expression levels remained inconsistent. However, suspension cultures were prepared from several callus lines and growth, and GFP accumulation levels were screened. Two cell lines, each carrying one of the two constructs, were selected for further experiments on the basis of good expression levels as well as suspension morphology and growth. In these lines, the GFP and GFP-HFBI represented up to 30% and 17% of TSP, respectively (Figure 1f).

HFBI fusion induces protein bodies in BY-2 suspension cells

Laser scanning confocal microscopy was applied to confirm the formation of hydrophobin-induced protein bodies in BY-2 suspension cells. ER-targeted free GFP was localized in the typical reticulate structure of ER (Figure 1a,c, Movie S1). By contrast, hydrophobin-fused GFP was found to induce formation of dense spherical protein bodies (Figure 1b,d, Movie S2). The protein bodies as well as the ER network were located in the periphery of the cells, whereas the central space was occupied by large vacuolar compartments (Figure 1e). During the late phase of the suspension culture, both free GFP and GFP-HFBI were located in large, sometimes irregularly shaped structures in some cells (data not shown). Fluorescence was also observed in vacuoles at this time.

Microscopic observation of the cell suspensions confirmed the heterogeneity of the population with respect to the fluorescence intensity of individual cells: high, moderate and nonexpressing cells were observed.

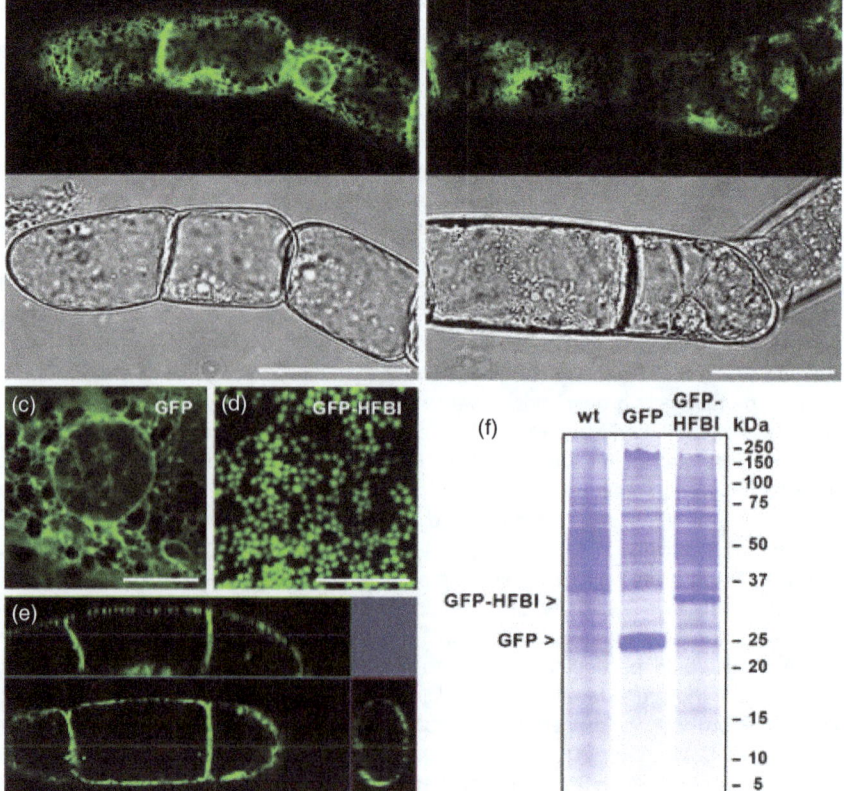

Figure 1 Expression of endoplasmic reticulum (ER)-targeted hydrophobin-fused GFP induces the formation of protein bodies in tobacco bright yellow 2 cells. (a,c) Free GFP is distributed in the typical web-like structure of ER, whereas (b,d) green fluorescent protein-hydrophobin fusion (GFP-HFBI) is located in small spherical protein bodies. A digital dissection of the Z-stack image (e) reveals the large vacuolar compartments restricting the ER and cytoplasm to the periphery of the cells (data shown only for a GFP-HFBI expressing cell). Scale bars represent 50 μm in (a) and (b) and 10 μm in (c) and (d). (f) Coomassie-stained sodium dodecyl sulphate-polyacrylamide gel electrophoresis from the corresponding cell suspensions. TSP from equal amounts of dry cell mass were loaded on the gel. Free GFP at 25 kD and GFP-HFBI at 35 kD. A cleavage product slightly smaller than 25 kD is visible in both lanes.

Growth of cell suspension in bioreactors

To evaluate scalability of the BY-2 cell suspension platform, propagation of the GFP-HFBI line was scaled up from 50 mL culture volumes in shake flasks first to 20 and finally to 600 L culture volumes in stirred tank bioreactors. Accumulation of biomass was comparable in all culture volumes (Figure 2a). Thus, detailed data are presented only for the 600 L cultivation. During the initial lag phase in growth, sucrose was hydrolysed to glucose (and fructose, not measured) and the level of dissolved oxygen (DO) decreased steadily, while the concentration of dissolved carbon dioxide (CO_2) increased (Figure 2b). Characteristically for the BY-2 cells, a drop in pH was observed during the first hours (Figure S1). The lag phase changed into exponential growth after 72 h. The agitation, aeration and vessel overpressure gradually increased according to the output of the DO cascade controller to maintain DO >30%, reaching their maximum values at 125 h (Figure S1). Thereafter, the level of DO decreased close to 0%, which may or may not have correlated

with oxygen deficiency in the culture (see Discussion). Approximately at the same time, the carbon sources were exhausted and cell growth reached a plateau, with culture dry weight (DW) peaking at 16–18 g/L (Figure 2a). Fresh weight (FW) and packed cell volume (PCV), however, continued to increase until termination of the culture, probably due to continuing uptake of water into the cell vacuoles.

Accumulation of GFP-HBFI fusion protein

Accumulation of GFP-HFBI fusion protein and total soluble protein (TSP) was monitored by offline sampling during the bioreactor operation. The proportion of TSP of the cell DW peaked during the phase of exponential growth and decreased after the stationary phase was reached (Figure 2c). However, the titre of GFP-HFBI in the culture suspension continued to increase until termination of the culture at 168 h. In the 600-L cultivation, GFP-HFBI titre reached a level of 0.30 ± 0.018 g/l, corresponding to 16.5% of TSP. Only minor cleavage of the fusion protein was observed from 120 h onwards (Figure 2d).

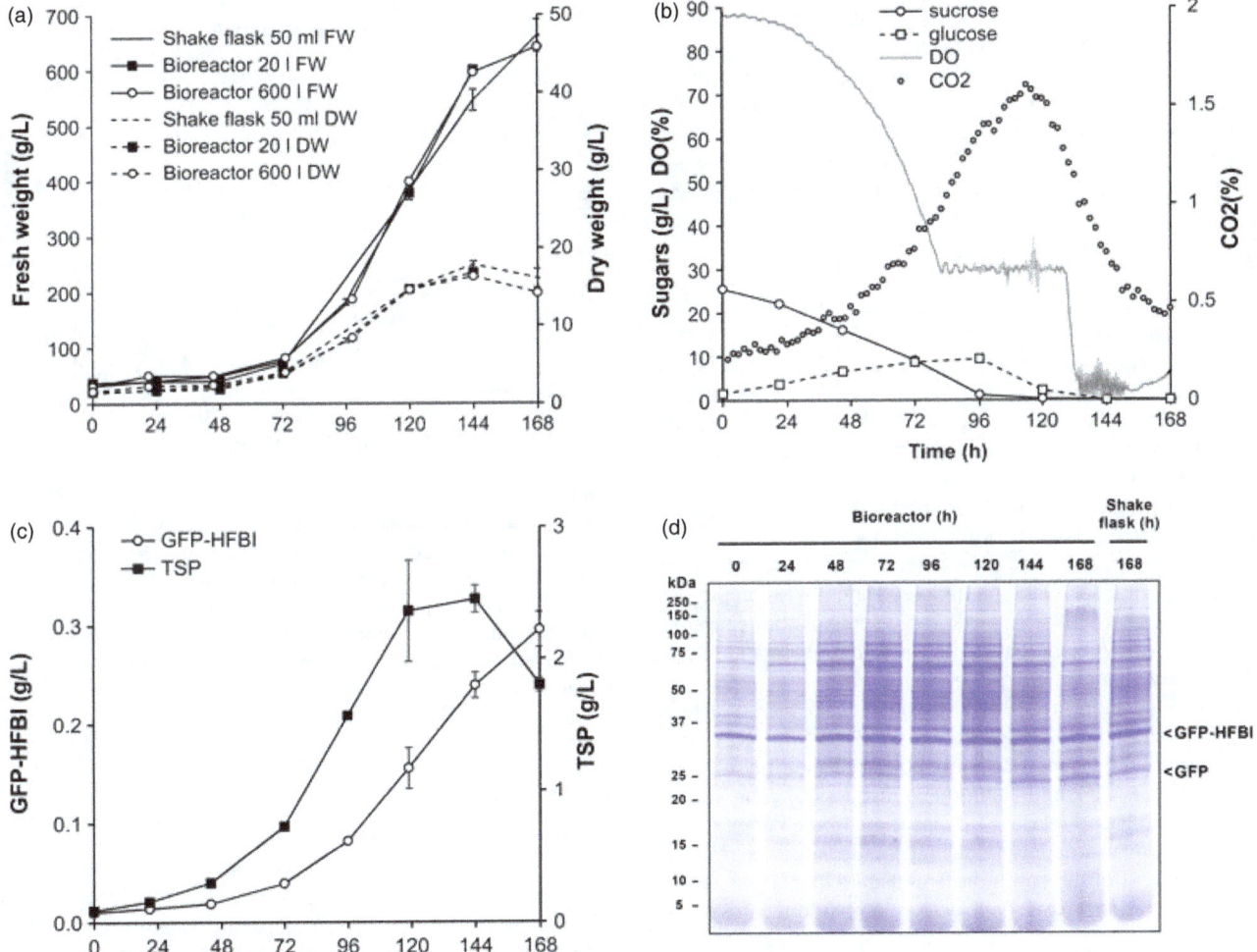

Figure 2 Suspension culture of tobacco bright yellow 2 cells is scalable in standard stirred tank bioreactors. (a) Accumulation of fresh and dry biomass in 20 and 600 L culture volumes in bioreactors ($n = 1$) compared with growth in a 50-mL culture volume in shake flask ($n = 3$, mean $\pm$ standard deviation). (b) Online measurements of dissolved oxygen (DO) and carbon dioxide (CO_2) and offline measured levels of sucrose and glucose in the 600 L culture. (c) Accumulation of intracellular green fluorescent protein-hydrophobin fusion (GFP-HFBI) and total soluble protein (TSP) in 600 L culture (technical repeats: n=3; mean $\pm$ standard deviation). (d) A Coomassie-stained sodium dodecyl sulphate-polyacrylamide gel electrophoresis indicating the stability of the fusion protein in 600 L culture in comparison with shake flask culture. The fusion protein (approximately 35 kD) and cleaved GFP fraction (25 kD), clearly visible at time points 144 and 168 h, are indicated with arrows. TSP from equal amounts of dry cell mass was loaded on the gel.

Biomass harvesting, lyophilization and protein extraction

The biomass propagated in the 600-L bioreactor was harvested from culture suspension using a filter press and lyophilized for storing. The water removal efficiency of the filter press was rather good: the biomass after filtration, amounting to 34% of original suspension mass, was quite dry to the touch. Because of the very high bulk density of the culture (centrifuged PCV 65%), centrifugation was not suitable as a means of separation.

The option to store dehydrated BY-2 biomass safely at room temperature and the effect of drying on protein extraction were evaluated by lyophilizing BY-2 suspension cells expressing GFP-HFBI, storing a sample at RT for 1 month and extracting the soluble proteins. The final DW of the biomass was only approximately 3.5% of the FW, reflecting the very high vacuolar volume of BY-2 cells. Neither the freeze drying procedure *per se* nor subsequent storage at RT for 1 month caused measurable degradation or cleavage of the target protein (Figure 3a).

A range of buffer volumes was tested for extraction of soluble proteins from the freeze-dried and powdered cell material (Figure 3d). High concentration, up to 4.4 ± 0.7 mg/mL of GFP-HFBI in crude extract, was reached using 5 mL extraction buffer for 1 g of dry cell powder without apparent reduction in recovery rate. This indicates good solubility of the fusion protein. However, low buffer volumes resulted in thick suspensions that were difficult to handle. Therefore, subsequent extractions were made using 40 volumes of extraction buffer per dry cell powder weight.

Scaled up purification by ATPS

Aqueous two-phase separation was applied to capture the GFP-HFBI fusion protein from cell extract. To assess scalability of the process, the initial surfactant extraction was conducted in volumes of 100 mL, 1 and 20 L using Triton X-114 (3% w/v) (Joensuu et al., 2010, 2012). Clear separation of the heavier surfactant phase carrying the GFP-HFBI was observed under UV-light after approximately 15 min in 100 mL, 30 min in 1 L and 1 h 30 min in 20 L volume (Figure 4a). In all separation volumes, the heavier phase comprised 35%–38% of the total volume.

The heavier phase was collected and the surfactant was removed by extraction with isobutanol, leaving the fusion protein in the buffer phase (back extract). The volume of the recovered back extract was 76% of that of the surfactant phase and <30% of the initial crude extract volume.

Visual observation of GFP-HFBI partitioning in the surfactant phase and subsequently to back extract (Figure 4a) was confirmed by sodium dodecyl sulphate-polyacrylamide gel electrophoresis (SDS-PAGE) and fluorometry (Figure 4b). The fusion protein was found to selectively concentrate in the back extract, and only residual amounts remained in the aqueous residual phase with most of the native proteins (Figure 4b,c). Concentration of the product was doubled in comparison with crude extract, representing more than 70% of TSP. Although the time required for phase separation increased with separation volume, efficiency of the ATPS was found not to be dependent on the volume (Figure 4c). The overall recovery rate after the ATPS was approximately 50%–60% of the total soluble GFP-HFBI in all extraction volumes (Figure 4c). Only minimal cleavage of the fusion protein was observed, although the whole process was carried out in room temperature.

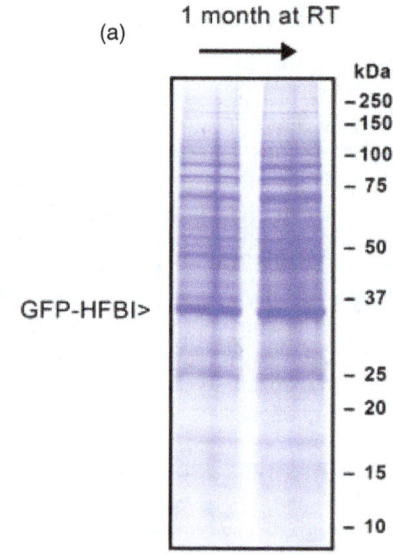

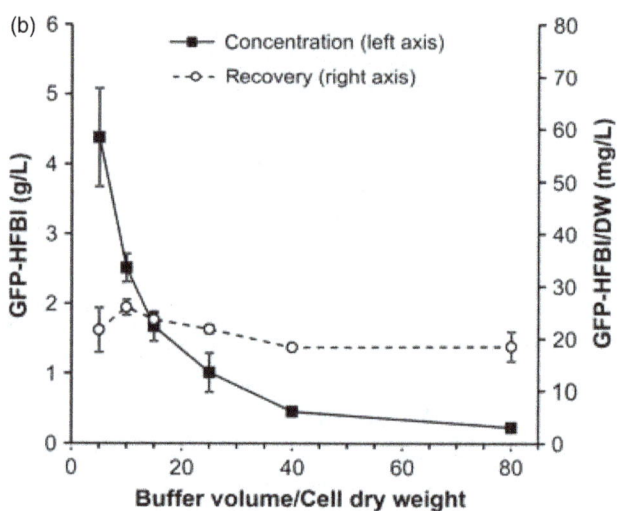

Figure 3 Extraction of the hydrophobin fusion protein from freeze-dried cell material. (a) Coomassie-stained sodium dodecyl sulphate-polyacrylamide gel electrophoresis of crude protein extract after lyophilization, and after storing, the dry material for 1 month at RT does not indicate apparent cleavage of the fusion protein. (b) Effect of the volume of extraction buffer on product concentration and recovery (n = 3, mean ± standard deviation). On the x-axis: volumes of buffer (mL)/cell material (g).

Discussion

Several issues including improvement in production rates, batch to batch consistency and efficient purification still need to be addressed before plant suspension cell platforms for production of high-value recombinant proteins can compete with already established systems on an industrial level. We set out to investigate the potential of hydrophobin fusion technology as a tool to facilitate both large-scale production and purification of recombinant proteins in tobacco BY-2 cell suspension.

Genetic instability and inconsistent yields are recognized as a major drawback in plant cell cultures (Doran, 2013; Xu et al., 2011). In this study too, no definite values could be given for expression levels due to the inconsistency in productivity. The

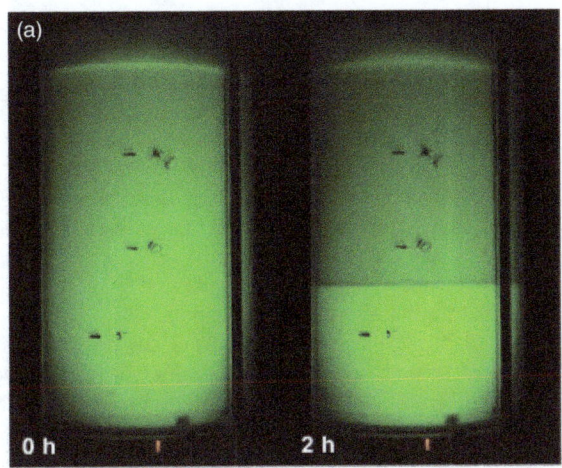

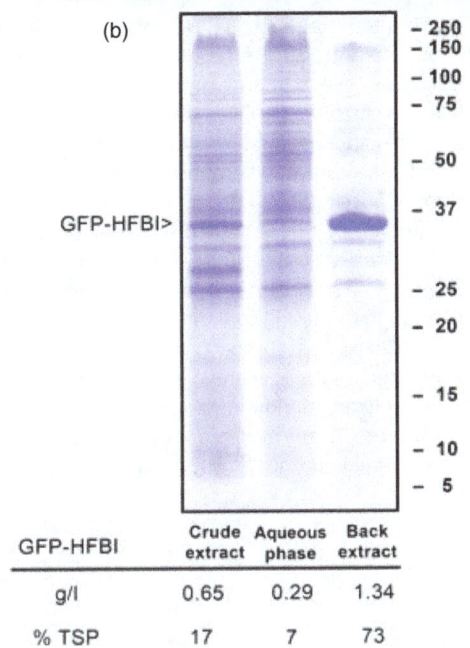

GFP-HFBI	Crude extract	Aqueous phase	Back extract
g/l	0.65	0.29	1.34
% TSP	17	7	73

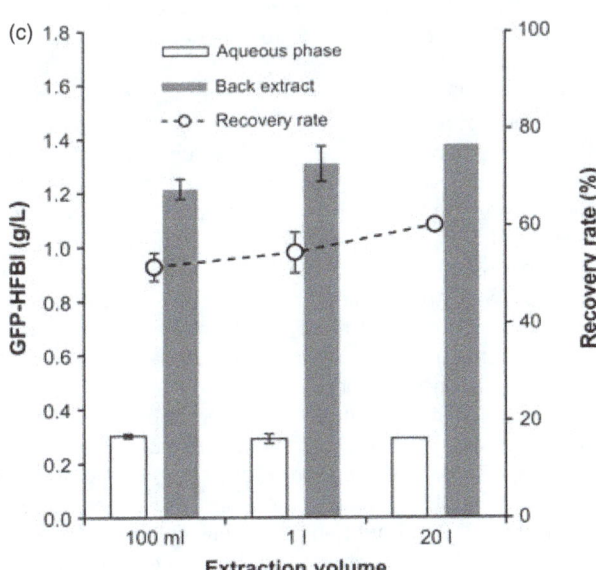

Figure 4 Purification of hydrophobin-fused green fluorescent protein (GFP) using surfactant-based aqueous two-phase separation (ATPS). (a) ATPS conducted in a volume of 20 L. Photographs were taken under UV-light to illustrate migration of the target protein to the surfactant phase. (b) Coomassie-stained sodium dodecyl sulphate-polyacrylamide gel electrophoresis showing selective recovery of the green fluorescent protein-hydrophobin fusion (GFP-HFBI) in the back extract, when most of the native proteins remain in the water phase. (c) The recovery rate and GFP-HFBI concentration in aqueous phase (residue) and in back extract in different separation volumes (100 mL: $n = 3$, mean $\pm$ standard deviation; 1 L: $n = 4$, mean $\pm$ standard deviation; 20 L: $n = 1$).

majority of the transgenic calli exhibited a mosaic or sectorial pattern of heterogenic intensity of visual fluorescence despite several passages manually selecting only the most fluorescent fractions. Microscopic examination of the cell suspensions derived from the calli confirmed this observation, revealing a heterogenic population of cells with varying fluorescence intensities. Furthermore, the levels of GFP expression decreased gradually over continuous passages in suspension culture. Similar heterogeneity has commonly been encountered in BY-2 cell lines expressing fluorescent proteins (Kirchhoff *et al.*, 2012; Nocarova and Fischer, 2009). Both genetic and epigenetic factors have been proposed as the cause for this variation (Doran, 2013). Preparation of monoclonal lines from the primary callus has been shown to reduce the heterogeneity significantly (Nocarova and Fischer, 2009) and improve the yield (Kirchhoff *et al.*, 2012). Although the expression levels of the monoclonal cell lines remained consistent over several months (Nocarova and Fischer, 2009) or even for up to 1 year (Kirchhoff *et al.*, 2012), they were not devoid of subsequent somaclonal variation or epigenetic changes, that is, gene silencing. Thus, the selection may ultimately need to be repeated. Nevertheless, generation of defined monoclonal cultures is required in order to meet the industrial demand for sufficient batch to batch consistency and specific characterization of production lines.

This is the first report confirming the formation of HFBI-induced protein bodies in BY-2 cells. The HFBI-fused GFP was localized in dense spherical structures closely resembling the protein bodies earlier reported in relation to HFBI in *N. benthamiana* (Joensuu *et al.*, 2010), in tobacco plants (Gutiérrez *et al.*, 2013) and on the other hand induced by ELP in BY-2 cells (Kaldis *et al.*, 2013).

Targeting of recombinant proteins to specific storage organelles using fusion tags, such as HFB, ELP or the zein-derived ZERA-peptide, has been a promising strategy to improve overall yields (Conley *et al.*, 2011; Khan *et al.*, 2012; Schmidt, 2013). Increased protein accumulation has been attributed to encapsulation of the recombinant protein from the proteolytic environment of the cytosol, thus protecting it from normal physiological turnover. Moreover, the encapsulation may also protect the host cell from the potentially toxic effects of the over-expressed protein (Joensuu *et al.*, 2010; Torrent *et al.*, 2009). However, the beneficial effect of the protein body formation on the yields in BY-2 cells could not be assessed reliably in this work due to the heterogeneity of the cell lines. Although brightly fluorescent protein bodies were observed in some of the GFP-HFBI-expressing suspension cells, some of the cells exhibited only weak or no fluorescence at all, and in these cells, the formation of protein bodies was not detected. As the yield was measured from the bulk suspension, it does not reliably represent that of the cells with obvious protein body formation. Furthermore, the yields of the suspension cultures fluctuated over time, hampering comparison between the lines. In order to reliably assess the effect of hydrophobin fusion on the yield, stable monoclonal cell lines need to be generated and compared as discussed above.

The detailed mechanism of protein body formation in relation to ELP, zein-derived peptides or HFB is not well understood, although it appears to be conserved in all eukaryotes (Torrent et al., 2009). All three fusion tags share hydrophobic or amphipathic properties and the tendency to self-assembly into stable aggregates in the ER (Conley et al., 2011; Khan et al., 2012). The variable levels of GFP-HFBI expression and consequently variable formation of protein bodies gives reason to assume that the concentration of the fusion protein in the ER may influence the self-assembly and that a certain threshold level of the fusion protein is required for protein body formation. Gutiérrez et al. (2013) reported similar conclusions in relation to HFBI-induced protein bodies in tobacco plants.

To assess the potential scalability of hydrophobin fusion technology-assisted protein production in BY-2 suspension cell platform, both pilot-scale suspension cultures and downstream processing were performed. Although several novel bioreactor systems have been proposed for propagation of plant suspension cells (Huang and McDonald, 2012; Kieran et al., 1997; Xu et al., 2011), we set out to evaluate the possibility of scaling up the propagation in standard steel stirred tank bioreactors designed for microbial cultivations.

The growth kinetics in various culture volumes and different bioreactors were found to be strikingly similar and well comparable to growth of the same cell line in shake flasks. Furthermore, the results obtained here are in line with the smaller-scale cultivations conducted with BY-2 suspension cells and reported previously (Holland et al., 2010; Schmale et al., 2006). Although the DO level of the pilot-scale culture was controlled by stirring speed, vessel overpressure and airflow, the level of DO decreased close to 0% at 125 h, coinciding with reaching the plateau of DW accumulation and peaking of pH. From this time point on, the culture might have suffered from lack of DO, possibly limiting the productivity. The equipment maxima of the DO cascade parameters (agitation, aeration, pressure) were far higher than those used in this cultivation of possibly mechanically sensitive plant cells. Another consideration was that it was not certain whether the observed low level of DO was in fact representative of the situation in the culture as a whole, or whether the DO probe was partially or completely covered by cell overgrowth in the conditions of low agitation. The fact that exhaust gas CO_2 was already decreasing before the start of the observed steep decrease in DO indicates that the latter explanation may be correct. Furthermore, on the basis of the offline curves, carbon source exhaustion also occurred at about the same time as the observed decrease in DO level, which is rather illogical from a biological perspective.

In this study, no growth inhibition due to shear stress was observed even in large culture volumes, even though the cell suspensions were cultured in bioreactors originally designed for microbial cultivations and equipped with Rushton turbines. This confirms the robustness of the BY-2 cell line and is in contrast to the general view of plant cells being highly sensitive to shear forces. Neither did the long cooling time of the media in

pilot-scale bioreactor after sterilization cause any growth retardation.

In our experiments, the volumetric productivity of the recombinant protein was very high, reaching up to 0.3 g/L. However, the high yields cannot be directly attributed to the stabilizing function of the fusion partner, as the free GFP also accumulated in high levels. Even further improvement in the productivity could be obtained by selection of elite monoclonal lines, as discussed above, and by media optimization: Holland et al. (2010) reported 10- to 20-fold increase in antibody yields by the use of extra nitrate in the culture media.

Pilot- or large-scale cultivations of BY-2 suspension cells for production of recombinant proteins have not been reported in the literature. Thus, we regard the results presented here as important evidence of robustness and scalability of the BY-2 platform for recombinant protein production in conventional stirred tank bioreactors. In fact, the very same laboratory that originally generated the tobacco BY-2 cell line in the early 1970s at Japan Tobacco and Salt Co. reported successful propagations of the wild-type cell suspension in bioreactors as large as 20 000 L (Noguchi et al., 1977). Despite the immense culture volume, the growth kinetics were similar to those in shake flask cultures and to the cultivations reported here.

Lyophilization of the filtered cell mass was found to allow convenient storage of the harvested material at room temperature as well as efficient cell disruption by milling. Removal of the intracellular water also provides significant concentration of the product. Although promising for small scale (<30 L) processing of high-value products, this approach may not be economical for handling large quantities of material due to limited capacity and the high cost of lyophilization in pilot scale. However, industrial-scale possibilities exist, for example, in the food industry, and in the case of high-value products, lyophilization may be considered feasible.

Surfactant-based ATPS is a convenient and low-cost method for first-step purification of hydrophobin fusion proteins from fungal cultures (Linder et al., 2004; Mustalahti et al., 2013; Selber et al., 2004), insect cells (Lahtinen et al., 2008) and plant tissues (Joensuu et al., 2010). Here, we applied ATPS in purification of hydrophobin-fused GFP from BY-2 cell extract. Separation in ambient room temperature (21–24 °C) using 3% (w/v) Triton X-114 (Bordier, 1981) resulted in a heavy phase comprising 35%–38% of the total volume. After the back extraction, the final volume was further reduced to <30% of the original volume. With these parameters, both good volume reduction and recovery rate (50–60%) were obtained. Purification of the same fusion protein from tobacco leaf extracts using Agrimul NRE 1205 as surfactant was reported to result in comparable levels of recovery (Joensuu et al., 2010). The work of Joensuu et al. (2010) further showed that optimization of the concentration of surfactant can be made to reach specific goals: more surfactant results in a better recovery rate, up to 90%, whereas less leads to higher product concentration.

The principles of how different sized hydrophobic molecules migrate in phases formed by different surfactants are not well understood. Thus, unpredictability and the requirement for empirical optimization remain the greatest challenges to development of the ATPS (Hatti-Kaul, 2001). By contrast, removal of surfactant by extracting with isobutanol has been shown to be very robust and has not been observed to cause denaturation of the target proteins (Joensuu et al., 2010; Linder et al., 2004). However, this step may also require optimization for less stable products. Capturing hydrophobin-fused proteins from BY-2 cell lysate by surfactant-based ATPS appears to be readily scalable for large separation volumes. Our results indicate that the separation volume has significant impact only on the time needed for phase separation, not on the resulting concentrations, fraction volumes or product concentration. These findings support previous reports of the scalability of ATPS. Selber et al. (2004) as well as Penttilä et al. (2008) showed that scaling a surfactant-based two-phase separation of recombinant proteins from fungal cultures from 10 mL to 1200 L volume did not change the yield or partitioning efficiency.

Conclusions

This study provides for the first time a proof of concept for applying hydrophobin fusion technology in tobacco BY-2 suspension cell platform. Hydrophobin-fused GFP accumulated in ER-derived protein bodies, and the recombinant protein was captured from the cell lysate by surfactant-based ATPS. Furthermore, we have shown that propagation of BY-2 suspension cells is readily scalable in standard stirred tank bioreactors. Further investigations have been initiated to assess the feasibility of the BY-2-hydrophobin platform for various other target proteins.

Experimental procedures

Constructs

The expression vectors for ER-targeted GFP and GFP-HFBI were previously described by Joensuu et al. (2010). Briefly, the coding sequences were placed under control of the dual-enhancer Cauliflower mosaic virus 35S promoter and A. tumefaciens nos terminator. A TEV protease cleavage site was located in between the GFP and HFBI moieties.

Transformation and maintenance of the BY-2 cultures

Transformation of the BY-2 cells was performed as described by De Sutter et al. (2005). The stock cultures were maintained as calli on modified MS-medium (Nagata and Kumagai, 1999) containing 1% agar and 25 mg/L kanamycin and were subcultured every 3–4 weeks by visually selecting the most fluorescent fractions under UV-light. Suspension cultures were maintained in 50 mL of the modified MS-medium supplied with 50 mg/L kanamycin and subcultured weekly by transferring 5% (v/v) of the culture to fresh media.

Confocal microscopy

Subcellular localization of GFP and GFP-HFBI and formation of protein bodies were visualized in suspension cells 7 days after subculturing. A Zeiss LSM 710 laser scanning confocal microscope (Carl Zeiss, Oberkochen, Germany) equipped with a 40× or a 63× water immersion objective lens was used. Excitation was performed with a 488-nm argon laser, and fluorescence was detected at 493–598 nm.

Bioreactor cultivations

All bioreactor operations were conducted in batch cultivation mode by inoculating at 5% (v/v) with 7-day-old suspension. The 20- and 30-L cultivations (Biostat C, Sartorius AG, Goettingen, Germany and IF 40, New Brunswick Scientific, Enfield, CT) were inoculated from bulked shake flask cultures. For the 600 L cultivation (BioFlo PRO, New Brunswick Scientific, Enfield, CT), the inoculum was grown in an intermediate step in the 30-L

bioreactor. The modified MS-medium (Nagata and Kumagai, 1999) was prepared and sterilized in the bioreactor for all cultivations. All cultivations were carried out in dark at 28 °C. Culture DO was controlled by stirring speed, airflow and vessel overpressure to maintain the DO concentration above a threshold of 30%; agitation was with standard Rushton turbines in all the bioreactors. In the pilot-scale cultivation (600 L), maximum agitation was set to 120 rpm (=2.3 m/s tip speed) and aeration to 200 L/min. The pH was monitored, but not controlled. No antifoam agent was added to the medium or during the cultivation procedure. All bioreactor cultivations were performed only once.

Packed cell volume was determined by sampling 10.0 mL of culture suspension in a conical tube and centrifuging at 3220 g for 10 min. The cell pellet was weighed to obtain FW and subsequently freeze-dried to obtain DW. The culture supernatant was stored at −20 °C and analysed later with a YSI 2900 Biochemistry Analyzer (YSI Life Sciences, Yellow springs, OH) to determine glucose and sucrose in the culture medium.

Downstream processing

In laboratory scale, the biomass was harvested by centrifugation and freeze-dried before cell disruption using steel beads and a Retsch mill (MM301, Haan, Germany). For pilot-scale downstream processing, the biomass was separated from the culture medium using a Larox filter press (PF 0.1 H2) and Aino T30 filter cloths, applying ca. 3–5 bar pressure. After primary filtration, the biomass cake was dried by applying a pressure of 8–10 bar via a rubber membrane over the biomass and removing the intercellular liquid thus released. The filtered cell mass was frozen and lyophilized. Cell disruption was performed with a Hosokawa Alpine (100 UPZ-lb) mill at 18 000 rpm.

Protein extraction and analysis

Disrupted cell powder was thoroughly mixed with extraction buffer (1 × phosphate buffered saline; 12 mM $Na_2HPO_4 \cdot 2H_2O$, 3 mM $NaH_2PO4 \cdot H_2O$, 150 mM NaCl), 1 mM EDTA, 100 mM sodium ascorbate and 0.4 μM leupeptine hemisulfate (Sigma-Aldrich, St. Louis, MO), and insoluble material was removed by centrifugation (Eppendorf Centrifuge 5810R, 3220 g, 10 min, RT). In pilot scale, centrifugation was carried out in 2-L bottles (Sorvall RC12BP, ca. 4000 g, 15 min, RT).

Concentration of TSP was measured using the Bradford assay (1976) with Bio-Rad reagent (Bio-Rad, Hercules, CA) and bovine serum albumin (BSA; Sigma-Aldrich) as standard. Extracted proteins were separated by SDS-PAGE on Bio-Rad Criterion-TGX and Mini-PROTEAN precast gels. The GFP concentration in TSP and ATPS samples was determined by fluorometry. Dilutions of 1 : 50, 1 : 100 or 1 : 200 were prepared in black microtiter plates (Microfluor 2; Thermo Fisher Scientific, Waltham, MA) as triplicates by addition of PBS containing 1% (w/v) BSA. The fluorescence of the diluted samples was determined at 485/527 nm using a VICTOR^2 plate reader (Perkin Elmer, Waltham, MA) at 12 nm bandwidth and 100 ms measurement time. Sample dilutions were compared to a standard curve constructed with purified GFP (BioVision, Milpitas, CA).

ATPS

For ATPS, the cell extract (100 mL, 1 or 20 L) was thoroughly mixed with TritonX-114 (3% w/v; Sigma-Aldrich) and left to separate at RT in a separation funnel or in a 20-L cylindrical glass vessel. When the phases were clearly separated, the surfactant phase was collected through the bottom valve and its volume was determined. The surfactant phase was mixed with an equal volume of isobutanol (Merck KGaA, Darmstadt, Germany), and phase separation was facilitated by centrifugation: in 100 mL scale in 50-mL tubes, in 1 L scale in 100-mL flasks (Eppendorf Centrifuge 5810R, 3220 g, 5 min, RT) and in 20 L scale in 2-L flasks (Sorvall RC12BP, ca. 4000 g, 5 min, RT). The product was recovered in the heavier aqueous phase. 100-mL separations were performed with three replicates, 1-L separations with four replicates and the 20-L separation only once. The recovery rate was calculated by dividing the total amount of GFP-HFBI in back extract by the total amount of GFP-HFBI in initial cell extract.

Acknowledgements

The authors gratefully thank Juha Tähtiharju, Merja Aarnio, Tuija Kössö and Riitta Suihkonen for their excellent assistance in pilot cultivations and downstream processing and Sirkka Kanervo, Tuuli Teikari, Jaana Rikkinen and Annika Majanen for technical assistance in cell culture work. This research was funded by Finnish Academy Grant 252442 and the VTT BizF_BY2UpScale 81598 funding. Support from COST action FA0804 Molecular Farming: plants as production platform for high-value proteins is also acknowledged.

References

Bordier, C. (1981) Phase separation of integral membrane proteins in Triton X-114 solution. J. Biol. Chem. **256**, 1604–1607.

Bortesi, L., Rademacher, T., Schiermeyer, A., Schuster, F., Pezzotti, M. and Schillberg, S. (2012) Development of an optimized tetracycline-inducible expression system to increase the accumulation of interleukin-10 in tobacco BY-2 suspension cells. BMC Biotechnol. **12**, 40 doi:10.1186/1472-6750-12-40.

Conley, A.J., Joensuu, J.J., Menassa, R. and Brandle, J.E. (2009) Induction of protein body formation in plant leaves by elastin-like polypeptide fusions. BMC Biol. **7**, 48.

Conley, A.J., Joensuu, J.J., Richman, A. and Menassa, R. (2011) Protein body-inducing fusions for high-level production and purification of recombinant proteins in plants. Plant Biotechnol. J. **9**, 419–433.

De Sutter, V., Vanderhaeghen, R., Tilleman, S., Lammertyn, F., Vanhoutte, I., Karimi, M., Inzé, D., Goossens, A. and Hilson, P. (2005) Exploration of jasmonate signalling via automated and standardized transient expression assays in tobacco cells. Plant J. **44**, 1065–1076.

Doran, P.M. (2013) Therapeutically Important Proteins From In Vitro Plant Tissue Culture Systems. Curr. Med. Chem. **20**, 1047–1055.

Fischer, R., Schillberg, S., Hellwig, S., Twyman, R.M. and Drossard, J. (2012) GMP issues for recombinant plant-derived pharmaceutical proteins. Biotechnol. Adv. **30**, 434–439.

Gutiérrez, S.P., Saberianfar, R., Kohalmi, S.E. and Menassa, R. (2013) Protein body formation in stable transgenic tobacco expressing elastin-like polypeptide and hydrophobin fusion proteins. BMC Biotechnol. **13**, 40. doi:10.1186/1472-6750-13-40.

Hakanpää, J., Szilvay, G.R., Kaljunen, H., Maksimainen, M., Linder, M. and Rouvinen, J. (2006) Two crystal structures of Trichoderma reesei hydrophobin HFBI—The structure of a protein amphiphile with and without detergent interaction. Protein Sci. **15**, 2129–2140.

Hatti-Kaul, R. (2001) Aqueous two-phase systems: a general overview. Mol. Biotechnol. **19**, 269–277.

Hellwig, S., Drossard, J., Twyman, R.M. and Fischer, R. (2004) Plant cell cultures for the production of recombinant proteins. Nat. Biotechnol. **22**, 1415–1422.

Holland, T., Sack, M., Rademacher, T., Schmale, K., Altmann, F., Stadlmann, J., Fischer, R. and Hellwig, S. (2010) Optimal nitrogen supply as a key to increased and sustained production of a monoclonal full-size antibody in BY-2 suspension culture. Biotechnol. Bioeng. **107**, 278–289.

Huang, T.K. and McDonald, K.A.. (2012) Bioreactor systems for in vitro production of foreign proteins using plant cell cultures. Biotechnol. Adv. **30**, 398–409.

Joensuu, J.J., Conley, A.J., Lienemann, M., Brandle, J.E., Linder, M.B. and Menassa, R. (2010) Hydrophobin fusions for high-level transient protein expression and purification in Nicotiana benthamiana. Plant Physiol. **152**, 622–633.

Joensuu, J.J., Conley, A.J., Linder, M.B. and Menassa, R. (2012) Bioseparation of recombinant proteins from plant extract with hydrophobin fusion technology. Methods Mol. Biol. **824**, 527–534.

Joseph, M., Ludevid, M.D., Torrent, M., Rofidal, V., Tauzin, M., Rossignol, M. and Peltier, J.B.. (2012) Proteomic characterisation of endoplasmic reticulum-derived protein bodies in tobacco leaves. BMC Plant Biol. **12**, 36. doi:10.1186/1471-2229-12-36.

Kaldis, A., Ahmad, A., Reid, A., Mcgarvey, B., Brandle, J., Ma, S., Jevnikar, A., Kohalmi, S.E. and Menassa, R. (2013) High-level production of human interleukin-10 fusions in tobacco cell suspension cultures. Plant Biotechnol. J. **11**, 535–545.

Khan, I., Twyman, R.M., Arcalis, E. and Stoger, E. (2012) Using storage organelles for the accumulation and encapsulation of recombinant proteins. Biotechnol. J. **7**, 1099–1108.

Kieran, P.M., MacLoughlin, P.F. and Malone, D.M. (1997) Plant cell suspension cultures: some engineering considerations. J. Biotechnol. **59**, 39–52.

Kirchhoff, J., Raven, N., Boes, A., Roberts, J.L., Russell, S., Treffenfeldt, W., Fischer, R., Schinkel, H., Schiermeyer, A. and Schillberg, S. (2012) Monoclonal tobacco cell lines with enhanced recombinant protein yields can be generated from heterogeneous cell suspension cultures by flow sorting. Plant Biotechnol. J. **10**, 936–944.

Lahtinen, T., Linder, M.B., Nakari-Setälä, T. and Oker-Blom, C. (2008) Hydrophobin (HFBI): a potential fusion partner for one-step purification of recombinant proteins from insect cells. Protein Expr. Purif. **59**, 18–24.

Linder, M.B., Qiao, M., Laumen, F., Selber, K., Hyytiä, T., Nakari-Setälä, T. and Penttilä, M.E. (2004) Efficient purification of recombinant proteins using hydrophobins as tags in surfactant-based two-phase systems. Biochemistry, **43**, 11873–11882.

Mustalahti, E., Saloheimo, M. and Joensuu, J.J. (2013) Intracellular protein production in Trichoderma reesei (Hypocrea jecorina) with hydrophobin fusion technology. New Biotechnol. **30**, 262–268.

Nagata, T. and Kumagai, F. (1999) Plant cell biology through the window of the highly synchronized tobacco BY-2 cell line. Methods Cell Sci. **21**, 123–127.

Nagata, T., Nemoto, Y. and Hasezawa, S. (1992) Tobacco BY-2 Cell Line as the "HeLa" Cell in the Cell Biology of Higher Plants. Int. Rev. Cytol. **132**, 1–30.

Nocarova, E. and Fischer, L. (2009) Cloning of transgenic tobacco BY-2 cells; An efficient method to analyse and reduce high natural heterogeneity of transgene expression. BMC Plant Biol. **9**, 44, doi:10.1186/1471-2229-9-44.

Noguchi, M., Matsumoto, T., Hirata, Y., Yamamoto, K., Katsuyama, A., Kato, A., Azechi, S. and Kato, K. (1977) Improvement of growth rates of plant cell cultures. In Plant Tissue Culture and Its Bio-technological Application, (Barz, W., Reinhard, E. and Zenk, M.H., eds), pp. 85–94. Springer-Verlag, Berlin.

Penttilä, M., Nakari-Setälä, T., Fagerström, R., Selber, K., Kula, M.R., Linder, M. and Tjerneld, F.. (2008) Process for partitioning of proteins. United States Patent 11/252,753.

Sack, M., Paetz, A., Kunert, R., Bomble, M., Hesse, F., Stiegler, G., Fischer, R., Katinger, H., Stoeger, E. and Rademacher, T. (2007) Functional analysis of the broadly neutralizing human anti-HIV-1 antibody 2F5 produced in transgenic BY-2 suspension cultures. FASEB J. **21**, 1655–1664.

Schiermeyer, A., Schinkel, H., Apel, S., Fischer, R. and Schillberg, S. (2005) Production of Desmodus rotundus salivary plasminogen activator a1 (DSPAa1) in tobacco is hampered by proteolysis. Biotechnol. Bioeng. **89**, 848–858.

Schinkel, H., Schiermeyer, A., Soeur, R., Fischer, R. and Schillberg, S. (2005) Production of an active recombinant thrombomodulin derivative in transgenic tobacco plants and suspension cells. Transgenic Res. **14**, 251–259.

Schmale, K., Rademacher, T., Fischer, R. and Hellwig, S. (2006) Towards industrial usefulness—cryo-cell-banking of transgenic BY-2 cell cultures. J. Biotechnol. **124**, 302–311.

Schmidt, S.R. (2013) Protein Bodies in Nature and Biotechnology. *Mol. Biotechnol.* **54**, 257–268.

Selber, K., Tjerneld, F., Collén, A., Hyytiä, T., Nakari-Setälä, T., Bailey, M., Fagerström, R., Kan, J., Van Der Laan, J., Penttilä, M. and Kula, M.R. (2004) Large-scale separation and production of engineered proteins, designed for facilitated recovery in detergent-based aqueous two-phase extraction systems. *Process Biochem.* **39**, 889–896.

Sun, Q.Y., Ding, L.W., Lomonossoff, G.P., Sun, Y.B., Luo, M., Li, C.Q., Jiang, L. and Xu, Z.F. (2011) Improved expression and purification of recombinant human serum albumin from transgenic tobacco suspension culture. *J. Biotechnol.* **155**, 164–172.

Torrent, M., Llompart, B., Lasserre-Ramassamy, S., Llop-Tous, I., Bastida, M., Marzabal, P., Westerholm-Pavinen, A., Saloheimo, M., Heifetz, P.B. and Ludevid, M.D. (2009) Eukaryotic protein production in designed storage organelles. *BMC Biol.* **7**, 5. doi:10.1186/1741-7007-7-5.

Walsh, G. (2010) Biopharmaceutical benchmarks 2010. *Nat. Biotechnol.* **28**, 917–924.

Xu, J., Ge, X. and Dolan, M.C. (2011) Towards high-yield production of pharmaceutical proteins with plant cell suspension cultures. *Biotechnol. Adv.* **29**, 278–299.

12

Simultaneous over-expressing of an acyl-ACP thioesterase (FatB) and silencing of acyl-acyl carrier protein desaturase by artificial microRNAs increases saturated fatty acid levels in *Brassica napus* seeds

Jin-Yue Sun[†], Joe Hammerlindl, Li Forseille, Haixia Zhang and Mark A. Smith*

National Research Council Canada, Saskatoon, SK, Canada

*Correspondence

email Mark.Smith@nrc-cnrc.gc.ca
[†]Present address: Institute of Agro-Food Science and Technology, Shandong Academy of Agricultural Sciences, Jinan, China.

Keywords: *Brassica napus*, saturated fatty acid, FatB, stearoyl-acyl carrier protein desaturase, artificial microRNAs, palmitic acid.

Summary

No temperate oilseeds crops are available that produce oil with a high saturated fatty acid content. To achieve such a profile, *Brassica napus* cv. DH12075 was engineered by simultaneous seed-specific over-expression of a native fatty acyl-ACP thioesterase B [BnFatB(2)] and artificial microRNA-mediated down-regulation of eight endogenous genes encoding putative stearoyl-ACP desaturases (BnSADs). Semi-quantitative RT-PCR analysis of transformed lines showed that the *BnFatB*(2) gene was highly over-expressed and the eight putative SAD genes were strongly down-regulated in developing seed demonstrating the successful application of microRNA as a tool for down-regulation of genes in the allotetraploid plant *B. napus*. Analysis of seed triacylglycerol (TAG) composition revealed that all lines contained high levels of palmitic acid (16:0) and moderately increased levels of stearic acid (18:0). Total saturated fatty acid content was increased from 7.4% in the control to 37.3–45.6% in the transformed lines, with *FatB* over-expression as the dominant trait. A twofold increase in 16:0 was observed in seed polar lipids. The melting point of oil from mature seeds was increased from −10 °C in DH12075 to 15 °C in the line with the highest saturated fatty acid content. TAG composition showed a shift from predominantly C54 TAG to C50 and C52 TAGs enriched in palmitic acid. Seedling establishment at low temperature was compromised in lines with high saturated fatty acid content. Results suggested that transcript encoding stearoyl-ACP desaturase in developing *B. napus* seeds is present in considerable excess of the level required for efficient desaturation of 18:0.

Introduction

As a result of their fatty acid composition, triacylglycerol (TAG) oils from temperate oilseeds crops are liquid, a property that limits their direct use in food applications such as margarine and baking shortenings where solid fats are required. For conversion from liquid oil to a spreadable or solid fat, these oils must either be blended or interesterified with other oils rich in saturated fatty acids, such as palm oil (*Elaeis* spp.), coconut oil (*Cocos nucifera*), hydrogenated vegetable oil or tallow (Chrysan, 2005). These approaches all have associated problems and limitations. For example, interesterification or blending with oils derived from tropical sources, particularly palm oil, is complicated by public perception concerning the sustainability and ecological impact of production (Laurance *et al.*, 2010). Use of hydrogenated oil can lead to the unintentional introduction of undesirable *trans*-fatty acids (TFAs), resulting from incomplete hydrogenation or double-bond isomerization (Sommerfeld, 1983). Under current dietary guidelines, it is recommended that consumption of TFAs from hydrogenated oils should be avoided (Hunter, 2006). A potential solution is to modify a temperate oilseed crop, through genetic engineering, to produce oil with a higher saturated fatty acid content. This oil could be used directly, or as a source of saturates for blending/interesterification. For *Brassica napus* (canola/rape-seed), a number of attempts have been made to achieve this goal

resulting in oils enriched in specific fatty acids including medium chain fatty acids, palmitic (hexadecanoic acid, 16:0) and stearic acids (octadecanoic acid, 18:0) (Stoll *et al.*, 2005).

The pathways of fatty acid biosynthesis and seed TAG assembly in oilseed species have been extensively studied, and recent reviews include those of Baud and Lepiniec (2010) and Bates *et al.* (2013). In summary, *de-novo* fatty acid biosynthesis occurs in the plastids in developing seeds. Synthesis is initiated by the condensation of acetyl-coenzyme A (acetyl-CoA) and malonyl-acyl carrier protein (malonyl-ACP), a multistep process yielding a 4-carbon acyl-ACP. In *B. napus*, repeated cycles of 2-carbon elongation result in the synthesis of palmitoyl-ACP (16:0-ACP). Release of 16:0 from ACP, catalysed by an acyl-ACP thioesterase, with subsequent export from the plastid and activation to 16:0-CoA, makes palmitic acid available for cytosolic processes including TAG assembly. Alternatively, newly formed 16:0-ACP can undergo a further two carbon elongation to stearoyl-ACP (18:0-ACP), with the condensation step being catalysed by the enzyme 3-ketoacyl-ACP synthase II (KASII). Desaturation between carbons 9 and 10, catalysed by stearoyl-ACP desaturase (SAD), then yields oleoyl-ACP (*cis*-9 octadecenoyl-ACP; 18:1-ACP). For further desaturation by the endoplasmic reticulum fatty acid desaturases, and for TAG assembly, 18:1 is released from ACP by a second acyl-ACP thioesterase with export from the plastid and activation to 18:1-CoA. As described by Stoll *et al.* (2005), acyl-ACP thioesterases, 3-ketoacyl-ACP synthase enzymes (KAS) and SAD have all been targeted in various studies aimed at increasing the saturated fatty acid composition of *B. napus* seed oil. As the terminal step in the fatty acid biosynthesis pathway, acyl-ACP thioesterases determine the chain length of acyl groups leaving the plastid for further metabolism in the cytosol. Higher plant acyl-ACP thioesterases can be divided into two distinct classes based on amino acid sequence. Referred to as FatA and FatB, these thioesterases primarily hydrolyse 18:1-ACP and C8-C16 saturated acyl-ACPs, respectively (Jones *et al.*, 1995; Mayer and Shanklin, 2007; Salas and Ohlrogge, 2002). The acyl specificity of FatB thioesterases has been used to advantage to engineer medium chain saturated fatty acid production and palmitic acid levels in *B. napus* (Dehesh *et al.*, 1996; Jones *et al.*, 1995; Voelker *et al.*, 1992). Thioesterase-mediated engineering of 18:0 in *B. napus* was more problematic, but was achieved by expression of a *Garcinia mangostana* (mangosteen) FatA1 acyl-ACP thioesterase, modified by site-directed mutagenesis to increase enzyme activity towards 18:0-ACP. Resulting plants accumulated up to 20% stearic acid in their seed oil compared to approximately 2% in control lines (Facciotti *et al.*, 1999). Manipulation of KAS activity in *B. napus* to achieve increased levels of palmitic acid has been less successful. Seed-specific expression of a KASIII (catalysing the first condensation step of fatty acid biosynthesis) from *Cuphea hookeriana* increased 16:0 levels from 4% to almost 9%, but severely impacted lipid synthesis and oil content (Dehesh *et al.*, 2001). Targeting desaturation of 18:0 to increase saturated fatty acid levels was one of the first examples of the successful engineering of a seed oil profile. Antisense-mediated suppression of SAD was used to increase 18:0 in *B. napus* from 2% to 40% (Knutzon *et al.*, 1992) and from 3.7% to 32% in a more recent report (Zarhloul *et al.*, 2006). Successful enhancement of saturated fatty acid content has also been reported for other oilseed species. For example, over-expression of endogenous *FatB1* or down-regulation of KASII has been used to increase palmitic acid levels in *Arabidopsis thaliana* (Dormann *et al.*, 2000; Pidkowich *et al.*, 2007), and hairpin RNA-mediated post-transcriptional gene silencing of SAD was used in the generation of high-stearic, high-oleic (HS-HO) cottonseed oil (*Gossypium hirsutum*) (Liu *et al.*, 2002).

Although *B. napus* engineered with increased saturated fatty acid content has not been a commercial success, market opportunities still exist for plant derived sources of 16:0 and 18:0, both for food and industrial uses. Temperate oilseed options for 18:0 include HS-HO sunflower (*Helianthus annuus*), now entering commercial production (Dubinsky and Garcés, 2011; Garcés *et al.*, 2012) and perhaps HS soybean (*Glycine max*) (Clemente and Cahoon, 2009), both produced by conventional breeding. Demand for 16:0 is largely met by palm oil (approximately 44% 16:0) or cottonseed oil (approximately 26% 16:0).

Our objective is to further examine the potential of *B. napus* as a platform for saturated fatty acid production, particularly for the temperate production of 16:0 to provide an alternative to palm oil. As the first step in our research, we achieved the simultaneous over-expression of a cDNA encoding a native FatB thioesterase and down-regulation of *SAD* gene expression, generating plants containing up to 46% saturated fatty acids. For precise silencing of specific genes encoding SAD, we used an artificial microRNA-mediated approach (amiRNA). The creation and analysis of these engineered *B. napus* plants is described.

Results

Isolation of *Brassica napus* cDNAs encoding FatB and SAD isoforms

To identify a *B. napus* gene encoding a palmitoyl-ACP-thioesterase (FatB) for over-expression, the GenBank database (http://www.ncbi.nlm.nih.gov/dbEST/index.html) was searched for known *FatB* genes from this species. A single sequence was identified, accession number DQ847275, and both the nucleotide sequence and deduced amino acid sequence (accession ABH11710) were used to search the *B. napus* ESTs database (http://brassicagenomics.ca). Assembly of identified ESTs indicated that, in addition to a sequence corresponding to DQ847275, a second transcript encoding a member of the *FatB* gene family in *B. napus* was also represented. A full-length cDNA representing the second *FatB* gene (designated *BnFatB(2)* in this work) was then amplified by RT-PCR using mRNA isolated from developing seeds of *B. napus* cv. DH12075.

The *BnFatB(2)* open reading frame (ORF) was 1248 bp in length, encoding a deduced precursor protein with a length of 415 amino acids (GenBank Accession KC202816). The coding regions of the two presently known *B. napus FatB* cDNAs, *BnFatB (2)* and DQ847275 are highly conserved, with 88% homology at the nucleotide level and 89% identity at the amino acid level. The deduced precursor proteins (compared in Figure 1) encoded by *BnFatB(2)* and DQ847275 have molecular masses of 46.2 and 46.0 kDa, respectively. Presence of a chloroplast transit peptide was predicted using ChloroP (Emanuelsson *et al.*, 1999; http://www.cbs.dtu.dk/services/ChloroP/) with a predicted cleavage site between amino acid residues Q91 and L92 in BnFatB(2), based on the structure of the mature Arabidopsis FatB (Mayer and Shanklin, 2007), resulting in a predicted mature polypeptide of 324 aa. The deduced mature proteins coded by *BnFatB(2)* and DQ847275 have molecular masses of both 36.7 kDa, and isoelectric point of 6.37 and 6.86, respectively. The three amino acid residues of the catalytic triad (N,H,C), essential for catalytic activity of FatB thioesterases, belonging to the conserved active-site motifs NQHVNN and YRRECG (Mayer and Shanklin, 2007; Yuan *et al.*, 1996), were found in BnFatB(2) mature polypeptide as N318, H320 and C355, respectively. Six amino acid residues (V164, K177, V201, M232, S265 and W312) related to the substrate specificity (Mayer and Shanklin, 2007) are also conserved. The novel *BnFatB(2)* cDNA was selected for over-expression in *B. napus*.

A similar strategy was used to identify ESTs encoding putative stearoyl-ACP desaturase homologues. The *B. napus* EST database was searched using previously identified *SAD* cDNAs, accession X74782, X97325, X63364 and AY642537. By sequence comparison (Table S1 and Figure S1), it was determined that accessions X63364 and AY642537 likely represent the same gene from two different *B. napus* cultivars. Assembly of overlapping ESTs identified a total of 8 sequences predicted to encode full-length *SAD* homologues or orthologues. Assembled sequences were Bn2745, Bn8595, Bn25810, Bn240, Bn47973, Bn47975, Bn47974 and Bn47976 and available as supplemental data (Data S1). Four sequences had high sequence homology to Arabidopsis *FAB2/SSI2* (At2g43710), encoding the chloroplast stearoyl-ACP desaturase. The remaining 4 sequences showed highest deduced amino acid similarity to Arabidopsis DES5 (At3g02630) and were designated BnDES5a to BnDES5d (Table S2). As *B. napus* is an allotetraploid (AACC), we

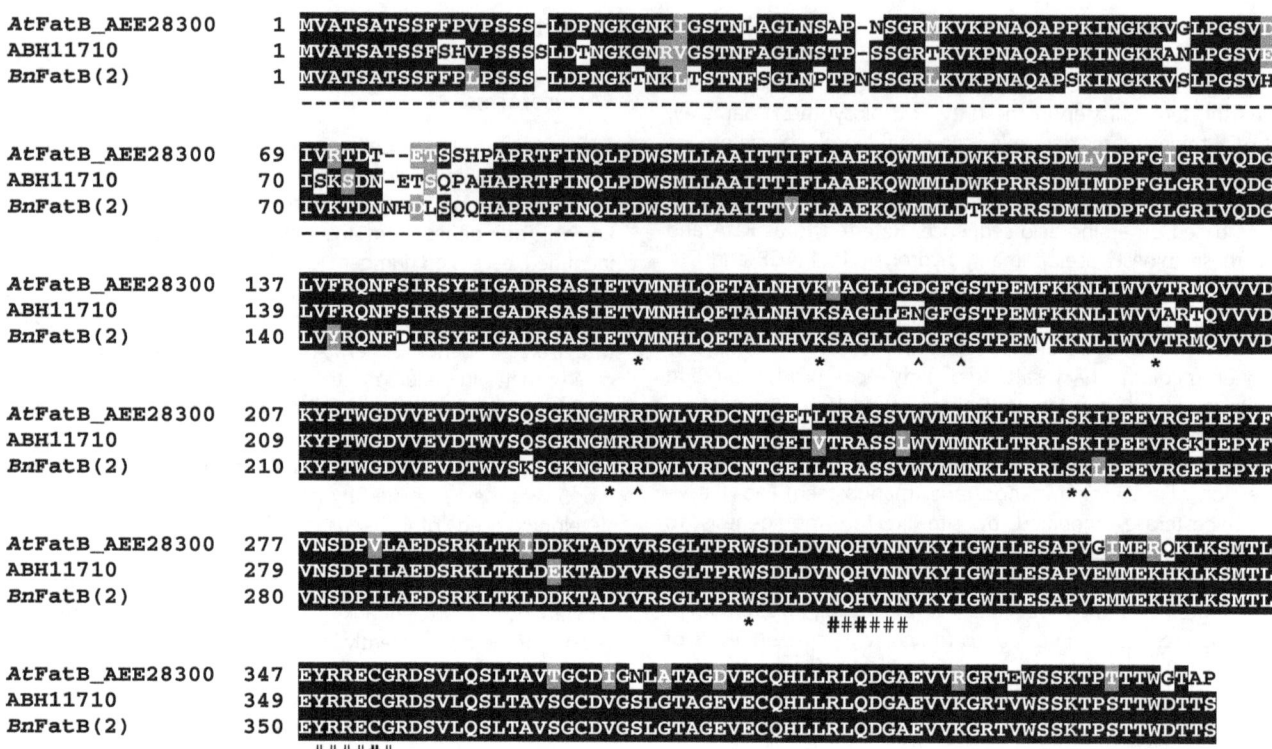

Figure 1 Deduced amino acid sequence comparison of *Arabidopsis thaliana* FatB (*At*FatB, AEE28300) and two FatBs from *Brassica napus*, (DQ847275 and BnFatB(2), KC202816). The predicted signal peptide (Mayer and Shanklin, 2007) is underlined. Conserved active-site motifs (Yuan *et al.*, 1996) are marked below with #, with the N, H and C residues of the catalytic triad marked in bold (#). Putative substrate determining residues identified by Mayer and Shanklin (2007) and marked * with additional previously identified residues marked ^ (Yuan *et al.*, 1996).

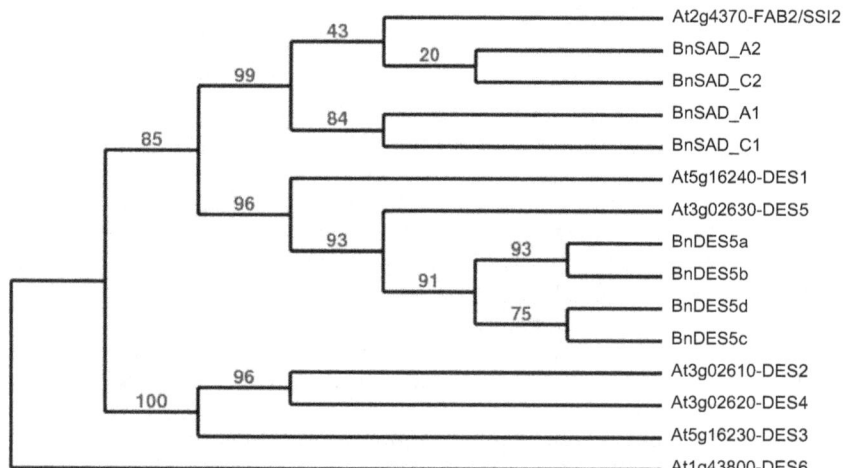

Figure 2 Cladogram with branch support values illustrating the relationship of the *Bn*SAD/DES deduced amino acid sequences, and the Arabidopsis stearoyl-ACP desaturase family (Kachroo *et al.*, 2007).

determined the likely genome of origin for the 4 FAB2/SSI2 homologues by comparison of cDNA and deduced amino acid sequences with available sequences from the diploid species *B. oleracea* (CC) and *B. rapa* (AA) using the NCBI GSS (Genome Survey Sequence) database (http://www.ncbi.nlm.nih.gov/projects/dbGSS/) and more recently the *B. rapa* genome sequence assembly using the *Brassica* database (BRAD, http://www.brassica.info/, Cheng *et al.*, 2011). Based on this analysis, sequences were designated *BnaA-SAD1*, *BnaA-SAD2*, *BnaC-SAD1* and *BnaC-SAD2* as shown in Table S2. A cladogram illustrating the relationship between the BnSAD/DES sequences

and the Arabidopsis stearoyl-ACP desaturase family is given in Figure 2. To determine the accuracy of our assembly of the EST sequences, a 1182 bp cDNA with 100% sequence identity to sequence *BnDES5b* was amplified by RT-PCR using mRNA isolated from developing seeds of *B. napus* cv. DH12075 (KF256138). The deduced amino acid sequence (Figure S2) of 393 residues had 94.4% identity to Arabidopsis Des5 (At3g02630), a predicted precursor-protein molecular mass of 44.8 kDa, and a putative N-terminal chloroplast transit peptide. A consensus binding motif [(D/E)X2H]2 characteristic of di-iron proteins (Shanklin and Cahoon, 1998) was also observed.

Creation and verification of transgenic *B. napus* lines with down-regulated *SAD* expression and up-regulated FatB activity

To simultaneously over-express a *FatB* thioesterase cDNA and down-regulate the four *B. napus* homologues of Arabidopsis *FAB2/SSI2* and the four *AtDES5* like genes, a binary vector was constructed. This contained the *FatB(2)* cDNA under control of a napin promoter for seed-specific expression, accompanied by two amiRNAi cassettes designed to target all 8 *SAD* family genes, also under control of a napin promoter. Design details are given in the experimental procedures section. After *Agrobacterium*-mediated plant transformation, a total of 52 transformants were recovered. All were screened by PCR, and the presence of the intact transformation cassette was confirmed in 33 individuals. Eight plants showed no cassette integration and were likely escapes. Initial characterization was conducted by GC-FAMES analysis of seed fatty acid composition (Figure 3). The majority of the plants exhibited a major increase in total saturated fatty acid content ranging from 7.4% in the control DH12075 line to 43.5% in line transgenic line 48, with palmitic acid being the predominant saturated fatty acid component. Lines with no detectable transgene cassette (plants 3, 9, 14, 27, 30) had saturated fatty acid composition similar to the untransformed control. Transgene locus number for each line was determined by germinating T1 seeds in the presence of kanamycin, and three lines (lines 39, 43 and 48) were selected for further analysis based on transgene locus number and the saturated fatty acid content of the T1 seeds. Kanamycin screening indicated that plant 39 was a single insertion line, plant 43 contained the transgene cassette in two loci, and plant 48 most likely contained the transgene cassette in 3 loci. Segregation for kanamycin sensitivity/resistance in the T1 seeds was 17/55, 5/85 and 2/118 for the three plants, respectively. Transgene copy number was also confirmed by the segregation ratio of 16:0 in single T1 seed, by GC analysis of FAMES. T1 seed was planted, and six homozygous plants (Bn39-17, Bn39-40, Bn39-46, Bn43-5, Bn48-93 and Bn48-115) were selected for detailed characterization. All 6 contained the intact transformation cassette, as evidenced by PCR with genomic DNA as template (Figure S3).

To demonstrate that the artificial microRNA was correctly expressed in the developing seed, we analysed the expression of the two amiRNAs by stem-loop RT-PCR in both 21 and 28 DPA developing seeds. The results from all six transgenic *B. napus* lines showed similar expression levels at 21 DPA and 28 DPA. Both BnSADamiR1 and BnSADamiR2 were highly expressed in 21 DPA developing seeds in the transgenic lines with no expression detected in the developing seeds of untransformed DH12075. Data from stem-loop RT-PCR analysis in 21 DPA developing seeds are shown in Figure S4.

To monitor changes in the expression levels of the target genes, we chose an RT-PCR-based approach, using a *B. napus* actin transcript as an internal control. Initially, the relative expression levels of *BnFatBs*, *BnSADs* and *BnACTIN* in 21 DPA developing seeds of untransformed *B. napus* DH12075 were analysed by RT-PCR using gene-specific primers and the same number of amplification cycles for all primer pairs. These results indicated that *BnaA-SAD1* and *BnaA-SAD2* (Figure S5, lanes 4 and 5) have the highest expression levels among the genes investigated, and the two *BnFatB* genes (Figure S5, lanes 2 and 3) were expressed at similar levels. Differences in amplification product size with each primer pair ensured that specific gene products could be efficiently distinguished. The semi-quantitative RT-PCR was then repeated for the control line and the 6 transgenic lines (Figure 4). Results demonstrated that the transcription level of the over-expressed *BnFatB2* gene in all the six transgenic lines was more than 10-fold higher compared to the endogenous expression seen in the untransformed DH12075, indicating successful over-expression. Expression of the endogenous *BnFatB* gene corresponding to the DQ847275 sequence, as determined by gene-specific primers, appeared unchanged in the transgenic lines. The expression levels of all four *BnSADs* and *BnDES5a-d* were all reduced in the six transgenic lines, but transcript of all genes was still detected. The three plants from line 39 showed the strongest down-regulation of *BnaA-SAD2*.

Determination of fatty acid composition and seed oil content

The fatty acid composition of leaves, young siliques and mature seeds from *B. napus* control and transgenic lines was determined by GC-FAME analysis. To verify that the modified fatty acid composition was restricted to the seed, the average levels of (16:0 + 18:0) and total saturated fatty acids in 7-day-old leaves of all transformed T0 plants were determined and found to be 13.9% ± 1.5% (± SD) and 23.2% ± 2.0% (± SD), respectively.

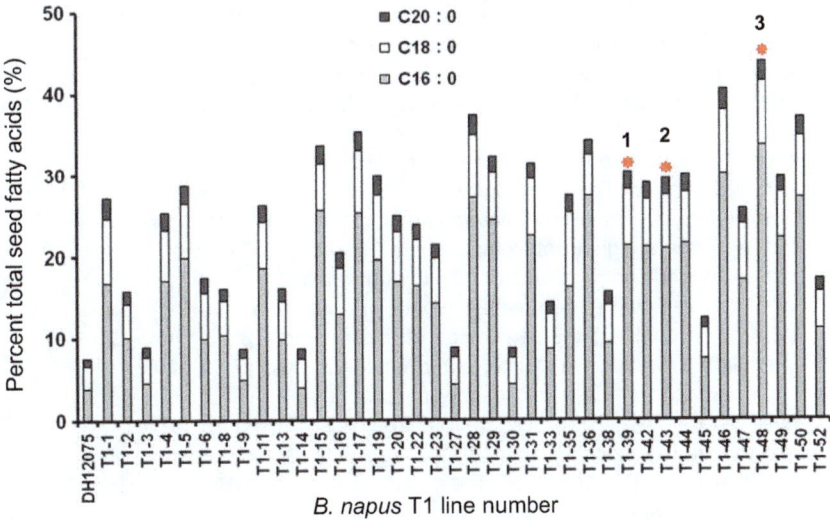

Figure 3 Saturated fatty acids as percentage of total seed fatty acids determined from pooled seeds of primary transgenic lines (T1 seed). Lines marked with a red star (lines 39, 43 and 48) were chosen for further study. Numbers above the columns indicate number of transgene loci in the selected lines.

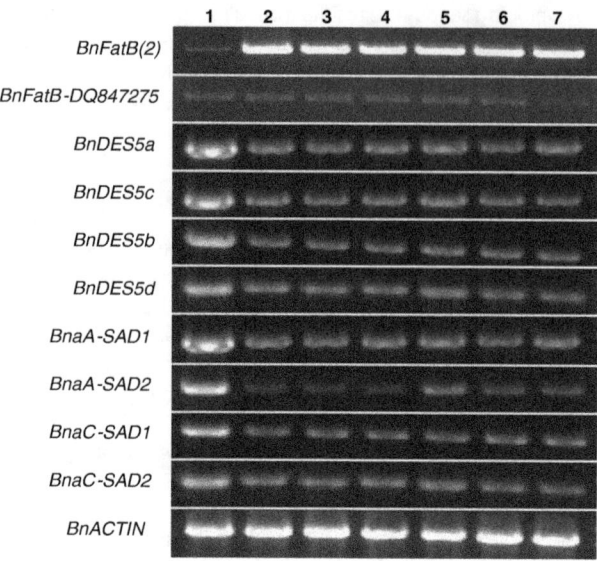

Figure 4 Semi-quantitative RT-PCR analysis of the expression of *BnFatBs* and *BnSADs/BnDES5s*. Total RNA of each line was extracted from 21 DPA developing seeds. All amplifications were tested in a linear range, 24 cycles for *BnFatBs*, *BnSAD/BnDES5s* and *BnACTIN* (control). 1, DH12075; 2, Bn39-17; 3, Bn39-40; 4, Bn39-46; 5, Bn43-5; 6, Bn48-93; and 7, Bn48-115.

No obvious difference was observed compared to levels in DH12075 untransformed controls grown at the same time [12.8% ± 0.8% (± SD) and 26.0% ± 0.7% (± SD), respectively]. Levels of (16:0 + 18:0) and total saturated fatty acids in five PDA siliques of the T0 plants were 20.8% ± 1.5% (± SD) and 31.3% ± 3.6% (± SD), respectively, with no obvious difference compared to that of DH12075 [21.9% ± 0.2% (± SD) and 28.2% ± 1.5% (± SD)]. The transformed lines clearly did not differ significantly from control plants in the fatty acid composition of leaf or young siliques, as would be expected with transgenes placed under control of the seed-specific napin promoter.

Seed lipid composition of T2 seeds from the six transgenic lines Bn39-17, Bn39-40, Bn39-46, Bn43-5, Bn48-93 and Bn48-115, on which detailed molecular characterization had been conducted, is given in Table 1. Total saturated fatty acid content was increased from approximately 7.4% in the control line to 45.6% in Bn48-93, with all lines maintaining the high saturated fatty acid profile seen in the screening of pooled T1 seed. Increase in saturated fatty acid content was largely accounted for by an increase in 16:0, with smaller relative increases seen in 18:0, 20:0 and 22:0. In line Bn39-17 for example, 16:0 increased over sevenfold compared to the control line, with a threefold increase in 18:0 and 20:0. The medium chain fatty acid myristic acid (14:0) was observed in all lines with increased saturated fatty acid content, but was not a significant component of the seed oil of the untransformed control. Levels of all monounsaturated fatty acids were reduced, with the greatest reduction seen in oleic acid (18:1$^{\Delta 9}$). In contrast, the percentage of linoleic acid (18:2$^{\Delta 9,12}$) was slightly increased in all lines except Bn48-93. Percentage of linolenic acid (18:3$^{\Delta 9,12,15}$) showed considerable variation between the lines, with those derived from the T0 plant BN39 not differing significantly from the control, whereas the other lines showed a small reduction. The minor n-7 fatty acids 16:1$^{\Delta 9}$ and 18:1$^{\Delta 11}$ were both reduced in the transgenic lines.

Table 1 Fatty acid composition of T3 seeds (Bn39-17, Bn39-40, Bn39-46 and Bn43-5) and T4 seeds (Bn48-93 and Bn48-115) from the six transgenic lines (± SD) and melting temperature of the extracted oil

Lines	% Fatty acid											Sat FA	Melting temp °C
	14:0	16:0	16:1	18:0	18:1D9	18:1D11	18:2	18:3	20:0	20:1D11	22:0		
DH12075	ND	4.2 ± 0.1	0.2 ± 0.0	2.1 ± 0.2	64.8 ± 1.3	2.5 ± 0.2	12.8 ± 0.6	10.4 ± 0.8	0.7 ± 0.1	1.3 ± 0.0	0.4 ± 0.0	7.4 ± 0.0	−10.0
Bn39-17	1.3 ± 0.2	28.0 ± 2.0	0.1 ± 0.0	6.8 ± 0.5	33.7 ± 4.0	1.7 ± 0.1	14.8 ± 0.8	9.0 ± 1.6	2.1 ± 0.1	0.9 ± 0.0	0.9 ± 0.1	39.0 ± 2.5	11.0
Bn39-40	1.4 ± 0.3	27.0 ± 0.9	0.1 ± 0.0	7.3 ± 0.4	30.5 ± 1.3	1.7 ± 0.1	15.0 ± 0.4	11.6 ± 0.4	2.3 ± 0.1	0.9 ± 0.0	1.0 ± 0.1	38.9 ± 1.1	10.0
Bn39-46	1.2 ± 0.1	26.5 ± 0.8	0.1 ± 0.0	7.7 ± 0.4	36.6 ± 1.7	1.9 ± 0.2	13.5 ± 0.9	8.2 ± 0.8	2.1 ± 0.1	0.8 ± 0.0	0.8 ± 0.1	38.3 ± 1.1	8.5
Bn43-5	0.9 ± 0.1	24.0 ± 0.9	0.1 ± 0.0	8.9 ± 0.5	40.0 ± 1.6	1.9 ± 0.2	13.1 ± 0.7	5.8 ± 0.9	2.6 ± 0.3	0.9 ± 0.0	1.0 ± 0.1	37.3 ± 0.1	7.0
Bn48-93	1.3 ± 0.2	31.1 ± 2.0	ND	11.0 ± 2.4	28.8 ± 2.2	1.1 ± 0.2	10.9 ± 2.3	5.6 ± 2.7	2.2 ± 0.3	ND	ND	45.6 ± 1.5	15.0
Bn48-115	1.0 ± 0.2	26.8 ± 2.7	0.1 ± 0.0	7.8 ± 0.9	35.1 ± 4.3	2.1 ± 0.2	15.0 ± 1.1	7.5 ± 1.0	2.1 ± 0.2	0.8 ± 0.1	0.8 ± 0.1	38.5 ± 3.0	12.0

ND, nondetectable.
Mean ± standard deviation for n = 3.

To examine whether saturated fatty acid content was increased in seed polar lipids, total lipid was extracted from mature seeds of DH12075 plants, and two transgenic lines (T3 seeds, plants Bn39-17-1 and Bn48-115-1) and separated into TAG and polar lipid fractions by TLC for analysis of fatty acid composition (Figure 5). In the DH12075 seeds, 16:0 accounted for 9.7% of total fatty acids in the polar lipid fraction. Levels increased approximately twofold to 20% in the transgenic lines. In contrast, the levels of 16:0 in TAG increased approximately sevenfold from 4.4% in the DH12075 seeds to 27.5% and 28.7% in the Bn39-17-1 and Bn48-115-1 seeds, respectively. The increase in 16:0 in the polar lipids and TAG of the transformed lines was accompanied by a significant decrease in 18:1, with a small increase in 18:2 and a slight decrease in 18:3.

Seed oil content was determined from pooled T2 seeds from the six transgenic lines and from control plants grown at the same time. Average oil content of the control line was 41.8% ± 1.4% (± SD) with an oil content of 48.3% ± 0.4% (± SD) recorded for line Bn43-5 (Figure 6a), as measured by NMR. Although line BN43-5 had considerably higher oil content compared to the control, lines Bn39-40 and Bn48-115 had slightly reduced oil content, and no clear pattern was obvious from these plants. We therefore returned to the T1 seeds for all of the transgenic lines and compared average oil content to 16:0 saturated fatty acid composition over the entire population (Figure 6b) Analysis of the results indicated that a weak correlation ($R^2 = 0.05$) was seen with increased 16:0 content appearing to correspond to a slight decrease in seed oil content. Taking data from the analysis of T2 seeds into consideration, is seems that the increase in saturated fatty acid does not have a major impact on oil content.

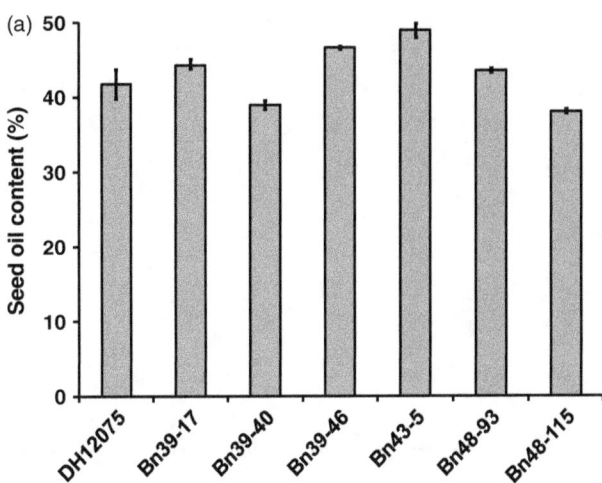

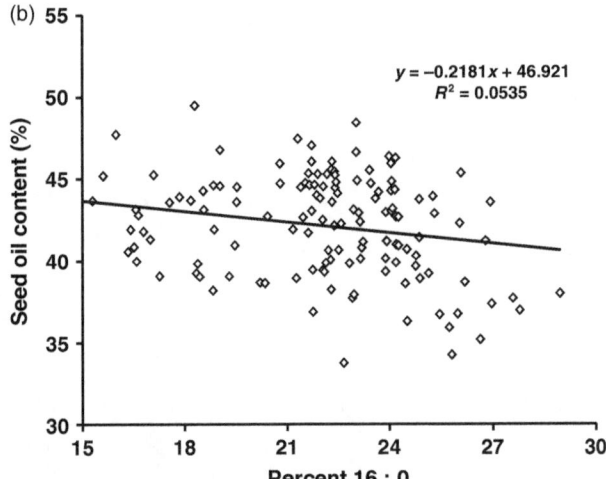

Figure 6 Seed oil content determined for *Brassica napus* line DH12075 and 6 transformed lines (a). Measurements were made in triplicate by NMR. Correlation between percentage fatty acid and percentage 16:0 in a population of *B. napus* lines transformed with vector pLS571. Points represent data for individual transformed lines, measured from pooled T1 seeds. Data (a) are mean ± SD for $n = 3$ or $n = 9$ (DH12075) and (b) mean ± SD for $n = 3$.

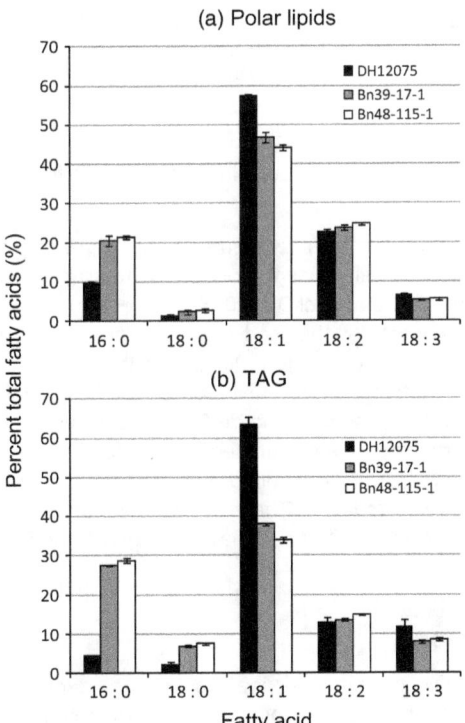

Figure 5 Fatty acid distribution in (a) seed polar lipids and (b) TAG of the control line DH12075 and representative transformed lines. Data are mean ± SD for $n = 3$.

Analysis of triacylglycerol by MALDI-TOF MS and physical properties of the seed oil

As the transgenic lines exhibited dramatically altered fatty acid profiles, we examined the seed triacylglycerols by Matrix-Assisted Laser Desorption/Ionization Time-of-Flight Mass Spectrometry (MALDI-TOF MS). Spectra were collected between 800 *m/z* and 1000 *m/z* for oil samples from line Bn48-115 and the control DH12075 line. Results (Figure 7) showed a shift from predominantly C54 TAG species (where the number indicates the total number of carbons in the fatty acid components) in DH12075 such as tri-oleate OOO (*m/z* = 907.8), OOL (*m/z* = 905.8), OOLn (*m/z* = 903.8), enriched in oleic acid (O), linoleic acid (L) and linolenic acid (Ln), to predominantly C50 and C52 TAGs in Bn48-115 enriched in palmitic acid (P) and stearic acid (S), such as POP (*m/z* = 855.8), PLP (*m/z* = 853.7), PLnP (*m/z* = 851.7), POO (*m/z* = 881.8), POL (*m/z* = 879.8), PLL (*m/z* = 877.7), POS (*m/z* = 883.8), PLS (*m/z* = 881.8), PLnS

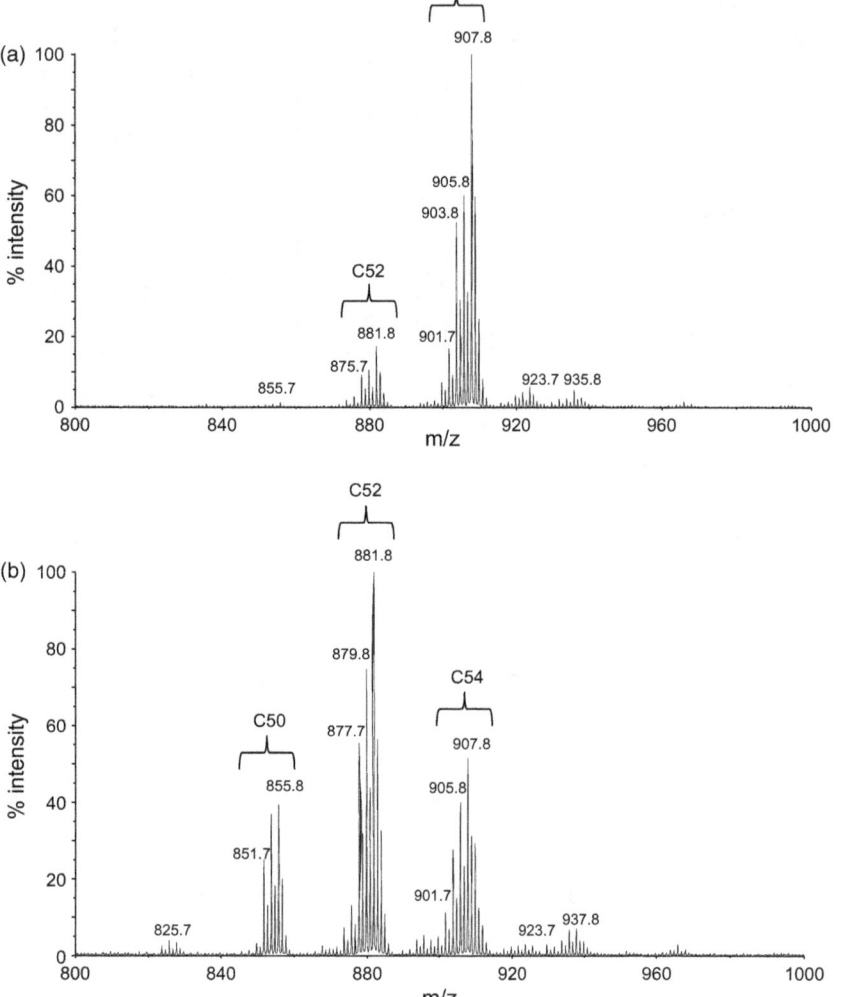

Figure 7 MALDI-TOF MS spectra (*m/z* 800–1000 region) of seed oil from *Brassica napus* line DH12075 (a) and transformed line Bn48-115 (b). Spectra were recorded in positive ion mode; the ions are sodium adducts ([M+Na]+). TAGs are labelled according to the total number of carbons in the fatty acid component.

(*m/z* = 879.8). The three most intense ions identified in the Bn48-115 TAG region, confirmed by MALDI-TOF MS/MS (data not shown), were attributed to POO (*m/z* 881.8), POL (*m/z* = 879.8) and PLP (*m/z* = 853.7) TAG species. For the control oil sample DH12075, the three dominant ions identified in the TAG region were identified as OOO (*m/z* = 907.8), OOL (*m/z* = 905.8) and OOLn (*m/z* = 903.8). The increase in saturated fatty acid content of the seed oil clearly corresponded to a major increase in TAG species containing a single palmitate moiety, and the appearance of substantial amounts of TAG containing two molecules of palmitate, TAG species that were largely absent in oil from the control seeds.

To assess the physical properties of the oil, melting point was determined for hexane-extracted oil prepared from seeds of the control plants and the six transgenic lines. Results (Table 1) indicated that the melting temperature of the oil from transgenic plants was greatly increased at 12.0 °C for Bn48-115 and 15.0 °C in Bn48-93, more than 20 °C higher than the melting temperature of −10 °C observed for oil from the control DH12075 seeds. Oil from the different transgenic lines chilled to 6 °C for 20 min is shown in Figure 8. All samples from the transgenic plants were solid at this temperature suggesting a potential use as a spreadable fat with zero trans-fatty acid content.

Seed germination and seedling growth at low temperature

When conducted at room temperature (22 °C), seed germination and seedling establishment for all 6 transgenic lines was comparable to the control DH12075, irrespective of the saturated fatty acid content of the seed oil. Plants grew normally under greenhouse conditions (Figure 9a). When the seeds were germinated at low temperature, inhibition of seedling establishment was observed. As shown in Figure 9b, seeds of the six lines were able to germinate when incubated at 6 °C for 13 days, but root extension was severely inhibited in most lines and greening of cotyledons was reduced. The most severely affected line was Bn48-93, the line with the highest saturated fatty acid content (45.6%).

Discussion

The objective of this work was to produce a prototype high saturate oil in a *B. napus* canola cultivar. This oil could provide an alternative to palm oil as a source of palmitic acid or be used for blending or interesterification to produce margarine or shortening. The use of canola, with around 23% polyunsaturated fatty acid (PUFA) content, would distinguish the oil from cotton

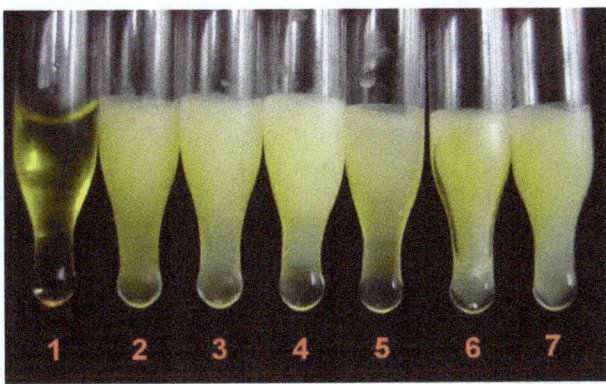

Figure 8 Oils extracted from pooled seeds of different transgenic lines chilled to 6 °C. 1, DH12075; 2, Bn39-17; 3, Bn39-40; 4, Bn39-46; 5, Bn43-5; 6, Bn48-93; and 7, Bn48-115.

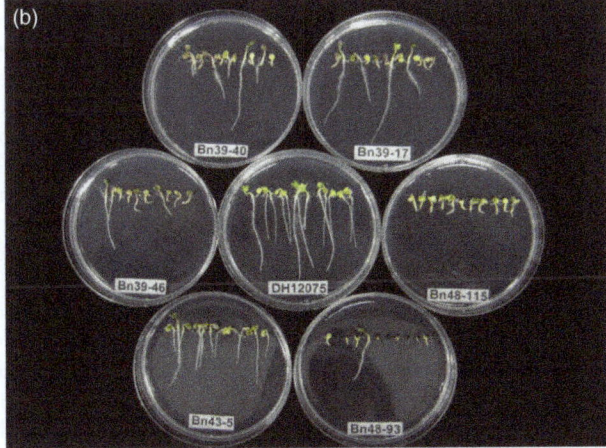

Figure 9 (a) DH12075 and transgenic lines grown under greenhouse conditions. (b) Seed germination and seedling establishment of transgenic *Brassica napus* lines at low temperature. Seeds germinated and grown at 6 °C for 13 days on A. T. medium.

seed oil, a source of palmitic acid in a high PUFA background. The approach was to use a combination of *FatB* over-expression to enhance palmitate release from ACP while simultaneously down-

regulating *SAD* expression to block desaturation of stearoyl-ACP. Based on previous work suggesting that FatB activity may in part be influenced by ACP structure (Salas and Ohlrogge, 2002), we chose to use a native *FatB* for the over-expression work and therefore cloned a cDNA encoding a novel *FatB* from *B. napus*. To down-regulate *SAD* expression by the highly gene-specific method of artificial microRNA-mediated gene silencing (Warthmann *et al.*, 2008), it was first necessary to identify all target genes in *B. napus*. Due to the nature of *Brassica napus*, an allotetrapoid ($2n = 4 \times = 38$, AACC) thought to have originated through spontaneous interspecific hybridization between *Brassica rapa* ($2n = 20$, AA) and *Brassica oleracea* ($2n = 18$ CC (Nagaharu, 1935; Parkin *et al.*, 1995), SAD activity was expected to be encoded by multiple genes. As no genome sequence was available for any *Brassica* species when the work described here was initiated, we searched all available *B. napus* EST resources using Arabidopsis sequences as reference and identified four genes encoding proteins with high sequence identity to Arabidopsis *FAB2/SSI2* (At2g43710) and four with high amino acid sequence identity to Arabidopsis *DES5* (At3g02630, Kachroo *et al.*, 2007). By RT-PCR (Figure S5), all eight genes were shown to be expressed in developing *B. napus* seeds. Previous work has shown that *B. napus* contains four genes encoding stearoyl-ACP desaturase (Slocombe *et al.*, 1994), cDNAs representing three of these genes were available in public databases, and we were able to clone a 4th. The Arabidopsis stearoyl-ACP desaturase gene family, however, contains seven members (Kachroo *et al.*, 2007). These are present as a tandem triplet on chromosome 3 (At3g02610, At3g02620 and At3g02630, designated *DES2*, *DES4* and *DES5*, respectively), a tandem pair on chromosome 5 (At5g16230 and At5g16240, designated *DES3* and *DES1*) and 2 single genes on chromosomes 1 and 2 (At1g43800 and At2g43710, *DES6* and *FAB2/SSI2*). The *FAB2/SSI2* locus (At2g43710) encodes the SAD largely responsible for the desaturation of 18:0 to 18:1 in plastids throughout the plant (Lightner *et al.*, 1994). Although the biological role of DES5 in Arabidopsis is currently unknown, the enzyme has been shown to catalyse the Δ9 desaturation of 18:0-ACP (Kachroo *et al.*, 2007). To ensure the highest probability of down-regulating SAD activity, our strategy was to target all 8 genes identified as being expressed in developing *B. napus* seed, not just the four *FAB2/SSI2* homologues.

With the recent release of the *B. rapa* (Chiifu-401) genome sequence (Wang *et al.*, 2011) complemented by syntenic gene analysis (Cheng *et al.*, 2012), it is now possible to examine the SAD gene family in the diploid species representing the A genome of *B. napus*. Like Arabidopsis, the *B. rapa* SAD gene family also contains seven genes. The *B. rapa* genes Bra000321 and Bra008631, located on chromosomes A03 and A05, respectively, are most likely orthologues of Arabidopsis *FAB2/SSI2*. Bra008631 and Bra008632 located in tandem on chromosome A10 appear to be orthologues of the Arabidopsis tandem pair *DES1* (At5g16240) and *DES3* (At5g16230). No orthologues of Arabidopsis *DES6* (At1g43800) were observed. By sequence comparison, the *B. rapa* genes Bra021427 and Bra001057 are orthologues of Arabidopsis *DES5*, whereas Bra039178 encodes a protein with high amino acid identity to both DES2 and DES4. Examination of the genomic sequence adjacent of the predicted Bra039178 ORF, using FGENESH+ (http://linux1.softberry.com/) uncovered an additional initiation codon 42 bp upstream, giving a predicted amino acid sequence of 409 residues for the encoded protein. Both potential translations result in a protein with higher amino acid identity to Arabidopsis *DES2* (90%) than *DES4* (85%). A previous detailed study of SAD loci in the *B. napus* genome and the dipoid A and C genomes using BAC sequencing (Cho *et al.*, 2010) presented evidence that the regions in those genomes corresponding to the Arabidopsis tandem triplicate of *SAD* genes At3g02610, At3g02620 and At3g02630 (*DES2*, *DES4* and *DES5* respectively) contain only a single gene. The conclusion was that the *Brassica* genomes contained no genes equivalent to At3g02610 and At3g02620 (*DES2* and *DES4*) and a single gene representing the At3g02630 (*DES5*) orthologue. The published *B. rapa* genome sequence indicates that there are 2 *DES5* orthologues on chromosomes A01 and A03, and a single *DES2* orthologue on chromosome A05 (Figure S6a), suggesting a complex rearrangement. No gene orthologous to of Arabidopsis *DES4* (At3g02620) was observed in the *B. rapa* genome sequence assembly, and this gene is also absent from *Arabidopsis lyrata* (Figure S6b). The presence of two orthologues of Arabidopsis *FAB2/SSI2* and two orthologues of *DES5* in *B. rapa* supports our assembly of ESTs representing four genes for *FAB2/SSI2* and 4 for *DES5* in *B. napus*. Further clarification is expected on release of the *B. oleracea* and *B. napus* genomes.

The transformed *B. napus* plants showed a major increase in saturated fatty acid content with 16:0 as the predominant fatty acid, suggesting that the phenotype observed was largely due to the activity of the over-expressed BnFatB2. The anticipated phenotype of high 16:0 in combination with high 18:0 levels was not observed. The moderate increase in 18:0 and its elongation products 20:0 and 22:0 suggests a slight decrease in SAD activity, although FatB activity may also contribute to this increase. We did not conduct over-expression of the *BnFatB(2)* cDNA alone in *B. napus*; however, ectopic expression of *BnFatB (2)* in Arabidopsis resulted in increased levels of 16:0 in the seed oil accompanied by a small increase in 18:0 (results not shown). As discussed previously, with the exception of the *FatB* from nutmeg (*Myristica fragrans*), only small increases in 18:0 were observed on *FatB* over-expression in *B. napus* (Hawkins and Kridl, 1998; Jones *et al.*, 1995; Voelker *et al.*, 1997) or Arabidopsis (Dormann *et al.*, 2000). Further evidence of efficient *SAD* down-regulation is a reduction in oil content, accompanied by a characteristic increase in 18:3 and VLCFAs levels, resulting from changes in flux through the fatty acid desaturation and elongation pathways (Cernac and Benning, 2004; Knutzon *et al.*, 1992). Significantly increased 18:3 was not seen in the transformed lines, but variation in 18:2 and 18:3 content between the lines may reflect alterations in flux between the pathways of fatty acid modification and TAG assembly.

Molecular analysis clearly demonstrated the presence of the silencing amiRNA and a significant reduction in expression of all eight target *SAD/DES5* genes. We examined amiRNA expression at a development time when seeds were actively synthesizing oil (Figure 4, 21 DPA), and it is clear that the transgenes were expressed at an appropriate time to influence oil composition. Previous modification of *B. napus* by antisense targeting of the *SAD* genes has successfully raised stearate levels to 40% of total seed fatty acids (Knutzon *et al.*, 1992; Zarhloul *et al.*, 2006). Although the previous studies did not report transcript levels, analysis of developing seeds by Western blot demonstrated an almost complete loss of SAD protein (Knutzon *et al.*, 1992) suggesting very efficient gene down-regulation. *SAD* transcript in *B. napus* is highly expressed in the developing seed (Slocombe *et al.*, 1992 and Figure S5), and our results suggest that the amiRNA-mediated silencing approach may not have achieved

sufficient silencing to significantly reduce SAD activity. Transcript encoding SAD in *B. napus* seeds appears to be present in considerable excess of the level required for efficient desaturation of 18:0.

In Arabidopsis, it is clear that although the SAD homologue FAB2/SSI2 is the primary enzyme responsible for 18:0 desaturation, other enzymes also contribute to 18:1 biosynthesis. The Arabidopsis *FAB2/SSI2* gene was first identified in an EMS-induced mutant population (James and Dooner, 1990). Although sequencing of the *FAD2/SSI2* gene from this mutant identified a point mutation resulting in a truncated, and likely inactive, protein (Kachroo *et al.*, 2001), the mutant still accumulates significant amounts of C18 unsaturated fatty acids and 20:1 in the seed oil. Subsequent identification and characterization of the additional six genes of the Arabidopsis *SAD/DES* gene family indicates that DES1 and DES5 catalyse the Δ9 desaturation of 18:0-ACP, whereas DES3 preferentially catalysed 16:0-ACP desaturation, all with very much lower *in-vitro* activity than FAB2/SSI2. The native expression of these genes was not able to compensate for the *fab2/ssi2* loss of function mutation, and T-DNA insertions in *DES1* and *DES4* did not alter 18:0 levels in any of the tissues analysed (Kachroo *et al.*, 2007). Without further knowledge of the extent of the *B. napus* SAD/DES gene family, and the activities of their encoded proteins, the contribution to 18:0 desaturation of enzymes other than the *SAD* and *DES5* homologues targeted here is unclear. A repeat of this work using an alternative silencing technique, or with the very effective antisense technique reported earlier (Knutzon *et al.*, 1992), may offer new insights into the results observed here. Artificial microRNA-mediated methods have been successfully applied to silence genes of lipid biosynthesis in Arabidopsis (Belide *et al.*, 2012), but there is no published systematic study of their effectiveness in *B. napus*. From our analysis, we were not able to determine any biological role for the *B. napus* DES5 gene products.

Unlike pure compounds, natural vegetable oils do not have true melting points as they are made up of complex mixtures of TAG molecules that pass through a gradual softening before becoming completely liquid (O'Brien, 1998). This is further complicated by the fact that oil crystals can exist in several polymorphic states (Fasina *et al.*, 2008). The melting point, as determined by the oil becoming completely liquid in phase, for the oils from the transformed plants ranged from 7 to 15 °C, compared to −10 °C for the control. This feature suggests that the oils may have some potential for direct use. From a nutritional aspect, low saturated fatty acid intake is currently considered desirable; however, the metabolic effects of saturated fatty acids differ according to chain length (German and Dillard, 2004), and many recent studies are now questioning the association between dietary saturated fatty acids and cardiovascular disease (Lawrence, 2013). Furthermore, the structure of plant derived 16:0 rich TAG may have some health benefits compared to oils where saturated fatty acids are introduced randomly by hydrogenation or interesterification. In most vegetable oils, saturated fatty acids are located in *sn*-1 and *sn*-3 positions of triacylglycerols, polyunsaturated fatty acids (specifically linoleic and linolenic acids) occupy *sn*-2 position (middle), and monoenoic acids are relatively evenly distributed in *sn*-1, *sn*-2 and *sn*-3 positions; longer-chain fatty acids (C20-C24) are apparently concentrated in the primary positions with some small preference for position *sn*-3; less-common fatty acids tend to be concentrated in position *sn*-3 (Christie *et al.*, 1991). Analysis of the oils by MALDI-TOF MS demonstrated that 16:0 was distributed throughout the TAG, with the predominant species being C52 TAGs (C16+C18+C18) and a significant amount of C50 TAG (C16+C18+C16). In humans, digestion of TAG by endogenous lipases generates free fatty acids and sn-2-monoacylglycerol (sn-2-MAG) (Mu and Hoy, 2004). Exclusion of saturates from the sn-2 position of TAG would, however, mean that digestion of the oil would generate sn-2 MAGs that were largely devoid of saturated fatty acids. The relatively poor intestinal absorption of 16:0 free fatty acid would further limit saturate uptake.

Of some concern, and a trait that requires further characterization, is the poor seedling establishment observed when seedlings were grown at low temperature. Oilseed germination and seedling establishment are correlated with the degradation of storage lipids catabolized by the action of lipases (Quettier and Eastmond, 2009; Theodoulou and Eastmond, 2012). Germinating seeds generally express multiple lipases, and no discrimination against saturated fatty acids has been reported, with most lipases now considered to be promiscuous (Kapoor and Gupta, 2012). Studies with sunflower (*Helianthus annuus*) seed lipases did not show a marked preference for any TAG species and TAG degradation in high and low saturated fatty acid varieties proceeded at a similar rate during germination (Fernández-Moya et al., 2000). Studies carried out with seeds of *Pinus edulis* indicated that the lipases were not specific for individual TAGs during germination, and levels of all TAGs were depleted similarly (Hammer and Murphy, 1994). Earlier work (Lin et al., 1986), however, suggested a correlation between lipase specificity and the fatty acid composition in the seeds of plants making unusual fatty acids including castor bean (*Ricinus communis*, ricinoleic acid), rapeseed (erucic acid) and elm (*Ulmus americana*, medium chain fatty acids). Germination and oil breakdown in the *B. napus* high 16:0 lines would likely result in enrichment of the seedling acyl-CoA pool with saturated fatty acids. Incorporation of these fatty acids into membrane lipids during seedling establishment may play a role in poor establishment at low temperature. Our analysis of seed polar lipids indicated that they already contained more than twice the percentage of saturated fatty acids found in the control line. The balance of saturated and unsaturated fatty acids in membrane lipids is an important determinant of membrane fluidity adaption to low temperature (Nishida and Murata, 1996). Further characterization to determine the reason for poor establishment and development of a solution to this problem is required if high saturated fatty acid canola is to become a viable crop. Although canola is a spring sown crop in Canada, night-time temperatures may be low enough to impact seedling establishment.

Looking beyond the experimental prototype and potential agronomic concerns described here, it is envisioned that high palmitic canola could coexist with other types if the crop was grown in an identity preserved system. Such a system is successfully established for high erucic rapeseed (*B. napus* HEAR) grown for industrial use. The future of high saturated fatty acid crops will necessarily be determined by market needs. The current low price of palm oil compared to other commodity vegetable oils suggests that demand for alternative sources of palmitic acid may be low. Co-engineering of these plants with enzymes that can modify palmitate to produce a higher value fatty acid, for example a monounsaturated fatty acid with a novel double-bond position, may therefore be an alternative.

Experimental procedures

Plant materials and growth condition

The *B. napus* cultivar used in this work was DH12075. Control and all transgenic lines were grown in pots with soil in a controlled greenhouse environment (16 h light/8 h dark, 24 °C/ 18 °C). Plants were bagged during bolting to ensure self-pollination, and flowers were tagged on opening (anthesis) to record the development stages of seeds.

Isolation of *BnFatB* and *BnDES5* cDNAs

The SV Total RNA Isolation System (Promega Corp. Madison, WI) was used for RNA isolation from 0.2 g of developing seeds collected at 20 days post anthesis (DPA). SuperScript II reverse transcriptase (Invitrogen Inc., Burlington, ON, Canada) was used for first-strand cDNA synthesis. To amplify a full-length open reading frame (ORF) encoding BnFatB(2), primer pairs BnEstFatbFullUp (5′-ATGGTGGCCACCTCTGCTACATCCTC) and BnEstFatbFullDo (5′-TTACGATGTAGTGTCCCAAGTCG) were designed based on the sequences of two expressed sequence tags (ESTs), GenBank accessions EL587436 and FG567880. PCR was conducted using the PCR Platinum® kit with Pfx DNA Polymerase (Invitrogen) using the following conditions: 94 °C for 3 min., then 32 cycles of 94 °C, 30 s; 55 °C, 30 s; 72 °C, 90 s, with a final extension at 72 °C for 10 min. To amplify an ORF encoding BnDES5b, primer pairs BnEstSADFullUp (5′-AAAATGG CGATGGCTATGAG) and BnEstSADFullDo (5′-TTAAGCCCTAAT CTCTCGATCA) were designed based on the sequences of three ESTs, EE536302, EE410520 and EE453671. TaKaRa Ex Taq™ DNA polymerase (EMD Millipore, Billerica, MA) was used for amplification under the following conditions: 95 °C for 3 min, then 35 cycles of 98 °C, 10 s; 68 °C, 100 s, with a final extension at 72 °C for 8 min. *Brassica napus* developing seeds cDNA was used as template for both PCR amplifications.

Artificial microRNA design and transformation vector construction

The 8 sequences encoding *BnaA-SAD1*, *BnaA-SAD2*, *BnaC-SAD1*, *BnaC-SAD2* and *BnDES5a* to *BnDES5d* were used for amiRNA design. Using website (http://wmd3.weigelworld.org/ cgi-bin/webapp.cgi), two amiRNA fragments were designed, BnSADamiR1 (target: *BnDES5a, BnDES5b, BnDES5c* and *BnDES5d*) and BnSADamiR2 (target: *BnaA-SAD1, BnaA-SAD2, BnaC-SAD1* and *BnaC-SAD2*). AmiRNAs are designed to resemble natural miRNAs using three criteria: they start with a U, they display 5′ instability relative to their amiRNA*, and their 10th nucleotide is either an A or an U. The strand with lower thermodynamic stability at its 5′ end (5′ instability) is preferentially incorporated into RISC. In designing amiRNA, the following rules are also applied: no mismatch between positions 2 and 12 of the amiRNA for all targets; one (or two) mismatches at the amiRNA 3′ end (pos.18-21); similar mismatch pattern for all intended targets; absolute hybridization energy between −35 and −38 kcal/mole. The primers for engineering amiRNAs and site-directed mutagenesis on precursors of endogenous miRNAs were designed by the artificial microRNA designer WMD (http://wmd3.weigelworld.org/cgi-bin/ webapp.cgi). Primer designing for *BnSAD/DES* amiRNAs was based on RS300 sequence (miR319a *A. thaliana*) gifted by Prof. Detlef Weigel, Max Planck Institute for Developmental Biology,

Germany. The cloning strategy and protocol for engineering artificial microRNAs using four oligonucleotide sequences (I–IV) and RS300 A and B into the endogenous miR319a precursor are available at the following website (http://wmd3.weigel-world.org/cgi-bin/webapp.cgi?page=Help;project=stdwmd). Figure S7 illustrates the procedure used to generate the amiRNAs. The sequence of oligonucleotides I–IV for BnSADamiR1 and BnSADamiR2 is given in Table S3. Sequences of BnSADamiRNAs and their targeted regions for *BnSAD* and *BnDES5* genes in *B. napus* are given in Table S4.

The two individually engineered BnSADamiR1 and BnSADamiR2 fragments in the miR319a precursor from pRS300 were fused with a napin promoter and a nos terminator (nos-T) to form two expression cassettes, Pnapin-BnSADamiR1-NOS-T and Pnapin-BnSADamiR2-NOS-T by over-lapping PCR. The *BnFatB(2)* ORF was inserted between the napin promoter and a nos-T terminator to form a third expression cassette, Pnapin-BnFatB-NOS-T. All three expression cassettes were integrated into the binary vector pHS723 (Hirji *et al.*, 1996) as shown in Figure S7. The successfully constructed vector was named plasmid pLS571. In each cassette napin, a seed-specific promoter was used to ensure expression only in the developing seed.

Agrobacterium-mediated transformation of *B. napus*

Plasmid pLS571 was introduced into *Agrobacterium* strain GV3101: pMP90 by electroporation. Colonies were picked from selection plates and grown overnight in LB medium with 50 mg/L each of kanamycin and gentamycin. Presence and integrity of the transformation vector was confirmed my miniprep and restriction enzyme digestion. *Agrobacterium*-mediated transformation of *B. napus* cotyledonary petiole explants from 5-day-old seedlings was carried out essentially as described by Moloney *et al.* (1989) except the explants were inoculated by bulk immersion in Agrobacterium suspension, and co-cultivation was carried out without plating the explants on medium as described in Lee (1996).

Transgene verification

Total genomic DNA was extracted from *B. napus* leaves using the following rapid DNA extraction method. Small pieces of fresh leaf were ground with a pestle in a sterile 1.5 mL microfuge tube containing 400 µL extraction buffer (200 mM Tris–HCl pH 7.5, 250 mM NaCl, 25 mM EDTA, 0.5% SDS). The ground material was vortexed and centrifuged at 16 000 *g* for 5 min. Three hundred microlitre of the supernatant was transferred into a microfuge tube with 300 µL isopropanol and mixed by inverting gently. The tubes were incubated for 5 min at room temperature and then centrifuged at 16 000 *g* for 5 min. The pellet was washed with 70% ethanol, air-dried and resuspended in 50 µL of sterile water. Primer pairs (Table S5) designed to confirm the presence of the different components of the transgene cassette were used as described in the text. PCR was conducted using Platinum® Taq DNA Polymerase (Invitrogen) for all amplifications.

Gene copy number determination for engineered *B. napus* plants

Two methods were employed to determine the number of transgene loci and zygosity of the engineered plants. Screening for resistance to kanamycin, conferred by the NPTII marker gene, was conducted by soaking the seeds in 250 mg/L kanamycin solution at room temperature for about 24 h under

continuous illumination until radicle emergence was observed. Seeds were transferred to soil, and kanamycin resistance was determined by scoring the emergence of green, undamaged cotyledons. In a second approach, seed fatty acid profiles for single seeds from the line of interest were determined by gas chromatography (GC) of fatty acid methyl esters (FAMES). Zygosity was estimated by calculating segregation ratios for 16:0 content.

Detection of BnSAD amiRNAs by stem-loop RT-PCR

To detect amiRNAs by step-loop PCR, amiRNA-specific primers (Table S6) were designed using the protocol of Varkonyi-Gasic *et al.* (2007). Total RNA was isolated from 21 and 28 DPA developing seeds using the TRIzol® Reagent (Invitrogen Inc.). Using 1 µL of total RNA as template, cDNA synthesis was conducted with stem-loop RT primers (BnSADamiR1RT and BnSADamiR2RT) and SuperScript II reverse transcriptase, as described in the manufacturer's protocol (Invitrogen Inc.). The PCR step of stem-loop RT-PCR was carried out as illustrated in Figure S8, using one microlitre of cDNA as template in a 25 µl PCR with Taq DNA Polymerase (Invitrogen). For detection of artificial microRNA BnSADamiR1 expression, PCR was performed with forward primer BnSADamiR1FW and reverse primer UniReverse, yielding PCR product BnSADamiR1PR with a size of 62 bp. For detection of artificial microRNA BnSADamiR2 expression, PCR was performed with forward primer BnSADamiR2FW and reverse primer UniReverse, giving PCR product BnSADamiR2PR with a size of 63 bp. PCR amplification conditions were as follows: 3 min of initial denaturation at 94 °C, 40 cycles at 94 °C for 20 s, 60 °C for 20 s, 72 °C for 20 s, followed by a final extension at 72 °C for 10 min. Three% agarose gel was used for separation of the PCR products.

Detecting *BnFatB* and *BnSAD/BnDES* gene expression by semi-quantitative RT-PCR

Total RNA was isolated from developing seeds at 21 and 28 DPA using the RNeasy Plant Mini kit (Qiagen Inc., Mississauga, ON, Canada). One microgram of total RNA was used as template for cDNA synthesis with oligo(dT)16 as reverse transcription primer and SuperScript II reverse transcriptase. RT-PCR was carried out using 1 µL of cDNA as template in a 25 µL PCR, with Taq DNA Polymerase (Invitrogen) for all amplification. Specific primers for each target *BnFatB* and *BnSAD/BnDES* gene were designed with Primer 3 software (http://frodo.wi.mit.edu/primer3/) as given in Table S7. *Brassica napus* actin (BnACTIN) (Accession No. AF111812) was used as a control. PCR amplification conditions for *BnFatBs*, *BnSAD/BnDES* transcripts and *BnACTIN* were as follows: 3 min of initial denaturation at 94 °C, 24 cycles at 94 °C for 30 s, 56 °C for 30 s, 72 °C for 90 s, followed by a final extension at 72 °C for 10 min.

Determination of seed fatty acid composition

For determination of total seed fatty acid composition, single or groups of 12 pooled seeds (from the same plant) were crushed and placed in a Pyrex screw-cap tube with 2 mL of 1 M HCl in methanol (Supelco, Bellefonte, PA) and 300 µL of hexane. The tubes were tightly capped and incubated at 80 °C overnight. After cooling, 2 mL of 0.9% NaCl was added, and FAMEs were recovered by collecting the hexane phase, with dilution as necessary. Gas chromatography of FAMEs was conducted using an Agilent 6890N GC equipped with a DB-23 capillary column (0.25 mm × 30 m, 0.25 mM thickness; J & W; Folsom, CA) and flame ionization detector, as described previously (Kunst *et al.*, 1992).

Seed oil content determination

Seed oil content was determined using a Maran Ultra benchtop NMR instrument (Oxford Instruments Molecular Biotools Ltd., Oxfordshire, UK), following the manual's procedure. About 0.5 g cleaned and weighted seeds from each plant were used for seed oil content analysis, with three replicates for each sample. Canola oil was used for calibration.

MALDI-TOF MS and MS/MS analysis

Two grams of seeds from transgenic lines Bn39-17, Bn39-40, Bn39-46, Bn43-5, Bn48-93, Bn48-115 and DH12075 were crushed in aluminium foil and transferred to a glass tube. Two millilitre of hexane was added, and the tubes were capped tightly, vortexed intensely and centrifuged to precipitate debris. The hexane was carefully transferred to a clean tube and evaporated under a steam of Nitrogen gas. Extracted oil samples for TAG analysis were dissolved in chloroform, and a modified pencil lead method (Black *et al.*, 2006) was used for MALDI-MS analysis. Pencil lead was scribbled on the MALDI plate, and sodium chloride solution was spotted on top to ensure that sodiated ions were the dominant ions in the mass spectra. Samples were analysed on an AB 4800 Matrix-Assisted Laser Desorption Ionization Time-of-Flight (MALDI-TOF) mass spectrometer (Applied Biosystems, LLC, Frederick, MD) equipped with a diode-pumped 355 nm Nd:YAG laser, and a laser intensity between 3000 and 3500 was used for MS data collection. The ion extraction delay time was set to 1000 ns, and positive ion, reflectron mode was used. All mass spectra were recorded as sums of 400 laser shots (800 ns) with *m/z* range from 500 to 2000.

Measurement of oil melting temperature

Twenty microlitre aliquots of the oil extracted for MALDI-TOF MS analysis were transferred to 200 µL glass GC vial inserts and frozen at −20 °C for 5 min to solidify the oil. The frozen tubes were transferred to a thermocycler set to 4 °C and left for 20 min for temperature equilibration. Subsequently, the temperature was gradually increased, in 0.5 °C intervals, from 4 to 18 °C, with 5 min at each temperature. At each interval, the tubes were quickly lifted from and returned to the thermocycler to observe oil phase and record melting temperature.

Fatty acid composition of TAG and polar lipids

To determine fatty acid composition, lipids extracted as described for MALDI-TOF MS analysis were separated by TLC on aluminium-backed silica gel G plates (Whatman Ltd., Maidstone, Kent, UK) using hexane:diethylether:acetic acid (140/70/3) as solvent. Lipid areas were identified by iodine staining of strips cut from plate edges and centre. Silica gel containing lipids was scraped from the plate and directly transesterified using 1 M HCl in methanol. For quantification, a lipid standard was added prior to transesterification. FAMES were determined by GC as described above.

Seed germination and seedling growth of transgenic lines at low temperature

To assess germination, harvested mature seeds were placed on plates with 1% Agar and home-made MS (Murashige and Skoog) medium (without sucrose) and/or in fresh water at different temperatures (4, 6 and 10 °C) and kept for 5–15 days in a growth chamber, either in darkness or with a 24 h photoperiod (300 µmol/m^2/s light intensity). To assess seedling establishment, seeds previously germinated at low temperature were allowed to continue to grow at the same temperature either in medium or in soil for a total of 13–27 days. Establishment was measured under a 24 h photoperiod (300 µmol/m^2/s light intensity). Germination rate, root length, hypocotyl length, plant fresh weight and leaf status were measured and recorded. Data were analysed and calculated using Microsoft Excel software. All experiments were performed in three replicates.

Acknowledgements

The work was conducted as part of the Bioactive Oils Program (BOP), with Funding from AVAC Ltd, Alberta, Canada (http://www.avacltd.com/) and the National Research Council of Canada (NRC). BOP is led by Dr Randall Weselake of the University of Alberta, to whom we owe special thanks. *Brassica napus* transformation was conducted by Mr Joe Hammerlindl and the plant transformation team. The authors also thank NRC Saskatoon DNA services for sequencing, the bioinformatics team, in particular Dr Chushin Koh, for bioinformatics support, and Drs Raju Datla and Edward Tsang for critical review of the manuscript. This is National Research Council of Canada publication number 55567. The authors declare no conflict of interest.

References

Bates, P.D., Stymne, S. and Ohlrogge, J. (2013) Biochemical pathways in seed oil synthesis. *Curr. Opin. Plant Biol.* **16**, 358–364.

Baud, S. and Lepiniec, L. (2010) Physiological and developmental regulation of seed oil production. *Prog. Lipid Res.* **49**, 235–249.

Belide, S., Petrie, J.R., Shrestha, P. and Singh, S.P. (2012) Modification of seed oil composition in *Arabidopsis* by artificial microRNA-mediated gene silencing. *Front Plant Sci.* **3**, 168.

Black, C., Poile, C., Langley, J. and Herniman, J. (2006) The use of pencil lead as a matrix and calibrant for matrix-assisted laser desorption/ionisation. *Rapid Commun. Mass Spectrom.* **20**, 1053–1060.

Cernac, A. and Benning, C. (2004) WRINKLED1 encodes an AP2/EREB domain protein involved in the control of storage compound biosynthesis in Arabidopsis. *Plant J.* **40**, 575–585.

Cheng, F., Liu, S., Wu, J., Fang, L., Sun, S., Liu, B., Li, P., Hua, W. and Wang, X. (2011) BRAD, the genetics and genomics database for *Brassica* plants. *BMC Plant Biol.* **11**, 136.

Cheng, F., Wu, J., Fang, L. and Wang, X. (2012) Syntenic gene analysis between *Brassica rapa* and other Brassicaceae species. *Front Plant Sci.* **3**, 198.

Cho, K., O'Neill, C.M., Kwon, S.J., Yang, T.J., Smooker, A.M., Fraser, F. and Bancroft, I. (2010) Sequence-level comparative analysis of the *Brassica napus* genome around two stearoyl-ACP desaturase loci. *Plant J.* **61**, 591–599.

Christie, W.W., Nikolova-Damyanova, B., Laakso, P. and Herslof, B. (1991) Stereospecific analysis of triacyl- sn -glycerols via resolution of diastereomeric diacylglycerol derivatives by high-performance liquid chromatography on silica. *J. Am. Oil Chem. Soc.* **68**, 695–701.

Chrysan, M.M. (2005) Margarines and spreads. In *Bailey's Industrial Oil and Fat Products*, 6th edn (Shahidi, F., ed), pp. 33–82. New York: John Wiley and Sons Inc.

Clemente, T.E. and Cahoon, E.B. (2009) Soybean oil: genetic approaches for modification of functionality and total content. *Plant Physiol.* **151**, 1030–1040.

Dehesh, K., Jones, A., Knutzon, D.S. and Voelker, T.A. (1996) Production of high levels of 8:0 and 10:0 fatty acids in transgenic canola by overexpression of Ch FatB2, a thioesterase cDNA from *Cuphea hookeriana*. *Plant J.* **9**, 167–172.

Dehesh, K., Tai, H., Edwards, P., Byrne, J. and Jaworski, J.G. (2001) Overexpression of 3-ketoacyl-acyl-carrier protein synthase IIIs in plants reduces the rate of lipid synthesis. *Plant Physiol.* **125**, 1103–1114.

Dormann, P., Voelker, T.A. and Ohlrogge, J.B. (2000) Accumulation of palmitate in Arabidopsis mediated by the acyl-acyl carrier protein thioesterase FATB1. *Plant Physiol.* **123**, 637–643.

Dubinsky, E. and Garcés, R. (2011) High-stearic/high-oleic sunflower oil: a versatile fat for food applications. *Inform*, **22**, 369–372.

Emanuelsson, O., Nielsen, H. and von Heijne, G. (1999) ChloroP, a neural network-based method for predicting chloroplast transit peptides and their cleavage sites. *Protein Sci.* **8**, 978–984.

Facciotti, M.T., Bertain, P.B. and Yuan, L. (1999) Improved stearate phenotype in transgenic canola expressing a modified acyl-acyl carrier protein thioesterase. *Nat. Biotechnol.* **17**, 593–597.

Fasina, O.O., Craig-Schmidt, M., Colley, Z. and Hallman, H. (2008) Predicting melting characteristics of vegetable oils from fatty acid composition. *LWT-Food Sci. Technol.* **41**, 1501–1505.

Fernández-Moya, V., Martínez-Force, E. and Garcés, R. (2000) Metabolism of triacylglycerol species during seed germination in fatty acid sunflower (*Helianthus annuus*) mutants. *J. Agric. Food Chem.* **48**, 770–774.

Garcés, R., Martinéz-force, E., Salas, J.J. and Bootello, M.A. (2012) Alternatives to tropical fats based on high-stearic sunflower oils. *Lipid Technol.* **24**, 63–65.

German, J.B. and Dillard, C.J. (2004) Saturated fats: what dietary intake? *Am. J. Clin. Nutr.* **80**, 550–559.

Hammer, M.F. and Murphy, J.B. (1994) Lipase activity and in vivo triacylglycerol utilization during *Pinus edulis* seed germination. *Plant Physiol. Biochem.* **32**, 861–867.

Hawkins, D.J. and Kridl, J.C. (1998) Characterization of acyl-ACP thioesterases of mangosteen (*Garcinia mangostana*) seed and high levels of stearate production in transgenic canola. *Plant J.* **13**, 743–752.

Hirji, R., Hammerlindl, J.K., Woytowich, A.E., Khachatourians, G.G., Datla, R.S.S., Keller, W.A. and Selvaraj, G. (1996) *Plasmid pHS723 and its derivative: plant transformation vectors that enable efficient selection and progeny analysis*. Fourth Canadian Plant Tissue Culture and Genetic Engineering Conference, June 1–4, Saskatoon, SK, Canada.

Hunter, J.E. (2006) Dietary *trans* fatty acids: review of recent human studies and food industry responses. *Lipids*, **41**, 967–992.

James, D.W. Jr and Dooner, H.K. (1990) Isolation of EMS-induced mutants in *Arabidopsis* altered in seed fatty acid composition. *Theor. Appl. Genet.* **80**, 241–245.

Jones, A., Davies, M. and Voelker, T.A. (1995) Palmitoyl-acyl carrier protein (ACP) thioesterase and the evolutionary origin of plant acyl-ACP thioesterases. *Plant Cell*, **7**, 259–371.

Kachroo, P., Shanklin, J., Shah, J., Whittle, E.J. and Klessig, D.F. (2001) A fatty acid desaturase modulates the activation of defense signaling pathways in plants. *Proc. Natl Acad. Sci. USA*, **98**, 9448–9453.

Kachroo, A., Shanklin, J., Whittle, E., Lapchyk, L., Hildebrand, D. and Kachroo, P. (2007) The Arabidopsis stearoyl-acyl carrier protein desaturase family and the contribution of leaf isoforms to oleic acid synthesis. *Plant Mol. Biol.* **63**, 257–271.

Kapoor, M. and Gupta, M.N. (2012) Lipase promiscuity and its biochemical applications. *Process Biochem.* **47**, 555–569.

Knutzon, D.S., Thompson, G.A., Radke, S.E., Johnson, W.B., Knauf, V.C. and Kridl, J.C. (1992) Modification of Brassica seed oil by antisense expression of a stearoyl-acyl carrier protein desaturase gene. *Proc. Natl Acad. Sci. USA*, **89**, 2624–2628.

Kunst, L., Taylor, D.C. and Underhill, E.W. (1992) Fatty acid elongation in developing seeds of *Arabidopsis thaliana*. *Plant Physiol. Biochem.* **30**, 425–434.

Laurance, W.F., Koh, L.P., Butler, R., Sodhi, N.S., Bradshaw, C.J.A., Neidel, J.D., Consunji, H. and Mateo Vega, J. (2010) Improving the performance of the roundtable on sustainable palm oil for nature conservation. *Conserv. Biol.* **24**, 377–381.

Lawrence, G.D. (2013) Dietary fats and health: dietary recommendations in the context of scientific evidence. *Adv. Nutr.* **4**, 294–302.

Lee, S.K. (1996) *Genetic transformation in broccoli and promoter tagging in Brassica species*. Dissertation, University of Saskatchewan. http://library2.usask.ca/theses/available/etd-10212004-000633/unrestricted/nq24027.pdf

Lightner, J., Wu, J. and Browse, J. (1994) A mutant of *Arabidopsis* with increased levels of stearic acid. *Plant Physiol.* **106**, 1443–1451.

Lin, Y.H., Yu, C. and Huang, A.H. (1986) Substrate specificities of lipases from corn and other seeds. *Arch. Biochem. Biophys.* **244**, 346–356.

Liu, Q., Singh, S.P. and Green, A. (2002) High-stearic and high-oleic cottonseed oils produced by hairpin RNA-mediated post-transcriptional gene silencing. *Plant Physiol.* **129**, 1732–1743.

Mayer, K.M. and Shanklin, J. (2007) Identification of amino acid residues involved in substrate specificity of plant acyl-ACP thioesterases using a bioinformatics-guided approach. *BMC Plant Biol.* **7**, 1 doi:10.1186/1471-2229-7-1.

Moloney, M.M., Walker, J.M. and Sharma, K.K. (1989) High efficiency transformation of *Brassica napus* using Agrobacterium vectors. *Plant Cell Rep.* **8**, 238–242.

Mu, H. and Hoy, C.-E. (2004) The digestion of dietary triacylglycerols. *Prog. Lipid Res.* **43**, 105–133.

Nagaharu, U. (1935) Genome analysis in Brassica with special reference to the experimental formation of *B. napus* and peculiar mode of fertilization. *Jpn. J. Bot.* **7**, 389–452.

Nishida, I. and Murata, N. (1996) Chilling sensitivity in plants and cyanobacteria: the crucial contribution of membrane lipids. *Annu. Rev. Plant Physiol. Plant Mol. Biol.* **47**, 541–568.

O'Brien, R. (1988) *Fats and Oils-Formulating and Processing for Applications*. Lancaster, PA: Technomic Publishing, pp. 694.

Parkin, I.A., Sharpe, A.G., Keith, D.J. and Lydiate, D.J. (1995) Identification of the A and C genomes of amphidiploids *Brassica napus* (oilseed rape). *Genome*, **38**, 1122–1131.

Pidkowich, M.S., Nguyen, H.T., Heilmann, I., Ischebeck, T. and Shanklin, J. (2007) Modulating seed B-ketoacyl-acyl carrier protein synthase II level converts the composition of a temperate seed oil to that of a palm-like tropical oil. *Proc. Natl Acad. Sci. USA*, **104**, 4742–4747.

Quettier, L.-A. and Eastmond, P.J. (2009) Storage oil hydrolysis during early seedling growth. *Plant Physiol. Biochem.* **47**, 485–490.

Salas, J.J. and Ohlrogge, J.B. (2002) Characterization of substrate specificity of plant FatA and FatB acyl-ACP thioesterases. *Arch. Biochem. Biophys.* **403**, 25–34.

Shanklin, J. and Cahoon, E.B. (1998) Desaturation and related modifications of fatty acids. *Annu. Rev. Plant Physiol. Plant Mol. Biol.* **49**, 611–641.

Slocombe, S.P., Cummins, I., Jarvis, R.P. and Murphy, D.J. (1992) Nucleotide sequence and temporal regulation of a seed-specific *Brassica napus* cDNA encoding a stearoyl-acyl carrier protein (ACP) desaturase. *Plant Mol. Biol.* **20**, 151–155.

Slocombe, S.P., Piffanelli, P., Fairbairn, D., Bowra, S., Hatzopoulos, P., Tsiantis, M. and Murphy, D.J. (1994) Temporal and tissue-specific regulation of a *Brassica napus* stearoyl-acyl carrier protein desaturase gene. *Plant Physiol.* **104**, 1167–1176.

Sommerfeld, M. (1983) *Trans* unsaturated fatty acids in natural products and processed foods. *Prog. Lipid Res.* **22**, 221–233.

Stoll, C., Luhs, W., Zarhloul, M.K. and Friedt, W. (2005) Genetic modification of saturated fatty acids in oilseed rape (*Brassica napus*). *Eur. J. Lipid Sci. Technol.* **107**, 244–248.

Theodoulou, F.L. and Eastmond, P.J. (2012) Seed storage oil catabolism: a story of give and take. *Curr. Opin. Plant Biol.* **15**, 322–328.

Varkonyi-Gasic, E., Wu, R., Wood, M., Walton, F.E. and Hellens, P.R. (2007) Protocol: a highly sensitive RT-PCR method for detection and quantification of microRNAs. *Plant Methods*, **3**, 1–12.

Voelker, T.A., Worrell, A.C., Anderson, L., Bleibaum, J., Fan, C., Hawkins, D.J., Radke, S.E. and Davies, H.M. (1992) Fatty acid biosynthesis redirected to medium chains in transgenic oilseed plants. *Science*, **257**, 72–74.

Voelker, T.A., Jones, A., Cranmer, A.M., Davies, H.M. and Knutzon, D.S. (1997) Broad-range and binary-range acyl-acyl-carrier-protein thioesterases suggest an alternative mechanism for medium-chain production in seeds. *Plant Physiol.* **114**, 669–677.

Wang, X., Wang, H., Wang, J., Sun, R., Wu, J., Liu, S., Bai, Y., Mun, J.H., Bancroft, I., Cheng, F., Huang, S., Li, X., Hua, W., Wang, J., Wang, X., Freeling, M., Pires, J.C., Paterson, A.H., Chalhoub, B., Wang, B., Hayward, A., Sharpe, A.G., Park, B.S., Weisshaar, B., Liu, B., Li, B., Liu, B., Tong, C., Song, C., Duran, C., Peng, C., Geng, C., Koh, C., Lin, C., Edwards, D., Mu, D.,

Shen, D., Soumpourou, E., Li, F., Fraser, F., Conant, G., Lassalle, G., King, G.J., Bonnema, G., Tang, H., Wang, H., Belcram, H., Zhou, H., Hirakawa, H., Abe, H., Guo, H., Wang, H., Jin, H., Parkin, I.A., Batley, J., Kim, J.S., Just, J., Li, J., Xu, J., Deng, J., Kim, J.A., Li, J., Yu, J., Meng, J., Wang, J., Min, J., Poulain, J., Wang, J., Hatakeyama, K., Wu, K., Wang, L., Fang, L., Trick, M., Links, M.G., Zhao, M., Jin, M., Ramchiary, N., Drou, N., Berkman, P.J., Cai, Q., Huang, Q., Li, R., Tabata, S., Cheng, S., Zhang, S., Zhang, S., Huang, S., Sato, S., Sun, S., Kwon, S.J., Choi, S.R., Lee, T.H., Fan, W., Zhao, X., Tan, X., Xu, X., Wang, Y., Qiu, Y., Yin, Y., Li, Y., Du, Y., Liao, Y., Narusaka, Y., Wang, Y., Wang, Z., Li, Z., Wang, Z., Xiong, Z., Zhang, Z. and Brassica rapa Genome Sequencing Project Consortium (2011) The genome of the mesopolyploid crop species *Brassica rapa*. *Nat. Genet.* **43**, 1035–1040.

Warthmann, N., Chen, H., Ossowski, S., Weigel, D. and Hervé, P. (2008) Highly specific gene silencing by artificial miRNAs in rice. *PLoS ONE*, **3**, e1829.

Yuan, L., Nelson, B.A. and Caryl, G. (1996) The catalytic cysteine and histidine in the plant acyl-acyl carrier protein thioesterases, *J. Biol. Chem.* **271**, 3417–3419.

Zarhloul, M.K., Stoll, C., Luhs, W., Syring-Ehemann, A., Hausmann, L., Topfer, R. and Freidt, W. (2006) Breeding high-stearic oilseed rape (*Brassica napus*) with high and low-erucic background using optimized promoter-gene constructs. *Mol. Breed.* **18**, 241–251.

13

A method for rapid production of heteromultimeric protein complexes in plants: assembly of protective bluetongue virus-like particles

Eva C. Thuenemann[1], Ann E. Meyers[2], Jeanette Verwey[3], Edward P. Rybicki[2] and George P. Lomonossoff[1,*]

[1]Department of Biological Chemistry, John Innes Centre, Norwich, UK
[2]Department of Molecular and Cell Biology, University of Cape Town, Rondebosch, South Africa
[3]Onderstepoort Biological Products SOC Ltd, Onderstepoort, South Africa

*Correspondence
email george.
lomonossoff@jic.ac.uk

Summary

Plant expression systems based on nonreplicating virus-based vectors can be used for the simultaneous expression of multiple genes within the same cell. They therefore have great potential for the production of heteromultimeric protein complexes. This work describes the efficient plant-based production and assembly of Bluetongue virus-like particles (VLPs), requiring the simultaneous expression of four distinct proteins in varying amounts. Such particles have the potential to serve as a safe and effective vaccine against Bluetongue virus (BTV), which causes high mortality rates in ruminants and thus has a severe effect on the livestock trade. Here, VLPs produced and assembled in *Nicotiana benthamiana* using the cowpea mosaic virus–based *HyperTrans* (CPMV-*HT*) and associated pEAQ plant transient expression vector system were shown to elicit a strong antibody response in sheep. Furthermore, they provided protective immunity against a challenge with a South African BTV-8 field isolate. The results show that transient expression can be used to produce immunologically relevant complex heteromultimeric structures in plants in a matter of days. The results have implications beyond the realm of veterinary vaccines and could be applied to the production of VLPs for human use or the coexpression of multiple enzymes for the manipulation of metabolic pathways.

Keywords: virus-like particle, bluetongue virus, cowpea mosaic virus, transient, heterologous expression, vaccine.

Introduction

The efficient production of complex heteromultimeric protein complexes, in which the various components are present in differing stoichiometries, represents a major challenge to plant expression methods. It requires the production of all the components of the complex within the same cell at appropriate levels. This has proved very difficult to achieve, especially with transient expression approaches using replicating virus vectors. Nonreplicating systems, such as the cowpea mosaic virus–based HyperTrans (CPMV-*HT*) system (Sainsbury and Lomonossoff, 2008; Sainsbury *et al.*, 2009), offer considerable potential benefits for the production of heteromultimeric complexes: it is possible to control the level of expression of the individual components, and multiple constructs can be expressed in the same cell without the problem of virus exclusion found with replicating systems. We therefore examined whether the CPMV-*HT* system can be used to efficiently produce virus-like particles (VLPs) of Bluetongue virus (BTV).

Bluetongue is a severe disease of ruminants, notably sheep and cattle, causing facial swelling, lameness and infertility amongst other symptoms and leading to mortality in some cases. The causal agent, the multicomponent dsRNA Bluetongue virus, is spread by an insect vector, *Culicoides sp.*, and occurs in its vector's habitat in temperate climates throughout much of the world (Carpenter *et al.*, 2009). BTV is the type member of genus *Orbivirus* in the family *Reoviridae*, with 26 known serotypes

(Maan *et al.*, 2012). When bluetongue first broke out in the United Kingdom in autumn of 2007, the disease was already rapidly spreading throughout continental Europe, causing high mortality rates in sheep and having a detrimental effect on the livestock trade through trade restrictions and loss of stock. The only effective weapon against the disease is control of the spread of BTV through rigorous vaccination programmes. Currently available commercial vaccines are based on both inactivated virus and live, attenuated strains and protect against a single serotype or multiple serotypes when provided as a cocktail (Savini *et al.*, 2008). However, the possibility of recombination between the live vaccine strain(s) and wild-type virus in infected animals, leading to the emergence of new infectious strains (Batten *et al.*, 2008), has motivated efforts to develop safer vaccines.

One approach in the development of an inherently safe vaccine has been the production of Bluetongue virus-like particles (VLPs) (French *et al.*, 1990; Roy *et al.*, 1992). BTV has a nonenveloped icosahedral structure, with four main structural proteins (VP3, VP7, VP5 and VP2) arranged in concentric shells around the segmented double-stranded RNA genome and minor structural and nonstructural proteins involved in virus replication. French *et al.* have shown that these four structural proteins, expressed in insect cells using a baculovirus expression system, assemble into virus-like particles devoid of nucleic acid (French *et al.*, 1990). The most internal of the structural proteins, VP3, is a 100-kDa protein, 120 copies of which form 60 dimers, which assemble into a *T* = 1 particle of 55 nm diameter (Grimes *et al.*, 1998) (schematic in

Figure 1c). During virus assembly, trimers of VP7 form an icosahedral shell of $T = 13$ symmetry on the scaffold provided by VP3, resulting in the production of core particles consisting of 120 copies of VP3 and 780 copies of VP7 (Prasad et al., 1992). Using a baculovirus expression system in insect cells, core-like particles (CLPs) of VP3 and VP7 have been produced (French and Roy, 1990). These are stable structures, although lacking 60 trimers of VP7 at the fivefold axes (Hewat et al., 1994), but do not induce a protective immune response in sheep. For this, a third shell consisting of 180 copies of VP2 and 360 copies of VP5 is needed. The major immunogenicity determinant, VP2, is a 102-kDa protein forming triskelion-shaped trimers on the surface of BTV (Hassan and Roy, 1999). Between these structures, 120 trimers of the 59-kDa VP5, which are involved in virus entry into cells, are positioned (Forzan et al., 2004). Inoculation of sheep with insect cell–expressed triple-shelled VLPs of BTV-10 containing all four structural proteins induces protective immunity against challenge with live virus of the same serotype. In addition, inoculation of sheep with similar structured BTV-1 VLPs produced in insect cells also induced neutralizing antibodies to BTV-1 and protected sheep challenged with live virus of the same serotype (Stewart et al., 2012). These results show that BTV VLPs have the potential to be used as an inherently safe vaccine (Roy et al., 1992). However, the high cost of production of such VLPs may be one barrier that has prevented wide-scale development of these vaccines. Over the past three decades, plants have become an increasingly popular alternative host for the heterologous expression of complex high-value proteins, providing cost-effective expression without the risk of contaminating animal pathogens. Transient expression systems in particular allow for very rapid expression, development and testing of new constructs in a matter of days (Rybicki, 2010).

Here, we report the plant-based high-level expression of assembled subcore-, core- and virus-like particles of BTV serotype 8 using the CPMV-*HT* and associated pEAQ vector system (Sainsbury and Lomonossoff, 2008; Sainsbury et al., 2009). Purified preparations of the VLPs, consisting of all four structural proteins, elicited an immune response in sheep and provided protective immunity against challenge with a South African BTV-8 field isolate. This demonstrates that CPMV-*HT* technology provides an economically viable method for producing complex VLPs, such as those of BTV, with the desired biological properties. It represents a significant advance in the use of plant-based systems for the production of complex biopharmaceuticals. The methods employed could also be applied to other situations where the expression of multiple proteins is required, such as the reconstruction of metabolic pathways.

Results

BTV coat proteins transiently expressed in *Nicotiana benthamiana* leaves assemble into subcore-, core- and virus-like particles

To determine whether BTV-8 structural proteins could be expressed in plant leaves, constructs containing each of the four structural proteins VP3, VP7, VP5 and VP2 were created. The genes encoding these proteins were codon-optimized for *N. benthamiana* expression and cloned between the CPMV untranslated regions (UTRs) in the pEAQ-*HT* vector (Figure 1b). Transient expression was achieved by infiltration of *N. benthamiana* leaves with cultures of *Agrobacterium tumefaciens* carrying these constructs, either individually or in various combinations. All constructs produced

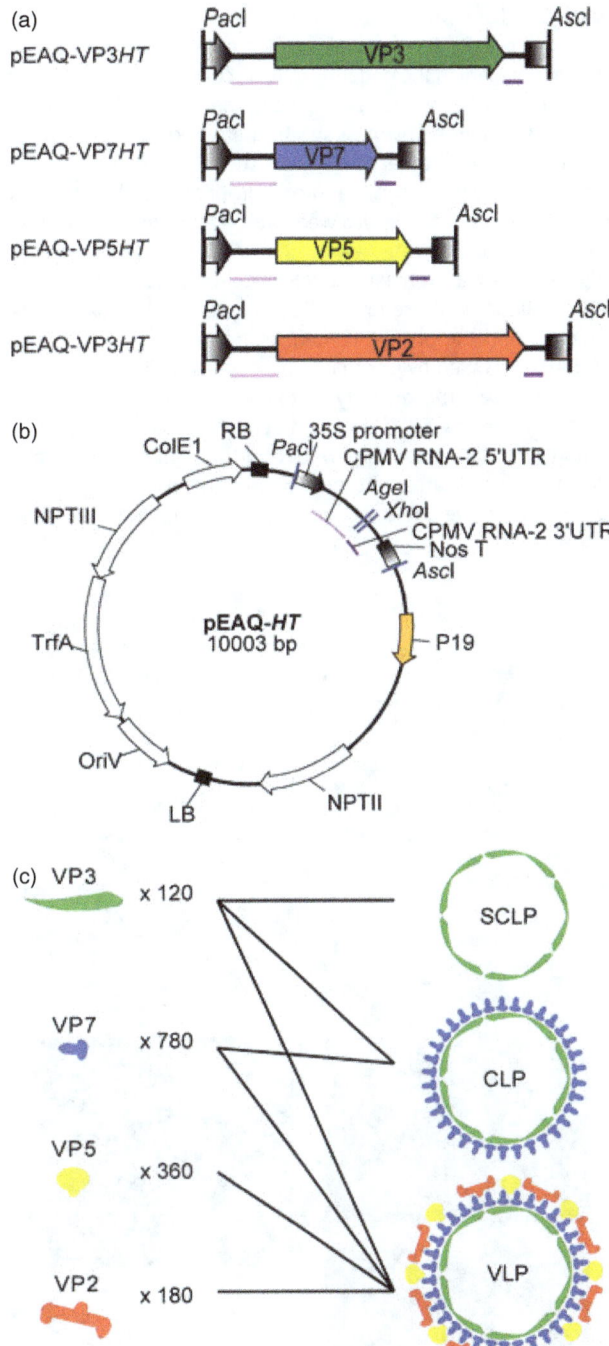

Figure 1 Schematic representations of constructs for plant-based expression of Bluetongue virus (BTV)-8 proteins and their assembly into particle structures. Maps of the cowpea mosaic virus–based *HyperTrans* (CPMV-*HT*) expression cassettes (a) and pEAQ-HT expression vector backbone (b) for expression of codon-optimized versions of BTV8 proteins VP3, VP7, VP5 and VP2. (c) Schematic of major BTV structural proteins, indicating how many copies of each structural protein come together to form subcore-, core- and virus-like particles (SCLP, CLP and VLP). pEAQ vectors contain an expression cassette for P19, a suppressor of gene silencing (Silhavy et al., 2002), required for high-level heterologous expression in plants.

chlorotic symptoms in the infiltrated leaf tissue. Individual expression of VP3 and VP5 produced necrotic symptoms, whilst in combination with the other proteins, no necrosis was observed

(Figure S1). Total protein extracts from leaf tissue infiltrated with the individual BTV-8 constructs in each case contained a distinct band, not present in the negative control, corresponding to the size of the relevant structural protein; the identity of these protein bands was confirmed by mass spectrometry (Figure S2).

To investigate whether the BTV-8 structural proteins expressed in leaf tissue could assemble into appropriate particulate structures, *N. benthamiana* leaves were either infiltrated with the VP3 gene alone, coinfiltrated with the VP3 and VP7 genes or infiltrated with all four structural protein genes. Plant extracts were fractionated by centrifugation on sucrose or iodixanol step gradients, and the protein content of fractions was analysed by SDS-PAGE followed by Coomassie blue staining (Figure 2a, Figure S3). In all cases, the majority of the BTV-specific proteins were found in the 45%–55% sucrose fractions (30%–40% iodixanol fractions), which were largely devoid of contaminating plant proteins. This indicates that the BTV proteins had assembled into high-molecular-weight structures. Furthermore, the cosedimen-

tation of the BTV proteins in extracts of leaves coinfiltrated with VP3 and VP7 or with VP3, VP7, VP2 and VP5 strongly suggested that the proteins had interacted to form supramolecular structures.

The complexes formed by the BTV proteins were analysed by transmission electron microscopy of the gradient fractions in which they were found. Expression of VP3 alone resulted in the formation of particles closely resembling BTV subcore-like particles (SCLPs) obtained by stripping CLPs of the VP7 layer (Loudon and Roy, 1991). These VP3 SCLPs had an average diameter of 54.9 ($\pm$1.3) nm and three distinct appearances (hexagonal, angular and round), depending on their orientation on the grid (Figure 2b). Coexpression of VP3 and VP7 yielded larger particles of 69.6 ($\pm$0.9) nm in diameter. These particles had a striking appearance consistent with that of BTV core particles and insect cell–produced CLPs (Figure 2c). The distinct difference in appearance compared with the SCLPs is likely due to the layer of VP7 trimers arranged in a concentric shell on top of the VP3 layer. Coexpression of all four structural proteins of BTV-8 resulted in a mixture of particle types, some of which appeared to be CLPs, whilst others were larger (Figure 2d). The larger particles had a less-structured appearance than CLPs, with some appearing to be assembly intermediates between CLPs and VLPs. Although these larger particles indicate that VP5 and VP2 can associate with CLPs in plant cells, the stoichiometry of the structural proteins (Figure 2a) is indicative of partial conversion of SCLPS and CLPs into fully assembled VLPs. Based on densitometric analysis (Figure S4) and electron microscopy, we estimate that these assembly intermediates constitute at least 50% of the particles produced by coinfiltration with four pEAQ-*HT* constructs.

Modulation of VP3 expression levels increases recovery of virus-like particles

To increase the yield of fully assembled VLPs in leaves expressing the four BTV-8 structural proteins, their relative expression levels were modulated. Due to the generally higher abundance of the smaller proteins, protein staining of a preparation of fully assembled virus produces bands of similar intensity for all four of the main structural proteins in an SDS-PAGE gel (Martin and Zweerink, 1972). Coexpression of the four BTV-8 structural proteins from separate pEAQ-*HT* constructs resulted in preparations with an over-representation of VP3, resulting in the presence of substantial numbers of subcore-like particles or assembly intermediates (Figure 2a).

To address this issue, we made use of the fact that multiple expression cassettes, each containing a gene of interest flanked by CPMV UTRs as well as the promoter and terminator, can be cloned into the vector pEAQexpress (Sainsbury et al., 2009) (Figure 3a). Thus, plasmids were created which were designed to express either the two core proteins (VP3 plus VP7) or the two outer shell proteins (VP5 plus VP2) from the same T-DNA. To control the level of CLP synthesis, two different versions of the VP3 + VP7 construct were produced. In addition to the codon-optimized genes under control of the *HyperTrans* UTRs for maximum expression (pEAQex-VP7*HT*-VP3*HT*), another (pEAQex-VP7*HT*-VP3wt) was made with codon-optimized genes but containing VP3 under control of the CPMV wild-type 5' UTR, that is, not containing the *HyperTrans* mutation. This latter construct was designed to down-regulate the synthesis of CLPs and shift the equilibrium towards VLPs when coexpressed with VP2 and VP5. Combining two proteins on the same T-DNA also halved the minimum number of different *Agrobacterium* T-DNA

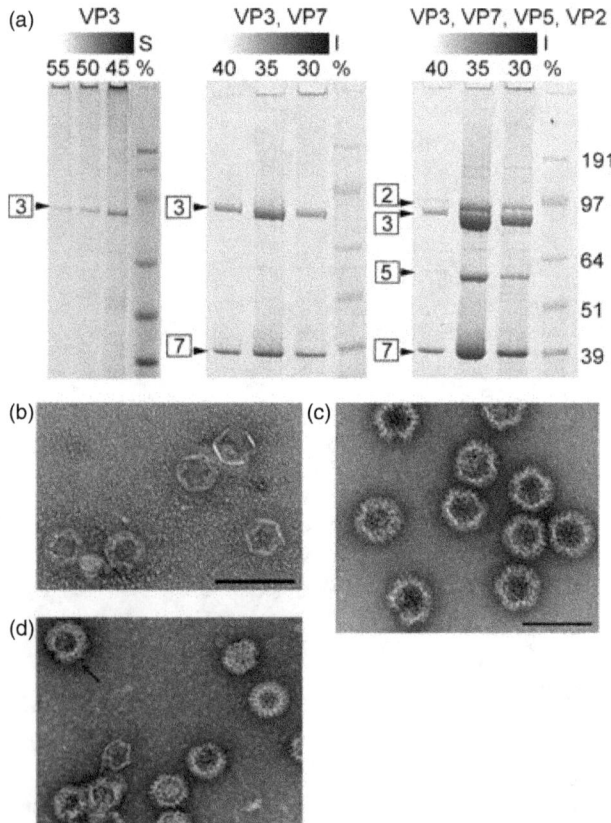

Figure 2 Assembly of BTV-8 structural proteins into particles. Leaves were infiltrated with pEAQ-*HT* constructs as indicated, harvested 8 dpi and extracted, and extracts subjected to sucrose (S) or iodixanol (I) density gradient centrifugation, as indicated. (a) Fractions were collected from the bottom and separated by denaturing SDS-PAGE followed by Coomassie blue staining. Marker sizes indicated on far right and location of viral proteins VP2, VP3, VP5 and VP7 (boxed) to the left of each gel. (b-d) Imaging of fractions from gradients of VP3 showing three different orientations of SCLPs (b), VP3 and VP7 showing CLPs (c), and VP3, VP7, VP5 and VP2 showing VLPs (arrow) and assembly intermediates (d). Scale bars, 100 nm.

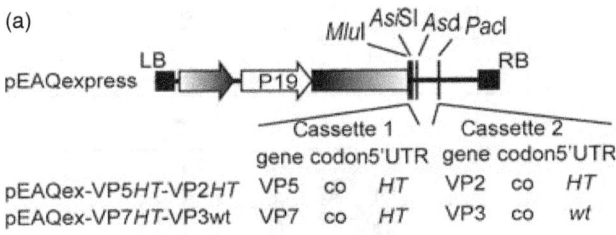

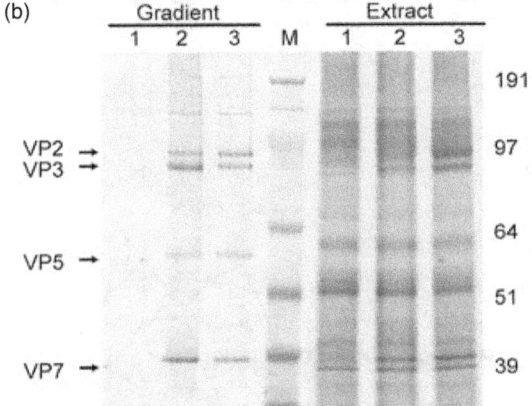

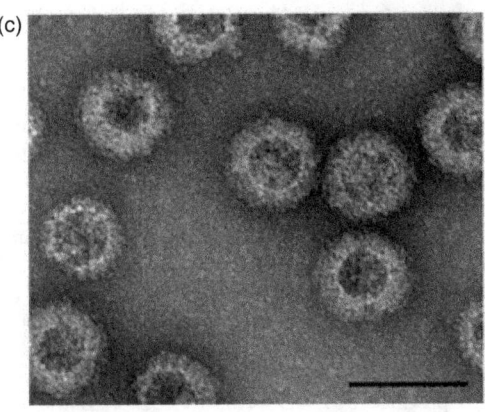

Figure 3 The stoichiometry of recovered VLPs can be influenced by changes in expression constructs. pEAQexpress vectors were made to incorporate two Bluetongue virus (BTV)-8 genes (VP5 and VP2, or VP7 and VP3) on the same construct to ensure coexpression. These constructs contained codon-optimized (co) genes and either *HyperTrans* (*HT*) or wild-type (wt) 5′UTRs. (a) Schematic representation of pEAQexpress vectors. (b) pEAQexpress vectors and pEAQ-*HT* vectors were coinfiltrated in combinations (1, 2 or 3) to codeliver all four BTV-8 genes. Leaves were harvested 8 dpi, and particles from clarified extracts purified on sucrose gradients. Clarified extracts (right) and peak gradient fractions (fraction 3; left) were run on denaturing SDS-PAGE and proteins stained with Coomassie blue. Infiltration combinations: (1) pEAQ-*HT* negative control; (2) pEAQ-VP7*HT* + pEAQ-VP3*HT* + pEAQ-VP5*HT* + pEAQ-VP2*HT*; (3) pEAQexpress-VP7*HT*-VP3wt + pEAQexpress-VP5*HT*-VP2*HT*; lane M: marker. Size markers are indicated on the right; location of BTV proteins is indicated on the left. (c) Purified particles arising from coinfiltration of pEAQex-VP5*HT*-VP2*HT* and pEAQex-VP7*HT*-VP3wt were imaged by TEM. Scale bar, 100 nm.

transfer events necessary to enable VLP production in any given cell from four to two. Restriction of the expression level of VP3 by the use of wild-type CPMV 5′ UTR produced VLP preparations with similar band intensities for all four BTV proteins on Coomassie blue-stained SDS-PAGE, indicative of high levels of

fully assembled VLPs (Figure 3b). Further gel analysis by densitometry revealed that, whilst levels of the outer layer proteins (VP2 and VP5) were similar in the two preparations, there was a marked reduction in band intensity of the core proteins (VP3 and VP7) when VP3 expression was controlled by the wild-type 5′UTR (Figure S4). This indicates that down-regulation of VP3 expression does not have a negative impact on VLP production, whilst reducing the burden of CLPs and SCLPs in the preparation. Visualization of these particles by TEM showed a far higher degree of VLP integrity in these preparations compared with coinfiltration of four separate codon-optimized constructs (Figure 3c). The particles deemed to be assembled VLPs had a mean diameter of 86.8 ($\pm$3.0) nm. Thus, for the larger-scale production of VLPs in plants, we utilized coinfiltration of *N. benthamiana* leaves with pEAQex-VP7*HT*-VP3wt and pEAQex-VP5*HT*-VP2*HT*. To further maximize the ratio of VLP to CLP/SCLP in the preparations, the most pure and stoichiometrically correct fractions of the density gradients were subjected to a second round of gradient centrifugation. The final yield of VLPs produced in this way was approximately 70 mg VLPs per kg leaf wet weight, as determined by protein staining and absorbance measured at a wavelength of 280 nm of the most stoichiometrically correct fractions. Total particulate BTV-8 protein yield was estimated to be more than 200 mg per kg leaf wet weight. To produce a preparation of CLPs, *N. benthamiana* leaves were infiltrated with pEAQex-VP7*HT*-VP3wt and CLPs isolated by two rounds of gradient centrifugation.

Plant-produced BTV-like particles induce an immune response in sheep

To test the immunogenicity of plant-produced BTV-8 VLPs, two sheep were injected with 20 μg of adjuvanted VLPs and received boost inoculations at 21 and 42 days post-initial inoculation. Serum was collected 18 days after each inoculation and on day 56. Serum samples from both sheep tested positive for BTV antibodies after the first boost using a Bluetongue virus antibody test kit (cELISA, VMRD Inc., Pullman, WA). Serum from day 56 (final bleed) was used to probe a Western blot of the original plant-produced particles (Figure 4). Signals for all four structural proteins could be identified. Degradation products were also recognized, indicating some loss of protein integrity in samples stored at 4 °C for 10 months. Prebleed serum used to probe a corresponding Western blot did not show any positive reaction with the proteins (data not shown). These results show that plant-produced BTV-8 VLPs stimulate an immune response in sheep, producing antibodies mainly against the major immunogenicity determinant (VP2) and the most abundant structural protein (VP7).

Plant-produced BTV-8 VLPs confer protective immunity in sheep

To test the efficacy of plant-produced BTV-8 VLPs as a potential vaccine against BTV-8, four groups of five sheep were injected with either 50 μg VLP, 200 μg CLP, 5×10^4 TCID$_{50}$/mL live, attenuated, monovalent BTV-8 used in the multivalent commercial vaccine (Onderstepoort Biological Products, South Africa) or PBS, in the presence of an adjuvant. On day 28, animals received a booster injection of the same composition. On day 63, animals were challenged with 1 mL infected sheep blood containing live BTV-8, and the clinical reactions were monitored for 2 weeks postchallenge. The severity of clinical bluetongue after challenge with the virulent virus was expressed in a numerical form as the clinical reaction index (CRI) as described by Huismans *et al.*

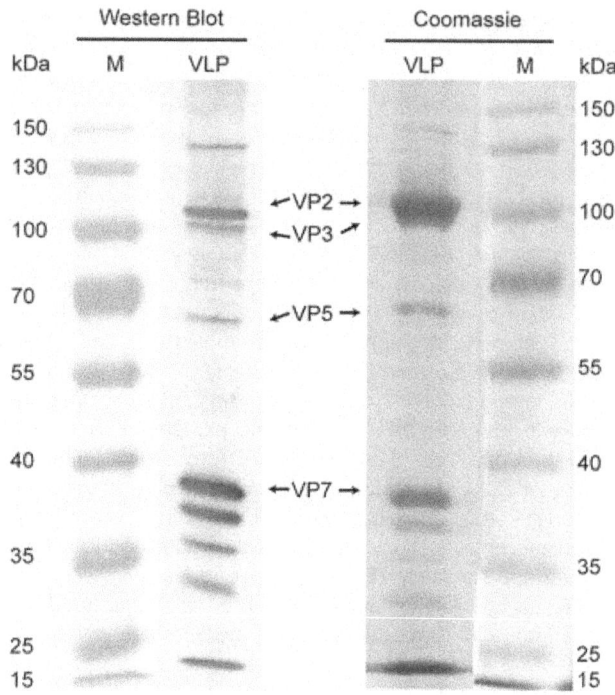

Figure 4 Immunogenicity of plant-made BTV-8 particles. Sheep were inoculated with 20 μg VLPs, boosted with 2 further doses after 21 and 42 days, and serum collected 56 days after the first inoculation. Bluetongue virus (BTV)-8 VLPs were run on SDS-PAGE and either stained by Coomassie blue or transferred by Western blotting. The Western blot was probed with the collected serum.

(1987a) (Figure 5a). The plant-produced BTV-8 VLP vaccine had the same protective efficacy as the live, attenuated, monovalent BTV-8 vaccine strain, with animals showing no clinical symptoms of bluetongue disease. Plant-produced CLPs were poorly protective, with a mean CRI of 10.8 compared with a mean CRI of 15.5 for the control group injected with PBS. Animal body temperature was recorded for 14 days postchallenge. None of the animals receiving VLP or live, attenuated vaccine developed an elevated temperature (>40.0 °C), in contrast with the CLP and PBS groups (Figure 5b). None of the animals died prematurely.

Serum neutralization tests were carried out weekly throughout the 91-day experiment (Table 1). Both the VLP group and the group vaccinated with the monovalent, live, attenuated BTV-8 vaccine strain showed high serum neutralization titres after day 28. However, whilst high levels of neutralizing antibodies were induced by the monovalent, attenuated BTV-8 vaccine as soon as 7 days after vaccination, plant-based VLPs only induced high antibody levels after booster injection, showing that a booster is necessary for protective efficacy of the plant-produced VLP vaccine. Neither the PBS control nor plant-produced CLPs induced a protective antibody response. However, plant-produced CLPs did appear to offer some protection against challenge, similar to insect cell–produced CLPs (Stewart et al., 2012).

Discussion

The results presented here demonstrate that it is possible to coexpress four structural proteins within the same plant cell using the CPMV-HT expression system and that correct assembly into complex structures (VLPs) with the appropriate immunological properties readily occurs within days of infiltration. The most

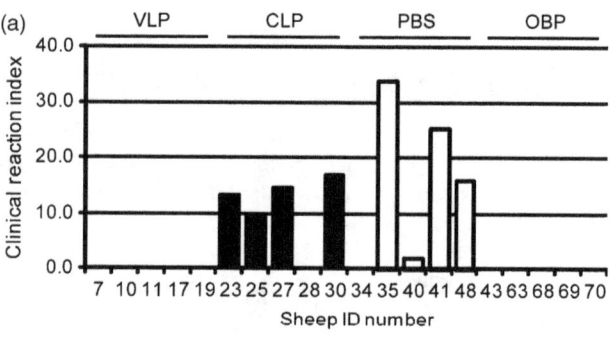

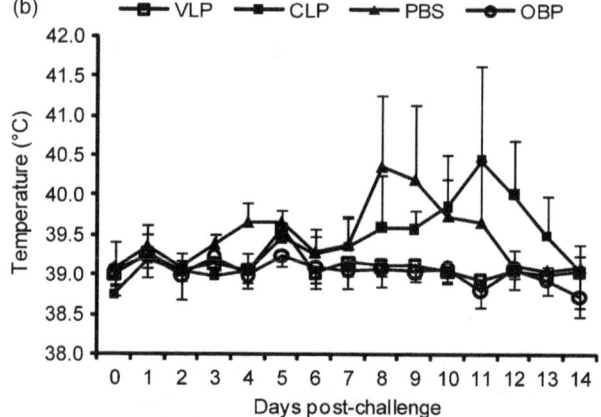

Figure 5 Protective efficacy of plant-made Bluetongue virus (BTV)-8 particles. A challenge experiment was performed by inoculating groups of 5 sheep with 50 μg plant-produced BTV-8 VLP, 200 μg plant-produced BTV-8 CLP, OBP monovalent, live, attenuated BTV-8 vaccine or PBS on day 0, boosting on day 28 and challenging with live BTV-8 on day 63. Animals were monitored for 14 days postchallenge and clinical reaction index determined (a). Mean recorded temperatures of each group of animals are presented with standard deviations indicated above (CLP and PBS) or below (VLP and OBP) each data point (b).

efficient way to produce BTV VLPs was to coexpress the core proteins (VP3 and VP7) on one plasmid and the outer proteins (VP2 and VP5) on another. As well as reducing the number of constructs that needed to be coinfiltrated, this approach also enabled the differential expression of the core and outer proteins. Down-regulation of VP3 synthesis through the use of the wild-type (wt) rather than the HT 5'UTR resulted in the production of fewer CLPs and shifted the equilibrium towards fully assembled VLPs. The ability to regulate translation levels of individual proteins within the same cell is a useful tool of the CPMV-HT expression system. In the case of BTV, it has been used to efficiently produce a protein complex where the individual components are present in varying stoichiometries; such an approach could also be applied to other complex viruses, such as human rotavirus. Furthermore, it could equally be used to reconstruct a metabolic pathway where the component enzymes are present at different levels. Thus, the results obtained with the production of BTV VLPs have implications outside the veterinary field.

The ability to express high levels of assembled BTV VLPs that can stimulate protective immunity in target animals (sheep) is both a very valuable proof of efficacy, as well as a further demonstration of the potential utility of plant-made pharmaceuticals. The timescale for production means that it is possible to respond quickly to outbreaks of emerging diseases. The present study made use of small-scale expression methodology (syringe

Table 1 Neutralizing antibody titres during challenge experiment

| | | Day | | | | | | | | | | | | | |
| | | Vaccination | | | | Boost | | | | | Challenge | | | | |
Group	Animal	0	7	14	21	28	35	42	49	56	63	70	77	84	91
VLP	7	0	0	8	8	16	256	256	256	256	512	128	512	512	4096
	10	0	0	0	4	16	256	256	256	256	64	64	128	128	64
	11	0	0	0	0	4	32	32	32	64	16	8	16	32	64
	17	0	0	0	0	4	256	256	256	64	128	64	64	128	4096
	19	0	0	0	2	4	256	256	256	64	16	8	16	64	128
CLP	23	0	0	0	0	2	0	0	0	8	0	0	256	1024	512
	25	0	0	0	0	0	0	0	0	0	0	0	2048	2048	4096
	27	0	0	0	0	0	0	0	0	0	0	0	32	512	2048
	28	0	0	4	0	8	0	0	0	0	0	0	256	1024	4096
	30	0	0	0	0	0	0	0	0	8	0	0	2048	2048	4096
PBS	34	0	0	0	0	0	0	0	0	0	0	0	32	0	512
	35	0	0	0	0	0	0	0	0	0	0	0	2048	1024	1024
	40	0	0	0	0	0	0	0	0	0	0	0	0	0	32
	41	0	0	0	0	0	0	0	0	0	0	0	0	0	512
	48	0	0	0	0	0	0	0	0	0	0	0	512	256	512
OBP vaccine strain	43	0	2	128	128	256	256	256	256	256	32	512	512	512	512
	63	0	2	256	256	256	256	256	256	256	128	256	1024	256	n/a
	68	0	64	256	256	256	256	256	256	256	128	64	512	512	512
	69	0	32	32	64	256	256	256	256	256	128	256	128	64	256
	70	0	16	64	34	256	256	256	256	256	64	128	64	256	256

Five sheep in each group were vaccinated with 50 μg plant-produced Bluetongue virus (BTV)-8 VLP, 200 μg plant-produced BTV-8 core-like particles (CLP), OBP monovalent, live, attenuated BTV-8 vaccine or PBS (control) on day 0, boosted with the same on day 28 and challenged with live BTV-8 on day 63. Serum samples were taken every 7 days, and neutralizing antibody titres were determined by serum neutralisation plaque reduction assay. Results are expressed as the reciprocal of the dilution factor causing a 50% reduction in cytopathic effect.

infiltration of individual leaves), but techniques for scaling up expression by the vacuum infiltration of whole plants have recently been developed for coexpression of multiple proteins and shown to give comparable yields to syringe infiltration (Vézina et al., 2009). Using such techniques, in combination with scalable downstream processing techniques such as size exclusion and ion-exchange chromatography, it should be possible to produce sufficient quantities of BTV-8 VLPs in plants for this to be a viable route to the production of recombinant subunit vaccines to protect animals against Bluetongue disease. This should be a valuable addition to the veterinary armamentarium, as such vaccines made by more conventional expression systems are not cost-effective for animal use. Furthermore, the methods utilized in this study could be applied to production of a wide variety of VLPs relevant to both veterinary and human diseases.

Experimental procedures

Constructs

Constructs for plant-based expression were based on the Netherlands NET2006/04 strain of BTV-8 (Maan et al., 2008). Gene sequences for VP2, VP3, VP5 and VP7 were codon-optimized for Nicotiana translation synthesized by GeneArt (Life Technologies, Grand Island, NY) with flanking AgeI and XhoI sites. These genes were inserted into pEAQ-HT (Sainsbury et al., 2009) by restriction cloning, producing pEAQ-VP2HT, pEAQ-VP3HT, pEAQ-VP5HT and pEAQ-VP7HT (Figure 1a). Expression cassettes of these vectors were excised using AscI/PacI sites and transferred to the compatible AsiSI/MluI or AscI/PacI sites of pEAQexpress to yield pEAQex-VP5HT-VP2HT, pEAQex-VP7HT-VP3HT. A vector for expression of VP3 from a construct lacking the HyperTrans mutations of the CPMV-HT system was made by substitution of the modified CPMV RNA-2 leader with an amplified wild-type leader. This vector's expression cassette was used to make pEAQex-VP7HT-VP3wt, containing codon-optimized genes of VP7 and VP3, but without the HyperTrans mutation in the VP3 expression cassette.

All expression constructs were transformed into Agrobacterium tumefaciens LBA4404 by electroporation and propagated at 28 °C in Luria–Bertani media containing 50 μg/mL kanamycin and 50 μg/mL rifampicin.

Plant growth and expression

Transient expression of the BTV-8 constructs was carried out by agroinfiltration of Nicotiana benthamiana leaves. Plants were grown in a greenhouse maintained at 23–25 °C and infiltrated 3–4 weeks after the seedlings were pricked out. The first three mature leaves of each plant were selected for infiltration. Agrobacterium tumefaciens strains were subcultured and grown overnight, pelleted and resuspended to OD_{600} = 0.3 in MMA (10 mM MES buffer, pH 5.6; 10 mM magnesium chloride; 100 μM acetosyringone) and then infiltrated into leaf intercellular spaces using a blunt-ended syringe. When n constructs were coinfiltrated, they were prepared to have an overall final OD = 0.3 * n, with equal concentrations of each strain. Tissue was harvested 8–9 days postinfiltration, a time span found to achieve expression of all four structural proteins of BTV10 in earlier experiments (data not shown). Leaf tissue was immediately extracted in three volumes of

VLP extraction buffer (50 mM Bicine, pH 8.4; 20 mM sodium chloride (NaCl), 0.1% (w/v) NLS sodium salt; 1 mM dithiothreitol; Complete Protease Inhibitor Cocktail (Roche, Welwyn Garden City, UK)) or CLP extraction buffer (as VLP extraction buffer but with 140 mM NaCl) by homogenization using a Waring (Torrington, CT) blender. Crude extracts were filtered through two layers of Miracloth (Merck Millipore, Darmstadt, Germany) and then centrifuged for 10 min at 4200 × g, 10 °C to remove cell debris.

Purification

Particles were purified by density gradient centrifugation using either OptiPrep iodixanol (Axis-Shield, Oslo, Norway) or sucrose step gradients. Iodixanol solutions (20%–50%) and sucrose solutions (30%–60%) were prepared to contain 50 mM Tris–HCl, pH 8.4, and 20 mM NaCl (for VLPs) or 140 mM NaCl (for CLPs). Step gradients of 3 mL 10% incrementing steps were overlayed with 24 mL of clarified leaf extract and centrifuged at 21,500 rpm (85,800 × g max.) and 10 °C for 3 h in a Surespin 630 rotor (Thermo Scientific Sorvall, Asheville, NC). Fractions of 1 mL were collected from the bottom, and 10 μL of each was analysed by denaturing SDS-PAGE (4%–12% NuPage, Life Technologies, Grand Island, NY) followed by protein staining with Instant Blue (Expedeon, Harston, UK). Particles were generally found in the densest fractions (45%–55% sucrose or 30%–40% iodixanol).

Particles produced for animal experiments were further purified and concentrated in a second round of gradient centrifugation by pooling the best fractions of the first gradient, gradually diluting them (50 mM Tris–HCl, pH 8.4, 20/140 mM NaCl) and reapplying to a second, smaller gradient. This was centrifuged at 21,500 rpm (79,100 × g max.) and 10 °C for 3 h in an SW-41Ti rotor (Beckman Coulter, High Wycombe, UK) and fractionated from the bottom.

Transmission electron microscopy

Particles from density gradient fractions were adsorbed onto plastic and carbon-coated copper grids, washed successively by floating on three droplets of water and then stained with 2% (w/v) uranyl acetate for 20 s. Grids were imaged using a FEI Tecnai G2 20 Twin TEM with bottom-mounted digital camera.

Particle sizing

Adobe Photoshop 6.0 software (Adobe Systems Inc., San Jose, CA) was used to measure particle diameters on images obtained from TEM using the 'Measure Tool'. Twelve particles of each type were measured and the mean diameter and standard deviation determined.

Mass spectrometry

Gel pieces were washed, treated with trypsin and extracted according to standard procedures adapted from the study by Shevchenko et al. (2007). The peptide solution resulting from the digest was spotted onto a PAC plate (Prespotted AnchorChip™ MALDI target plate, Bruker Daltonics; Coventry, UK), and the spots were washed briefly with 10 mM ammonium phosphate and 0.1% TFA according to the manufacturer.

After drying, the samples were analysed by MALDI-TOF on a Bruker Ultraflex TOF/TOF. The instrument was calibrated using the prespotted standards (ca. 200 laser shots). Samples were analysed using a method optimized for peptide analysis, and spectra were summed from ca. 30 × 15 laser shots. Data were processed in FlexAnalysis (Bruker) and submitted for a database search using an in-house Mascot Server 2.2 (Matrixscience; London, UK) on the sptrembl20100119 or sptrembl20090901 database with taxonomy set to Viruses. For the search, the enzyme was set to trypsin with maximum one missed cleavage using a peptide mass tolerance of 30 ppm. Carbamidomethyl (C) was used as fixed and oxidation (M) as variable modification. Protein scores >72 were considered significant ($P < 0.05$).

Seroconversion of sheep injected with plant-produced VLPs

One milliliter of blood was taken from each of two BTV-free sheep before they were each injected subcutaneously with 20 μg plant-produced VLPs in a ratio of 1 : 1 with Freund's incomplete adjuvant. They were boosted with 20 μg VLPs at days 21 and 42 and bled 14 days after each inoculation with the final bleed taken at day 56. Serum was tested for BTV antibodies using a Bluetongue virus antibody test kit, cELISA (VMRD Inc., Pullman, WA). Dilutions of the BTV-8 VLPs originally used for the inoculations were separated on a 10% denaturing SDS polyacrylamide gel and blotted onto nylon membranes (transblotter) for Western blotting. Membranes were probed with 1/2000, 1/5000 and 1/10 000 dilutions of sheep serum from the final bleed and subsequently with a 1/10 000 dilution of anti-goat/anti-sheep alkaline phosphatase-conjugated secondary antibody (Sigma-Aldrich, Gillingham, UK). BTV-8 proteins were visualized using 5-bromo-4-chloro-3-indolyl phosphate/nitro blue tetrazolium (Roche, Welwyn Garden City, UK) substrate. A similar Western blot membrane was probed with prebleed sheep serum diluted 1 : 5000 and subsequently treated as described above.

Immunization and challenge of sheep

The immunization and challenge experiments were approved by the Onderstepoort Biological Products Animal Ethics Committee.

Twenty 1-year-old, BTV-naive merino sheep were kept in insect-proof isolation stables and divided into 4 groups and subcutaneously injected into the inner thigh with 1 mL of the following: 50 μg VLP with 50% (v/v) Montanide ISA70 VG (Seppic; SEPPIC, Puteaux, France) as an adjuvant, 200 μg CLP with 50% (v/v) Montanide ISA70 VG, PBS with 50% (v/v) Montanide ISA70 VG and 5 × 10^4 TCID$_{50}$/mL live, attenuated, monovalent BTV-8 (commercial vaccine seed stock for BTV-8, Onderstepoort Biological Products; South Africa). On day 28, each group was boosted with the same. Serum samples were taken on days 0, 7, 14, 21 and 28 postvaccination as well as after booster vaccination. Animals were monitored for reactions at the injection site, and temperature reactions were monitored for 14 days after injection. This was carried out to ensure that the inoculum was safe and secondly to monitor for possible other infections that could affect the results. On day 56, each group was challenged with an intravenous 1-mL injection of virulent BTV-8 sheep blood. The clinical reactions were monitored for the first 14 days postchallenge. The severity of clinical bluetongue after challenge with the virulent virus was expressed in a numerical form as the clinical reaction index (CRI; Huismans et al., 1987b), which was obtained by adding the following three scores (a + b + c):

● Febrile reaction – The cumulative total fever readings above 40 °C from days 3 to 14.
● Clinical lesion score – Lesions of the nose, mouth and feet are scored on a scale of 0–4.
● Fatality – an additional four points are added if death occurs within 14 days.

This CRI represents an estimate of bluetongue disease that takes into account the duration of any temperature increase

(>40 °C) and the extent, types and location of lesions (nose, mouth, hoof, etc.).

Bluetongue virus (BTV)-specific neutralizing antibodies were measured according to the procedure of the serum neutralisation test as described in the Office International des Epizooties (OIE) Manual of diagnostic tests and vaccines for terrestrial animals (2009 version; http://www.oie.int/fileadmin/Home/eng/Health_standards/tahm/2.01.03_BLUETONGUE.pdf). Antibody titres are expressed as the reciprocal of the serum dilution causing a 50% reduction in cytopathic effect and are calculated using the Spearman–Karber method.

Acknowledgements

We thank Louis Maartens and Stephan Swanepoel from Deltamune (Pretoria) for help with the sheep antiserum experiments, and Gerhard Saalbach (John Innes Centre) for help with mass spectrometry. E.C.T. acknowledges funding from a Marie Curie Early Stage Training Fellowship MEST-CT-2005-019727 and the Trustees of the John Innes Foundation. This work was supported, in part, by the EU FP7 'PLAPROVA' project (Grant Agreement No. KBBE-2008-227056). At the John Innes Centre, the work was also supported by BB/J004561/1 from BBSRC and the John Innes Foundation.

References

Batten, C.A., Maan, S., Shaw, A.E., Maan, N.S. and Mertens, P.P.C. (2008) A European field strain of bluetongue virus derived from two parental vaccine strains by genome segment reassortment. *Virus Res.* **137**, 56–63.

Carpenter, S., Wilson, A. and Mellor, P.S. (2009) *Culicoides* and the emergence of bluetongue virus in northern Europe. *Trends Microbiol.* **17**, 172–178.

Forzan, M., Wirblich, C. and Roy, P. (2004) A capsid protein of nonenveloped Bluetongue virus exhibits membrane fusion activity. *Proc. Natl Acad. Sci. USA*, **101**, 2100–2105.

French, T.J. and Roy, P. (1990) Synthesis of bluetongue virus (BTV) corelike particles by a recombinant baculovirus expressing the two major structural core proteins of BTV. *J. Virol.* **64**, 1530–1536.

French, T.J., Marshall, J.J.A. and Roy, P. (1990) Assembly of double-shelled, viruslike particles of bluetongue virus by the simultaneous expression of four structural proteins. *J. Virol.* **64**, 5695–5700.

Grimes, J.M., Burroughs, J.N., Gouet, P., Diprose, J.M., Malby, R., Zientara, S., Mertens, P.P.C. and Stuart, D.I. (1998) The atomic structure of the bluetongue virus core. *Nature*, **395**, 470–478.

Hassan, S.S. and Roy, P. (1999) Expression and functional characterization of bluetongue virus VP2 protein: Role in cell entry. *J. Virol.* **73**, 9832–9842.

Hewat, E.A., Booth, T.F. and Roy, P. (1994) Structure of correctly self-assembled bluetongue virus-like particles. *J. Struct. Biol.* **112**, 183–191.

Huismans, H., Vandijk, A.A. and Els, H.J. (1987a) Uncoating of parental bluetongue virus to core and subcore particles in infected L-cells. *Virology*, **157**, 180–188.

Huismans, H., van der Walt, N.T., Cloete, M. and Erasmus, B. (1987b) Isolation of a capsid protein of bluetongue virus that induces a protective immune response in sheep. *Virology*, **157**, 172–179.

Loudon, P.T. and Roy, P. (1991) Assembly of five bluetongue virus proteins expressed by recombinant baculoviruses - inclusion of the largest protein VP1 in the core and virus-like particles. *Virology*, **180**, 798–802.

Maan, S., Maan, N.S., Ross-Smith, N., Batten, C.A., Shaw, A.E., Anthony, S.J., Samuel, A.R., Darpel, K.E., Veronesi, E., Oura, C.A.L., Singh, K.P., Nomikou, K., Potgieter, A.C., Attoui, H., van Rooij, E., van Rijn, P., De Clercq, K., Vandenbussche, F., Zientara, S., Breard, E., Sailleau, C., Beer, M., Hoffman, B., Mellor, P.S. and Mertens, P.P.C. (2008) Sequence analysis of bluetongue virus serotype 8 from the Netherlands 2006 and comparison to other European strains. *Virology*, **377**, 308–318.

Maan, N.S., Maan, S., Belaganahalli, M.N., Ostlund, E.N., Johnson, D.J., Nomikou, K. and Mertens, P.P. (2012) Identification and differentiation of the twenty-six bluetongue virus serotypes by RT-PCR amplification of the serotype-specific genome segment 2. *PLoS ONE* **7**, e32601.

Martin, S.A. and Zweerink, H.J. (1972) Isolation and characterization of two types of bluetongue virus particles. *Virology*, **50**, 495–506.

Prasad, B.V.V., Yamaguchi, S. and Roy, P. (1992) 3-Dimensional structure of single-shelled bluetongue virus. *J. Virol.* **66**, 2135–2142.

Roy, P., French, T. and Erasmus, B.J. (1992) Protective efficacy of virus-like particles for bluetongue disease. *Vaccine*, **10**, 28–32.

Rybicki, E.P. (2010) Plant-made vaccines for humans and animals. *Plant Biotechnol. J.* **8**, 620–637.

Sainsbury, F. and Lomonossoff, G.P. (2008) Extremely high-level and rapid transient protein production in Plants without the use of viral replication. *Plant Physiol.* **148**, 1212–1218.

Sainsbury, F., Thuenemann, E.C. and Lomonossoff, G.P. (2009) pEAQ: versatile expression vectors for easy and quick transient expression of heterologous proteins in plants. *Plant Biotechnol. J.* **7**, 682–693.

Savini, G., MacLachlan, N.J., Sanchez-Vinaino, J.M. and Zientara, S. (2008) Vaccines against bluetongue in Europe. *Comp. Immunol. Microbiol. Infect. Dis.* **31**, 101–120.

Shevchenko, A., Tomas, H., Havliš, J., Olsen, J.V. and Mann, M. (2007) In-gel digestion for mass spectrometric characterization of proteins and proteomes. *Nat. Protoc.* **1**, 2856–2860.

Silhavy, D., Molnár, A., Lucioli, A., Szittya, G., Hornyik, C., Tavazza, M. and Burgyán, J. (2002) A viral protein suppresses RNA silencing and binds silencing-generated, 21- to 25-nucleotide double-stranded RNAs. *EMBO J.* **21**, 3070–3080.

Stewart, M., Dovas, C.I., Chatzinasiou, E., Athmaram, T.N., Papanastassopoulou, M., Papadopoulos, O. and Roy, P. (2012) Protective efficacy of Bluetongue virus-like and subvirus-like particles in sheep: Presence of the serotype-specific VP2, independent of its geographic lineage, is essential for protection. *Vaccine*, **30**, 2131–2139.

Vézina, L.-P., Faye, L., Lerouge, P., D'Aoust, M.-A., Marquet-Blouin, E., Burel, C., Lavoie, P.-O., Bardor, M. and Gomord, V. (2009) Transient co-expression for fast and high-yield production of antibodies with human-like *N*-glycans in plants. *Plant Biotechnol. J.* **7**, 442–455.

14

The *Vitis vinifera* C-repeat binding protein 4 (*VvCBF4*) transcriptional factor enhances freezing tolerance in wine grape

Richard L. Tillett[†], Matthew D. Wheatley[†], Elizabeth A. R. Tattersall[‡], Karen A. Schlauch, Grant R. Cramer and John C. Cushman*

Department of Biochemistry and Molecular Biology, University of Nevada, Reno, NV, USA

*Correspondence
email
jcushman@unr.edu
[†]These authors contributed equally to this work.
[‡]Present address: Elizabeth A. R. Tattersall, Department of Biology, Western Nevada College, Bently Hall 102, 1680 Bently Parkway South, Minden, NV 89423, USA.

Keywords: CBF transcription factor, freezing, cold tolerance, dwarf, wine grape, *Vitis vinifera*.

Summary

Chilling and freezing can reduce significantly vine survival and fruit set in *Vitis vinifera* wine grape. To overcome such production losses, a recently identified grapevine C-repeat binding factor (CBF) gene, *VvCBF4*, was overexpressed in grape vine cv. 'Freedom' and found to improve freezing survival and reduced freezing-induced electrolyte leakage by up to 2 °C in non-cold-acclimated vines. In addition, overexpression of this transgene caused a reduced growth phenotype similar to that observed for CBF overexpression in *Arabidopsis* and other species. Both freezing tolerance and reduced growth phenotypes were manifested in a transgene dose-dependent manner. To understand the mechanistic basis of *VvCBF4* transgene action, one transgenic line (9–12) was genotyped using microarray-based mRNA expression profiling. Forty-seven and 12 genes were identified in unstressed transgenic shoots with either a >1.5-fold increase or decrease in mRNA abundance, respectively. Comparison of mRNA changes with characterized CBF regulons in woody and herbaceous species revealed partial overlaps, suggesting that CBF-mediated cold acclimation responses are widely conserved. Putative *VvCBF4*-regulon targets included genes with functions in cell wall structure, lipid metabolism, epicuticular wax formation and stress-responses suggesting that the observed cold tolerance and dwarf phenotypes are the result of a complex network of diverse functional determinants.

Introduction

Vitis vinifera or wine grape was domesticated more than 7000 years ago and continues to the present day to produce one of the world's most important fruit crops (Arroyo-Garcia et al., 2006; This et al., 2006). Cultivation of *V. vinifera* and other *Vitis spp.* encompasses ~8 million hectares of land worldwide, more than any other cultivated fruit (Vivier and Pretorius, 2002). *Vitis vinifera* cultivars grow well in temperate, semi-arid climates that can sometimes experience freezing or subfreezing temperatures each winter.

As a deciduous perennial, *Vitis spp.* acquire freezing tolerance in advance of annual freezes, when shoots mature into overwintering canes and buds enter dormancy, cued by chilling temperatures and/or shortened day length (Sreekantan et al., 2010). Cooled gradually, *V. vinifera* cultivars can tolerate sustained winter temperatures as low as −15 °C without injury, whereas wild North American and Asian species can tolerate exotherms of −35 to −40 °C (Mullins et al., 1992; Fennell, 2004). As vines exit dormancy each spring, freezing vulnerability returns quickly (Fennell, 2004). Breaking buds and newly emergent green tissues can suffer injury at just −2.5 °C (Fuller and Telli, 1999). Moreover, damage to floral primordia of primary and secondary buds can drastically reduce crop yields (Fennell, 2004).

Cold acclimation, the process whereby plants sense low temperatures and activate mechanisms to increase tolerance to chilling or freezing stress, is a complex response involving multiple biochemical pathways (Nakashima and Yamaguchi-Shinozaki, 2006; Sreenivasulu et al., 2007). Central to the early cold-response pathway is the C-repeat binding factor (CBF)/dehydration-responsive element binding (DREB) family of transcription factors, of which the overexpression of a single *Arabidopsis* CBF gene family member is sufficient to impart an improved stress tolerance phenotype in *Arabidopsis* (Stockinger et al., 1997; Gilmour et al., 1998; Jaglo-Ottosen et al., 1998; Liu et al., 1998), canola (*Brassica napus*) (Jaglo-Ottosen et al., 2001), tomato (Hsieh et al., 2002; Lee et al., 2003) and potato (Pino et al., 2007). In addition to cold tolerance, constitutive, ectopic expression of *AtCBF2* or *AtCBF3* has been shown to delay leaf senescence and extend plant longevity in *Arabidopsis* presumably to enable winter survival until spring for lifecycle completion (Sharabi-Schwager et al., 2010).

C-repeat binding factor/DREB genes have been identified in monocots and eudicots alike (Benedict et al., 2006; El Kayal et al., 2006; Nakashima and Yamaguchi-Shinozaki, 2006; Badawi et al., 2007; Campoli et al., 2009), many of which have been characterized to have similar or conserved functions. For example, overexpression of CBF genes from cotton (Huang et al., 2009), rice (Dubouzet et al., 2003), birch (Welling and Palva, 2008), perennial ryegrass (Zhao and Bughrara, 2008) and wine grape (*V. vinifera*) cv. Koshu (Takuhara et al., 2011) or wild grape (*V. riparia*) (Siddiqua and Nassuth, 2011) in transgenic *Arabidopsis* improved freezing tolerance in a manner similar to *AtCBF* overexpression. Ectopic expression of two *Eucalyptus CBF* genes in transgenic *Eucalyptus* plants resulted in improved cold tolerance (Navarro *et al.*, 2011). Constitutive overexpression of *AtCBF1* (*AtDREB1b*) in transgenic *V. vinifera* cv. Centennial Seedless resulted in about a 2 °C improvement in cold resistance as measured by electrolyte leakage and improved vine survival at −4 °C (Jin *et al.*, 2009). Lastly, constitutive overexpression of a peach (*Prunus persica*) *CBF* gene in apple (*Malus* × *domestica*) not only improved cold hardiness, but also resulted in short day length–related growth cessation and leaf senescence (Wisniewski *et al.*, 2011).

The transcription factors encoded by the first three *CBF/DREB* family members described for *V. vinifera* and *V. riparia* were found to regulate genes that respond to low temperature, drought stress and exogenous ABA application (Xiao *et al.*, 2006). These three transcription factor genes (*VvCBF1*, *VvCBF2* and *VvCBF3*) showed increased mRNA expression in young compared with mature vegetative tissues upon exposure to freezing and drought stresses (Xiao *et al.*, 2006). A fourth member of the *Vitis CBF/DREB1* gene family, *VvCBF4*, was also identified for both *V. vinifera* and *V. riparia* (Xiao *et al.*, 2008). This transcription factor gene is unique among the *Vitis CBF* gene family in both its expression profile and sequence (Xiao *et al.*, 2008). Expression of *VvCBF4* mRNA was sustained for several days following induction of a 4 °C cold stress, in contrast to the transient expression of *VvCBF1–3* transcripts. In addition, *VvCBF4* was induced similarly in both young and mature tissues. More recently, *VvCBF4* transcripts have been shown to be induced after 4 h cold (4 °C) stress in leaf, stem and flower of *V. vinifera* cv. Koshu (Takuhara *et al.*, 2011). VvCBF4:green fluorescent protein (GFP) fusions localized to the nucleus, and *VvCBF4* expression was shown to induce beta-glucuronidase (GUS) expression under the control of the *rd29a* promoter *in trans* when co-infiltrated in tobacco leaves (Xiao *et al.*, 2008). *CBF* gene transcripts with similarly sustained cold-induction patterns have been reported in other woody perennial species, such as birch (Welling and Palva, 2008), poplar (Benedict *et al.*, 2006) and *Eucalyptus* (Navarro *et al.*, 2009). These CBF transcription factors and their target genes have been suggested to play a role in the freezing tolerance of overwintering woody perennials. Additional members of the *AP2/ERF* gene superfamily in *Vitis vinifera* display diverse expression patterns in both vegetative and reproductive tissues (Licausi *et al.*, 2010).

In this report, the overexpression of the *VvCBF4* gene is shown to impart a reduced growth phenotype as well as confer improved freezing survival in nonacclimated vines, similar to the effects of CBF overexpression in *Arabidopsis* and other species. Microarray transcriptional profiling of a *VvCBF4* overexpressor identified 47 and 12 genes with either a >1.5-fold increase or decrease in mRNA abundance, respectively, in young, unstressed shoots. These observed changes in mRNA expression suggest that the CBF regulon is widely conserved among woody and herbaceous species and that modulations in cell wall structure, lipid metabolism, epicuticular wax formation, and various stress-responses might participate in the acquisition of freezing tolerance in wine grape.

Results

VvCBF4 is most similar to CBFs of woody perennials

VvCBF4 was identified previously (Xiao *et al.*, 2008) and so-named simply because of the order of its discovery. To iden-

tify its relationship with other *CBF* genes in *Vitis*, and eudicots in general, phylogenetic analyses were performed (Figure 1). Phylogenetic analysis of all four *Vitis CBF* genes with the full-length amino acid sequences of *CBFs* from other dicotyledonous plants (and a non-CBF, AP2-domain protein AtERD10 as an outgroup gene) confirmed that among the *Vitis* CBFs, the *VvCBF4* gene product shares the greatest similarity with *Arabidopsis* CBFs as well as two CBF gene products from another woody species, *Populus trichocarpa*, PtDREB70 and PtDREB71 (Figure 1a). The amino acid sequence alignment of the full-length CBF-encoding genes of *Arabidopsis*, *P. trichocarpa* and *V. vinifera* (Figure 1b) showed that the *VvCBF4* gene product has 100% conservation with the *Arabidopsis* CBF consensus sequences that flank the AP2 region (N-flanking: PKKR/PRAGRxKFxETHRP, C-flanking: DSAWR), which are required for CBF function (Canella *et al.*, 2010). In contrast, the proteins encoded by the three other *Vitis CBF* genes (*VvCBF1*, *VvCBF2* and *VvCBF3*) as well as those encoded by the *PtDREB66* and *PtDREB67* genes share a variant of the established consensus between positions 1 and 4 of the N-flanking consensus (H/KKR/NK instead of PKKR/P). However, only the three other *Vitis CBF* (*VvCBF1*, *VvCBF2* and *VvCBF3*) gene products contain a 21–25-amino acid stretch between the DSAWR motif and the C-terminus found in neither the *Arabidopsis* and *Populus* CBF gene products nor that of VvCBF4. Additionally, the gene products of *VvCBF1*, *VvCBF2* and *VvCBF3* as well as those of the *PtDREB66* and *PtDREB67* genes share a variation (e.g., LWN(E)D(H/E)) in the generally conserved LWSY amino acid sequence motif at their C-terminus.

Quantitative real-time RT-PCR (qRT-PCR)

To quantify expression of the cauliflower mosaic virus (CaMV) 35S::*VvCBF4* transgene in three independently transformed lines of *Vitis* rootstock 'Freedom,' quantitative real-time, reverse-transcriptase PCR (qRT-PCR) was performed. Whereas the ORFs of native and transgenic *VvCBF4* transcript are identical, the native 3'UTR was replaced by a 35S CaMV terminator region in the 35S::*VvCBF4* construct. Thus, primers were designed to detect either 'native' or transgenic ('tg') *VvCBF4* transcripts exclusively from these regions (Table S1, Figure S1).

qRT-PCR experiments demonstrated that both native and tg primers were specific to the mRNAs for which they were designed. In each of the three independent transgenic lines, 9–3, 9–12 and 9–1, tg transcripts were readily detected, whereas no amplification of tg could be detected in control line 8–6 (Figure 2). *VvCBF4* overexpression resulted in no observable changes in native transcript abundance. One-way ANOVA of native *VvCBF4* abundance showed no significant difference in genotype ($F > 0.05$), which remained low regardless of the line tested (Figure 2).

In the 35S::*VvCBF4* transformed line 9–12, tg transcript accumulated to concentrations 20-fold greater than native transcript levels. Another line, 9–3, expressed 35S::*VvCBF4* less strongly than 9–12, with only 71% of the transgene abundance of 9–12, or 14-fold greater than native transcript. The tg mRNA abundance difference between 9–3 and 9–12 was statistically significant as determined by Student's *t*-test with Bonferroni correction ($P = 0.048$). Transgene expression in line 9–1, however, was highly variable among biological replicates, with expression ranging between 72% and 360% of 9–12 transgene relative abundance (or between 14- and 72-fold of native transcript). Owing to this large variability, comparison of expression in 9–1 with the other transgenic lines failed to reveal any significant expression differences in contrast to 9–3 or 9–12.

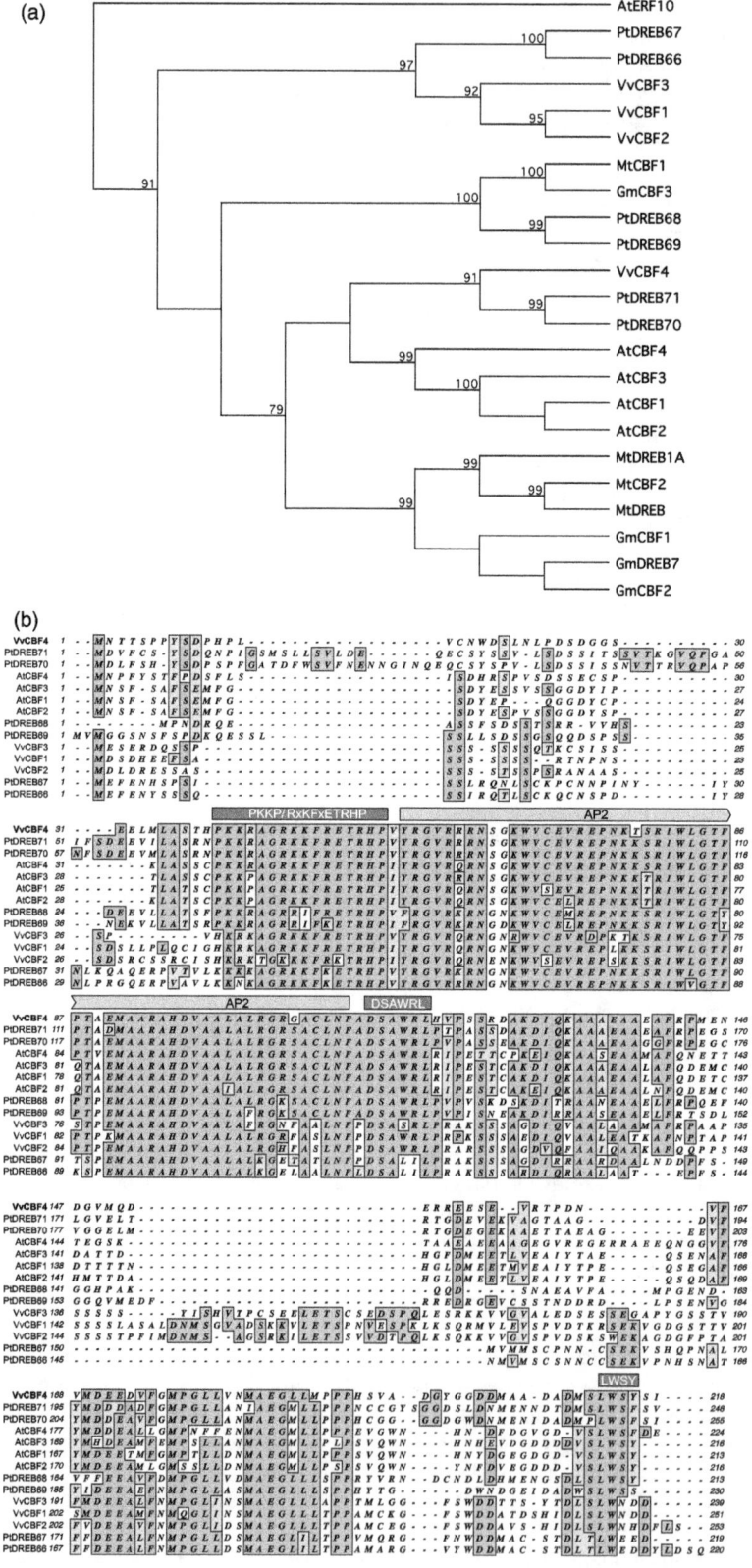

Figure 1 Comparative sequence analysis of C-repeat binding factor (CBF) protein family members in eudicot species. (a) Phylogenetic analysis of CBF/DREB proteins found in eudicot species. CBF proteins were identified from *Arabidopsis thaliana*, *Vitis vinifera*, *Populus trichocarpa*, *Glycine max* and *Medicago truncatula* – species with published draft genomes. *AtERF10* is included as an outgroup root. (b) Amino acid alignment of *A. thaliana*, *P. trichocarpa* and *V. vinifera CBF* transcription factors. Identical and similar amino acids are coloured dark and light gray, respectively. The conserved APETALA 2 (AP2) domain and the CBF-conserved AP2 N-flanking PKKP/RAGRxKFxETRHP, AP2 C-flanking DSAWRL and LWSY motifs are labelled with boxes above the alignment. Accession numbers are listed in the Experimental Procedures section.

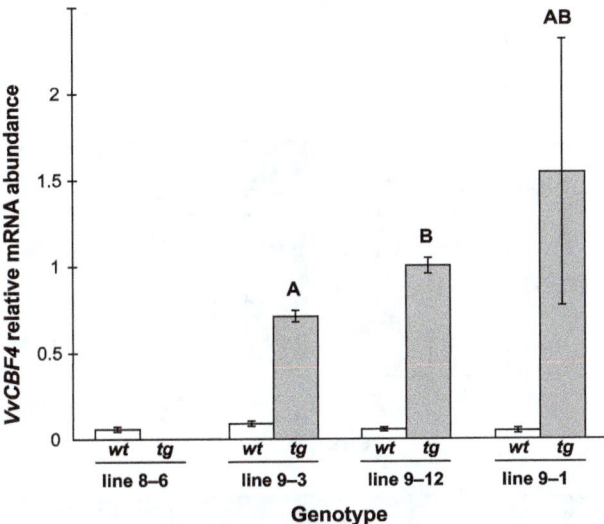

Figure 2 Overexpression of 35S::*VvCBF4* in transformed *Vitis* cv. 'Freedom' shoots. qRT-PCR of native and transgenic *VvCBF4* transcripts for control (line 8–6) and three independently transformed 35S::*VvCBF4*-overexpressing lines (9–3, 9–12, 9–1). Wild-type *VvCBF4* (wt) transcript relative abundance; transgenic *VvCBF4* (tg) transcript relative abundance. Transcript abundances were normalized to an actin reference gene. Error bars indicate ± standard error (SE). Significantly different abundance was determined using the Student's *t*-test with Bonferroni correction, *P* < 0.05; *n* = 3. Significant pairwise differences in tg expression are indicated with letters (a, b).

VvCBF4 overexpression reduces shoot elongation

The constitutive overexpression of some CBF/DREB genes using the 35S-CaMV promoter is known to result in dwarfing effects in species such as *Arabidopsis thaliana* (Liu *et al.*, 1998; Gilmour *et al.*, 2000), *Medicago sativa* (Zhang *et al.*, 2005) and *Lolium perenne* L. (Zhao and Bughrara, 2008). To assess the effect of 35S::*VvCBF4* transgene expression on plant phenotype, the shoot elongation rate (SER) was measured at 3-day intervals in young vines over an 18-day period for control line 8–6 and the three selected 35S::*VvCBF4*-overexpressing lines (Figure 3a,b). By one-way ANOVA, genotype accounted for highly significant observed differences in SER (*F* < 0.0001). Compared to control line 8–6, which had an average SER of 12.5 mm/day, SER was reduced in all 35S::*VvCBF4*-transformed vines. Line 9–3 SER was reduced slightly, although significantly, to 10.9 mm/day (*P* < 0.05). Lines 9–12 and 9–1 vines displayed even more dramatic reductions in SER, with rates of 7.4 and 4.5 mm/day, respectively (both were significantly different from control line 8–6, each *P* < 0.001) (Figure 3b).

In lines 9–12 and 9–1, lengths of stem sections, or internodes, were also reduced greatly (Figure 3c). Over the course 18 days of growth observation, the longest internodes on individual 9–12 vines (e.g., the single longest stem section from each of eight vines, averaged) were 36.8 mm, or 66% of the lengths achieved by 8–6 vines. Similarly, line 9–1 grew to maximum lengths of 33.1 mm (59% of control). These reduced node lengths were highly significant (*P* < 0.01).

The two most dwarfed lines, 9–12 and 9–1, differed in the number of nodes and the attachment site of leaves produced

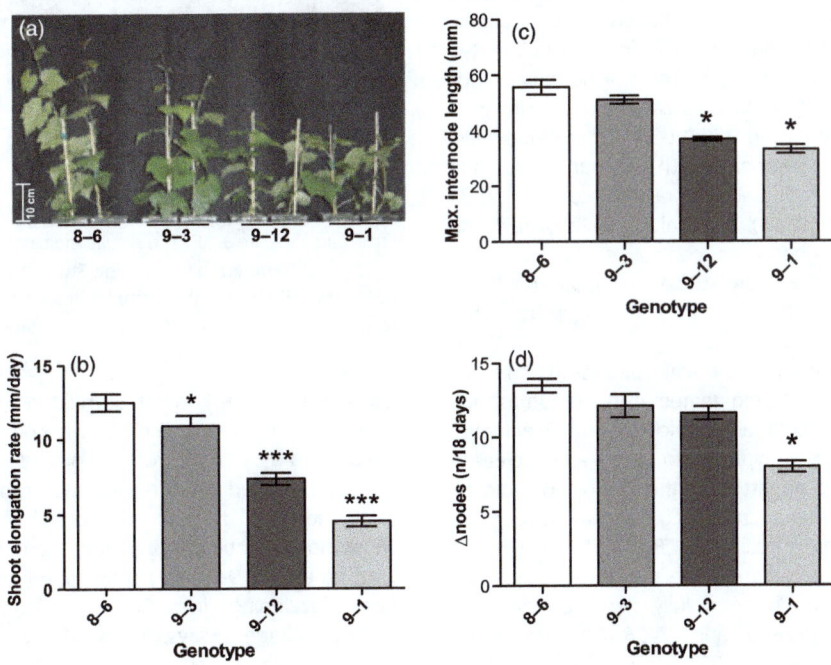

Figure 3 The C-repeat binding factor (CBF)-induced dwarfing effect is dependent on relative level of *VvCBF4* overexpression in 35S::*VvCBF4 Vitis* cv. 'Freedom.' (a) Image showing two representative vines each of control line 8–6 and 35S::*VvCBF4* overexpressing lines 30 days after transplantation to soil. (b) Shoot elongation rate for line 8–6 (control) and 35S::*VvCBF4*-overexpressing lines (9–3, 9–12, 9–1). Each bar represents the mean shoot elongation rate of eight vines measured over 18 days, every third day. Error bars indicate standard error of the mean; *n* = 8. Growth for each 35S::*VvCBF4*-overexpressing line was significantly different from control line 8–6 (repeated measures ANOVA). **P* < 0.05; ***P < 0.001. (c) Average length of the maximum (Max.) internode length per plant. (d) Total increase in number of nodes (Δ nodes) after 18 days of observation for control line 8–6 and 35S::*VvCBF4*-overexpressing lines (*n* = 8). Error bars indicate standard error of the mean; n = 8. *significant difference between control (line 8–6) and *VvCBF4*-overexpressing lines based on the Student's *t*-test with Bonferroni correction, *P* < 0.05; *n* = 8.

during this 18-day period (Figure 3d). Line 9–12 vines initiated 11.6 (SD ± 2.1) leaves on average, 86% as many leaves as did 8–6, which produced an average of 13.5 leaves (SD ± 1.3). This difference was not statistically significant. In contrast, the other dwarfed line, 9–1, produced an average of only 8.0 (SD ± 1.1) leaves in the same period. Genotypic differences were significant (one-way ANOVA by genotype, $P < 0.001$), and line 9–1 (and only line 9–1) differed from each of the other three lines (Bonferroni-corrected pairwise tests, $P < 0.01$).

Taken together with the *VvCBF4* transgene expression data for each of these transformed lines, the growth data indicated that a dose–response relationship exists between *VvCBF4* transcript abundance and the dwarf phenotype, with high-overexpressing line 9–12 showing a more reduced SER than the medium-overexpressing line 9–3. Line 9–1 growth reduction was the most severe, although the high variability of *VvCBF4* transgene expression (Figure 2) limited the correlation with the growth phenotype in this line. Overall, these results are consistent with reports for other species in which constitutive overexpression of a CBF/DREB gene under the control of a constitutive promoter resulted in dwarf or reduced aerial biomass phenotypes (Kasuga *et al.*, 1999; Gilmour *et al.*, 2000; Benedict *et al.*, 2006; Pino *et al.*, 2007; Achard *et al.*, 2008a; Navarro *et al.*, 2011).

*Vv*CBF4 overexpression increases freezing survival

As *VvCBF4* transcripts were previously reported to undergo increased relative abundance during the application of 4 °C chilling in Chardonnay (Xiao *et al.*, 2008), the effect that *VvCBF4* overexpression had on freezing survival rates was investigated. Freezing stress resistance was assessed in 35S::*VvCBF4* transgenic vines by whole-plant survival analysis of nonacclimated vines. The 50% lethal temperature (LT_{50}) for control line 8–6 vines was determined experimentally to be at or near −2 °C for 24 h (data not shown), so these conditions were chosen for survival testing between control and 35S::*VvCBF4* transgenic lines. Following nine replicate experiments with 10 young vines/line in each experiment, the freezing survival rate for control line 8–6 averaged only 29% (Figure 4). Survival of 35S::*VvCBF4* line 9–12 was significantly higher than control, at 52% survival ($P < 0.01$). Survival of lines 9–1 and 9–3 were 39% and 43%, respectively, which was not significantly different from the control line ($P > 0.05$).

Cold acclimation in plants is a dynamic process that results from exposure to low nonfreezing temperatures (Thomashow, 1999). Following a given cold acclimation period, electrolyte leakage assays are employed typically to assay any increased cold- or freezing-tolerance imparted to the cell by the numerous biochemical changes that occur during acclimation (Gilmour *et al.*, 2000). Because overexpression of CBF/DREB genes might mimic the effects of cold acclimation, we performed electrolyte leakage assays on leaf discs cut from fully expanded leaves of nonacclimated control line 8–6 and *VvCBF4*-overexpressing vines. Line 9–3, which expressed ~70% as much *VvCBF4* mRNA as line 9–12 (Figure 2), exhibited a change in leakage only at −6 °C (Figure 5a). Line 9–12 showed a 2 °C greater resistance to electrolyte leakage than the control line with these differences appearing at −6 and −7 °C (Figure 5b). At −8 °C, line 9–12 leaf discs had begun leaking and no further differences could be observed. No significant changes between line 9–1 and control line 8–6 were observed (Figure 5c). The lack of measurably significant enhancement of survival or electrolyte

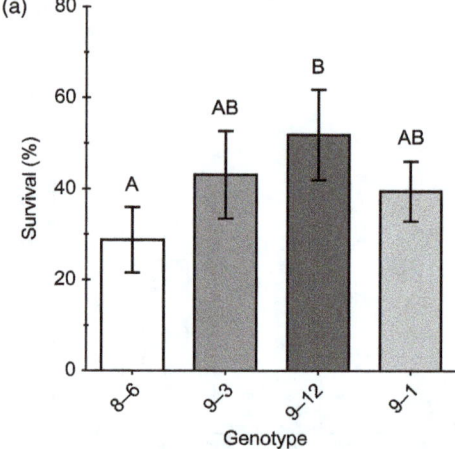

Figure 4 35S::*VvCBF4* overexpression enhances freezing survival in transformed line 9–12. (a) Percentage survival for mock-transformed line 8–6 (control) and 35S::*VvCBF4*-overexpressing lines (9–3, 9–12, 9–1) following exposure to freezing at −2 °C for 24 h with 14 days recovery at 22 °C. Each bar represents the mean of nine replicate experiments with 10 individual vines used for each genotype in each experiment. Error bars indicate ± SE. The results indicated by different letters are significantly different based on the Student's *t*-test with Bonferroni correction; ($P < 0.01$). (b) Exemplar images of freezing survival after recovery from one of the nine experimental trials.

leakage in line 9–1 might be related to the severity of growth retardation in this line (Figure 3) in combination with the propagation and assay conditions employed. Sterile cuttings from line 9–1 typically had shorter and fewer roots and had a lower survival rate when transplanted to soil compared with the other lines under identical propagation conditions, suggesting a lowered fitness for 9–1 vines propagated on the same time scale as faster-developing vines. Based upon freezing survival and electrolyte leakage assays, line 9–12 was selected for detailed genotypic evaluation.

Global changes in gene expression in 35S::*Vv*CBF4 overexpressor

To investigate the molecular mechanisms involved in the observed enhancement of freezing tolerance in *VvCBF4* overexpressor line 9–12, differences in relative mRNA abundance changes in this line were compared with those of empty vector control line 8–6 using microarray transcript profiling. RNA was

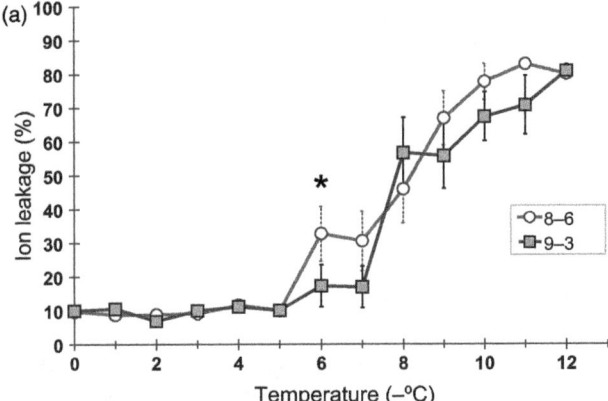

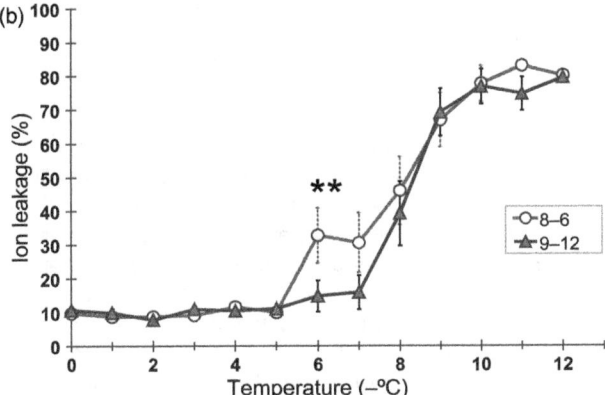

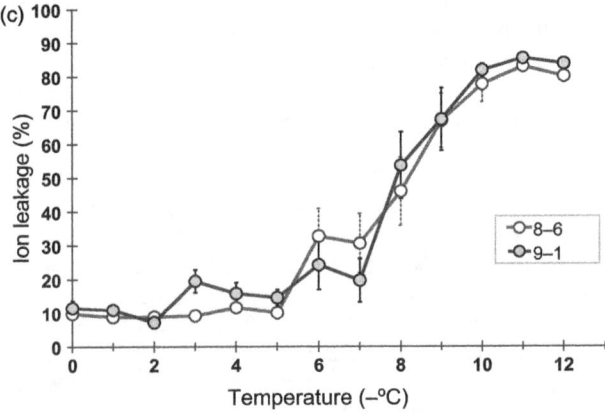

Figure 5 35S::*VvCBF4*-transformed 'Freedom' lines exhibit less electrolyte leakage upon freezing stress. Percentage leaf-disc electrolyte leakage for line 8–6 (control) and 35S::*VvCBF4*-overexpressing lines (a) 9–3, (b) 9–12 and (c) 9–1 following exposure to subfreezing temperatures as indicated. Each point represents the mean electrolyte leakage of leaf discs from 11. Error bars indicate ± SE. *Significant difference between the 35S::*VvCBF4*-transformed line and control line 8–6 at the marked temperature (one-way ANOVA with *post hoc* Student's *t*-test with ordered differences test, $P < 0.05$; $n = 11$).

extracted from whole aerial portions of young vines of line 9–12 and control line 8–6 (four biological replicates each), which had been propagated and harvested 30 day after transplantation to soil. RNA was verified to be of high quality with no degradation using an Agilent 2100 Bioanalyzer (Figure S2). After cRNAs were hybridized to Affymetrix® *Vitis* GeneChip® microarrays and probeset intensities normalized by robust multiaverage (RMA),

principal component analysis (PCA) of the RMA probe intensities found 99.4% of the variation among all samples to be explained by two components, p1 and p2 (Figure 6). Along these two axes, the microarray data can be seen to segregate by genotype, indicating that expression of the transgene is associated strongly with the observed phenotypic differences.

In *VvCBF4*-overexpressing line 9–12, 48 probesets were identified to have 1.5-fold or greater abundance than in control line 8–6, (significance of $P < 0.05$) using t statistics corrected for multiple comparisons (Benjamini and Hochberg, 1995). A total of 47 unique transcripts increased in abundance in the *VvCBF4* overexpressor, which included an acid phosphatase class B (XP_002273448) that was specified by two independent probesets that showed 2.0-fold (probeset ID 1621892_a_at) and 1.6-fold (probeset ID 1617422_at) increase, respectively (Table 1). Changes in increased relative expression ranged from 1.5-fold to 5.6-fold. An additional 12 transcripts exhibited significantly decreased relative abundance in line 9–12, compared with control line 8–6 (Table 1). Changes in decreased relative expression ranged from −1.5-fold to −37.1-fold. The complete list of transcripts that displayed significantly different patterns of expression is presented in Table S2.

Validation of transcript abundance by qRT-PCR

To validate the observed transcript abundance changes obtained using the *Vitis* GeneChip® microarray, qRT-PCR was performed on a set of nine genes that were selected at random from among those genes exhibiting either significantly increased or decreased transcript abundance using gene-specific primer pairs (Table S1). Linear regression analysis of the $\log_2$-transformed values from the microarray analysis with those from qRT-PCR showed a goodness of fit (R^2) of 0.81, confirming that the observed changes in transcript abundance were accurate (Figure 7).

Comparison of woody CBF regulons

Transcriptome changes observed because of ectopic 35S::*VvCBF4* expression were compared with those identified in annual (leaf) and perennial (stem) tissues of 35S::*AtCBF1* regulons in the woody species *Populus* (Benedict *et al.*, 2006). The closest protein homologues within previously reported CBF regulons were compared using BLAST homology searches. Comparing 35S::*VvCBF4* transgene expression with that in 35S::*AtCBF1* poplar revealed 10 genes with similar trends of increased and one gene, a chitinase class IV C gene, with decreased transcript abundance (Table 2). Three additional genes shared significant expression differences between the two transgenic woody species, but displayed varying directionality of expression (Table 2). These results indicate that at least part of the CBF regulon is shared between these two diverse woody species.

Comparison of herbaceous CBF regulons

Next, the ectopic 35S::*VvCBF4* mRNA expression profile was compared with that from various independent reports of ectopic expression of *At*CBF or closely related genes including the 35S::*AtCBF1*, 35S::*AtCBF2* 35S::*AtCBF3* regulons (Fowler and Thomashow, 2002; Maruyama *et al.*, 2004, 2009; Vogel *et al.*, 2005; Sharabi-Schwager *et al.*, 2010) and the 35S::*DDF1* regulon (Magome *et al.*, 2008) in the herbaceous species *Arabidopsis* (Table 3). A total of nine and four shared genes showed similar increases or decreases, respectively, in relative transcript abundance with two genes, a pectin methylesterase inhibitor (PMEI) and a GDSL lipase being common to four and five independent

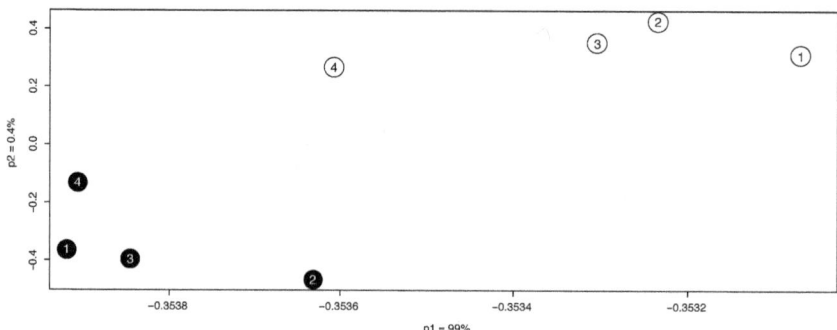

Figure 6 Principal component analysis of microarray probeset variation. Robust Multichip Average (RMA) normalized intensities were analysed with ~16 000 probesets on Affymetrix® *Vitis* GeneChip® microarrays for the RNA transcripts of 35S::*VvCBF4*-overexpressing line 9–12 (black) and empty vector-transformed *Vitis* rootstock 'Freedom' control line 8–6 (white). Whole aerial portions (stem + five leaves) of nonstressed young vines were used for RNA extraction, and tissue from three different vines was pooled for RNA isolation and microarray hybridization (*n* = 4 arrays per genotype). Principal components 1 and 2 (p1, p2) account for 99% and 0.4% of the variation among the eight microarrays.

studies, respectively. Another five genes were shared among different experiments, but had only partially similar transcript expression trends between 35S::*VvCBF4* and the different *AtCBF*- or *DDF1f*-overexpressing *Arabidopsis* lines. Three genes showed consistent transcript expression trends with one or more independent experiments: a chitinase class IV gene, a thaumatin-like gene and a pepsin-like aspartic protease (Table 3). Lastly, six genes showed disparate transcript expression trends between 35S::*VvCBF4* expression profiles and among the different *AtCBF*- or *AtDDF1f*-overexpressing Arabidopsis lines (Table 3). The amount of overlap in target genes observed for 35S::*VvCBF4* and these various *Arabidopsis* 35S::*AtCBF* (*DDF1f*) transgenic lines is comparable to the degree of overlap observed when 35S::*AtCBF1*-driven expression in Poplar was compared with the *AtCBF3* regulon in *Arabidopsis*, wherein 12 genes showed similar increased transcript abundance (Benedict *et al.*, 2006). These results indicate that CBF regulons share common targets in both woody and herbaceous species.

Two-dimensional difference in gel electrophoresis (2D-DIGE) analysis

To quantify the effect of 35S::*VvCBF4* transgene expression on relative protein abundance, 2D-DIGE analysis was performed. Seventy-seven spots showed significant differences (one-way ANOVA, $P \leq 0.05$) in abundance in 35S::*VvCBF4* line 9–12 when compared with control line 8–6 across four biological replicates (data not shown). Of these, 29 spots, of which 17 and 12 proteins were increased or decreased in protein abundance, respectively, were identified by MALDI-TOF/TOF analysis (Table S4).

Discussion

In this study, the role of a wine grape CBF transcription factor in conferring improved freezing tolerance was confirmed in the leaves of *V. vinifera* cv. Freedom. *VvCBF4* is unique among the four known *CBF/DREB* genes found in grapevine, sharing only 48%, 45% and 45% homology with *VvCBF1*, *VvCBF2* and *VvCBF3*, respectively (Xiao *et al.*, 2008). In contrast, *VvCBF4* shares 58% sequence homology with *A. thaliana CBF1*, one of four known *Arabidopsis CBF/DREB1* genes. Among all genes found in the fully sequenced plant genomes, *VvCBF4* gene product has the greatest amino acid similarity the *Populus*

trichocarpa PtDREB70 and *PtDREB 71* gene products (Figure 1). *PtDREB70* and *PtDREB 71*, which were previously identified as *PtCBF2* and *PtCBF1* by Benedict *et al.* (2006), are cold-inducible *CBF* genes with peak mRNA expression at 9 and 6 h after cold-exposure, respectively. Previously, transient expression assays of *VvCBF4* constructs in tobacco leaves confirmed that *VvCBF4* was capable of inducing the transcription of reporter genes via CRT cis-elements and that *Vitis* CBF4:GFP localized to the nucleus (Xiao *et al.*, 2008). These observations, combined with the long duration of cold induction (2–5 days) of its transcript, which is even longer in the cold-hardy *Vitis riparia* species (Xiao *et al.*, 2008), suggested that *VvCBF4* is likely to play an important role in adaptation to cold or freezing conditions. In *Eucalyptus*, another woody species, differences in the duration of cold induction of CBF genes between cold-hardy and cold-sensitive species have also been reported (Navarro *et al.*, 2009).

The constitutive overexpression of *VvCBF4* enhanced freezing survival (Figure 4) and reduced electrolyte leakage under freezing conditions by about 2 °C (Figure 5) in two of three *VvCBF4*-expressing lines tested. The observed improvements in freezing tolerance are comparable with those observed in transgenic poplar ectopically expressing *AtCBF1*, which showed an improvement in freezing tolerance of 1.5 °C in stems and 3 °C in leaves, as assessed by electrolyte leakage assays (Benedict *et al.*, 2006). These results were also consistent with the improved freezing tolerance afforded by ectopic, constitutive expression of two endogenous *CBF* genes in transgenic *Eucalyptus* (Navarro *et al.*, 2011). Overexpression of *AtCBF1* (*AtDREB1b*) under the control of the constitutive CaMV 35S promoter in transgenic *V. vinifera* cv. Centennial Seedless resulted in about a 2 °C improvement in cold resistance as assessed by electrolyte leakage assays, reduced vine wilting and improved vine survival at −4 °C after 12 h (Jin *et al.*, 2009). Ectopic expression of a *VvCBF4* gene from cv. Koshu under the control of the CaMV 35S promoter conferred cold tolerance in transgenic *Arabidopsis* (Takuhara *et al.*, 2011). Lastly, these results were also consistent with the 2–3 °C or 4–6 °C increase in freezing tolerance exhibited by different transgenic apple trees expressing a peach *PpCBF1* gene compared with untransformed control trees as determined by electrolyte leakage assays (Wisniewski *et al.*, 2011). Taken together, these results indicate that *Vitis spp.* employ CBF-mediated response regulons, in part, to modulate

Table 1 Transcripts with significantly different expression in *VvCBF4* overexpressor line (9–12) compared with control line (8–6)

Annotation	Fold ratio	adj. *P*-value	Probeset ID	Protein ID	Gene ID
Increased abundance					
Pectin methylesterase Inhibitor (PMEI)	5.629	0.026	1606429_at	XP_002264167	LOC100251832
AAA-type ATPase domain (chaperone-like)	4.724	0.004	1610816_at	XP_002275572	LOC100252698
Calmodulin-like	2.621	0.019	1617516_at	XP_002284268	LOC100252031
Sulphite exporter TauE/SafE	2.532	0.048	1609884_at	XP_002267318	LOC100253061
Leucine-rich repeat extensin	2.413	0.026	1620976_at	XP_002277227	LOC100252056
Hypothetical protein	2.321	0.045	1612536_s_at	XP_002283137	LOC100255628
Gibberellin-regulated protein (GASA5)	2.272	0.049	1617881_at	XP_002280219	LOC100259439
Phytepsin aspartyl protease	2.252	0.048	1616668_at	XP_002276363	LOC100266485
Lipid transfer protein (LTP3)	2.193	0.049	1608175_at	XP_002275107	LOC100265591
Lactoylglutathione lyase/glyoxalase I	2.148	0.048	1619235_at	XP_002271396	LOC100266132
Flavonoid 3′,5′-hydroxylase (F3′5′H)/CYP96A10	2.089	0.026	1612325_at	XP_002279531	LOC100266173
Serine carboxypeptidase III	2.069	0.05	1620729_at	XP_002271855	LOC100242723
Unknown protein	2.052	0.031	1613368_at	XP_002283053	LOC100243457
Acid phosphatase class B	2.024	0.01	1621892_a_at	XP_002273448	LOC100246316
Xyloglucan endo-transglycosylase (XET)	2.008	0.038	1617739_at	XP_002274520	LOC100232906
Rho GDP-dissociation inhibitor 2	1.931	0.05	1611669_at	XP_002263904	LOC100260088
RD22-C	1.931	0.048	1621818_at	XP_002284286	LOC100264522
DUF1070 domain AGP41-like	1.927	0.05	1622530_at	XP_002283086	LOC100249036
Lipase GDSL	1.88	0.039	1607341_at	XP_002283363	LOC100242887
DNA-3-methyladenine glycosidase I	1.876	0.044	1621976_at	XP_002263612	LOC100256507
Lipase GDSL	1.829	0.049	1620618_at	XP_002271851	LOC100263626
Stellacyanin-like	1.824	0.019	1613509_at	XP_002280885	LOC100246223
Pepsin-like aspartic protease	1.808	0.045	1611138_s_at	XP_002265771	LOC100246744
Glycosyl hydrolase 17-like	1.784	0.031	1622656_at	XP_002285661	LOC100257244
Glycerol-3-phosphate acyltransferase 4-like	1.775	0.031	1612479_at	XP_002275348	LOC100243093
BSL1-like serine/threonine phosphoesterase	1.772	0.031	1615748_at	XP_002270638	LOC100249353
Pollen proteins Ole e I	1.761	0.027	1610746_at	XP_002272595	LOC100265200
Embryo-specific 3	1.757	0.026	1614771_at	XP_002266473	LOC100242179
Lipase GDSL	1.752	0.031	1607744_at	XP_002272970	LOC100249332
Zinc finger (C3HC4-type RING)	1.729	0.031	1612240_at	XP_002262825	LOC100263579
Polyketide cyclase/dehydratase (MLP28-like)	1.724	0.045	1607196_at	XP_002284578	LOC100262819
Pectate lyase	1.717	0.048	1616158_at	XP_002285340	LOC100242302
Unknown protein	1.696	0.031	1617384_at	XP_002281002	LOC100261133
Similar to thaumatin-like VVTL1	1.689	0.049	1614746_at	XP_002284403	LOC100259225
Exordium-like 3	1.689	0.048	1614951_at	XP_002264723	LOC100266367
Acid phosphatase class B	1.686	0.046	1617433_at	XP_002273448	LOC100246316
DUF1070 domain AGP20-like	1.655	0.019	1619401_at	XP_002280494	LOC100250031
FRA8-like	1.632	0.048	1608896_at	XP_002275679	LOC100243163
ROP6-like	1.569	0.031	1615962_at	XP_002269907	LOC100256456
LACERATA-like CYP450	1.564	0.048	1621973_at	XP_002275806	LOC100247907
GDP-fucose protein O-fucosyltransferase	1.562	0.045	1613171_at	XP_002267185	LOC100260861
Calcium-binding EF-hand-containing MSS3-like	1.554	0.048	1612996_at	XP_002266359	LOC100254364
Tubulin beta-1 chain	1.536	0.045	1607001_at	XP_002275306	LOC100259087
Glucose–methanol–choline(GMC) oxidoreductase	1.535	0.05	1622345_at	XP_002282510	LOC100266705
Unknown protein	1.529	0.035	1610862_at	XP_002274279	LOC100260563
Ferredoxin-related	1.52	0.049	1610753_at	XP_002281459	LOC100257849
Plant Basic Secretory Protein	1.518	0.031	1615434_at	XP_002283729	LOC100257403
Protein transport Sec61 subunit beta	1.507	0.048	1622732_at	XP_002276029	LOC100247887
Decreased abundance					
Chlorophyllase 2 (CHL2)	−1.514	0.035	1616275_at	XP_002279285	LOC100252835
Unknown protein (DUF1279, thylakoid)	−1.521	0.045	1619538_at	XP_002271410	LOC100258627
Galactinol synthase 4-like	−1.618	0.048	1608907_s_at	XP_002265947	LOC100260266
Copper amine oxidase	−1.675	0.045	1608650_at	XP_002263349	LOC100257527
ABC transporter MRP4-like	−1.703	0.049	1617849_at	XP_002265012	LOC100253698
Secretory peroxidase	−2.095	0.048	1621431_at	XP_002269918	LOC100257005
Chitinase class IV C (CHI4C)	−2.205	0.048	1617192_at	XP_002275516	LOC100232911

Table 1 Continued

Annotation	Fold ratio	adj. *P*-value	Probeset ID	Protein ID	Gene ID
Ubiquitin-protein ligase (PUB23-like)	−2.256	0.04	1606741_at	XP_002267438	LOC100250551
Nicotinamidase 1	−2.264	0.048	1610169_at	XP_002270896	LOC100250570
CYP94 family protein	−2.491	0.026	1609552_at	XP_002278009	LOC100267102
Alpha-glucan phosphorylase (PHS2-like)	−3.04	0.043	1614707_at	XP_002280732	LOC100251865
AtBAG6-like	−37.14	0.049	1614614_at	XP_002279584	LOC100256846

cold acclimation responses. If the protection against frost damage observed here for shoot tissues were pertinent to floral primordia of primary and secondary buds (Fennell, 2004), then this engineering strategy might serve to improve crop yields by reducing frost damage to fruit-bearing structures.

35S::*Vv*CBF4 overexpressor results in dwarf phenotypes

Interestingly, the 35S::*Vv*CBF4 expressing lines displayed a dwarf phenotype characterized by slower shoot elongation rates, shorter internodes and fewer interodes per plant in a transgene dose-dependent manner (Figure 3). This observation is well known from constitutive, ectopic expression of *AtCBF1*, *AtCBF2* and *AtCBF3* genes in *Arabidopsis*, which results in dwarf phenotypes with smaller leaves (Liu *et al.*, 1998; Kasuga *et al.*, 1999; Gilmour *et al.*, 2004; Sharabi-Schwager *et al.*, 2010). Although direct measurements were not made to assess changes in either cell numbers or cell size in the transgenic *Vitis* lines, the dwarf phenotype observed here is likely to be the result of reduced cell expansion or elongation rather than cell division as observed in *EguCBF1a* overexpressing *Eucalyptus* (Navarro *et al.*, 2011).

In the woody species poplar, ectopic *AtCBF1* expression resulted in a slowing of growth rate in plants <6 weeks of age (Benedict *et al.*, 2006). After this age, tree growth rates returned to normal. Similarly, constitutive ectopic expression of two endogenous CBF genes in *Eucalyptus* resulted in reduced growth of microcuttings and leaf area, with *EguCBF1a* having a more pronounced effect than *EguCBF1b* (Navarro *et al.*, 2011). Similarly, ectopic expression of a peach *CBF* gene in transgenic apple, a woody perennial species, resulted in reduced leaf size, but increased leaf dry weight, increased anthocyanin accumulation in cold-acclimated leaves, reduced shoot growth, and onset of dormancy as exhibited by terminal bud set and basipetal leaf senescence triggered by exposure to short day length (SD) or cold (4 °C) conditions (Wisniewski *et al.*, 2011). In contrast, constitutive expression of a *VvCBF4* gene from *cv*. Koshu in transgenic *Arabidopsis* apparently did not result in dwarfing, but only limited morphometric data were reported (Takuhara *et al.*, 2011).

In wine grape, the observed reduction in growth rates observed for the 35S::*Vv*CBF4 transgenic lines might be desirable in a production setting because it would be expected to reduce vine vigour, thereby reducing labour costs normally associated with pruning and leaf-pulling, activities customarily practiced in vineyards to improve fruit and wine quality via increased sunlight exposure of the fruit (Matus *et al.*, 2009). *CBF* gene overexpression is also known to delay flowering in some instances depending upon the *CBF* gene family member (Gilmour *et al.*, 2004; Sharabi-Schwager *et al.*, 2010). Delays in flowering time in wine grape might help avoid late spring frost damage to floral primordial and improve berry yield in some areas (Fennell, 2004). However, such delays might also limit berry-ripening potential in areas with short growing seasons.

Gene expression changes in the 35S::*Vv*CBF4 overexpressor

The 47 transcripts with significantly increased relative abundance in *Vv*CBF4-overexpressing grape vines spanned a wide range of molecular functions (Table 1). Five transcripts encoded proteins that catalyse the polymerization or depolymerization of cell wall structural components. For example, increased expression of the proteinaceous pectin methylesterase inhibitor (PMEI; 5.6-fold; 1606429_at) might act to reduce pectin depolymerization-based cell wall loosening leading to growth inhibition consistent with a dwarf phenotype (Figure 3), although the exact *in vivo* role of this inhibitor remains unclear (Jolie *et al.*, 2010). In another example, increased transcript abundance of a leucine-rich repeat extensin (LRX; 2.4-fold; 1620976_at), which are thought to reinforce and stabilize cell wall polysaccharide structure by cross-linking (Baumberger *et al.*, 2003; Ringli, 2010), might also be consistent with dwarfing as increased activity of this enzyme would be expected to limit cell size. Increased expression of an xyloglucan endotransglycosylase (XET; 2.0-fold;

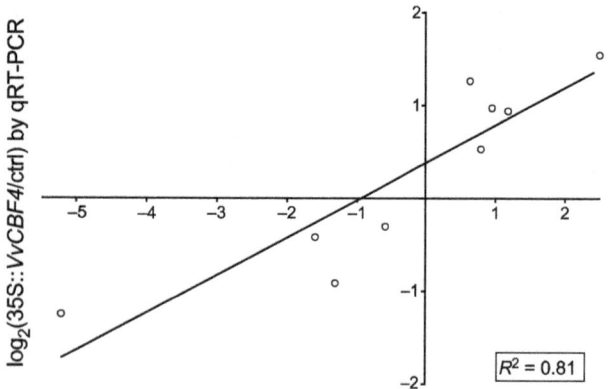

Figure 7 Verification of microarray results by real-time qRT-PCR. Log$_2$-transformed values of Affymetrix® *Vitis* GeneChip® signal intensities (*x*-axis) and real-time PCR expression ratios (*y*-axis) for 35S::*Vv*CBF4/control of nine differentially expressed microarray probe sets (open circles). The linear regression had a goodness of fit R^2 = 0.81.

Table 2 Comparison of changes in gene expression in *VvCBF4*-overexpressing grapevine with *AtCBF1*-overexpressing Poplar

In 35S::*Vv*CBF4-grapevine					In 35S::*AtCBF1*-Poplar (as reported by Benedict *et al.*, 2006)		
Vitis vinifera gene description	*V. vinifera* RefSeq ID	*Vitis* Shoot FC	Poplar Leaf FC	Poplar Stem FC	Poplar gene ID (best *At* match)*	Poplar gene description	BLAST e-value
Reported transcript changes similar in Vitis and Poplar CBF overexpressors							
Exordium-like 3	XP_002264723	1.7	−1.4	1.8	At2g17230.1	Exordium-like 5	3.4E-92
Glycosyl hydrolase 17-like	XP_002285661	1.8	2.3	−1.1	At5g55180.2	O-Glycosyl hydrolase family 17	2.0E-18
Calcium-binding EF-hand MSS3-like	XP_002266359	1.6	1.0	1.8	At5g39670.1	Calcium-binding EF-hand family protein	9.0E-16
Xyloglucan endo-transglycosylase (XET)	XP_002274520	2.0	1.6	2.0	At3g23730.1	Xyloglucan endo-transglucosylase/hydrolase 16	2.6E-87
Unknown protein	XP_002283053	2.1	1.1	1.9	At4g35320.1	Unknown protein	4.7E-30
Leucine-rich repeat extensin	XP_002277227	2.4	1.8	−1.1	At1g28290.1	Arabinogalactan protein 31 (AGP31)	9.4E-30
Rare Cold-Inducible 2B (RCI2B)	XP_002279333	1.4	1.9	1.9	At3g05880.1	Low temperature and salt responsive protein (LTI6A)	5.0E-22
Lipase GDSL (1)	XP_002271851	1.8	2.3	1.1	At5g45950.1	GDSL-like Lipase/Acylhydrolase (1)	2.3E-70
Lipase GDSL (2)	XP_002272970	1.8	2.3	1.2	At1g29660.1	GDSL-like Lipase/Acylhydrolase (2)	1.1E-56
Lipase GDSL (3)	XP_002283363	1.9	2.8	1.1	At1g71691.2	GDSL-like Lipase/Acylhydrolase (3)	6.1E-55
Chitinase class IV C (CHI4C)	XP_002275516	−2.2	−1.2	−2.0	At3g54420.1	Chitinase (ATEP3, ATCHITIV)	1.1E-111
Reported transcript changes differ between Vitis and Poplar CBF overexpressors							
Pectate lyase	XP_002285340	1.7	−2.0	−2.1	At4g24780.1	Pectin lyase-like	7.1E-202
Lactoylglutathione lyase/glyoxalase I	XP_002271396	2.1	−2.5	1.1	At1g15380.1	Lactoylglutathione lyase/glyoxalase I	9.0E-45
ABC transporter MRP4-like	XP_002265012	−1.7	−1.2	2.4	At3g21250.2	Multidrug resistance-associated protein 6 (MRP6)	3.7E-297

FC, fold change.
Arabidopsis gene locus identifiers used as the sole designators for UPSC-KTH 13K Poplar microarray probes designed by Benedict *et al.*, 2006 (Tables 1a–c).

1617739_at), which can participate in cell wall strengthening and reduce cell wall extension *in vitro* (Maris *et al.*, 2009), might also be in agreement with the observed dwarf phenotype. However, the exact role of this gene product is difficult to predict based on sequence information alone given the large size of this gene family and should be investigated by direct experimentation as other xyloglucan-specific enzymes have been reported to stimulate cell expansion (Maris *et al.*, 2009). Increased abundance of two DUF1070 transcripts predicted to encode short arabinogalactosylated proteins (AGP; 1.7- and 1.9-fold; 1622530_at, 1619401_at) might also contribute to dwarfing as overexpression of an *Arabidopsis* arabinogalactan protein gene *AtAGP18* resulted in plants with smaller rosettes and shorter stems and roots (Zhang *et al.*, 2011). Poplar, acclimated to cold for 7 days, or overexpressing *AtCBF1*, also displayed increased abundance of an *AGP31* gene (Table 2) (Benedict *et al.*, 2006). The elevated mRNA expression of LRX and EXGT genes was also observed in *AtCBF1*-overexpressing Poplar (Table 2), and that of the PMEI was conserved in *AtCBF2* and *AtCBF3* overexpressing *Arabidopsis* (Table 3) indicative of conservation of *CBF* gene regulons in both woody and herbaceous species. Lastly, a gibberellin-regulated gibberellic acid-stimulated Arabidopsis (GASA) *GASA5*-like gene (1617881_at) increased in abundance 2.27-fold (Table 1). In *Arabidopsis*, *GASA5* gene expression is stimulated by GA signalling and subsequent GASA5 activity negatively regulates GA-related gene expression and inflorescence stem growth, making it both GA-regulated and GA-regulating (Zhang *et al.*, 2009). Transcripts for a related *GASA1* gene have also been observed

to be elevated in *AtCBF2* overexpressing *Arabidopsis* (Table 3); however, the exact role of this gene remains unclear at this time. Additional experiments are required to determine whether the expression of the *Vitis* GASA5-like gene is influenced by GA and its product is involved in GA-signalling.

Another class of genes showing increased relative transcript abundance in the *VvCBF4* gene overexpressing line included those involved in lipid metabolism and epicuticular wax formation or modification. A homolog of the lipid transfer protein 3 (*LTP3*-like) gene increased in abundance 2.2-fold (1608175_at). Although the function of this gene is unclear, various functions have been assigned to lipid transfer genes including lipid exchange between membranes or as lipid sensors or chaperones (D'Angelo *et al.*, 2008). The *Vitis LTP3*-like gene is most similar to the *Arabidopsis LTP3* (At5g59320) gene, which encodes a PR-14 family lipid transfer protein that is located in the apoplast and cell wall and whose transcript abundance is regulated by water deficit and ABA (Huang *et al.*, 2008).

Three GDSL-motif lipase genes increased in abundance between 1.7- and 1.9-fold (1607744_at, 1607341_at, 1620618_at). In plants, members of the GDSL-lipase family are known to be involved in various types of lipid and acyl-group metabolism, including lipase, lysophospholipase, esterase, thioesterase and arylesterase activities (Akoh *et al.*, 2004). GDSL lipases comprise a large gene family in land plant species, including grapevine, which contains more than 90 GDSL-lipase genes (Volokita *et al.*, 2010). Some GDSL-lipase genes are required for proper cuticle formation. For example, in rice, the GDSL-lipase gene *wilted dwarf and lethal* 1 (*WLD1*) is required

Table 3 Comparisons of changes in gene expression in *VvCBF4*-overexpressing grapevine with multiple *C-repeat binding factor (CBF)/DREB1f*-overexpressing *Arabidopsis*

	In 35S::*VvCBF4*-grapevine		In CBF-overexpressing *Arabidopsis*								
Vitis vinifera RefSeq ID	*V. vinifera* gene description	*Vitis* CBF4	CBF1* FC	CBF2† FC	CBF2‡ FC	CBF3§ FC	DDF1¶ FC	*Arabidopsis* Gene ID	*Arabidopsis* gene description	BLAST e-value
Transcript abundance trends similar between 35S::VvCBF4-grapevine and CBFs in Arabidopsis										
XP_002264167	Pectin methylesterase Inhibitor (PMEI)	5.6		5.6	2.9	7.3		At5g62350.1	Invertase/PMEI	1.9E-52
				4.8	3.9	4.6		At5g62360.1	Invertase/PMEI	6.9E-42
			4.4			3.9		At3g47380.1	Invertase/PMEI	1.7E-35
XP_002280219	Gibberellin-regulated protein (GASA5)	2.3		3.9				At1g75750.1	GASA1	1.6E-19
XP_002271855	Serine carboxypeptidase III	2.1		2.5				At2g22980.1	Serine carboxypeptidase-like 13	2.3E-39
XP_002283363	Lipase GDSL (#1)	1.9		3.0				At1g29660.1	GDSL-like Lipase/Acylhydrolase (#2)	1.1E-56
XP_002271851	Lipase GDSL (#2)	1.8	32.4	11.0	258.5		64.0	At2g24560.1	GDSL-like Lipase/Acylhydrolase (#3)	1.0E-68
XP_002272970	Lipase GDSL (#3)	1.8		3.8		2.1		At1g29670.1	GDSL-like Lipase/Acylhydrolase (#4)	1.7E-59
						2.3		At4g30140.1	CDEF1 | GDSL-like	5.4E-42
XP_002280885	Stellacyanin-like	1.8				2.2		At4g27520.1	Early nodulin-like 2	1.5E-16
XP_002285340	Pectate lyase	1.7		2.5	4.5			At3g09540.1	Pectin lyase-like superfamily	1.4E-39
XP_002275806	LACERATA-like CYP450	1.6					4.0	At2g27690.1	CYP94C1	1.0E-84
XP_002279285	Chlorophyllase 2 (CHL2)	−1.5		−9.8	−5.1	−2.8	−5.3	At1g19670.1	Chlorophyllase 1 (CHL1)	3.2E-70
XP_002263349	Copper amine oxidase	−1.7		−2.7				At4g12280.1	Copper amine oxidase	3.6E-129
XP_002269918	Secretory peroxidase	−2.1					−8.6	At5g06730.1	Peroxidase superfamily	2.7E-85
XP_002278009	CYP94 family protein	−2.5					−5.3	At5g63450.1	CYP94B1	1.3E-118
Transcript abundance trends partially similar between 35S::VvCBF4-grapevine and CBFs in Arabidopsis										
XP_002275516	Chitinase class IV C (CHI4C)	−2.2					−4.0	At2g43590.1	Chitinase family protein	2.2E-91
			52.1	29.0	40.4	19.7	3.0	At2g43620.1	Chitinase family protein	1.0E-69
XP_002284403	Similar to thaumatin-like VVTL1	1.7				−4.2	−3.2	At1g75040.1	Pathogenesis-related gene 5	1.5E-64
				4.5	5.0	2.0		At4g36010.1	Pathogenesis-related thaumatin	1.1E-59
XP_002265771	Pepsin-like aspartic protease	1.8				−2.7		At5g10770.1	Eukaryotic aspartyl protease family	7.8E-41
					−3.6	−4.6		At5g10760.1	Eukaryotic aspartyl protease family	5.4E-34
				5.5				At3g52500.1	Eukaryotic aspartyl protease family	2.1E-30
				10.0	3.0		3.7	At3g54400.1	Eukaryotic aspartyl protease family	1.9E-26
XP_002265947	Galactinol synthase 4-like	−1.6	17.7		6.1		4.6	At1g60470.1	Galactinol synthase 4	5.5E-158
				3.5				At2g47180.1	Galactinol synthase 1	3.6E-149
				−4.4		18.3		At1g56600.1	Galactinol synthase 2	3.2E-142
			88.4	62.7	346.5	50.6	21.1	At1g09350.1	Galactinol synthase 3	1.3E-138
XP_002284578	Polyketide cyclase/dehydratase (MLP28-like)	1.7				2.4	−3.2	At1g70850.3	MLP34 | MLP-like protein 34	5.2E-44
							−17.1	At1g70880.1	Polyketide cyclase/dehydrase	8.4E-34
Transcript abundance trends dissimilar between 35S::VvCBF4-grapevine and CBFs in Arabidopsis										
XP_002265012	ABC transporter MRP4-like	−1.7			2.2			At3g62700.1	Multidrug resistance-associated protein 10	0.0E+00
XP_002273448	Acid phosphatase class B	2.0		−13.3			−5.7	At5g24780.1	Vegetative storage protein (VSP1)	8.0E-42
XP_002279531	Flavonoid 3′,5′-hydroxylase CYP96A10	2.1				−2.4	−19.7	At5g52320.1	CYP96A4	2.5E-138
XP_002280732	Alpha-glucan phosphorylase (PHS2-like)	−3.0				3.2		At3g29320.1	Glycosyl transferase, family 35	0.0E+00
						3.9		At3g46970.1	Alpha-glucan phosphorylase 2 (PHS2)	0.0E+00
XP_002266359	Calcium-binding EF-hand MSS3-like	1.6				−2.2		At2g43290.1	Calcium-binding EF-hand (MSS3)	4.5E-76
						−2.2		At3g59440.1	Calcium-binding EF-hand family	3.7E-70
XP_002274520	Xyloglucan endo-transglycosylase (XET)	2.0				−2.5		At5g65730.1	XTH6	1.0E-87
						−2.9		At4g30270.1	XTH24	1.8E-79
							−11.3	At2g18800.1	XTH21	1.6E-75
				−2.8		−2.4		At3g44990.1	XTH31	7.6E-54

FC, fold change.

*35S::AtCBF1 in *Arabidopsis* (Fowler and Thomashow, 2002).

†35S::AtCBF2 (Sharabi-Schwager *et al.*, 2010).

‡35S::AtCBF2 (Vogel *et al.*, 2005).

§35S::AtCBF3 (Maruyama *et al.*, 2009).

¶35S::DDF1 (Magome *et al.*, 2008).

GASA, gibberellic acid-stimulated Arabidopsis.

for correct cuticle formation; *wld1* knockouts contain abnormal wax crystals and exhibit rapid water loss as a result of aberrant cuticle (Park *et al.*, 2010). GDSL-lipase genes might also participate in bacterial and fungal pathogen resistance through eliciting local and systemic resistance (Kwon *et al.*, 2009). Homologues of three GDSL-motif lipase genes also exhibited increased transcript abundance in *AtCBF1* and *CBF*-overexpressing Poplar and *Arabidopsis*, respectively (Tables 2 and 3).

The relative mRNA expression of genes involved in biosynthesis of the epicuticular wax also increased, including a glycerol-3-phosphate acyltransferase 4-like gene (*GPAT4*; 1.8-fold; 1612479_at), a LACERATA-like CYP450 (1.6-fold; 1621973_at) and a glucose-methanol-choline (GMC) oxidoreductase (1.5-fold; 1622345_at). The best matches in *Arabidopsis* for each of these transcripts were elevated in *A. thaliana* transgenic plants overexpressing the wax-inducing *WIN1/SHN1* AP2 domain-containing ERF transcription factor genes (Kannangara *et al.*, 2007). The increased expression of epicuticular wax biosynthetic genes is consistent with the observed enhancement of epicuticular wax deposition in transgenic *Eucalyptus* (*E. urophylla* × *E. grandis*) overexpressing an endogenous *CBF1a* gene (Navarro *et al.*, 2011). Although additional experiments are needed to determine whether and how ectopic expression of *VvCBF4* might have affected cuticle formation in grapevine, transgenic vines retained a normal appearance.

A set of stress-responsive genes also showed increased transcript abundance in the *VvCBF4*-overexpressing line including genes encoding the RESPONSIVE TO DESICCATION 22 (*VvRD22*) gene (1.9-fold; 1621818_at), a polyketide cyclase/dehydratase (major latex protein 28-like) (1.7-fold; 1607196_at), thaumatin-like *VvTL1* (1.7-fold; 1614746_at), stress-related basic secretory protein (1.5-fold; 1615434_at) and an unknown stress-regulated transcript (2.1-fold; 1613368_at) genes, which have each been reported previously to increase in mRNA abundance under abiotic stress conditions either in grape or in *Arabidopsis* (Kuwabara *et al.*, 1999; Fowler and Thomashow, 2002; Hanana *et al.*, 2008; Lytle *et al.*, 2009). Transcripts coding for the DNA repair enzyme DNA-3-methyladenine glycosidase I (1.9-fold; 1621976_at) and the abiotic stress-responsive methylglyoxal detoxification enzyme lactoylglutathione lyase/glyoxalase I (2.1-fold; 1619235_at) also increased (Santerre and Britt, 1994; Yadav *et al.*, 2008). Two other stress-regulated transcripts with increased mRNA expression with likely roles in signal transduction included a calmodulin (CaM)-like gene (2.6-fold; 1617516_at), which has also been shown to increase in expression in cold-stressed *Arabidopsis* (Fowler and Thomashow, 2002), and a calcium-binding EF-hand-containing *MSS3*-like transcript (1.6-fold; 1612996_at). Lastly, the mRNA expressing an AAA-type ATPase domain (chaperone-like) gene (1610816_at) increased in abundance 4.7-fold (Table 1). A homolog of this gene is known to respond to salt stress in *Arabidopsis* (Fujita *et al.*, 2005).

In contrast to the relatively large numbers of genes with elevated transcript abundance, only twelve genes showed a significant decrease in mRNA abundance in the *VvCBF4* overexpressor. Most notable was a BAG-domain (BCL-2-associated athanogene) gene that showed a 37-fold decrease in relative transcript abundance (Table 1). This nuclear localized, calmodulin-binding domain containing gene is most similar to *AtBAG6*, which when overexpressed induces programmed cell death in transformed yeast and *Arabidopsis* (Kang *et al.*, 2006). *AtBAG6* is also thought to coordinate stress-induced hormone signalling and might play a role in limiting pathogen coloniza-

tion (Doukhanina *et al.*, 2006). Additional experiments will be necessary to determine the exact role of this gene in *Vitis*.

Lack of correlation between mRNA and protein expression

Interestingly, none of the proteins identified as being significantly overexpressed in the *VvCBF4*-overexpressing line 9–12 (Table S4) matched the differentially expressed transcripts described here. This result is in contrast to the correlation observed between transcript and protein abundance changes described for heat-stressed, 35S::*DREB2C* overexpressing *Arabidopsis* plants, wherein a total of 10 different proteins showed significant changes in protein abundance (six and four proteins showed increased or decreased abundance, respectively) compared with wild-type plants (Lee *et al.*, 2009). However, no significant differences were apparent if the transgenic plants were not heat-treated. One possible explanation for these different results is that heat stress might be necessary for transcription factor activation as described for *DREB2A*, which was activated by dehydration, high salinity, ABA or cold treatments, presumably by post-translational modification (Liu *et al.*, 1998). Indeed, phosphorylation of a stress-inducible *DREB2A* transcription factor from *Pennisetum glaucum* was shown to prevent DNA binding (Agarwal *et al.*, 2007). However, such post-translational modifications are not always necessary as in the case of *DREB1A*-related transcription factors (Liu *et al.*, 1998). In another example, comparison of nontransgenic and transgenic potato expressing 35S::*AtDREB1A* using 2D-DIGE showed increased expression of only patatin, a major storage protein and decreased expression of lipoxygenase and starch synthase (Nakamura *et al.*, 2010). Thus, there is precedent for ectopic expression of *CBF*-family transgenes having a relatively small effect on the proteome.

One possible explanation for the lack of correlation between transcript and protein abundance is that *VvCBF4* might require post-translational modification for activation. Additional experiments would be required to test this hypothesis. Studies in various plant species, including wine grape berries, have shown that there are only moderate-to-weak correlations ($r = 0.50$ or less) between mRNA and protein depending on the study (Watson *et al.*, 2003; Gion *et al.*, 2005; Liu *et al.*, 2006; Faurobert *et al.*, 2007; Grimplet *et al.*, 2009b). Additional possible explanations for this weak correlation include the observation that mRNA abundance changes are typically presented as fold changes and not in absolute terms, so this might favour the reporting of low-abundance transcripts. Also, alternatively spliced transcripts might not be detected by microarray or qRT-PCR analytical methods; however, such events might result in changes in relative protein expression.

Conclusions

In conclusion, ectopic expression of the *VvCBF4* resulted in improved freezing survival in nonacclimated vines to a degree comparable with that observed for *CBF* overexpression in *Arabidopsis* and other woody species within *Eucalyptus*, *Malus*, *Populus* and *Vinifera* genera. The gene expression profiling results presented here clearly show that wine grape possesses an evolutionarily conserved CBF regulon that is widely conserved among cold-adapted herbaceous species as well as other woody species. Improvement in freezing tolerance should be useful in reducing late spring frost damage to floral primordia of primary and secondary buds in *Vitis*. The observed dwarf phenotype might be advantageous to improve berry ripening by reducing canopy vigour and allowing more sun exposure to developing berries. However, to avoid possible undesirable effects of dwarfing, such as reductions in sink tissue (i.e., berries) production typically associated with constitutive expression of CBF transcription factors, future experiments should focus on the development of transgenic vines expressing this or other VvCBF gene family members under the control of abiotic stress-inducible promoters. VvCBF4-overexpressing lines might also be predicted to be more tolerant to salinity and water deficit stress. Additional experiments are in progress to verify these possibilities.

Experimental procedures

VvCBF4 sequence analysis

Phylogenetic analysis, based on minimum evolution, was conducted on the full-length protein sequences of CBF gene family members from eudicots for which complete genome drafts are available. The polypeptides aligned included four CBF genes from Arabidopsis: AtCBF1 (NP_567721.1), AtCBF2 (NP_567719.1), AtCBF3 (NP_567720.1), AtCBF4 (NP_200012.1); four CBFs from Glycine max: GmCBF1 (AAQ02703.1), GmCBF2 (ACB45077.1), GmCBF3 (ACA63936.1) and GmDREB7 (ABQ42206.1); six CBFs from Populus trichocarpa: PtDREB66 (XP_002313656.1), PtDREB67 (XP_002328068.1), PtDREB68 (XP_002299565.1), PtDREB69 (XP_002298067.1), PtDREB70 (XP_002318846.1) and PtDREB71 (XP_002321877.1); four CBF genes from Medicago truncatula: MtCBF1 (ABX80062.1), MtDREB1A (ABG75914.1), MtCBF2 (ABX80063.1) and MtDREB (ABB72792.1); and four CBF genes from Vitis vinifera: VvCBF1 (AAR28673.1), VvCBF2 (AAR28677.1), VvCBF3 (XP_002267961.1) and VvCBF4 (XP_002280097.1). The Arabidopsis AP2-AE ERF10 (NP_171876.1) was included as a non-CBF/DREB rooting outlier. Alignments were generated in the MacVector (ver. 12.0; MacVector, Inc., Cary, NC) software suite, using ClustalW alignment. The phylogeny tree was constructed using neighbour-joining tree building, with Poisson-corrected distances, 1000 bootstrap replicates and postbuild outlier rooting.

CBF4 overexpression construct

The full-length VvCBF4 open reading frame (ORF, GenBank accession: DQ497624) was PCR-amplified from genomic DNA using the forward primer 5′-CACCATGAATACTACTTCTCCACCATATTCC-3′ and reverse primer 5′- CTAAATAGAGTAACTCCATAATGA-CATGTC-3′. The PCR product was cloned into pENTR/D-TOPO and then into the pH2GW7 GATEWAY-type expression vector using LR clonase to recombine the VvCBF4 ORF between sites attr1 and attr2, downstream of the CaMV 35S constitutive promoter. Empty vector control line (e.g., '8–6') vectors were constructed by the removal of the GATEWAY cassette from pH2GW7 with ApaI/SstI digestion, blunt-end formation via T4 polymerase 3′ exonuclease activity, and re-ligation with T4 ligase (see Figure S1 for details). Both the VvCBF4 expression vector and the empty vector control were verified by DNA sequencing and restriction digestion (Tattersall, 2006). Expression constructs were then introduced into Agrobacterium tumefaciens strain EHA105 by electroporation.

Vitis vinifera cv. Freedom transformation

Grape vine transformation was performed by the Ralph M. Parsons Plant Transformation Facility, University of California, Davis,

CA. Briefly, immature anthers from the Vitis hybrid rootstock cv. 'Freedom' were used as a source of tissue for embryogenic callus. The embryogenic callus was inoculated with an overnight suspension of Agrobacterium tumefaciens containing the expression cassettes outlined earlier and adjusted to an O.D.$_{600}$ of 0.075. Callus tissue was plated onto a sterile No. 1 Whatman filter paper, which was placed on top of coculture medium consisting of Woody Plant Media (WPM) (Lloyd and McCown, 1981) supplemented with 10 mg/L picloram, 2 mg/L thidiazuron (TDZ), 500 mg/L activated charcoal, 1000 mg/L casein and 200 μM acetosyringone. Agrobacterium was added dropwise to the callus until thoroughly moistened. After 15 min, the callus was blotted dry using a second piece of sterile No. 1 Whatman filter paper and incubated in the dark at 23 °C. After 48–72 h, the callus was transferred to WPM supplemented with 10 mg/L picloram, 2 mg/L thidiazuron (TDZ), 500 mg/L activated charcoal, 1000 mg/L casein, 400 mg/L carbenicillin, 250 mg/L cefotaxime and 25 mg/L hygromycin sulphate. Callus was incubated in the dark at 26 °C and transferred to fresh medium every 21–28 days. Hygromycin-resistant embryogenic callus colonies formed after 3–4 months. These colonies were harvested and transferred to germination medium consisting of WPM medium supplemented with 20 g/L sucrose, 500 mg/L activated charcoal, 1000 mg/L casein, 150 mg/L timentin, 0.5 mg/L benzylaminopurine (BAP) and 0.1 mg/L naphthalene acetic acid (NAA). Elongating embryos were transferred to rooting media consisting of one-half strength Murashige and Skoog minimal organics medium (Murashige and Skoog, 1962) supplemented with 15 g/L sucrose, 0.01 mg/L NAA, 150 mg/L timentin and 25 mg/L hygromycin. Rooted plantlets were acclimated to soil in a growth chamber under 16 h photoperiod at 26 °C and then transferred to greenhouse for further development.

Vine propagation

Green cuttings (2–3 internodes in length) from well-established woody vines were sterilized with 50% (v/v) household bleach and 0.02% (v/v) Triton X-100 for 30 s and placed in autoclaved, sterile 77 × 77 × 97 mm (W × L × H) Magenta GA-7 boxes (Magenta Corp., Chicago, IL) containing 80 mL of 0.8% (w/v) plant tissue culture agar (#A111; Phytotechnology Laboratories, Shawnee Mission, KS) with Murashige and Skoog modified basal medium w/Gamborg vitamins (#M404; Phytotechnology Laboratories), 1.5% sucrose, pH 5.6–5.7 and 300 μM indoleacetic acid (IAA) to promote rooting (Murashige and Skoog, 1962; Gamborg et al., 1968). Vines were allowed to develop roots in a Percival Scientific growth chamber (Model # CU-32L; Perry, IA) under 18 h of fluorescent light (200 μmol/m^2 s^{-1} PPFD) at 25 °C and 6 h darkness at 20 °C. Upon development of root primordia (~14 days), individual vines were transferred to a second Magenta box containing media prepared as above, except without IAA, to promote normal plant growth for about 10 days. Upon establishment of normal growth, cuttings from these vines were further propagated in Magenta boxes (one cutting/box) for subsequent use in stress phenotyping assays, electrolyte leakage assays, mRNA and protein expression assays and growth phenotyping. These cuttings were handled identically as the 'mother vines' (omitting the bleaching step, as they were already sterile) and were allowed ~24 days to develop roots in Magenta boxes, followed by transplantation to 4' square plastic pots (Part # TSD4; McConkey Co., Garden Grove, CA) containing Metromix® 200 soil (Sun Gro Horticulture, Bellevue, WA). Vines were allowed an additional 2 weeks to adapt to soil and low humidity before abiotic stress assays were performed. Acclimation was accomplished by placing the newly potted vines in a tray covered with a clear plastic dome to maintain high humidity. After 5 days, a dome having a 50% vented surface area replaced the full dome. After an additional 5 days, the vented dome was removed completely. Twice during the 2-week time period, vines were given ¼ strength modified Hoagland's solution with full-strength iron and micronutrients (600 µM KNO_3, 400 µM $Ca(NO_3)_2 \cdot H_2O$, 200 µM $NH_4H_2PO_4$, 100 µM $MgSO4 \cdot 7H_2O$, 280 µM Fe-EDTA, 50 µM KCl, 25 µM H_3BO_3, 2 µM $MnSO_4 \cdot H_2O$, 2 µM $ZnSO_4 \cdot 7H_2O$, 0.5 µM $CuSO_4 \cdot 5H_2O$, and 0.5 µM H_2MoO_4) (Hoagland and Arnon, 1950).

Freezing stress assays

For freezing stress assays, groups of vines (n = 10) from the empty vector control line (8–6) and three 35S::VvCBF4 overexpressor lines (9–1, 9–3 and 9–12) were placed in a prechilled Percival Scientific growth chamber (model # AR75L) for 24 h at −2 °C. The built-in humidifier was turned off and water drained from the reservoir to avoid ice build-up and damage to the growth chamber. Twenty kilograms of frozen ice packs and a small fan were added to the growth chamber to maintain maximum temperature equilibrium throughout the chamber during stress application. Plant position within the growth chamber was randomized. Assays began and ended at 10 AM, with 18 h light (200 µmol/m²/s light) beginning at 6 AM and 6 h dark beginning at 10 PM. Vines were watered 2 days post-stress and allowed to recover under normal growth conditions (25 °C day/20 °C night temperature) under the same light conditions as above for 2 weeks, at which point survival was scored by the presence of new growth. Freezing stress assays were performed on multiple groups of vines, as described, and Student's t-test corrected for multiple comparisons was used to identify statistically significant genotypic differences in freezing tolerance (adjusted P-value <0.05).

Electrolyte leakage assay

The freezing tolerance of the three transgenic V. vinifera cv. 'Freedom' lines overexpressing VvCBF4 and one control line was also assayed by leaf electrolyte leakage. Transgenic V. vinifera lines 9–1, 9–3, 9–12, and the empty vector control line, 8–6, were grown in pots for 6 weeks (150–200 µmol/m²/s PPFD, 16 h light at 25 °C/8 h dark at 20 °C) without any cold acclimation. Identical-sized leaf discs (6 mm diameter) were punched from individual fully expanded leaves and were equilibrated in 200 µL of Nanopure water (Millipore, Inc., Bedford, MA) in 1.5-mL microfuge tubes for 1 h. For freezing treatment, we manually adjusted a Neslab RTE-4 (Thermo Fisher Scientific, Inc., Waltham, MA) refrigerated circulating antifreeze bath to achieve 1 °C/30 min decrement from 0 to −16 °C. Tubes were floated in the refrigerated water bath and exposed to the freezing conditions. At each 1 °C/30-min interval, tubes were removed and gently thawed at 4 °C for 18 h. Leaf discs and incubation solutions were transferred to another tube containing 4 mL of Nanopure water. After shaking overnight, conductivity was measured using an Orion 4Star Portable Conductivity Meter with the Orion 013010MD Conductivity Cell

(Thermo Fisher Scientific, Inc.). Samples were autoclaved for 20 min at 121 °C, and the conductivity was re-measured. The level of electrolyte leakage induced by freezing was determined as the percentage of conductivity before autoclaving versus conductivity because of total leakage after autoclaving. Student's t-test corrected for multiple comparisons was used to identify significant genotypic differences in electrolyte leakage (adjusted P-value <0.05).

Growth phenotyping

The shoot elongation rate of 2-week-old soil-rooted plantlets of VvCBF4 overexpressor lines and empty vector control lines (n = 8) was measured to identify possible alterations (reductions) in growth. The shoots were marked with acrylic nail polish just below the first above-soil internode, and all internodal lengths were measured using dial callipers (Mitutoyo America Corp., model no. 505-675-66, Aurora, IL) every third day. Total shoot length and SER were determined over 21 days. Student's t-test corrected for multiple comparisons was used to identify significant genotypic differences in growth rates and internode length (adjusted P-value <0.05).

RNA extraction

Entire aerial portions (stems and young leaves) of grapevines that had been growing in soil for 2 weeks were collected in 50-mL conical tubes and immediately frozen in liquid nitrogen. Total RNA was isolated using a three-step process. First, total RNA was extracted and applied to an RNeasy Midi column (Qiagen, Inc., Valencia, CA) with 2% polyethylene glycol (PEG, mw = 20 000; Sigma Aldrich, Inc., St. Louis, MO) added to the RLT buffer, followed by on-column DNase digestion. The resultant RNA was then subjected to phenol:chloroform:isoamyl–alcohol (25:24:1) extraction (Sambrook and Russell, 2001). Lastly, the RNA extract was further purified using an RNeasy mini column (Qiagen, Inc.) with 2% PEG (Tattersall et al., 2005). Total RNA was quantified and 260/280 ratios determined using a NanoDrop® ND-1000 spectrophotometer (Thermo Fisher Scientific, Inc.). RNA integrity was confirmed by electrophoresis on formaldehyde-containing 1.5% agarose gels.

Quantitative real-time (reverse transcription) PCR (qRT-PCR)

cDNA was synthesized using the high-capacity cDNA reverse transcription kit (Applied Biosystems, Inc., Foster City, CA) according to manufacturer's instructions, using uniform 2 µg RNA per reaction volume reverse transcription reactions. Gene-specific primers for qRT-PCRs were selected using the Primer-BLAST tool at NCBI (http://www.ncbi.nlm.nih.gov/tools/primer-blast) using RefSeq V. vinifera transcripts as input, screened against all other V. vinifera RefSeq sequences, and the following Primer3 (Rozen and Skaletsky, 2000) settings: T_m range 58–60 °C, product size = 50–150 bp, primer size = 13–25 nt, max poly-X = 3, and G/C content = 30%–80%. Primer pair selection against a GC clamp, such that no more than two of the last five 3' nucleotides were G or C, was conducted per qRT-PCR instrument recommendations. qRT-PCRs were prepared using Fast SYBR® Green Master Mix and performed using the ABI PRISM® 7500 Sequence Detection System (Applied Biosystems, Inc.) using four biological replicate samples and normalized to an endogenous actin 7 control gene (NCBI locus ID, LOC 100232968) (Reid et al., 2006). Gene-specific primer pairs and products used are summarized in Table S1. Relative quantitation of qRT-PCR outputs was performed using the Pfaffl method, an elaborated $\Delta\Delta Ct$ method that employs observed primer efficiencies (rather than assuming $2^{-\Delta\Delta Ct}$) when comparing different primer pairs/products (Pfaffl, 2001).

Microarray-based mRNA expression profiling

For microarray experiments, RNA was re-quantified using RiboGreen® fluorescent nucleic acid stain (Invitrogen Life Technologies, Inc., Carlsbad, CA) and read using a Labsystems Fluoroskan Ascent fluorescence plate reader (Thermo Fisher Scientific, Inc.). Sample RNA integrity was determined with an Agilent 2100 Bioanalyzer microfluidics station (Agilent Technologies, Inc., Santa Clara, CA). Total RNA (300 ng) was loaded onto a capillary electrophoretic column to determine major (rRNA) band sizes and quality (see Figure S2). Complementary DNA (cDNA) was generated from mRNA using a GeneChip® T7-Oligo(dT) Promoter Primer Kit containing the T7 polymerase promoter sequence and oligo(dT) priming region (5'-GGCCAGTGAATTGTAATACGACTCACTATAGG-GAGGCGG-(dT)$_{24}$-3') (Affymetrix®, Santa Clara, CA) and SuperScript™ II Reverse Transcriptase (Invitrogen Life Technologies, Inc.) according to the manufacturer's instructions (Affymetrix, 2009). Biotinylated complementary RNAs (cRNAs) were synthesized in vitro from four biological replicates using T7 RNA polymerase in the presence of biotin-labelled UTP/CTP, then purified, fragmented and hybridized to GeneChip® 16K Vitis vinifera (Grape) Genome Arrays ver. 1.0 (Affymetrix®). The hybridized arrays were washed and stained with streptavidin–phycoerythrin and biotinylated antistreptavidin antibody using an Affymetrix® 3000 7G Scanner, Hybridization, and Fluidics System. Scanner and image data were collected and processed on a GeneChip® Workstation using Affymetrix® GCOS software.

The images of all arrays were examined, and no obvious scratches or spatial variations were observed. The 'present' call rates were also consistent across the eight arrays, ranging from 66% to 70% (mean rate = 69%). Raw hybridization intensity values were processed by robust multiarray average (RMA) (Irizarry et al., 2003), using the R package affy (Gautier et al., 2004). Specifically, expression values were extracted from raw CEL files by first applying the RMA model of probe-specific correction to PM (perfect match) probes. Corrected probe values were then normalized via quantile normalization, and a median value was computed for the PM probeset. Resulting RMA expression values were log$_2$-transformed. Distributions of the RMA-expression values of all arrays (four biological replicates) were visualized by two-dimensional PCA.

A t-test was performed for each RMA-PM probeset to determine which genes were differentially expressed between the transgenic line (9–12) and empty vector control-transformed (8–6) genotypes. Multiple testing correction was performed on the t statistic of each probeset to minimize false discovery rate (Benjamini and Hochberg, 1995). Genes that were differentially expressed (±1.5-fold) at adjusted P-value <0.05 were identified and assigned functional annotation derived from VitisNet (Grimplet et al., 2009a) and the PlexDB microarray annotations http://www.plexdb.org/modules/PD_probeset/annotation.php? genechip=Grape (Wise et al., 2007) and presented in Table 1. The complete list of significantly differentially expressed genes is presented in Table S2. In this table, the light red shading indicates increased abundance, light green shading indicates decreased abundance, and genes not significant by false discovery were shaded in light gray. Microarray data were deposited in the NCBI GEO database under series GSE29948 viewable at: http://www.ncbi.nlm.nih.gov/geo/query/acc.cgi?acc=GSE29948.

Genes identified by microarray analysis that showed significantly increased or decreased transcript abundance in the 35S::VvCBF4 overexpressor were compared by sequence homology against other CBF/DREB1f regulon genes from previous studies as listed in Tables 2 and 3. The translated protein sequences of the genes or transcripts associated with each increased/decreased probeset were compared using basic local alignment search tool (BLAST), with the NCBI BLAST+ software (Camacho et al., 2009) using the command line BLASTP algorithm, word size 3, open penalty 11, extend penalty 1, window size 40 and maximum e-value cut-off 1×10^{-15}.

Protein extraction

Whole aerial portions of young vines grown in soil for 3 weeks (control line 8–6 plant height of ~16 cm; 35S::VvCBF4, line 9–12 plant height of ~10 cm) were collected and frozen immediately in liquid nitrogen and stored at −80 °C until further use. Four biological replicates from each treatment were individually ground in liquid nitrogen with a mortar and pestle, followed by extraction of total proteins with a phenol-based protocol optimized for grape (Vincent et al., 2006), based upon protocols previously developed for recalcitrant plant tissues (Hurkman and Tanaka, 1987; Saravanan and Rose, 2004). From each of the eight samples, 5 g of frozen, ground tissue was added to 10 mL of protein extraction buffer (0.5 M Tris–HCl pH 7.5, 0.7 M sucrose, 50 mM EDTA, 0.1 mM KCl, 2 mM PMSF and 2% (v/v) β-mercaptoethanol in water) containing 1 Complete™ Protease Inhibitor Cocktail Tablet per 10 mL of buffer (Roche Applied Science, Inc., Indianapolis, IN) in a 50-mL BD Falcon™ tube and vortexed for 30 s followed by a 10-min incubation at 4 °C. Next, 10 mL of Tris-saturated phenol (pH 7.9) was added to the mixture vortexed for 30 s, followed by a 30-min incubation at 4 °C with inversion of the samples every 10 min. Samples were then centrifuged at 3650 **g** and 4 °C for 30 min in a Beckman Allegra™ 6R centrifuge (Beckman Coulter Inc., Brea, CA). The upper phenol phase was then removed to a new 50-mL BD Falcon™ tube and an equal volume of fresh protein extraction buffer added. This was followed by vortexing for 30 s and incubation for 30 min at 4 °C with inversion of the samples every 10 min. Samples were again centrifuged at 3650 **g** and 4 °C for 30 min, and the phenol phase for each sample was again removed to a new 50-mL BD Falcon™ tube. To precipitate proteins, five volumes of cold (−20 °C) methanolic ammonium acetate (0.1 M ammonium acetate in methanol) were added to each Falcon tube, followed by placement of samples for 3 h at −20 °C with inversion every 10 min. Next, the 50-mL BD Falcon™ tubes were spun at 3650 **g** at 4 °C for 30 min using the Beckman Allegra™ centrifuge to pellet the proteins. Following centrifugation, the supernatants were discarded and 5 mL cold (−20 °C) methanol was added to each tube to wash the pellet. The samples were vortexed for 30 s and placed for 1 h at −20 °C with inversion every 10 min, followed by centrifugation at 3650 **g** at 4 °C for 30 min. The supernatant was discarded, and three additional wash, vortex and centrifugation steps were performed in the same manner as the methanol wash, all with ice-cold (−20 °C) acetone.

Following the final acetone wash, wet pellets were transferred to 2-mL Eppendorf tubes and placed on ice until 2D-DIGE analysis.

2D-DIGE analysis

The acetone-wetted protein pellets were centrifuged at 13 000 *g* at 0 °C for 10 min and washed an additional two times with acetone/water (4:1) before being chilled to −20 °C. The final pellets were allowed to dry in open tubes for 10 min, including 1 min of drying at 3000 *g* in a Speed Vac System (Thermo Fisher Scientific Inc.). Individual dried pellets were resuspended in 200 μL DIGE Reaction Buffer (7 M urea, 2 M thiourea, 4% CHAPS, 30 mM Tris, pH 8.74). The tubes were vortexed frequently a minimum of 10 times and then sonicated for a total of 10 min using 30 s pulses in a water bath sonicator (model no. FS30; Fisher Scientific, Pittsburgh, PA) over a period of 2 h. Samples were then centrifuged at 13 000 *g* at 22 °C for 10 min. The supernatant from each sample was removed to a clean 1.7-mL microfuge tube and assayed for protein concentration using an EZQ™ Protein Quantification Kit with ovalbumin as the standard (Bio-Rad Laboratories, Inc., Hercules, CA). Samples were stored overnight at −20 °C.

Following thawing on ice, additional aliquots of DIGE reaction buffer were added to each sample to bring the final protein concentration to 1.33 mg/mL. Sixty microlitres of each sample (80 μg protein) was then pipetted into a 0.5-mL microfuge tube. Each of three Cy-dyes (e.g., Cy2, Cy3, Cy5; GE Healthcare, Inc., Piscataway, NJ), which were solutions of 5 nmol dye in 5 μL dimethylformamide (DMF), were diluted 5 × (1 μL Cy dye in 4 μL DMF) prior to use. Samples were then subjected to a random dye-swap scheme for normalization of differences in Cy-dye fluorescence intensity (Table S3). A 1.15-μL aliquot of Cy3 and Cy5 was added to each sample. For an internal control, 30 μL of resuspended protein from each sample was mixed in a single tube and 4.6 μL Cy2 was added. At this point, each of the eight sample tubes contained approximately 80 μL total protein and 230 pmoles total Cy dyes (Cy3 and Cy5), whereas the pooled sample control tube contained 331 μg protein and 920 pmol Cy2. The tubes were quickly vortexed and then placed on ice, centrifuged briefly, then placed back on ice. The ice bucket was then covered with aluminium foil and stored in a dark cabinet for 30 min for Cy dye/protein binding. At the end of the 30 min incubation, 2 μL of 10 mM aqueous lysine was added to each tube. The tubes were vortexed and then put back on ice in the dark for an additional 10 min. Four 1.7-mL microfuge tubes were numbered gel 1–4, corresponding to the gel numbers given in Table S3. The contents of the incubated protein/Cy dye mixes were then transferred to the labelled 'gel' tubes, completed by washing each 0.5-mL tube with 50 μL 2 × DIGE Reaction buffer (7 M urea, 2 M thiourea, 4% CHAPS). To each of the 'gel' tubes, the following were added: 60 μL pooled sample, 137 μL 2 × DIGE reaction buffer, 4 μL 0.1% bromophenol blue, 2.35 μL carrier ampholytes pI = 3–10 (Bio-Rad Laboratories, Inc.), and 47.0 μL of 10.5 M DTT solution. The mixtures were then centrifuged at 13 000 *g* and 22 °C for 10 min. From each of the four gel tube mixtures, 450 μL of supernatant was applied to a 24-cm 4–7 immobilized pH gradient (IPG) strip (GE Healthcare, Inc.). The strip was rehydrated passively overnight (for 22 h) at 20 °C. Isoelectric focusing (IEF) was performed as follows: active rehydration at 50 V for 4 h, a linear increase to 200 V in 1 h, a linear increase to 500 V in 1 h, a linear increase to 1000 V in 1 h, a

linear increase to 10 000 V in 2 1/2 h, and maintenance of steady voltage from 10 000 V to 70 000 Vh (Volt-hours) were conducted (∼7½ h). IPG strips were removed and stored at −80 °C until the second dimension was run.

IEF strips were thawed at room temperature (10 min) followed by equilibration in equilibration base buffer (6 M urea, 30% v/v glycerol, 2 M Tris–HCl pH 8.8, and 2% w/v SDS) containing 1% w/v DTT for 20 min to reduce proteins, and then 2.5% w/v iodoacetamide for 20 min to alkylate proteins. The strips were then loaded onto 12.0% (v/v) 26 × 20 × 0.1 cm polyacrylamide gels (Jule Inc., Milford, CT). The gels had been treated with Bind-Silane (γ-methacryloxypropyltrimethoxysilane) and had reference markers attached on the short plate. Electrophoresis was performed using a Protean® Plus Dodeca Cell (Bio-Rad Laboratories, Inc.) with standard Tris–glycine–SDS buffer (25 mM Tris-base, 0.5 M glycine, 0.1% v/v SDS) under the following conditions: 40 V for 2 h followed by 100 V for 24 h at 10 °C.

Prior to imaging, the larger glass plate from each of the gel cassettes was removed. A Typhoon Trio™ Imaging System (part no. 63-0055-87; GE Healthcare, Inc.) was used for image acquisition using the blue (488 nm) laser. Immediately following imaging, each gel was placed in destain solution (7% v/v acetic acid, 10% v/v methanol) and shaken gently for 72 h. Images captured with the Typhoon were analysed using DeCyder™ 2D Software ver. 7.0 (GE Healthcare, Inc.) for protein quantification. Sample labelling and CyDye swapping schema are summarized in Table S3.

Protein identification

For spot picking and protein identification, two gels (replicates #1 and #4) were selected for Sypro® Ruby staining. The destain solution was removed from the gels, and ∼110 mL of Sypro® Ruby Staining solution (Invitrogen, Inc., Carlsbad, CA) in ∼960 mL of fresh destain solution was added. Gels were gently shaken for 48 h followed by washing once with fresh destain solution and once with water. Spot excision was performed using the EXQuest™ spot cutter (Bio-Rad Laboratories, Inc.) followed by trypsin digestion according to the protocol developed by Rosenfeld *et al.* (Rosenfeld *et al.*, 1992) using the Investigator™ ProPrep™ protein digestion kit (Genomic Solutions, Ann Arbor, MI). The trypsin-digested fragments were analysed using an ABI 4700 Proteomics Analyzer™ (Applied Biosystems) matrix-assisted laser desorption/ionization (MALDI) time-of-flight/time-of-flight (TOF/TOF) mass spectrometer (MS). A 0.5-mL aliquot of matrix solution with 5 mg/mL alpha-cyano-4-hydroxycinnamic acid (CHCA) (Sigma-Aldrich, Inc.) and 10 mM ammonium phosphate (Sigma-Aldrich, Inc.) in 0.2% formic acid was cospotted with 0.5 mL of sample (Zhu and Papayannopoulos, 2003). The data were acquired in reflector mode from a mass range of 700–4000 Da, and 1200 laser shots were averaged for each mass spectrum. Each sample was internally calibrated if both the 842.51 and 2211.10 Da ions from trypsin autolysis were present. The eight most intense ions from the MS analysis not present on the exclusion list were subjected to MS/MS analysis. To this end, the mass range was from 70 to precursor ion size with a precursor window of 1–3 Da using an average of 2500 laser shots for each spectrum. The resulting mass data were then used to search the NCBI *nr* database (ver. 10_09_2009; 9 694 989 sequences) and the contigs from *Vitis* Gene Index (ver. 18_9_2009, 23 493 sequences) using automated MASCOT V.2.1 software (http://www.matrix-science.com). Peptide tolerance was 20 ppm, one missed cleavage was allowed and MS/MS tolerance was 0.8 Da. The possibility of matching multiple translated isoforms was examined by manual analysis of peptides present within the sequences. All MS analyses were performed in cooperation with the Nevada Proteomic Center at the University of Nevada, Reno. The complete list of differentially expressed proteins is presented in Table S4.

Acknowledgements

This work was supported by funding from the National Science Foundation NSF (DBI-0217653) and the University of Nevada Agricultural Experiment Station (to GRC and JCC). The authors thank David Tricoli and Kim Carney of the Ralph M. Parsons Plant Transformation Facility, Davis, CA for performing the *Vitis* transformations. The authors thank Craig Osborn of the Nevada Genomic Center for performing microarray services, Rebecca Woolsey, Kathy Schegg and David Quilici of the Nevada Proteomics Center for support and for performing MS analyses, and Rebecca Albion and Kitty Spreeman for invaluable technical support. We thank Mary Ann Cushman for her critical reading of the manuscript. This publication was also made possible by NIH Grant Number P20 RR-016464 from the INBRE Program of the National Center for Research Resources through its support of the Nevada Genomics, Proteomics and Bioinformatics Centers.

References

Achard, P., Gong, F., Cheminant, S., Alioua, M., Hedden, P. and Genschik, P. (2008a) The cold-inducible CBF1 factor-dependent signaling pathway modulates the accumulation of the growth-repressing DELLA proteins via its effect on gibberellin metabolism. *Plant Cell*, **20**, 2117–2129.

Affymetrix (2009) GeneChip® Expression Analysis Technical Manual.

Agarwal, P., Agarwal, P., Nair, S., Sopory, S. and Reddy, M. (2007) Stress-inducible DREB2A transcription factor from *Pennisetum glaucum* is a phosphoprotein and its phosphorylation negatively regulates its DNA-binding activity. *Mol. Genet. Genomics*, **277**, 189–198.

Akoh, C., Lee, G., Liaw, Y., Huang, T. and Shaw, J. (2004) GDSL family of serine esterases/lipases. *Prog. Lipid Res.* **43**, 534–552.

Arroyo-Garcia, R., Ruiz-Garcia, L., Bolling, L., Ocete, R., Lopez, M., Arnold, C., Ergul, A., Soylemezoglu, G., Uzun, H., Cabello, F., Ibanez, J., Aradhya, M., Atanassov, A., Atanassov, I., Balint, S., Cenis, J., Costantini, L., Goris-Lavets, S., Grando, M., Klein, B., McGovern, P., Merdinoglu, D., Pejic, I., Pelsy, F., Primikirios, N., Risovannaya, V., Roubelakis-Angelakis, K., Snoussi, H., Sotiri, P., Tamhankar, S., This, P., Troshin, L., Malpica, J., Lefort, F. and Martinez-Zapater, J. (2006) Multiple origins of cultivated grapevine (*Vitis vinifera* L. ssp. *sativa*) based on chloroplast DNA polymorphisms. *Mol. Ecol.* **15**, 3707–3714.

Badawi, M., Danyluk, J., Boucho, B., Houde, M. and Sarhan, F. (2007) The CBF gene family in hexaploid wheat and its relationship to the phylogenetic complexity of cereal CBFs. *Mol. Genet. Genomics*, **277**, 533–554.

Baumberger, N., Doesseger, B., Guyot, R., Diet, A., Parsons, R., Clark, M., Simmons, M., Bedinger, P., Goff, S., Ringli, C. and Keller, B. (2003) Whole-genome comparison of leucine-rich repeat extensins in *Arabidopsis* and rice. A conserved family of cell wall proteins form a vegetative and a reproductive clade. *Plant Physiol.* **131**, 1313–1326.

Benedict, C., Skinner, J., Meng, R., Chang, Y., Bhalerao, R., Huner, N., Finn, C., Chen, T. and Hurry, V. (2006) The CBF1-dependent low temperature signalling pathway, regulon and increase in freeze tolerance are conserved in *Populus* spp. *Plant Cell Environ.* **29**, 1259–1272.

Benjamini, Y. and Hochberg, Y. (1995) Controlling the false discovery rate: A practical and powerful approach to multiple testing. *J. Royal Stat. Soc. Ser. B*, **57**, 289–300.

Camacho, C., Coulouris, G., Avagyan, V., Ma, N., Papadopoulos, J., Bealer, K. and Madden, T. (2009) BLAST+: architecture and applications. *BMC Bioinformatics*, **10**, 421.

Campoli, C., Matus-Cádiz, M., Pozniak, C., Cattivelli, L. and Fowler, D. (2009) Comparative expression of Cbf genes in the Triticeae under different acclimation induction temperatures. *Mol. Genet. Genomics*, **282**, 141–152.

Canella, D., Gilmour, S., Kuhn, L. and Thomashow, M. (2010) DNA binding by the *Arabidopsis* CBF1 transcription factor requires the PKKP/RAGRxKFxETRHP signature sequence. *Biochim. Biophys. Acta*, **1799**, 454–462.

D'Angelo, G., Vicinanza, M. and De Matteis, M. (2008) Lipid-transfer proteins in biosynthetic pathways. *Curr. Opin. Cell Biol.* **20**, 360–370.

Doukhanina, E., Chen, S., van der Zalm, E., Godzik, A., Reed, J. and Dickman, M. (2006) Identification and functional characterization of the BAG protein family in *Arabidopsis thaliana*. *J. Biol. Chem.* **281**, 18793–18801.

Dubouzet, J., Sakuma, Y., Ito, Y., Kasuga, M., Dubouzet, E., Miura, S., Seki, M., Shinozaki, K. and Yamaguchi-Shinozaki, K. (2003) *OsDREB* genes in rice, *Oryza sativa* L., encode transcription activators that function in drought-, high-salt- and cold-responsive gene expression. *Plant J.* **33**, 751–763.

El Kayal, W., Navarro, M., Marque, G., Keller, G., Marque, C. and Teulieres, C. (2006) Expression profile of CBF-like transcriptional factor genes from *Eucalyptus* in response to cold. *J. Exp. Bot.* **57**, 2455–2469.

Faurobert, M., Mihr, C., Bertin, N., Pawlowski, T., Negroni, L., Sommerer, N. and Causse, M. (2007) Major proteome variations associated with cherry tomato pericarp development and ripening. *Plant Physiol.* **143**, 1327–1346.

Fennell, A. (2004) Freezing tolerance and injury in grapevines. *J. Crop. Improv.* **10**, 201–235.

Fowler, S. and Thomashow, M.F. (2002) *Arabidopsis* transcriptome profiling indicates that multiple regulatory pathways are activated during cold acclimation in addition to the CBF cold response pathway. *Plant Cell*, **14**, 1675–1690.

Fujita, Y., Fujita, M., Satoh, R., Maruyama, K., Parvez, M.M., Seki, M., Hiratsu, K., Ohme-Takagi, M., Shinozaki, K. and Yamaguchi-Shinozaki, K. (2005) AREB1 Is a transcription activator of novel ABRE-dependent ABA signaling that enhances drought stress tolerance in *Arabidopsis*. *Plant Cell*, **17**, 3470–3488.

Fuller, M. and Telli, G. (1999) An investigation of the frost hardiness of grapevine (*Vitis vinifera*) during bud break. *Ann. Appl. Biol.* **135**, 589–595.

Gamborg, O., Miller, R. and Ojima, K. (1968) Nutrient requirements of suspension cultures of soybean root cells. *Exp. Cell Res.* **50**, 151–158.

Gautier, L., Cope, L., Bolstad, B. and Irizarry, R. (2004) affy—Analysis of Affymetrix GeneChip data at the probe level. *Bioinformatics*, **20**, 307–315.

Gilmour, S., Zarka, D., Stockinger, E., Salazar, M., Houghton, J. and Thomashow, M. (1998) Low temperature regulation of the *Arabidopsis* CBF family of AP2 transcriptional activators as an early step in cold-induced COR gene expression. *Plant J.* **16**, 433–442.

Gilmour, S., Sebolt, A., Salazar, M., Everard, J. and Thomashow, M. (2000) Overexpression of the *Arabidopsis* CBF3 transcriptional activator mimics multiple biochemical changes associated with cold acclimation. *Plant Physiol.* **124**, 1854–1865.

Gilmour, S., Fowler, S. and Thomashow, M. (2004) *Arabidopsis* transcriptional activators CBF1, CBF2, and CBF3 have matching functional activities. *Plant Mol. Biol.* **54**, 767–781.

Gion, J., Lalanne, C., Le Provost, G., Ferry-Dumazet, H., Paiva, J., Chaumeil, P., Frigerio, J., Brach, J., Barre, A., de Daruvar, A., Claverol, S., Bonneu, M., Sommerer, N., Negroni, L. and Plomion, C. (2005) The proteome of maritime pine wood forming tissue. *Proteomics*, **5**, 3731–3751.

Grimplet, J., Cramer, G., Dickerson, J., Mathiason, K., Van Hemert, J. and Fennell, A. (2009a) VitisNet: "Omics" integration through grapevine molecular networks. *PLoS ONE*, **4**, e8365.

Grimplet, J., Wheatley, M., Jouira, H., Deluc, L., Cramer, G. and Cushman, J. (2009b) Proteomic and selected metabolite analysis of grape berry tissues under well-watered and water-deficit stress conditions. *Proteomics*, **9**, 2503–2528.

Hanana, M., Deluc, L., Fouquet, R., Daldoul, S., Leon, C., Barrieu, F., Ghorbel, A., Mliki, A. and Hamdi, S. (2008) Identification and characterization of 'rd22' dehydration responsive gene in grapevine (*Vitis vinifera* L.). *C. R. Biol.* **331**, 569–578.

Hoagland, D. and Arnon, D. (1950) *The Water Culture Method of Growing Plants Without Soil, in Circular No 347*, Berkeley, California, USA: California Agricultural Experiment Station.

Hsieh, T., Lee, J., Yang, P., Chiu, L., Charng, Y., Wang, Y. and Chan, M. (2002) Heterology expression of the *Arabidopsis* C-repeat/dehydration response element binding factor 1 gene confers elevated tolerance to chilling and oxidative stresses in transgenic tomato. *Plant Physiol.* **129**, 1086–1089.

Huang, D., Wu, W., Abrams, S. and Cutler, A. (2008) The relationship of drought-related gene expression in *Arabidopsis thaliana* to hormonal and environmental factors. *J. Exp. Bot.* **59**, 2991–3007.

Huang, J., Yang, M., Liu, P., Yang, G., Wu, C. and Zheng, C. (2009) GhDREB1 enhances abiotic stress tolerance, delays GA-mediated development and represses cytokinin signalling in transgenic *Arabidopsis*. *Plant Cell Environ.* **32**, 1132–1145.

Hurkman, W. and Tanaka, C. (1987) The effects of salt on the pattern of protein synthesis in barley roots. *Plant Physiol.* **83**, 517–524.

Irizarry, R.A., Hobbs, B., Collin, F., Beazer-Barclay, Y.D., Antonellis, K.J., Scherf, U. and Speed, T.P. (2003) Exploration, normalization, and summaries of high density oligonucleotide array probe level data. *Biostatistics*, **4**, 249–264.

Jaglo-Ottosen, K., Gilmour, S., Zarka, D., Schabenberger, O. and Thomashow, M. (1998) *Arabidopsis* CBF1 overexpression induces COR genes and enhances freezing tolerance. *Science*, **280**, 104–106.

Jaglo-Ottosen, K., Kleff, S., Amundsen, K., Zhang, X., Haake, V., Zhang, J., Deits, T. and Thomashow, M. (2001) Components of the *Arabidopsis* C-repeat/dehydration-responsive element binding factor cold-response pathway are conserved in *Brassica napus* and other plant species. *Plant Physiol.* **127**, 910–917.

Jin, W., Dong, J., Hu, Y., Lin, Z., Xu, X. and Han, Z. (2009) Improved cold-resistant performance in transgenic grape (*Vitis vinifera* L.) overexpressing cold-inducible transcription factors AtDREB1b. *HortScience*, **44**, 35–39.

Jolie, R., Duvetter, T., Van Loey, A. and Hendrickx, M. (2010) Pectin methylesterase and its proteinaceous inhibitor: a review. *Carbohydr. Res.* **345**, 2583–2595.

Kang, C., Jung, W., Kang, Y., Kim, J., Kim, D., Jeong, J., Baek, D., Jin, J., Lee, J., Kim, M., Chung, W., Mengiste, T., Koiwa, H., Kwak, S., Bahk, J., Lee, S., Nam, J., Yun, D. and Cho, M. (2006) AtBAG6, a novel calmodulin-binding protein, induces programmed cell death in yeast and plants. *Cell Death Differ.* **13**, 84–95.

Kannangara, R., Branigan, C., Liu, Y., Penfield, T., Rao, V., Mouille, G., Hofte, H., Pauly, M., Riechmann, J. and Broun, P. (2007) The transcription factor WIN1/SHN1 regulates cutin biosynthesis in *Arabidopsis thaliana*. *Plant Cell*, **19**, 1278–1294.

Kasuga, M., Liu, Q., Miura, S., Yamaguchi-Shinozaki, K. and Shinozaki, K. (1999) Improving plant drought, salt, and freezing tolerance by gene transfer of a single stress-inducible transcription factor. *Nat. Biotechnol.* **17**, 287–291.

Kuwabara, C., Arakawa, K. and Yoshida, S. (1999) Abscisic acid-induced secretory proteins in suspension-cultured cells of winter wheat. *Plant Cell Physiol.* **40**, 184–191.

Kwon, S., Jin, H., Lee, S., Nam, M., Chung, J., Kwon, S., Ryu, C. and Park, O. (2009) GDSL lipase-like 1 regulates systemic resistance associated with ethylene signaling in *Arabidopsis*. *Plant J.* **58**, 235–245.

Lee, J., Prasad, V., Yang, P., Wu, J., Ho, T.D., Charng, Y. and Chan, M. (2003) Expression of *Arabidopsis* CBF1 regulated by an ABA/stress inducible promoter in transgenic tomato confers stress tolerance without affecting yield. *Plant Cell Environ.* **26**, 1181–1190.

Lee, K., Han, K., Kwon, Y., Lee, J., Kim, S., Chung, W., Kim, Y., Chun, S., Kim, H. and Bae, D. (2009) Identification of potential DREB2C targets in *Arabidopsis thaliana* plants overexpressing DREB2C using proteomic analysis. *Mol. Cells*, **28**, 383–388.

Licausi, F., Giorgi, F., Zenoni, S., Osti, F., Pezzotti, M. and Perata, P. (2010) Genomic and transcriptomic analysis of the AP2/ERF superfamily in *Vitis vinifera*. *BMC Genomics*, **11**, 719.

Liu, Q., Kasuga, M., Sakuma, Y., Abe, H., Miura, S., Yamaguchi-Shinozaki, K. and Shinozaki, K. (1998) Two transcription factors, DREB1 and DREB2, with an EREBP/AP2 DNA binding domain separate two cellular signal transduction pathways in drought- and low-temperature-responsive gene expression, respectively, in *Arabidopsis*. *Plant Cell*, **10**, 1391–1406.

Liu, Y., Lamkemeyer, T., Jakob, A., Mi, G., Zhang, F., Nordheim, A. and Hochholdinger, F. (2006) Comparative proteome analyses of maize (*Zea mays* L.) primary roots prior to lateral root initiation reveal differential protein expression in the lateral root initiation mutant *rum1*. *Proteomics*, **6**, 4300–4308.

Lloyd, G. and McCown, B. (1981) Commercially feasible micropropagation of Mountain laurel, *Kalmia latifolia*, by the use of shoot tip culture. *Proc. Int'l. Plant Prop. Soc.* **30**, 421–427.

Lytle, B.L., Song, J., de la Cruz, N.B., Peterson, F.C., Johnson, K.A., Bingman, C.A., Phillips Jr, G.N. and Volkman, B.F. (2009) Structures of two *Arabidopsis thaliana* major latex proteins represent novel helix-grip folds. *Proteins*, **76**, 237–243.

Magome, H., Yamaguchi, S., Hanada, A., Kamiya, Y. and Oda, K. (2008) The DDF1 transcriptional activator upregulates expression of a gibberellin-deactivating gene, GA2ox7, under high-salinity stress in *Arabidopsis*. *Plant J.* **56**, 613–626.

Maris, A., Suslov, D., Fry, S., Verbelen, J. and Vissenberg, K. (2009) Enzymic characterization of two recombinant xyloglucan endotransglucosylase/hydrolase (XTH) proteins of *Arabidopsis* and their effect on root growth and cell wall extension. *J. Exp. Bot.* **60**, 3959–3972.

Maruyama, K., Sakuma, Y., Kasuga, M., Ito, Y., Seki, M., Goda, H., Shimada, Y., Yoshida, S., Shinozaki, K. and Yamaguchi-Shinozaki, K. (2004) Identification of cold-inducible downstream genes of the *Arabidopsis* DREB1A/CBF3 transcriptional factor using two microarray systems. *Plant J.* **38**, 982–993.

Maruyama, K., Takeda, M., Kidokoro, S., Yamada, K., Sakuma, Y., Urano, K., Fujita, M., Yoshiwara, K., Matsukura, S., Morishita, Y., Sasaki, R., Suzuki, H., Saito, K., Shibata, D., Shinozaki, K. and Yamaguchi-Shinozaki, K. (2009) Metabolic pathways involved in cold acclimation identified by integrated analysis of metabolites and transcripts regulated by DREB1A and DREB2A. *Plant Physiol.* **150**, 1972–1980.

Matus, J., Loyola, R., Vega, A., Peña-Neira, A., Bordeu, E., Arce-Johnson, P. and Alcalde, J. (2009) Post-veraison sunlight exposure induces MYB-mediated transcriptional regulation of anthocyanin and flavonol synthesis in berry skins of *Vitis vinifera*. *J. Exp. Bot.* **60**, 853–867.

Mullins, M., Bouquet, A. and Williams, L. (1992) *Biology of the Grapevine*, Cambridge, New York: Cambridge University Press.

Murashige, T. and Skoog, F. (1962) A revised medium for rapid growth and bio assays with tobacco tissue cultures. *Physiol. Plant.* **15**, 473–497.

Nakamura, R., Satoh, R., Nakamura, R., Shimazaki, T., Kasuga, M., Yamaguchi-Shinozaki, K., Kikuchi, A., Watanabe, K. and Teshima, R. (2010) Immunoproteomic and two-dimensional difference gel electrophoresis analysis of *Arabidopsis dehydration response element-binding protein 1A* (*DREB1A*)-transgenic potato. *Biol. Pharm. Bull.* **33**, 1418–1425.

Nakashima, K. and Yamaguchi-Shinozaki, K. (2006) Regulons involved in osmotic stress-responsive and cold stress-responsive gene expression in plants. *Physiol. Plant.* **126**, 62–71.

Navarro, M., Marque, G., Ayax, C., Keller, G., Borges, J., Marque, C. and Teulieres, C. (2009) Complementary regulation of four *Eucalyptus* CBF genes under various cold conditions. *J. Exp. Bot.* **60**, 2713–2724.

Navarro, M., Ayax, C., Martinez, Y., Laur, J., El Kayal, W., Marque, C. and Teulieres, C. (2011) Two EguCBF1 genes overexpressed in *Eucalyptus* display a different impact on stress tolerance and plant development. *Plant Biotech. J.* **9**, 50–63.

Park, J., Jin, P., Yoon, J., Yang, J., Jeong, H., Ranathunge, K., Schreiber, L., Franke, R., Lee, I. and An, G. (2010) Mutation in Wilted Dwarf and Lethal 1 (WDL1) causes abnormal cuticle formation and rapid water loss in rice. *Plant Mol. Biol.* **74**, 91–103.

Pfaffl, M. (2001) A new mathematical model for relative quantification in real-time RT-PCR. *Nucleic Acids Res.* **29**, e45.

Pino, M., Skinner, J., Park, E., Jeknic, Z., Hayes, P., Thomashow, M. and Chen, Y. (2007) Use of a stress inducible promoter to drive ectopic AtCBF expression improves potato freezing tolerance while minimizing negative effects on tuber yield. *Plant Biotech. J.* **5**, 591–604.

Reid, K., Olsson, N., Schlosser, J., Peng, F. and Lund, S. (2006) An optimized grapevine RNA isolation procedure and statistical determination of reference genes for real-time RT-PCR during berry development. *BMC Plant Biol.* **6**, 27.

Ringli, C. (2010) The hydroxyproline-rich glycoprotein domain of the *Arabidopsis* LRX1 requires Tyr for function but not for insolubilization in the cell wall. *Plant J.* **63**, 662–669.

Rosenfeld, J., Capdevielle, J., Guillemot, J. and Ferrara, P. (1992) In-gel digestion of proteins for internal sequence-analysis after 1-dimensional or 2-dimensional gel-electrophoresis. *Anal. Biochem.* **203**, 173–179.

Rozen, S. and Skaletsky, H. (2000) Primer3 on the WWW for general users and for biologist programmers. *Methods Mol. Biol.* **132**, 365–386.

Sambrook, J. and Russell, D. (2001) *Molecular Cloning: A Laboratory Manual*, Cold Spring Harbor, NY: Cold Spring Harbor Laboratory.

Santerre, A. and Britt, A. (1994) Cloning of a 3-methyladenine-DNA glycosylase from *Arabidopsis thaliana*. *Proc. Natl Acad. Sci. USA*, **91**, 2240–2244.

Saravanan, R. and Rose, J. (2004) A critical evaluation of sample extraction techniques for enhanced proteomic analysis of recalcitrant plant tissues. *Proteomics*, **4**, 2522–2532.

Sharabi-Schwager, M., Lers, A., Samach, A., Guy, C. and Porat, R. (2010) Overexpression of the CBF2 transcriptional activator in *Arabidopsis* delays leaf senescence and extends plant longevity. *J. Exp. Bot.* **61**, 261–273.

Siddiqua, M. and Nassuth, A. (2011) *Vitis* CBF1 and *Vitis* CBF4 differ in their effect on *Arabidopsis* abiotic stress tolerance, development and gene expression. *Plant Cell Environ.* **34**, 1345–1359.

Sreekantan, L., Mathiason, K., Grimplet, J., Schlauch, K., Dickerson, J. and Fennell, A. (2010) Differential floral development and gene expression in grapevines during long and short photoperiods suggests a role for floral genes in dormancy transitioning. *Plant Mol. Biol.* **73**, 191–205.

Sreenivasulu, N., Sopory, S. and Kavi Kishor, P. (2007) Deciphering the regulatory mechanisms of abiotic stress tolerance in plants by genomic approaches. *Gene*, **388**, 1–13.

Stockinger, E., Gilmour, S. and Thomashow, M. (1997) *Arabidopsis thaliana* CBF1 encodes an AP2 domain-containing transcriptional activator that binds to the C-repeat/DRE, a *cis*-acting DNA regulatory element that stimulates transcription in response to low temperature and water deficit. *Proc. Natl Acad. Sci. USA*, **94**, 1035–1040.

Takuhara, Y., Kobayashi, M. and Suzuki, S. (2011) Low-temperature-induced transcription factors in grapevine enhance cold tolerance in transgenic *Arabidopsis* plants. *J. Plant Physiol.* **168**, 967–975.

Tattersall, E. (2006) *Changes in Gene Expression in Response to Abiotic Stress in Grapevine (Vitis vinifera)*, p 154, Reno, NV: University of Nevada.

Tattersall, E., Ergul, A., AlKayal, F., Deluc, L., Cushman, J. and Cramer, G. (2005) Comparison of methods for isolating high-quality RNA from leaves of grapevine. *Am. J. Enol. Vitic.* **56**, 400–406.

This, P., Lacombe, T. and Thomas, M. (2006) Historical origins and genetic diversity of wine grapes. *Trends Genet.* **22**, 511–519.

Thomashow, M. (1999) Plant cold acclimation: freezing tolerance genes and regulatory mechanisms. *Annu. Rev. Plant Physiol. Plant Mol. Biol.* **50**, 571–599.

Vincent, D., Wheatley, M. and Cramer, G. (2006) Optimization of protein extraction and solubilization for mature grape berry clusters. *Electrophoresis*, **27**, 1853–1865.

Vivier, M. and Pretorius, I. (2002) Genetically tailored grapevines for the wine industry. *Trends Biotech.* **20**, 472–478.

Vogel, J., Zarka, D., Van Buskirk, H., Fowler, S. and Thomashow, M. (2005) Roles of the CBF2 and ZAT12 transcription factors in configuring the low temperature transcriptome of *Arabidopsis*. *Plant J.* **41**, 195–211.

Volokita, M., Rosilio-Brami, T., Rivkin, N. and Zik, M. (2010) Combining comparative sequence and genomic data to ascertain phylogenetic relationships and explore the evolution of the large GDSL-lipase family in land-plants. *Mol. Biol. Evol.* **28**, 551–565.

Watson, B., Asirvatham, V., Wang, L. and Sumner, W. (2003) Mapping the proteome of barrel medic (*Medicago truncatula*). *Plant Physiol.* **131**, 1104–1123.

Welling, A. and Palva, E. (2008) Involvement of CBF transcription factors in winter hardiness in birch. *Plant Physiol.* **147**, 1199–1211.

Wise, R., Caldo, R., Hong, L., Shen, L., Cannon, E. and Dickerson, J. (2007) BarleyBase/PLEXdb. *Methods Mol. Biol.* **406**, 347–363.

Wisniewski, M., Norelli, J., Bassett, C., Artlip, T. and Mascarisin, D. (2011) Ectopic expression of a novel peach (*Prunus persica*) CBF transcription factor in apple (*Malus x domestica*) results in short-day induced dormancy and increased cold hardiness. *Planta*, **233**, 971–983.

Xiao, H., Siddiqua, M., Braybrook, S. and Nassuth, A. (2006) Three grape CBF/DREB1 genes respond to low temperature, drought and abscisic acid. *Plant Cell Environ.* **29**, 1410–1421.

Xiao, H., Tattersall, E., Siddiqua, M., Cramer, G. and Nassuth, A. (2008) CBF4 is a unique member of the CBF transcription factor family of *Vitis vinifera* and *Vitis riparia*. *Plant Cell Environ.* **31**, 1–10.

Yadav, S.K., Singla-Pareek, S. and Sopory, S. (2008) An overview on the role of methylglyoxal and glyoxalases in plants. *Drug Metabol. Drug Interact.* **23**, 51–68.

Zhang, J., Broeckling, C., Blancaflor, E., Sledge, M., Sumner, L. and Wang, Z. (2005) Overexpression of WXP1, a putative *Medicago truncatula* AP2 domain-containing transcription factor gene, increases cuticular wax accumulation and enhances drought tolerance in transgenic alfalfa (*Medicago sativa*). *Plant J.* **42**, 689–707.

Zhang, S., Yang, C., Peng, J., Sun, S. and Wang, X. (2009) GASA5, a regulator of flowering time and stem growth in *Arabidopsis thaliana*. *Plant Mol. Biol.* **69**, 745–759.

Zhang, Y., Yang, J. and Showalter, A. (2011) AtAGP18 is localized at the plasma membrane and functions in plant growth and development. *Planta* **233**, 675–683.

Zhao, H. and Bughrara, S. (2008) Isolation and characterization of cold-regulated transcriptional activator LpCBF3 gene from perennial ryegrass (*Lolium perenne* L.). *Mol. Genet. Genomics* **279**, 585–594.

Zhu, X. and Papayannopoulos, I. (2003) Improvement in the detection of low concentration protein digests on a MALDI TOF/TOF workstation by reducing alpha-cyano-4-hydroxycinnamic acid adductions. *J. Biomol. Tech.* **14**, 298–307.

15

Transglutamination allows production and characterization of native-sized ELPylated spider silk proteins from transgenic plants

Nicola Weichert[1], Valeska Hauptmann[1], Matthias Menzel[2], Kai Schallau[1], Philip Gunkel[1], Thomas C. Hertel[3], Markus Pietzsch[3], Uwe Spohn[2] and Udo Conrad*[1]

[1]*Leibniz Institute of Plant Genetics and Crop Plant Research, Stadt Seeland/Ortsteil, Gatersleben, Germany*
[2]*Fraunhofer Institute for Mechanics of Materials, Halle (Saale), Germany*
[3]*Institute of Pharmacy, Faculty of Sciences I, Martin Luther University Halle-Wittenberg, Halle (Saale), Germany*

*Correspondence
email conradu@ipk-gatersleben.de

Keywords: spider silk, transgenic tobacco, ELPylation, transglutaminase, atomic force microscopy, nanoindentation.

Summary

In the last two decades it was shown that plants have a great potential for production of specific heterologous proteins. But high cost and inefficient downstream processing are a main technical bottleneck for the broader use of plant-based production technology especially for protein-based products, for technical use as fibres or biodegradable plastics and also for medical applications. High-performance fibres from recombinant spider silks are, therefore, a prominent example. Spiders developed rather different silk materials that are based on proteins. These spider silks show excellent properties in terms of elasticity and toughness. Natural spider silk proteins have a very high molecular weight, and it is precisely this property which is thought to give them their strength. Transgenic plants were generated to produce ELPylated recombinant spider silk derivatives. These fusion proteins were purified by Inverse Transition Cycling (ITC) and enzymatically multimerized with transglutaminase *in vitro*. Layers produced by casting monomers and multimers were characterized using atomic force microscopy (AFM) and AFM-based nanoindentation. The layered multimers formed by mixing lysine- and glutamine-tagged monomers were associated with the highest elastic penetration modulus.

Introduction

Molecular Farming generally means large-scale production of proteins and other compounds for pharmaceutical and technical use in plants or in plant cells. In the last decade an extended number of different complete antibodies, antibody derivatives of several types, vaccines and other therapeutic proteins for medical and veterinary use has been produced in plants (Fischer and Schillberg, 2004; Koprowski, 2005; Ma *et al.*, 2005; Phan *et al.*, 2013; Sharma and Sharma, 2009; Stoger *et al.*, 2005; Streatfield, 2005). Plant-based expression systems can provide complex, correctly folded and post-translationally modified proteins (Sharma and Sharma, 2009). Relatively low cost, safety and scalability are specific advantages of plant expression systems, but it has to be taken into account, that the cost of downstream processing steps as protein extractions, protein recovery and protein purification are generally similar in all recombinant production systems and that such cost can cover more than 80% of the overall processing costs (Hassan *et al.*, 2008; Yusibov and Rabindran, 2008). Therefore, high cost and inefficient downstream processing are a main technical bottleneck for the broader use of plant-based production technology especially for protein-based products for technical use as fibres or biodegradable plastic (Scheller and Conrad, 2005) and for biomedical applications (Radtke *et al.*, 2011). Here, high-performance fibres as spider silk are a prominent example. During evolution, spiders developed many rather different proteinaceous silk materials (Craig, 2004; Hinman *et al.*, 1992). These fibres are superior to man-made fibres in terms of elasticity and toughness (Gosline *et al.*, 1999; Vollrath and Knight, 2001). The mechanical properties of dragline silk are extraordinary: dragline silk is five times stronger than steel and three times tougher then *p*-aramide, the most promising artificial fibre produced by humans. Dragline silk consists of two spidroins, major ampullate spidroin 1 (MaSp1) and major ampullate spidroin 2 (MaSp2; Xu and Lewis, 1990; Hinman *et al.*, 1992). The molecular structure of spider silks is characterized by protein crystals, interrupted by less organized protein chains (Kluge *et al.*, 2008). Different repetitive primary structure modules, so-called 'ensemble repeats', are combined to secondary structures which properties could be assigned to different spider silks (Hayashi *et al.*, 1999; Romer and Scheibel, 2008). Crystalline β-sheets essentially contribute to the high tensile strength of spider silks. β-Sheets are folded by interactions between amino acid sequences with multiple repeats of alanine, glycine-alanine or glycine-alanine-serine. These sequences are typical for MaSp1 and MaSp2 silks of *Nephila clavipes* (Hinman *et al.*, 1992; Xu and Lewis, 1990) as well as for fibroin-3 and fibroin-4 of *Araneus diadematus* (Scheibel, 2004). Non-crystalline regions of spider silk proteins contain either GPGXX repeats (β-spirals) or helical GGX repeats (Kluge *et al.*, 2008) which provide elasticity. Natural spider silk proteins show a very high molecular weight from at least 250 kDa to several hundred kDa (Ayoub *et al.*, 2007; Sponner *et al.*, 2005). Major ampullate (MA) fibroin genes have very large exons of about 9000 bp (Romer and Scheibel, 2008) resulting in a corresponding protein of about 330 kDa per exon. Currently only two complete spider silk protein sequences are known: the flagelliform protein (FLAG) from *Nephila clavipes* (Hayashi and Lewis, 2000) and a MA protein of *Latrodectus hesperus* (Ayoub and Hayashi, 2008; Ayoub et al., 2007). For expression experiments in general 3'-sequences of different spider silk–encoding genes have been used (Beckwitt and Arcidiacono, 1994; Guerette et al., 1996; Hayashi and Lewis, 1998; Hinman et al., 1992; Scheibel, 2004; Xu and Lewis, 1990). Therefore, strategies have been developed to construct recombinant silk-like proteins that strongly resemble MA or flagelliform silk proteins (Scheibel, 2004). Expression experiments have been performed in *Escherichia coli* (Arcidiacono et al., 1998; Fahnestock and Irwin, 1997; Huemmerich et al., 2004; Prince et al., 1995), in *Pichia pastoris* (Fahnestock and Bedzyk, 1997), in plants (Hauptmann et al., 2013; Menassa et al., 2004; Scheller et al., 2001, 2004), in mammalian cells (Lazaris et al., 2002), in transgenic mice (Karatzas et al., 2005) and in the *Baculovirus* system (Miao et al., 2006). Spider silk production in *E. coli* has been optimized by combinations of synthetic modules with different authentic spider silk sequences (Huemmerich et al., 2004), but proteins of native size could not be produced by this technique. Remarkably, dragline silk-like proteins of native size have been produced in *E. coli* by metabolic engineering and the fibres spun from this material show mechanical properties comparable to those of native spider silk (Xia et al., 2010). All spiders investigated so far produce high molecular weight dragline silk proteins, therefore, it is anticipated that the size of spider silk proteins is a key factor for toughness and tensile strength (Ayoub et al., 2007; Sponner et al., 2005; Xia et al., 2010). Large dragline silk proteins contain several repeated motifs which facilitate multiple intramolecular and intermolecular interactions. Smaller proteins could lead to more chain end caused gaps during the spinning (Xia et al., 2010). Transgenic plants could be an option for scaling up spider silk production, but only if economical downstream processing and purification strategies can be developed (Kluge et al., 2008; Romer and Scheibel, 2008). Here, the use of elastin-like peptide (ELP) fusions and purification by membrane-based Inverse Transition Cycling (mITC) are suitable options (Ge et al., 2006; Phan and Conrad, 2011); Floss et al. (2010b). The high repetitivity of spider silk sequences causes genetic instability and mRNA instability in plants and, therefore, the size of open reading frames is limited. Very high molecular weight spider silk proteins were produced by intein-based *in vivo* post-translational multimerization leading to linear multimers of very variable size (Hauptmann et al., 2013). A further method of post-translational multimerization *in vitro* is the use of tranglutamination by enzymatic connection of lysines and glutamines. For this purpose a soluble recombinant transglutaminase, recently described and characterized in detail, was used (Marx et al., 2007, 2008). A further challenge to be addressed is the adaptation of a ELPylation based purification method (Phan and Conrad, 2011) for spider silk-ELP fusions. This includes selection of the most suitable ITC temperature and sodium chloride concentration as well as removal of sodium chloride after the ITC process. The latter process is presumptive for the small-scale production of layers that could be used to characterize the material. We casted layers of cross-linked MaSp1-ELP fusion proteins and characterized them by topographical analyses and by atomic force microscopy (AFM)-based nanoindentation to investigate the mechanical properties of the purified material in terms of elastic modulus. We proposed to deposit it as small and smooth layer spots with diameters smaller than 2 mm and thicknesses between 1 and 5 μm to mechanically characterize small amounts

of the genetically engineered proteins (<1 mg) from plants. Then, at least for isotropic materials the AFM-based nanoindentation opens up an effective way to determine the Young's modulus E, from which the shear modulus $G = 0.5E/(1 + v)$, the bulk modulus $\kappa = E/3(1-2v)$ and the compressibility $\beta = 1/\kappa$ could be calculated or at least roughly estimated taking into consideration the Poisson's ratio v of lateral contraction to longitudinal elongation. It should be noted that the obtained Young's moduli can be tightly correlated with those determined by Dynamic Mechanical Analysis (Cappella et al., 2005) based on $E = (E'^2 + E''^2)$ with the elastic storage modulus E' and the plastic loss modulus E''. Thus, plant-produced multimeric spider silk protein variants were characterized according their mechanical properties. The results achieved show that post-translational multimerization could be a useful option for biobased material development.

Results

Design of ELPylated MaSp1-derivatives

A major ampullate spidroin 1 derived gene (*MaSp1*), originated from the appropriate *N. clavipes* sequence (Xu and Lewis, 1990), was designed to enable *in vitro* post-translational multimerization by a recombinantly produced microbacterial transglutaminase from *Streptomyces mobaraensis* (rMTG). For this purpose an appropriate lysine-tag (K-tag, GSGMKETAAARFERNHMDSGS) or a glutamine-tag (Q-tag, GSGMAETAAAA FERQHMDSGS), respectively, adapted from Tanaka et al. (2004), was inserted at the N-terminal site of the MaSp1 sequence. ELP was applied at C-terminal site of MaSp1 as a mean for both expression enhancement and purification by ITC to recover the recombinant fusion proteins (Floss et al., 2010b). The resulting cassettes K-MaSp1-100xELP (lysine-tagged) and Q-MaSp1-100xELP (glutamine-tagged) encoded ELPylated spider silk fusion proteins with 1258 amino acid residues, respectively (Figures 1 and S1).

Transgenic expression of ELPylated spider silk proteins in tobacco

The expression profiles of the primary transgenic T_0 plants carrying either the Q-MaSp1-100xELP or the K-MaSp1-100xELP construct were visualized by Western blotting, which showed that the target protein was accumulated in tobacco leaves (data not shown). Two transgenic lines were selected out of 16 (Q-MaSp1-100xELP) and 24 (K-MaSp1-100xELP) recombinant protein expressing T_0 plants. Leaf extracts from T_1 progeny of these two lines, respectively, were analyzed by Western blotting showing the expected size of 102 kDa (Figure 2). These plants were used as a source for purification of Q-MaSp1-100xELP and K-MaSp1-100xELP monomers.

Purification of ELPylated monomers by mITC

The Q-MaSp1-100xELP and K-MaSp1-100xELP monomers as well as 100xELP were purified by mITC which was described by Phan and co-workers (Phan and Conrad, 2011) and adapted to purify plant-expressed highly repetitive spider silk proteins by introduction of a 1 h heat incubation step at the beginning of the procedure. Exemplarily, the purification of lysine-tagged MaSp1-ELP monomers is presented in Figure 3a. As expected, in the centrifugation supernatants and in the filtrate the MaSp1-ELP fusion protein was detectable, but not in either the centrifugation pellets or in the filtration flow through. Enrichment of the fusion protein was achieved by elution from the cellulose acetate filter (Figure 3a,b). The reproducibility of the adapted mITC method

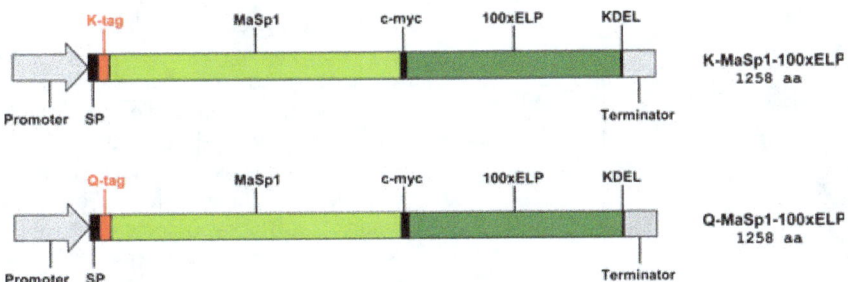

Figure 1 Schematic representation of expression cassettes of tagged MaSp1-100xELP fusion proteins for heterologous expression in tobacco. MaSp1: designed major ampullate spidroin 1. 100xELP: 100 repeats of the elastin-like peptide VPGXG (where X represents any amino acid instead of proline). SP: legumin B4 signal peptide. KDEL: ER retention signal. c-myc: detection tag. K-tag and Q-tag facilitated transglutaminase effected cross-linking. The expression of the transgene was driven by the CaMV 35S promoter.

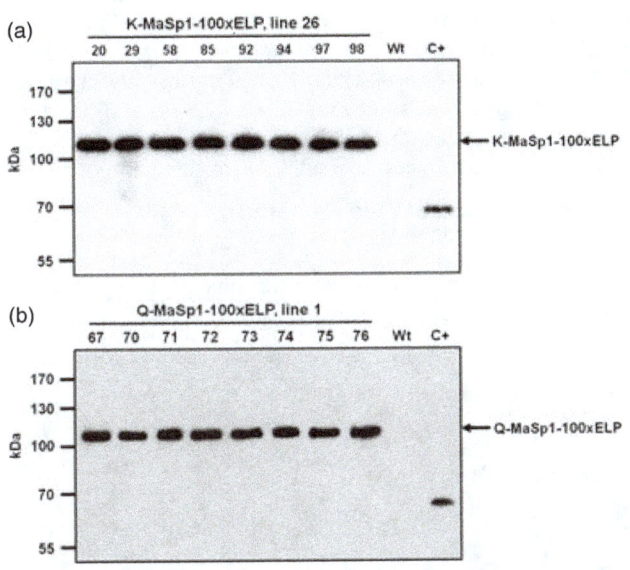

Figure 2 Expression of the fusion proteins K-MaSp1-100xELP (a) and Q-MaSp1-100xELP. (b) Leaf extracts of stable transgenic tobacco plants were separated by 10% SDS-PAGE and analyzed by Western blot. The numbers represent independent T_1-plants of the selected transgenic T_0 lines 1 and 26. WT: wild type *N. tabacum* cv. SNN. C+: c-myc Western blot standard [1 ng anti-TNFα-ELP (Conrad *et al.*, 2011)].

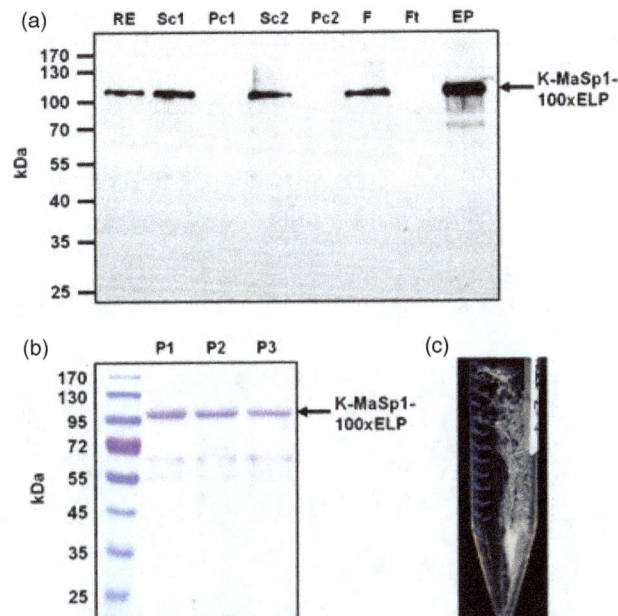

Figure 3 Purification of K-MaSp1-100xELP. (a) Western blot analysis of a complete mITC procedure. Samples (RE: raw extract; Sc: centrifugation supernatant; Pc: centrifugation pellet; F: polyethersulfone filtrate; Ft: flow through; EP: protein eluted from cellulose acetate (CA) filter) were separated by 10% SDS-PAGE and analyzed by Western blotting. (b) Robustness of the ITC procedure. Separation (10% SDS-PAGE) of 30 μL aliquots of dissolved fusion protein, followed by Coomassie Blue staining. P1–P3: three independent replicated purifications of K-MaSp1-100xELP. (c) Desalted and lyophilized K-MaSp1-100xELP protein.

was proven by repeating the procedure three times with same quantities of leaf material resulting in comparable amounts of purified protein (Figure 3b). The yield of desalted and purified fusion proteins was determined by weighing: 65–75 mg of lyophilized MaSp1-ELP fusion protein were produced per kg of leaf material (Figure 3c), comparable yields were achieved for purified Q-MaSp1-ELP.

Cross-linking of tagged MaSp1-100xELP fusion proteins by transglutaminase

The purified monomers Q-MaSp1-100xELP and K-MaSp1-100xELP were successfully cross-linked by recombinant microbial transglutaminase (rMTG) treatment *in vitro*, producing much larger protein molecules. The cross-linking experiments involved either the Q- or the K-MaSp1-100xELP on their own, or a mixture of both. In all three approaches, the molecular weight of the product rose to over 250 kDa (Figure 4). Whereas the two monomers carried an N-terminal glutamine-

or lysine-tag, their sequences also included a number of internal glutamine residues, and each featured a lysine residue at c-myc and at its C-terminus (Figure S1), implying that not just the N-terminal Q- or K-residue was targeted by the transglutaminase. Therefore, also internal glutamines and lysines than the N-terminal Q or K could be used by the transglutaminase. We did not expect a linear end-to-end dimerization. Nevertheless, nearly native-sized spider silk protein variants were produced by this strategy. The application of a rather simple and scalable purification strategy as ITC allowed the isolation of sufficient cross-linked material at an appropriate concentration to enable the production and the mechanical analysis of spider silk-ELP layers.

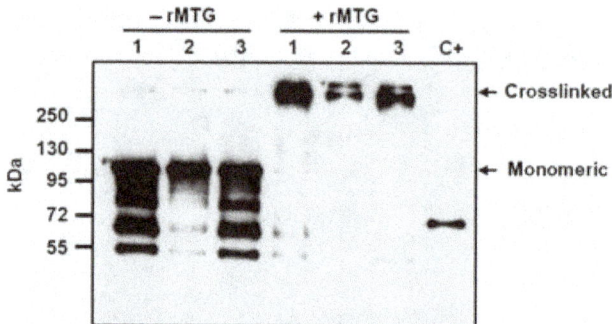

Figure 4 *In vitro* cross-linking of purified K-MaSp1-100xELP and Q-MaSp1-100xELP by recombinant microbial transglutaminase (rMTG). 1: Q/K-MaSp1-100xELP (equimolar mixture of Q-MaSp1-100xELP and K-MaSp1-100xELP). 2: Q-MaSp1-100xELP. 3: K-MaSp1-100xELP. −rMTG: without transglutaminase. +rMTG: with transglutaminase. C+: c-myc Western blot standard [1 ng anti-TNFα-ELP (Conrad *et al.*, 2011)]. Separation on a 10% SDS-polyacrylamide gel, Western blot and detection by anti-c-myc antibody.

Surface topography of casted spider silk protein layers

Layers of Q- and K-tagged MaSp1-ELP, cross-linked Q/K-MaSp1-ELP, silkworm silk fibroin and low density polyethylene (LDPE) Riblene® FL30 foil were characterized by high resolution AFM phase image analysis and compared to phase images of pure 100xELP (Figure 5a–f). All casted spider silk protein layers appeared homogenous on surface with very low roughnesses smaller than 10.5 nm. They were partially crystalline also including some amorphous regions which were identified from their more darkly colored appearance. The lysine-tagged monomer showed indications of formation of a fine fibrous superstructure, however, the glutamine-tagged one appeared more globular and amorphous. The deposited cross-linked Q/K MaSp1-100xELP had a roughness on the nanoscale, which was very similar to that of Q-MaSp1-100xELP. In general, no preferable direction for the orientation of fine structures on the surface was observed. All materials should be able to take a load equally from any direction in space, an essential precondition for the subsequent nanoindentation measurements. The layer thicknesses of the recombinant ELPylated spider silk proteins and 100xELP were measured by AFM profilometry (Table 1). The silkworm silk fibroin and the LDPE layers were thicker than 10 μm. Therefore, the fundamental requirement for nanoindentation experiments that the indentation depth should be smaller than 10% of the layer thickness, was fulfilled (Berg and Grau, 1993; Bückle, 1965). In Table 1 the mean and the squared surface roughnesses which were acceptable for AFM-based nanoindentation were summarized.

Elastic properties of recombinant spider silk fusion proteins

The AFM-based nanoindentation experiments to measure the elastic properties of recombinant spider silk proteins require a

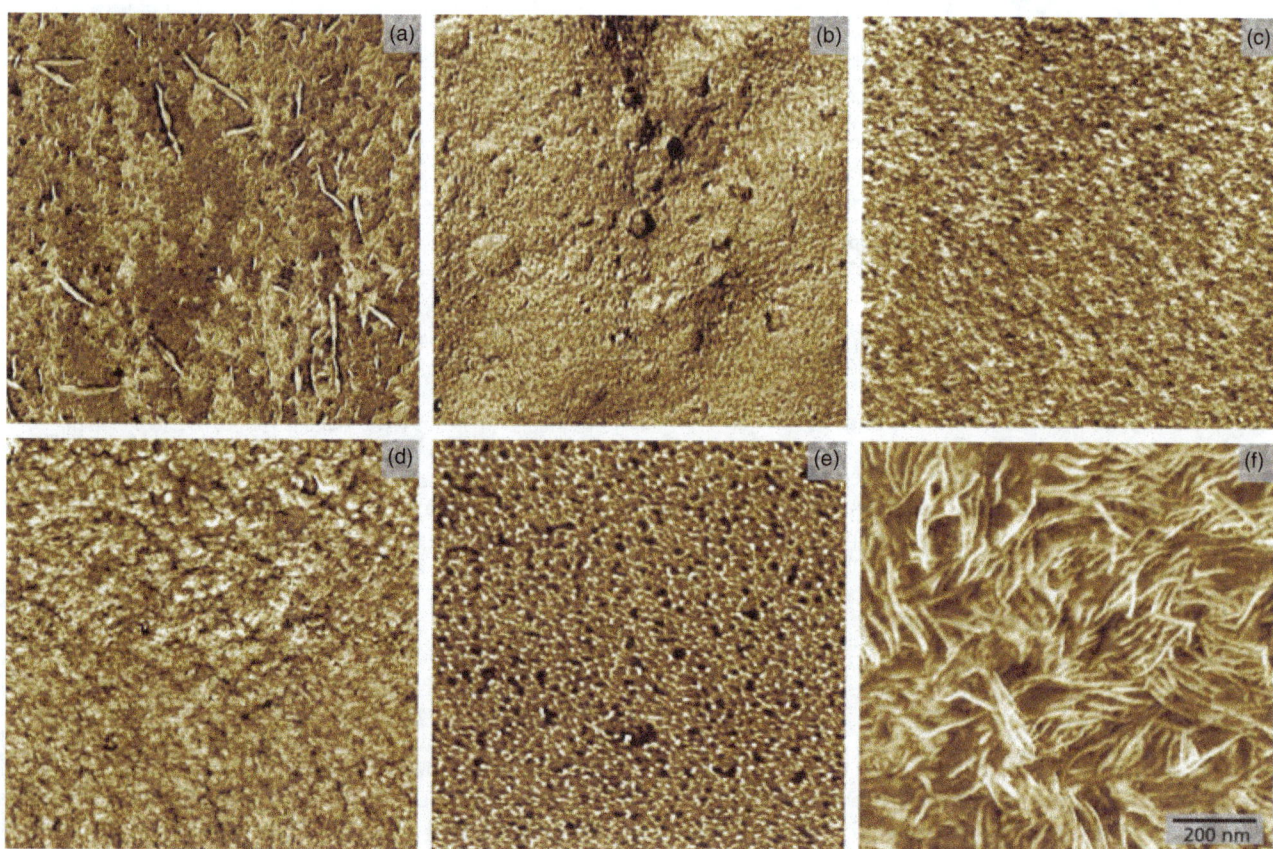

Figure 5 High resolution AFM phase images of ELPylated spider silk protein layers by intermittent tapping mode. Representative 1 x 1 μm (x-y) AFM images shown were generated using an SSS-NCHR™ cantilever (2 nm-tip). (a) K-MaSp1-100xELP monomer. (b) Q-MaSp1-100xELP monomer. (c) Q/K-MaSp1-100xELP, cross-linked. (d) 100xELP. (e) Reference silkworm silk fibroin layer. (f) Reference low density polyethylene (LDPE) Riblene® FL30 foil. Silkworm fibroin was prepared as described previously (Jin and Kaplan, 2003)

Table 1 AFM determination of surface roughness of recombinant spider silk fusion protein layers, according to DIN EN ISO 25178. Data were generated from a defined 5×5 μm² grid

Sample	Thickness of the sample [d] d [μm]	Mean surface roughness [s_a] (5×5 μm²) s_a [nm]	Squared surface roughness [s_q] (5×5 μm²) s_q [nm]
100xELP	5.53	5.0	6.5
K-MaSp1-100xELP	4.72	1.9	2.6
Q-MaSp1-100xELP	1.86	7.4	9.3
Q/K-MaSp1-100xELP, cross-linked	2.24	8.5	10.3
Silkworm silk fibroin (ref)	>10	2.1	2.6
LDPE Riblene® FL30 (ref)	>200	17	22

Mean surface roughness: $S_a = \frac{1}{A} \int_A /z(x,y)/dxdy$;

Squared surface roughness: $S_a = \frac{1}{A} \int\int_A /z^2(x,y)/dxdy$;

ref: reference.

homogenous layer surface. The topography of the recombinant protein layers was, therefore, recorded by AFM in a defined two-dimensional grid (5×5 μm² and 20×20 μm²). The determined thicknesses (1.9–5.5 μm) of the ELPylated spider silk protein layers, as well as that of 100xELP were several 100-fold greater than their surface roughnesses, which ranged from 1.9 to 8.5 nm for all prepared layers (Table 1, Figure S3). The homogeneity of the ELPylated spider silk protein layers (roughness, squared roughness, cross-sectional profiles) was sufficient to perform highly reproducible nanoindentation measurements, based on a penetration depth of 10–15 nm. Despite fulfilling the Buckle rule (Bückle, 1965) and the homogeneity requirement as well as having low roughness, the small penetration depths restricted the interpretation of the mechanical properties to the near-surface range. The shape of the used cantilever tip was shown before and after a series of nanoindentations demonstrating its advantageous mechanical stability in comparison to other tips with attached spheres (Figure S2).

All these control data mentioned in this chapter ensure the reliability of the mechanical property measurements.

The elastic penetration moduli E were determined from three independent series of indentation experiments involving from 3100 to 3300 measurements for each protein material and 1089 for LDPE (Table 2). LDPE and silkworm silk fibroin were used as reference materials to check the cantilever calibration. The E value of LDPE was the same as measured by Jee and co-workers (Jee and Lee, 2010): 0.16 GPa and 0.19 GPa using the AFM-based nanoindentation but different methods of evaluation. The E modulus of silkworm silk fibroin was different to literature data (Bai et al., 2006), but these data were produced by classical nanoindentation. For comparison with data from plant-produced spider silk fusion proteins we used measuring data from silkworm silk produced with the same method of AFM-based nanoindentation with the same cantilevers.

The confidence intervals of the evaluated data were narrow enough to statistically distinguish all materials from each other (Figure 6). The various polymers differed significantly with respect to their elastic moduli E according to the differential t-test (Noack and Schulze, 1980) with

$$t = \frac{E_{m,i} - E_{m,j}}{S_d} \left(\frac{N_i N_j}{N_i + N_j} \right)^{0.5} \tag{1}$$

$$S_d^2 = \left(\sum (E_i - E_{m,i})^2 + \sum (E_j - E_{m,j})^2 \right) / (N_i + N_j - 2) \tag{2}$$

Table 2 Elastic penetration moduli of recombinant spider silk fusion protein layers obtained by AFM-based nanoindentation. $E = E_a \pm t_{0.05}{}^{N-1} \sigma/N^{0.5}$; σ – standard deviation. Experimental conditions: temperature = 23.4 °C, relative humidity = 38.0%. Cantilever DT-NCHR #1, system sensitivity = 20.49 nm/V, spring constant = 67.67 N/m, tip radius = 111.7 nm, setpoint control = 500 nN, relative setpoint = 1500 nN, z-closed loop active, z-length = 1 μm, retract delay = 0.01 s, extend speed = 2 μm/s, grid size = 25×25 μm², data points 33×33 per run with an orthogonal and lateral inter-sampling point distance of at least 758 nm

Sample	Elastic indentation modulus E E [GPa]	Total number of measurements N
$100 \times$ ELP	2.74 ± 0.02	3267
K-MaSp1-100 $\times$ ELP	2.78 ± 0.02	3267
Q-MaSp1-100 $\times$ ELP	3.10 ± 0.03	3172
Q/K-MaSp1-100 $\times$ ELP, cross-linked	3.29 ± 0.03	3267
Silkworm silk fibroin (ref)	3.01 ± 0.02	3212
LDPE Riblene® FL30 (ref)	0.16 ± 0.01	1089

GPa, Gigapasqual; ref, reference; ELP, elastin-like peptide.

Differential t-tests were performed for the comparison of all E_i values with all E_j values with $i \neq j$ (i and j are the indices of the corresponding proteins in test). If all calculated t-values were $t > t_{0.05, N1+N2-2}$, then the arithmetically averaged values $E_{m,i}$ and $E_{m,j}$ of the elastic moduli were significantly different. N_i and N_j were the numbers of repeated measurements of the values E_i and E_j, respectively. The confidence intervals were defined by $\pm\Delta = \pm st/N^{0.5}$ with the standard deviation s, the t-value for the probability of 95% of the number N of repeated measurements which did not significantly overlap in any case.

The highest E value of 3.29 ± 0.03 was determined for the cross-linked Q/K-MaSp1-100xELP layer when compared with the other characterized recombinant polymers. Although the E values of the K-MaSp1-100x-ELP and Q-MaSp1-100xELP monomers clearly differed from each other, both monomers had a greater elastic penetration modulus than 100xELP (2.74 ± 0.02; Table 2).

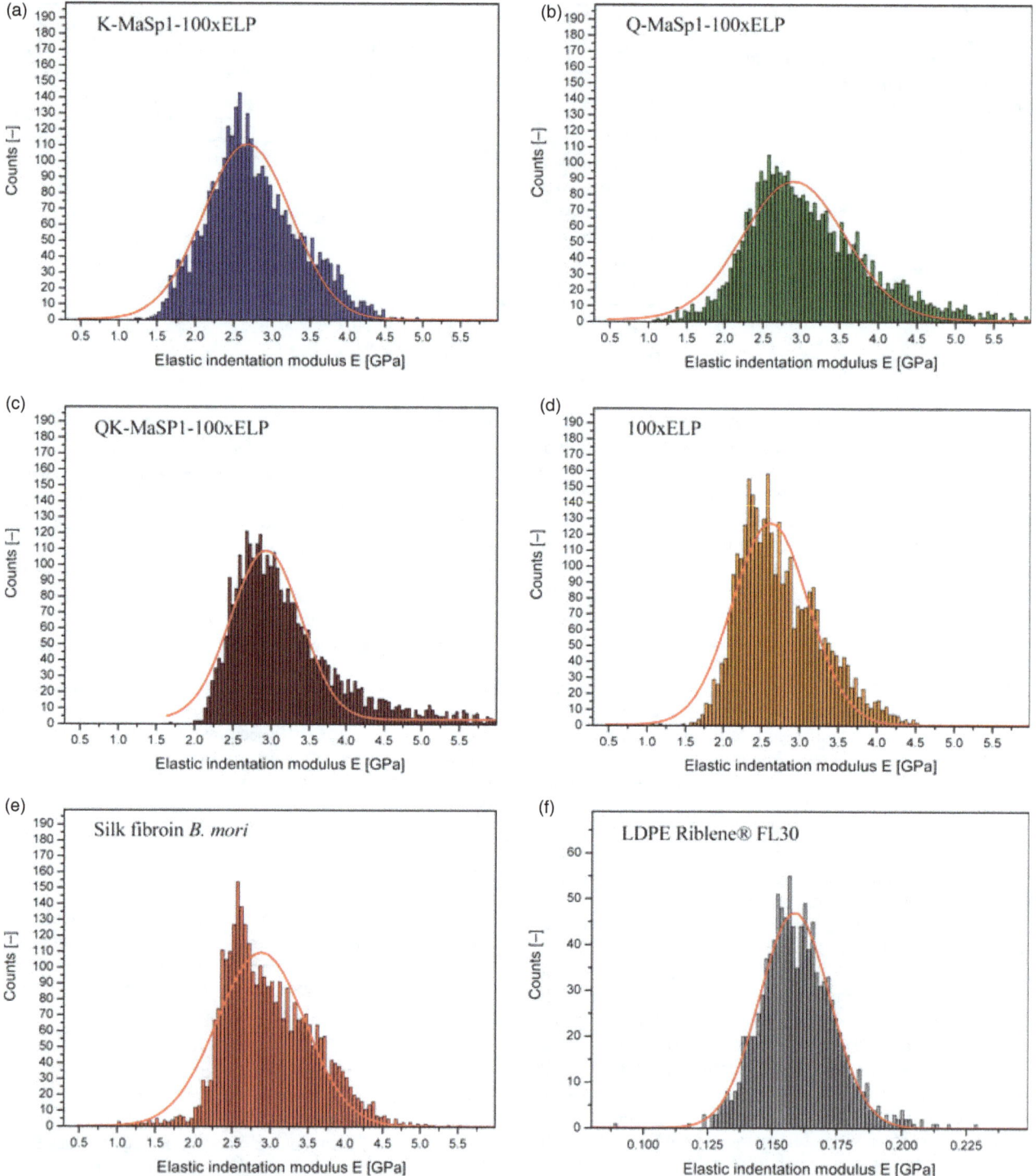

Figure 6 Frequency distribution pattern of elastic penetration moduli E of recombinant ELPylated spider silk proteins (a–c), 100xELP (d), silkworm silk fibroin (e) and low density polyethylene (LDPE) Riblene® FL30 (f). The confidence intervals were calculated from evaluated data of three independent series of indentations (~1000 measurements per single run) using differential *t*-test.

Discussion

Natural spider silk proteins have a rather high molecular weight spanning from at least 250 kDa to several hundreds of kDa (Ayoub *et al.*, 2007; Sponner *et al.*, 2005). The extended size of spider silk proteins seems to be a key factor for the superior mechanical properties of spider silk fibres (Ayoub *et al.*,

2007; Sponner *et al.*, 2005; Xia *et al.*, 2010). The highly repetitive nature of the spider silk amino acid sequences and the need to providing sufficient tRNA and precursor synthesis for dominant amino acids such as glycine, limit the production of high molecular weight spider silk proteins in fast growing microorganisms such as *E. coli*. The latter problems may be overcome by optimizing glycyl-tRNA and glycine synthesis (Xia *et al.*, 2010). Spider silk proteins could be also produced efficiently in transgenic plants, but there was a size limit of about 100 kDa (Scheller *et al.*, 2001). High number of repetitions in the coding sequences caused genetic instability in *Agrobacterium tumefaciens*, which was used as a transformation vehicle to produce stable transformants and for transient expression in plants. The use of post-translational multimerization would solve this problem, because the size of the open reading frames as well as the size of the transcripts would be limited. Post-translational multimerization *in vivo* by intein-mediated *trans*-splicing allowed the production of native-sized spider silk proteins in plants. This *in planta* process resulted in the production of very large molecules connected end-to-end (Intein C-Intein N) by a linear mode. Multimer mixes of the spider silk protein FLAG of different size spanning from 70 to 500 kDa could be produced by this method (Hauptmann *et al.*, 2013). Post-translational multimerization *in vitro* was a further option to achieve large-sized spider silk multimers from plants. We chose highly repetitive MaSp1 sequences for these experiments and used a bacterial transglutaminase for the interconnection of lysine and glutamine. One essential presumption was the purification of monomers before the enzymatic interconnection. ELPylation is a useful tool for both purification of spider silk-ELP fusion proteins and expression enhancement *in planta* (Floss *et al.*, 2010b; Scheller *et al.*, 2004). Here we show, that K- and Q-tagged MaSp1-ELP fusions could be successfully produced in tobacco leaves by ER retention, a commonly accepted mode for enhancement of plant-based production of proteins (Floss *et al.*, 2010b). Purification of the monomers was achieved by mITC, developed for the plant-based production of avian flu antigens (Phan and Conrad, 2011) and adapted to spider silk proteins. The major modification of the method was the inclusion of a boiling step at the beginning, as MaSp1 derivatives remain soluble under these conditions (Scheller *et al.*, 2001). For further applications, this modification has to be proved effective for the relevant spider silk derivatives. Nearly complete multimerization by transglutaminase results in spider silk-ELP fusion proteins in excess of 250 kDa in size. Interestingly, we achieved comparable results using either Q-tagged fusion proteins, K-tagged fusion proteins or a mixture of both versions. At least in case of the first two multimerization methods, other glutamines and lysines (glutamines are present in the coding sequence, whereas two lysines are present in the c-myc-tag and in the KDEL signal) were used for transglutaminase-based multimerization. This provides an alternative to the end-to-end *trans*-splicing *via* inteins and could lead to a higher degree of cross-linking. The degree of cross-linking should have a significant influence on the elasticity and the elastic modulus E. According to Flory (1953) and Nowatzki *et al.* (2008) for homogeneous rubber-like materials equation (3) with the density of the polymeric material ρ_o, the polymer volume fraction $X_{p,V}$, the Poisson's ratio v, the relative molecular weight of the polymer M and the average molecular weight M_C between the cross-links can give a very first approximation of the effect of cross-linking on the elastic modulus E:

$$E = \frac{2\rho_o X_{p,V}(1-v)RT}{M_C} \cdot \left(1 - \frac{2M_C}{M}\right) \qquad (3)$$

The right factor is the fraction of elastically active cross-links decreasing with M_C.

It could be assumed that the high E modulus of the Q/K-MaSp1-100xELP layer is caused by a significant degree of cross-linking. Transglutaminase-catalyzed cross-links of proteins causes an increase in tensile strength and elongation to break of films as well as these cross-links decreased the film permeability (Mariniello *et al.*, 2013). Further investigations are necessary to confirm our results by achieving a higher degree of cross-linking, e.g. by a longer reaction time in presence of transglutaminase.

Another important difference was the notable size of the fused proteins wherein all proteins yielded are larger than 250 kDa. After the transglutaminase reaction the resulting proteins were re-purified and even these large proteins were highly soluble in water at low-salt conditions. This solubility could be due to the ELP component. These properties enabled us to produce layers of sufficient thickness and rather low squared roughness values to perform highly reproducible nanoindentation measurements. The layers of Q-tagged, K-tagged and Q/K-multimerized MaSp1-ELP showed a uniform structure. K-tagged monomers contrast with Q-tagged monomers in the formation of fine superstructures (Figure 5). The K-tagged monomers contain a hydrophilic tag-region whereas the Q-tag is hydrophobic such as the majority of the MaSp1 and the ELP sequences (Figure S4). This difference in hydrophobicity at the N-terminus could strongly influence the folding/assembly process and could explain the differences mentioned above. Further experiments as modifications of the tag-sequences as well as transfer to the C-terminus or insertion of tag-sequences into the MaSp1 open reading frame and production of new transgenic lines are necessary to investigate this problem more in detail.

The AFM images (Figure 5a–d) demonstrate that homogeneous layers can be produced from highly purified and ELP-tagged recombinant spider silk proteins. The extremely low roughness between 1.9 and 10 nm (Table 1), and the achieved thickness of the deposited protein layers between 1.8 and 5 μm (Figure S3) allowed precise measurements of elastic penetration modulus by AFM-based nanoindentation. The critical influence of relatively small amounts of salts could be excluded by the applied protein purification. The E values of all recombinant spider silk layers were similar to analogous layers prepared from the native silk fibroin of *Bombyx mori* (Table 2) and they were in the typical range for glassy amorphous polymers (E = 2–4 GPa), e.g. Polystyrene, Polymethacrylate and Nylon 6 (Jee and Lee, 2010; Kaliappan, 2007). The question that remains is whether these large-sized and enzymatically cross-linked MaSp1-ELP-multimers also show a higher stiffness. The layered Q/K-MaSp1-100xELP multimers formed by transglutaminase catalyzed cross-linking of mixed lysine- and glutamine-tagged monomers were associated with the highest elastic penetration modulus which was slightly but significantly higher than the E moduli both of K-MaSp1-100xELP and Q-MaSp1-100xELP. It should be noted that E of Q/K-MaSp1-100xELP is also significantly higher than that of the layers of native silk fibroin despite in the same order of magnitude. The toughness of this kind of proteins layers could be increased by the proposed enzyme catalyzed cross-linking as it was achieved with other protein-based materials (Liu *et al.*, 2009; McDermott *et al.*, 2004).

At this point it is important to note again, that the Young's moduli could be correlated to other moduli determined by Dynamic Mechanical Analysis as well as to compression moduli that described important properties for the use of such materials (Cappella *et al.*, 2005; Hrouz *et al.*, 1978). Following these arguments we expected higher toughness and stiffness of fibres and layers derived from multimerized plant-based spider silk protein derivatives. Thus, our experiments demonstrated that spider silk-ELP derivatives, with their potential for further use in nanotechnology and medicine, could be produced in plants, purified, modified and mechanically tested according to important parameters. A nanomechanically and reliably working screening procedure was developed to characterize small amounts of genetically engineered silk and other proteins, which could be relevant for technical and medical applications. Bioresorbable materials could be produced as a promising alternative for applications, such as promotion of nerve regeneration. Spider silk-based materials provide crucial unique characteristics for nerve regeneration guidance as long-term degradable fibres of sufficient tensile strength, toughness, elasticity and thermal stability (Radtke et al., 2011). Schwann cells adhered to spider silk and showed cell proliferation (Allmeling et al., 2006). Additionally, spider silk layers have been shown to enhance survival of chondrocytes and to prevent dedifferentiation of these cells crucial for cartilage formation (Scheller et al., 2004). Plant-based production and purification by ITC are in principle scalable and, therefore, may potentially meet the demand for production of gram-amounts of product that will allow further exploration of this strategy in the development of novel protein-based materials.

Experimental procedures

Construction of tagged spider silk genes and generation of stably transformed tobacco plants

Pre-hybridized oligonucleotids encoding on the one hand a glutamine-tag (5'-primer: 5'-GATCCGGCTCTGGAATGGCTGAA ACGGCCGCAGCGGCTTTCGAAAGACAGCATATGGATTCTG-3'; 3'-primer: 5'-GATCCAGAATCCATATGCTGTCTTTCGAAAGCCG CTGCGGCCGTTTCAGCCATTCCAGAGCCG-3') and on the other hand a lysine-tag (5'-primer: 5'-GATCCGGCTCTGGAATGAAGGA AACGGCCGCAGCGAGATTCGAAAGAAACCATATGGATTCTG-3'; 3'-primer: 5'-GATCCAGAATCCATATGGTTTCTTCGAATCTCGCT GCGGCCGTTTCCTTCATTCCAGAGCCG-3') were individually ligated into the BamHI digested vector MaSp1-100xELP-pRTRA (Schallau, 2008). Expression was driven by an ubiquitous Cauliflower Mosaic Virus (CaMV) 35S Promoter and terminated by a CaMV 35S terminator (Franck et al., 1980; Gardner et al., 1981), further including the legumin B4 (LeB4) signal peptide (Baumlein et al., 1986), a glutamine- or lysine-tag, the spider silk gene MaSp1 (derived from Nephila clavipes MaSp1 (Xu and Lewis, 1990), a c-myc-tag (Munro and Pelham, 1987), respectively 100 copies of ELP, the KDEL ER retention signal for enhancing correct folding and high accumulation (Wandelt et al., 1992). For stable transformation of tobacco (Nicotiana tabacum cv. SNN), the expression cassettes were digested by HindIII and then transferred individually into the binary vector pCB301-Kan (Gahrtz and Conrad, 2009) which was subsequently electroporated into Agrobacterium tumefaciens C85C1 (pGV2260; Deblaere et al., 1985). Transgenic tobacco plants were generated by leaf disk transformation (Horsch, et al., 1985).

SDS-PAGE and Western blot

Frozen leaf material was ground in a Mixer Mill MM 300 (Retsch® GmbH, Haan, Germany) and extracted for 10 min at 95 °C in SDS sample buffer (72 mM Tris, 10% glycerol, 2% SDS, 5% 2-mercaptoethanol, 0.0025 mM bromophenol blue, pH 6.8).

After centrifugation (19 000 g, 30 min, 4 °C) the concentration of total soluble protein (TSP) was determined using the Bradford assay (Bio-Rad, Munich, Germany). Extracts were separated by reducing SDS-PAGE, stained by Coomassie® Brillant Blue R-250 (SERVA Electrophoresis GmbH, Heidelberg, Germany) or electro-transferred to a nitrocellulose membrane (Whatman™, Maidstone, UK) in 25 mM Tris, 0.1% w/v SDS, 192 mM glycine and 20% v/v methanol. Processing of the membranes and the detection of recombinant proteins using an anti-c-myc monoclonal antibody (Conrad et al., 1998) were performed as described previously (Floss et al., 2010a). Purified recombinant ELPylated proteins were visualized by Coomassie® Brillant Blue R-250 after SDS-PAGE.

Purification of ELPylated spider silk proteins using ITC

Both glutamine- or lysine-tagged MaSp1-100xELP monomers and 100xELP were purified via mITC (Phan and Conrad, 2011), adapted for purification of ELPylated spidroins. Frozen and grounded transgenic tobacco leaf material (80 g) was added to 260 mL 85 °C-heated stirred 50 mM Tris–HCl (pH 8.0) and then boiled for 1 h. The suspension was cleared by centrifugation (75 600 g, 30 min, 4 °C). Sodium chloride was added at 4 °C to the supernatant to a final concentration of 2 M NaCl, the cold extract was re-centrifuged (75 600 g, 30 min, 4 °C) and then passed through a polyethersulfone membrane, pore size 0.22 μm, (Corning®, St. Louis, MO) at 4 °C. The pre-treated extract was warmed up to 20 °C and the precipitate was captured by vacuum filtration through a 0.2 μm cellulose acetate membrane (Sartorius Stedim, Göttingen, Germany). After washing the membrane twice (2 M NaCl, 20 °C) to remove contaminating proteins, the ELPylated proteins were eluted with ice-cold Millipore-Q water, concentrated by lyophilization (ALPHA2-4LSD; Christ, Osterode, Germany) and desalted via gel filtration (Sephadex G-50; Sigma-Aldrich, St. Louis, MO) in LC-MS grade deionized water. Centrifugation-based ITC (cITC) was performed as described (Scheller et al., 2004).

Preparation of recombinant microbial transglutaminase

The S2P variant of rMTG from S. mobaraensis was expressed as Pro-TG in E. coli BL21 in a histidine-tagged form as described previously (Sommer et al., 2011). After cell disruption and activation of Pro-TG with Proteinase K the transglutaminase was purified by Ni-IMAC (Sommer et al., 2012). By the use of dialysis of active fractions against 50 mM Tris/HCl-buffer, the TG precipitated. The rMTG (S2P) was re-solubilized in 50 mM Tris/HCl (pH 8.0), 300 mM NaCl. The activity of rMTG was assayed according to the colorimetric hydroxamate procedure (Folk and Cole, 1966).

Cross-linking of tagged monomeric spider silk proteins

Lyophilized monomers were solubilized in 50 mM Tris–HCl, pH 8.0 to a final concentration of 1 mg/mL. An equimolar solution of K-MaSp1-100xELP and Q-MaSp1-100xELP was treated with 0.5 U of rMTG (Sommer et al., 2012) per μL protein solution at 11 °C with gentle shaking. The resulting cross-linked ELPylated proteins were purified by cITC and desalted via gel filtration (Sephadex G-50; Sigma-Aldrich).

Casting of protein layers for AFM imaging and AFM-based nanoindentation

Lyophilized proteins were solubilized in LC-MS grade deionized water and desalted by gel filtration (Sephadex G-50; Sigma-

Aldrich). Layering was achieved by applying 6 mL of desalted protein solution (1 mg/mL) in overlapping sets of 20 µL droplets onto a borosilicate glass slide (5 × 25 × 1 mm), and drying the slide at room temperature *in vacuo* (Vacuum Concentrator 5301; Eppendorf, Hamburg, Germany) between droplet applications to achieve layers of sufficient thickness.

Atomic force microscopy

The surface characteristics of the protein layers were investigated by AFM operated in the intermittent tapping mode providing both their topological and phase shift imaging by using the set-up Nanowizard®II (JPK Instruments, Berlin, Germany). Protein layer thickness and roughness were analyzed in standard contact mode. Layer thickness was determined by measuring the step heights between layer surfaces and the surface of the substrate after specific scratching procedure (Figure S3). Roughness of the layer surfaces was recorded using a scan size of 5 × 5 µm² and 20 × 20 µm². A silicon cantilever SSS-NCHR™ (NanoWorld AG, Neuchâtel, Switzerland) with a force constant of about 40 N/m and a tip radius of ~2 nm was used for high resolution imaging. For larger scan widths, a Si_3N_4-cantilever MLCT (Bruker Corporation, Billerica, MA) with lowest force constant of 0.01 N/m was used.

AFM-based nanoindentation

Atomic force microscopy-based nanoindentation (Nanowizard®II, JPK Instruments, Berlin, Germany) was applied to record and evaluate load penetration curves as a course of the load in dependence on the penetration depth to measure the elastic penetration modulus of the protein layers. The temperature and the relative humidity were adjusted at constant values for 23.4 °C and 38.0%, respectively to prevent system drift errors. Three indentation series were performed including all samples within single runs and checked according aberrations among each other. A diamond-coated cantilever DT-NCHR #1 (NanoWorld AG) was used and calibrated by the thermal noise method (Sader *et al.*, 1999). A spring constant of 67.7 ± 0.8 N/m was determined. Scanning electron micrographs of the cantilever tips were taken to evaluate the exact tip radius both before and after indentation (Figure S2). A tip radius of 112 ± 2 nm was measured. Elastic penetration moduli were calculated from the load penetration curve according to an advanced Hertzian model for spherical indenter geometry (Johnson, 1985; Lin *et al.*, 2007).

$$F = \frac{E}{1 - v^2} \left[\frac{a^2 + R^2}{2} \ln \frac{R+a}{R-a} - aR \right]$$
$$\delta = \frac{a}{2} \ln \frac{R+a}{R-a}$$

(4)

The penetration depth was represented by δ, the Poisson's ratio by v ($v = 0.3$), the radius of sphere by R, the contact circle by a, the load by F and the elastic penetration modulus by E.

With the exception of LDPE more than 3000 penetration measurements (scan size: 25 × 25 µm², 33 × 33 data points per run) were performed for each layer sample with an orthogonal and lateral intersampling point distance of at least 758 nm. The confidence intervals were calculated as described above. The significance of the differences of elastic moduli was clearly confirmed by calculations based on equations (1) and (2) of the corresponding *t*-tests. There was no overlapping of the confidence intervals of the different materials. Silk fibroin of the silkworm *B. mori* which was prepared as described previously (Jin and Kaplan, 2003) and also LDPE Riblene® FL30 (POLIMERI

EUROPA SpA, Milano, Italy) were used as reference materials to check cantilever calibration.

Acknowledgements

We thank Christine Helmold for excellent technical assistance and Heiko Weichert for support with freeze-drying. Special thanks to Jeremy N. Timmis for helpful comments on the manuscript. We also acknowledge the fruitful discussions with colleagues participating in COST action FA0804 'Molecular Farming'. The AFM-based investigation was supported by the grant KF 2330204FR0 of the ZIM program of the Federal Ministry of Economy and Technology, F.R.G. The work was also funded by the grant FKZ 22037511 of the Fachagentur Nachwachsende Rohstoffe e. V., supported by the Federal Ministry of Food, Agriculture and Consumer Protection, Germany. The technical and scientific support of JPK Instruments is acknowledged. We, the authors, disclose any potential sources of conflict of interest.

References

Allmeling, C., Jokuszies, A., Reimers, K., Kall, S. and Vogt, P.M. (2006) Use of spider silk fibres as an innovative material in a biocompatible artificial nerve conduit. *J. Cell Mol. Med.* **10**, 770–777.

Arcidiacono, S., Mello, C., Kaplan, D., Cheley, S. and Bayley, H. (1998) Purification and characterization of recombinant spider silk expressed in *Escherichia coli. Appl. Microbiol. Biotechnol.* **49**, 31–38.

Ayoub, N.A. and Hayashi, C.Y. (2008) Multiple recombining loci encode MaSp1, the primary constituent of dragline silk, in widow spiders (Latrodectus: Theridiidae). *Mol. Biol. Evol.* **25**, 277–286.

Ayoub, N.A., Garb, J.E., Tinghitella, R.M., Collin, M.A. and Hayashi, C.Y. (2007) Blueprint for a high-performance biomaterial: full-length spider dragline silk genes. *PLoS ONE*, **2**, e514.

Bai, J.M., Ma, T., Chu, W., Wang, R.Z., Silva, L., Michal, C., Chiao, J.C. and Chiao, M. (2006) Regenerated spider silk as a new biomaterial for MEMS. *Biomed. Microdevices*, **8**, 317–323.

Baumlein, H., Wobus, U., Pustell, J. and Kafatos, F.C. (1986) The legumin gene family – structure of a B type gene of *Vicia faba* and a possible legumin gene specific regulatory element. *Nucleic Acids Res.* **14**, 2707–2720.

Beckwitt, R. and Arcidiacono, S. (1994) Sequence conservation in the C-terminal region of spider silk proteins (Spidroin) from *Nephila clavipes* (Tetragnathidae) and *Araneus bicentenarius* (Araneidae). *J. Biol. Chem.* **269**, 6661–6663.

Berg, G. and Grau, P. (1993) Influence of the substrate hardness on the validity of Buckles Rule. *Cryst. Res. Technol.* **28**, 989–994.

Bückle, H. (1965) *Mikrohärteprüfung und ihre Anwendung.* Stuttgart: Berliner Union.

Cappella, B., Kaliappan, S.K. and Sturm, H. (2005) Using AFM force-distance curves to study the glass-to-rubber transition of amorphous polymers and their elastic-plastic properties as a function of temperature. *Macromolecules*, **38**, 1874–1881.

Conrad, U., Fiedler, U., Artsaenko, O. and Philips, J. (1998) Single-chain Fv antibodies expressed in plants. In *Methods in Biotechnology – Recombinant Proteins from Plants: Production and Isolation of Clinically Useful Compounds* (Cunningham, C. and Porter, A., eds), pp. 103–127. Totowa, NJ: Humana Press.

Conrad, U., Plagmann, I., Malchow, S., Sack, M., Floss, D.M., Kruglov, A.A., Nedospasov, S.A., Rose-John, S. and Scheller, J. (2011) ELPylated anti-human TNF therapeutic single-domain antibodies for prevention of lethal septic shock. *Plant Biotechnol. J.* **9**, 22–31.

Craig, C.L. (2004) Broad patterns of speciation are correlated with the evolution of new silk proteins in spiders but not in the lepidoptera. *Biomacromolecules*, **5**, 739–743.

Deblaere, R., Bytebier, B., De Greve, H., Deboeck, F., Schell, J., Van Montagu, M. and Leemans, J. (1985) Efficient octopine Ti plasmid-derived vectors for

Agrobacterium-mediated gene transfer to plants. *Nucleic Acids Res.* **13**, 4777–4788.

Fahnestock, S.R. and Bedzyk, L.A. (1997) Production of synthetic spider dragline silk protein in *Pichia pastoris*. *Appl. Microbiol. Biotechnol.* **47**, 33–39.

Fahnestock, S.R. and Irwin, S.L. (1997) Synthetic spider dragline silk proteins and their production in *Escherichia coli*. *Appl. Microbiol. Biotechnol.* **47**, 23–32.

Fischer, R. and Schillberg, S. (2004) *Molecular Farming*. Weinheim: Wiley VCH Verlag GmbH & Co. KGaA.

Flory, P.J. (1953) *Principles of Polymer Chemistry*. Ithaca, New York: Cornell University Press.

Floss, D.M., Mockey, M., Zanello, G., Brosson, D., Diogon, M., Frutos, R., Bruel, T., Rodrigues, V., Garzon, E., Chevaleyre, C., Berri, M., Salmon, H., Conrad, U. and Dedieu, L. (2010a) Expression and immunogenicity of the mycobacterial Ag85B/ESAT-6 antigens produced in transgenic plants by elastin-like peptide fusion strategy. *J. Biomed. Biotechnol.* **2010**, 274346.

Floss, D.M., Schallau, K., Rose-John, S., Conrad, U. and Scheller, J. (2010b) Elastin-like polypeptides revolutionize recombinant protein expression and their biomedical application. *Trends Biotechnol.* **28**, 37–45.

Folk, J.E. and Cole, P.W. (1966) Transglutaminase – mechanistic features of active site as determined by kinetic and inhibitor studies. *Biochim. Biophys. Acta*, **122**, 244–264.

Franck, A., Guilley, H., Jonard, G., Richards, K. and Hirth, L. (1980) Nucleotide sequence of cauliflower mosaic virus DNA. *Cell*, **21**, 285–294.

Gahrtz, M. and Conrad, U. (2009) Immunomodulation of plant function by in vitro selected single-chain Fv intrabodies. *Methods Mol. Biol.* **483**, 289–312.

Gardner, R.C., Howarth, A.J., Hahn, P., Brownluedi, M., Shepherd, R.J. and Messing, J. (1981) The complete nucleotide sequence of an infectious clone of cauliflower mosaic virus by M13mp7 shotgun sequencing. *Nucleic Acids Res.* **9**, 2871–2888.

Ge, X., Trabbic-Carlson, K., Chilkoti, A. and Filipe, C.D.M. (2006) Purification of an elastin-like fusion protein by microfiltration. *Biotechnol. Bioeng.* **95**, 424–432.

Gosline, J.M., Guerette, P.A., Ortlepp, C.S. and Savage, K.N. (1999) The mechanical design of spider silks: from fibroin sequence to mechanical function. *J. Exp. Biol.* **202**, 3295–3303.

Guerette, P.A., Ginzinger, D.G., Weber, B.H. and Gosline, J.M. (1996) Silk properties determined by gland-specific expression of a spider fibroin gene family. *Science*, **272**, 112–115.

Hassan, S., van Dolleweerd, C.J., Ioakeimidis, F., Keshavarz-Moore, E. and Ma, J.K. (2008) Considerations for extraction of monoclonal antibodies targeted to different subcellular compartments in transgenic tobacco plants. *Plant Biotechnol. J.* **6**, 733–748.

Hauptmann, V., Weichert, N., Menzel, M., Knoch, D., Paege, N., Scheller, J., Spohn, U., Conrad, U. and Gils, M. (2013) Native-sized spider silk proteins synthesized in planta via intein-based multimerization. *Transgenic Res.* **22**, 369–377.

Hayashi, C.Y. and Lewis, R.V. (1998) Evidence from flagelliform silk cDNA for the structural basis of elasticity and modular nature of spider silks. *J. Mol. Biol.* **275**, 773–784.

Hayashi, C.Y. and Lewis, R.V. (2000) Molecular architecture and evolution of a modular spider silk protein gene. *Science*, **287**, 1477–1479.

Hayashi, C.Y., Shipley, N.H. and Lewis, R.V. (1999) Hypotheses that correlate the sequence, structure, and mechanical properties of spider silk proteins. *Int. J. Biol. Macromol.* **24**, 271–275.

Hinman, M., Dong, Z., Xu, M. and Lewis, R.V. (1992) Spider silk: a mystery starting to unravel. *Results Probl. Cell Differ.* **19**, 227–254.

Horsch, R.B., Fry, J.E., Hoffmann, N.L., Eichholtz, D., Rogers, S.G. and Fraley, R.T. (1985) A simple and general method for transferring genes into plants. *Science*, **227**, 1229–1231.

Hrouz, J., Ilavsky, M., Havlicek, I. and Dusek, K. (1978) Comparison of penetration, tensile and compression moduli of elasticity of poly(n-alkyl acrylate) networks in rubberlike state. *Collect. Czech. Chem. Commun.* **43**, 1999–2007.

Huemmerich, D., Helsen, C.W., Quedzuweit, S., Oschmann, J., Rudolph, R. and Scheibel, T. (2004) Primary structure elements of spider dragline silks and their contribution to protein solubility. *Biochemistry*, **43**, 13604–13612.

Jee, A.Y. and Lee, M. (2010) Comparative analysis on the nanoindentation of polymers using atomic force microscopy. *Polym. Test.* **29**, 95–99.

Jin, H.J. and Kaplan, D.L. (2003) Mechanism of silk processing in insects and spiders. *Nature*, **424**, 1057–1061.

Johnson, K.L. (1985) *Contact Mechanics*. Cambridge, UK: Cambridge University Press.

Kaliappan, S.K. (2007) *Characterization of Physical Properties of Polymers Using AFM Force-Distance Curves*. Siegen, Germany: University of Siegen.

Karatzas, C.N., Chretien, N., Duguay, F., Bellemare, A., Zhou, J.F., Rodenhiser, A., Islam, S.A., Turcotte, C., Huang, Y. and Lazaris, A. (2005) High-toughness spider silk fibers spun from soluble recombinant silk produced in mammalian cells. In *Biotechnology of Biopolymers* (Steinbuchel, A. and Doi, Y., eds), pp. 945–965. Weinheim, Germany: Wiley-VCH Verlag GmbH & Co. KGaA.

Kluge, J.A., Rabotyagova, O., Leisk, G.G. and Kaplan, D.L. (2008) Spider silks and their applications. *Trends Biotechnol.* **26**, 244–251.

Koprowski, H. (2005) Vaccines and sera through plant biotechnology. *Vaccine*, **23**, 1757–1763.

Kyte, J. and Doolittle, R.F. (1982) A simple method for displaying the hydropathic character of a protein. *J. Mol. Biol.* **157**, 105–132.

Lazaris, A., Arcidiacono, S., Huang, Y., Zhou, J.F., Duguay, F., Chretien, N., Welsh, E.A., Soares, J.W. and Karatzas, C.N. (2002) Spider silk fibers spun from soluble recombinant silk produced in mammalian cells. *Science*, **295**, 472–476.

Lin, D.C., Dimitridas, E.K. and Horkay, F. (2007) Robust strategies for automated AFM force curve analysis-I. Non adhesive indentation of soft, inhomogeneous materials. *J Biomech Eng-T Asme*, **129**, 430–440.

Liu, Y., Kopelman, D., Wu, L.Q., Hijji, K., Attar, I., Preiss-Bloom, O. and Payne, G.F. (2009) Biomimetic sealant based on gelatin and microbial transglutaminase: an initial in vivo investigation. *J. Biomed. Mater. Res. B.* **91B**, 5–16.

Ma, J.K., Barros, E., Bock, R., Christou, P., Dale, P.J., Dix, P.J., Fischer, R., Irwin, J., Mahoney, R., Pezzotti, M., Schillberg, S., Sparrow, P., Stoger, E., Twyman, R.M. and European Union Framework 6 Pharma-Planta Consortium. (2005) Molecular farming for new drugs and vaccines. Current perspectives on the production of pharmaceuticals in transgenic plants. *EMBO Rep.* **6**, 593–599.

Mariniello, L., Porta, R., Sorrentino, A., Giosafatto, C.V., Rossi Marquez, G., Esposito, M. and Di Pierro, P. (2013) Transglutaminase-mediated macromolecular assembly: production of conjugates for food and pharmaceutical applications. *Amino Acids*, doi: 10.1007/s00726-013-1561-6.

Marx, C.K., Hertel, T.C. and Pietzsch, M. (2007) Soluble expression of a pro-transglutaminase from *Streptomyces mobaraensis* in *Escherichia coli*. *Enzyme Microb. Tech.* **40**, 1543–1550.

Marx, C.K., Hertel, T.C. and Pietzsch, M. (2008) Purification and activation of a recombinant histidine-tagged pro-transglutaminase after soluble expression in *Escherichia coli* and partial characterization of the active enzyme. *Enzyme Microb. Tech.* **42**, 568–575.

McDermott, M.K., Chen, T.H., Williams, C.M., Markley, K.M. and Payne, G.F. (2004) Mechanical properties of biomimetic tissue adhesive based on the microbial transglutaminase-catalyzed crosslinking of gelatin. *Biomacromolecules*, **5**, 1270–1279.

Menassa, R., Zhu, H., Karatzas, C.N., Lazaris, A., Richman, A. and Brandle, J. (2004) Spider dragline silk proteins in transgenic tobacco leaves: accumulation and field production. *Plant Biotechnol. J.* **2**, 431–438.

Miao, Y., Zhang, Y., Nakagaki, K., Zhao, T., Zhao, A., Meng, Y., Nakagaki, M., Park, E.Y. and Maenaka, K. (2006) Expression of spider flagelliform silk protein in *Bombyx mori* cell line by a novel Bac-to-Bac/BmNPV baculovirus expression system. *Appl. Microbiol. Biotechnol.* **71**, 192–199.

Munro, S. and Pelham, H.R. (1987) A C-terminal signal prevents secretion of luminal ER proteins. *Cell*, **48**, 899–907.

Noack, S. and Schulze, G. (1980) Statistical evaluation of analytical measurements - approximation of the integral limits of the R-distribution, T-distribution, and F-distribution. *Fresen. Z. Anal. Chem.* **304**, 250–254.

Nowatzki, P.J., Franck, C., Maskarinec, S.A., Ravichandran, G. and Tirrell, D.A. (2008) Mechanically tunable thin films of photosensitive artificial proteins: preparation and characterization by nanoindentation. *Macromolecules*, **41**, 1839–1845.

Phan, H.T. and Conrad, U. (2011) Membrane-based inverse transition cycling: an improved means for purifying plant-derived recombinant protein-elastin-like polypeptide fusions. *Int. J. Mol. Sci.* **12**, 2808–2821.

Transglutamination allows production and characterization of native-sized ELPylated spider silk proteins...

167

Phan, H.T., Floss, D.M. and Conrad, U. (2013) Veterinary vaccines from transgenic plants: highlights of two decades of research and a promising example. *Curr. Pharm. Des.* **19**, 5601–5611.

Prince, J.T., McGrath, K.P., DiGirolamo, C.M. and Kaplan, D.L. (1995) Construction, cloning, and expression of synthetic genes encoding spider dragline silk. *Biochemistry*, **34**, 10879–10885.

Radtke, C., Allmeling, C., Waldmann, K.H., Reimers, K., Thies, K., Schenk, H.C., Hillmer, A., Guggenheim, M., Brandes, G. and Vogt, P.M. (2011) Spider silk constructs enhance axonal regeneration and remyelination in long nerve defects in sheep. *PLoS ONE*, **6**, e16990.

Romer, L. and Scheibel, T. (2008) The elaborate structure of spider silk: structure and function of a natural high performance fiber. *Prion*, **2**, 154–161.

Sader, J.E., Chon, J.W.M. and Mulvaney, P. (1999) Calibration of rectangular atomic force microscope cantilevers. *Rev. Sci. Instrum.* **70**, 3967–3969.

Schallau, K. (2008) *Herstellung von Spinnenseidenproteinen in Tabaksamen.* Halle (Saale), Germany: Martin-Luther-Universität Halle-Wittenberg, Universitäts-und Landesbibliothek Sachsen-Anhalt.

Scheibel, T. (2004) Spider silks: recombinant synthesis, assembly, spinning, and engineering of synthetic proteins. *Microb. Cell Fact.* **3**, 14.

Scheller, J. and Conrad, U. (2005) Plant-based material, protein and biodegradable plastic. *Curr. Opin. Plant Biol.* **8**, 188–196.

Scheller, J., Guhrs, K.H., Grosse, F. and Conrad, U. (2001) Production of spider silk proteins in tobacco and potato. *Nat. Biotechnol.* **19**, 573–577.

Scheller, J., Henggeler, D., Viviani, A. and Conrad, U. (2004) Purification of spider silk-elastin from transgenic plants and application for human chondrocyte proliferation. *Transgenic Res.* **13**, 51–57.

Sharma, A.K. and Sharma, M.K. (2009) Plants as bioreactors: recent developments and emerging opportunities. *Biotechnol. Adv.* **27**, 811–832.

Sommer, C., Volk, N. and Pietzsch, M. (2011) Model based optimization of the fed-batch production of a highly active transglutaminase variant in *Escherichia coli. Protein Expres. Purif.* **77**, 9–19.

Sommer, C., Hertel, T.C., Schmelzer, C.E. and Pietzsch, M. (2012) Investigations on the activation of recombinant microbial pro-transglutaminase: in contrast to proteinase K, dispase removes the histidine-tag. *Amino Acids*, **42**, 997–1006.

Sponner, A., Schlott, B., Vollrath, F., Unger, E., Grosse, F. and Weisshart, K. (2005) Characterization of the protein components of *Nephila clavipes* dragline silk. *Biochemistry*, **44**, 4727–4736.

Stoger, E., Ma, J.K., Fischer, R. and Christou, P. (2005) Sowing the seeds of success: pharmaceutical proteins from plants. *Curr. Opin. Biotechnol.* **16**, 167–173.

Streatfield, S.J. (2005) Plant-based vaccines for animal health. *Rev. Sci. Tech.* **24**, 189–199.

Tanaka, T., Kamiya, N. and Nagamune, T. (2004) Peptidyl linkers for protein heterodimerization catalyzed by microbial transglutaminase. *Bioconjug. Chem.* **15**, 491–497.

Vollrath, F. and Knight, D.P. (2001) Liquid crystalline spinning of spider silk. *Nature*, **410**, 541–548.

Wandelt, C.I., Khan, M.R., Craig, S., Schroeder, H.E., Spencer, D. and Higgins, T.J. (1992) Vicilin with carboxy-terminal KDEL is retained in the endoplasmic reticulum and accumulates to high levels in the leaves of transgenic plants. *Plant J.* **2**, 181–192.

Xia, X.X., Qian, Z.G., Ki, C.S., Park, Y.H., Kaplan, D.L. and Lee, S.Y. (2010) Native-sized recombinant spider silk protein produced in metabolically engineered *Escherichia coli* results in a strong fiber. *Proc. Natl Acad. Sci. USA*, **107**, 14059–14063.

Xu, M. and Lewis, R.V. (1990) Structure of a protein superfiber: spider dragline silk. *Proc. Natl Acad. Sci. USA*, **87**, 7120–7124.

Yusibov, V. and Rabindran, S. (2008) Recent progress in the development of plant derived vaccines. *Expert Rev. Vaccines*, **7**, 1173–1183.

16

A dsRNA-binding protein MdDRB1 associated with miRNA biogenesis modifies adventitious rooting and tree architecture in apple

Chun-Xiang You[†], Qiang Zhao[†], Xiao-Fei Wang, Xing-Bin Xie, Xiao-Ming Feng, Ling-Ling Zhao, Huai-Rui Shu and Yu-Jin Hao*

National Key Laboratory of Crop Biology, National Research Center for Apple Engineering and Technology, College of Horticulture Science and Engineering, Shandong Agricultural University, Tai-An, Shandong, China

Correspondence

email haoyujin@sdau.edu.cn
[†]These authors contributed equally to this study.
Accession numbers: *MdDRB1* (GenBank accession number KC857648)

Keywords: apple, double-strand RNA-binding protein, miRNAs, adventitious rooting, leaf curvature, tree architecture.

Summary

Although numerous miRNAs have been already isolated from fruit trees, knowledge about miRNA biogenesis is largely unknown in fruit trees. Double-strand RNA-binding (DRB) protein plays an important role in miRNA processing and maturation; however, its role in the regulation of economically important traits is not clear yet in fruit trees. EST blast and RACE amplification were performed to isolate apple *MdDRB1* gene. Following expression analysis, RNA binding and protein interaction assays, *MdDRB1* was transformed into apple callus and *in vitro* tissue cultures to characterize the functions of MdDRB1 in miRNA biogenesis, adventitious rooting, leaf development and tree growth habit. MdDRB1 contained two highly conserved DRB domains. Its transcripts existed in all tissues tested and are induced by hormones. It bound to double-strand RNAs and interacted with AtDCL1 (Dicer-Like 1) and MdDCL1. Chip assay indicated its role in miRNA biogenesis. Transgenic analysis showed that *MdDRB1* controls adventitious rooting, leaf curvature and tree architecture by modulating the accumulation of miRNAs and the transcript levels of miRNA target genes. Our results demonstrated that *MdDRB1* functions in the miRNA biogenesis in a conserved way and that it is a master regulator in the formation of economically important traits in fruit trees.

Introduction

MicroRNAs (miRNAs) play multiple important regulatory roles in plants (Jones-Rhoades *et al.*, 2006). They are generally 20–24 nucleotides in length and control post-transcriptional mRNA stability, translation or target epigenetic modification to specific regions of the genome by complementarily binding to target nucleic acids (Brodersen and Voinnet, 2009). Mature miRNAs silence or down-regulate their target genes at post-transcriptional or translational level, and are therefore involved in the regulations associated with plant development, growth (Mathieu *et al.*, 2009), nutrient allocation (Buhtz *et al.*, 2008) and stress responses (Zhou *et al.*, 2008).

In plants, miRNA biogenesis and functional pathways have been partly described. The miRNA genes are firstly transcribed as long pri- and pre-miRNA transcripts that form foldback structures (Bartel, 2004). Following transcription, RNase III-like protein DCL1 is thought to catalyse the processing of pri-miRNAs and pre-miRNAs, subsequently producing a wide variety of miRNAs that control the expression of various important genes, such as transcription factors and development-related genes (Park *et al.*, 2002; Reinhart *et al.*, 2002). Genetic evidence suggests that the double-stranded RNA (dsRNA)-binding domain (dsRBD) of DCL1 is essential for its *in vivo* function (Jacobsen *et al.*, 1999; Schauer *et al.*, 2002). In addition, other double-stranded RNA-binding proteins, such as HYL1 and HEN1, are involved in miRNA biogenesis and the resultant RNA silencing as crucial components of the RNA-induced silencing complex (RISC).

HYPONASTIC LEAVES 1 (HYL1) is the first double-stranded RNA-binding protein identified as a miRNA biogenesis protein that participates in miRNA accumulation in coordination with DCL1, HEN1 and other RISC components. The *hyl1* null mutant exhibits reduced miRNA levels and developmental defects such as reduced leaf size and severe leaf hyponasty, slow growth and reduced plant height, late flowering and reduced fertility, as well as multiple lateral shoots, giving a bushy phenotype. Homozygous *hyl1* mutant plants also showed reduced sensitivity to exogenous cytokinin and auxin, as well as additional auxin phenotypes that include reduced root gravitropism and apical dominance (Liu *et al.*, 2011; Lu and Fedoroff, 2000).

In *Arabidopsis*, HYL1 is also called dsRNA-binding protein1 (DRB1). It has four other homologues, that is, DRB2, DRB3, DRB4 and DRB5. In the SAM region, DRB2 plays a regulatory role in miRNA biogenesis and therefore adds an additional layer of gene regulatory complexity in this developmentally important tissue. DRB2 and DRB4 are necessary for an appropriate accumulation of p4-siRNAs (Eamens *et al.*, 2012; Pélissier *et al.*, 2011). DRB3 and DRB5 also function in the same noncanonical miRNA pathway, just as DRB2 does (Eamens *et al.*, 2012). Furthermore, similar to the cooperation of DCL1 and HYL1 in modulating miRNA biogenesis, DRB4 interacts with DCL4, and facilitates DCL4 activity, and forms a protein complex to regulate the biogenesis of ta-siRNA and participate in defending against *Turnip yellow mosaic virus* infection (Jakubiec *et al.*, 2012). APC/C-mediated degradation of DRB4 involves in RNA silencing (Marrocco *et al.*, 2012).

In addition, the rice homologue of HYL1 is OsRBP that contains the typical dsRNA-binding domains. Its expression pattern suggests that it may be associated with embryo and seed development. *OsLMS* encodes a predicted protein containing a carboxyl-terminal domain (CTD) phosphatase domain and two double-stranded RNA-binding motifs (dsRBM), which is similar to the *Arabidopsis FIERY2/CPL1* gene, and involve in many plant processes such as stress response and development (Tang *et al.*, 2002; Undan *et al.*, 2012). In Chinese cabbage, BcpLH, as a HYL1 homologue, plays a role in upward and inward curvature of folding leaves in Chinese cabbage (Yu, 2005), indicating its involvement in the regulation of economically important traits in horticulture crops.

Recently, many mature miRNAs or pre-miRNAs have been isolated in fruit trees such as grapevine (Pantaleo *et al.*, 2010), citrus (Xu *et al.*, 2010), apple (Gleave *et al.*, 2008) and strawberry (Ge *et al.*, 2012); however, limited is known about the RISC components and their roles in economically important traits of fruit trees. In this study, MdDRB1, an apple homologue of *Arabidopsis* DRB1/HYL1, was isolated and identified. Its functions in miRNA biogenesis and consequently in the regulation of adventitious rooting and tree architecture were characterized.

Results

Molecular cloning, sequence identity and expression analysis of MdDRB1 in apple

To clone *MdDRB1* gene, we firstly blasted the genome database using arabidopsis *AtDRB1/AtHYL1* (At1g09700) as information probe. However, it was found that there is an assembly error at the homologous locus (Fig. S1). Then, we blasted through the GenBank database for homologous ESTs of *AtHYL1*. EST CN880177 showed the highest similarity to *AtDRB1* and therefore was chosen as core sequence for further BLAST search through the apple EST database. Subsequently, the obtained ESTs were aligned with DNAMAN software. The resultant contig was then used as probe for another round of BLAST search for more overlapped EST sequences. The BLAST search and alignment were repeated till the new contig could not be prolonged anymore.

Consequently, a 580-bp homologous fragment was acquired. After confirmed with sequencing, this fragment was used for primer design for RACE amplification. Finally, a full-length cDNA in 1853 bp was obtained and confirmed with sequencing. Hereafter, it was named as *MdDRB1* (GenBank accession number KC857648). *MdDRB1* cDNA contained an open reading frame (ORF) of 1479 bp in length as well as 3'- and 5'-UTRs of 137 and 237 bp, respectively. Its ORF encoded a deduced protein containing 493 amino acid residues with a predicated molecular weight of 52.2 kD and a PI of 5.40.

The predicted MdDRB1 protein contains two typical dsRNA-binding motifs, that is, Ds-RBM1 and Ds-RBM2. In addition, MdDRB1 protein also contained two repeats of HDAGAVLLPY and HDVAAVLLPC, which appeared two times in the C-terminal end (Figure 1a). Compared with other DRB proteins, MdDRB1 is closely related to the Chinese cabbage BcpLH with 75.31% amino acid identity over Ds-RBMs and 31.1% identity to the entire protein, followed by *Arabidopsis* HYL1 protein with 72.96% similarity over Ds-RBMs and 33.7% to the entire protein.

The amino acid sequences of 16 *Arabidopsis* dsRBM-containing proteins and 9 apple DRB proteins were aligned using Clustal X

1.83, and the tree was constructed using MEGA 4.0 with the neighbour-joining (NJ) method with 1000 bootstrap replicates. As a result, the phylogenic tree showed MdDRB1 in the same clade as BcpLH and AtHYL11, but far from the clade as other dsRNA-binding proteins such as AtDCL (Figure 1b).

Semi-quantitative RT-PCR was conducted to analyse the expression pattern of *MdDRB1* gene. The results indicated that *MdDRB1* transcripts were constitutively present in all organs tested. Generally, *MdDRB1* was expressed at a higher level in the root than others. In addition, the expressions of *MdDRB1* gene were analysed in *in vitro* tissue cultures treated with IAA and 6-BA. The result showed that its expression levels were noticeably enhanced by auxin and cytokinin (Figure 1c), indicating its involvement in the responses to auxin and cytokinin in apple.

MdDRB1 protein binds to dsRNAs and interacts with DCL1s

To examine if MdDRB1 protein binds dsRNA, mobility shift assays were conducted using probes of dsRNA, ssRNA, dsDNA and ssDNA that were designed and synthesized according to the *18S* RNA sequence. The result showed that MdDRB1 protein bound only to dsRNA probe, but not to ssRNA, dsDNA and ssDNA probes (Figure 2a). Therefore, MdDRB1 protein preferentially binds dsRNA.

To verify if MdDRB1 interacts with RISC component such as AtDCL1 and MdDCL1 proteins, yeast two-hybrid assays were carried out. Protein MdDRB1 was used as bait in the vector pGBT9, while each of C-terminal polypeptides of AtDCL1 and MdDCL1, that is, AtDCL1(ΔN) and MdDCL1(ΔN) as prey in the vector pGAD424. The result showed that positive α-gal activity was observed in yeast containing either pGBT9-MdDRB1 plus pGAD424-AtDCL1(ΔN) or pGBT9-MdDRB1 plus pGAD424-MdDCL1(ΔN) grown on -T/-L/-H/-A screening medium, but not in those containing pGBT9-MdDRB1 plus the empty pGAD424 vector (Figure 2b,c), indicating that MdDRB1 protein interacted with DCL1 proteins at N-terminal.

Mis-expression of *MdDRB1* influences miRNA processing in apple callus

To characterize the function of *MdDRB1*, transgenic apple callus mix expressing *MdDRB1* transcripts was obtained using agro-mediated transformation with sense and antisense *MdDRB1* genes, respectively. As a result, *MdDRB1* transcripts were increased in sense callus, while decreased in antisense callus (Figure 3a). Subsequently, the antisense transgenic apple callus was used to examine the accumulation levels of miRNAs with miRNA microarray assay using a plant miRNA chip including a total of 427 miRNA corresponding probes. The result showed that 169 miRNAs were abnormally accumulated in antisense transgenic callus compared with the pBI control. Among them, 95 miRNAs such as miR153, miR156, miR157, miR165, miR399, were remarkably down-regulated, while the other 74 such as miR159, miR168, miR172, miR393, miR398, miR408 up-regulated, indicating the involvement of *MdDRB1* in miRNA formation in apple cells (Figure 3b,c; see Data S1 online).

Semi-quantitative stem-loop RT-PCRs were performed to confirm the abnormal accumulation of four miRNAs including miR153, miR165, miR398 and miR408 in transgenic apple calluses. The result showed that miRNAs miR165 and miR153 were down-regulated in antisense callus, but up-regulated in sense callus. In contrast, miR398 and miR408 were up-regulated

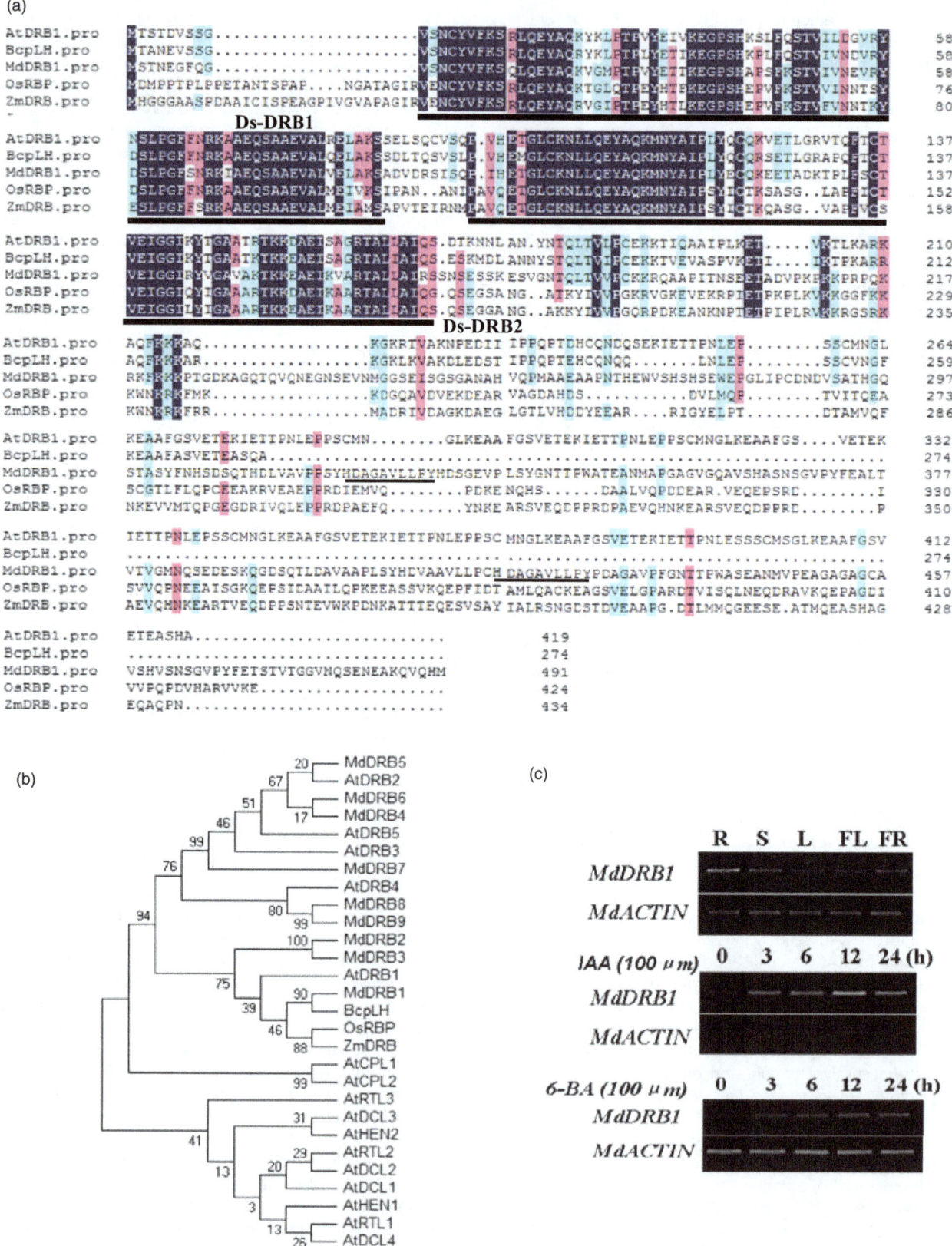

Figure 1 Sequence identity and expression analysis of MdDRB1. (a) Multiple sequence alignment of MdDRB1 with other double-strand RNA-binding (DRB) proteins in plants. The underlined amino acid residues indicate two Ds-DRB motifs (black lines) and two repeats (red lines), respectively. Gaps to optimize alignments are designated by dots. The consensus amino acid identity among all organisms is highlighted with black colour. Amino acids are numbered on the right side of the sequence. (b) Phylogenetic relationship of in MdDRB1 with other DRB proteins. The numbers beside the branches indicate the bootstrap values that support the adjacent node. (c) Expression analysis of *MdDRB1* gene in different organs, as well as in response to IAA and 6-BA treatments. The transcript levels of *MdACTIN* gene were used as loading control. R, root; S, shoot; L, leaf; FL, flower; FR, fruit.

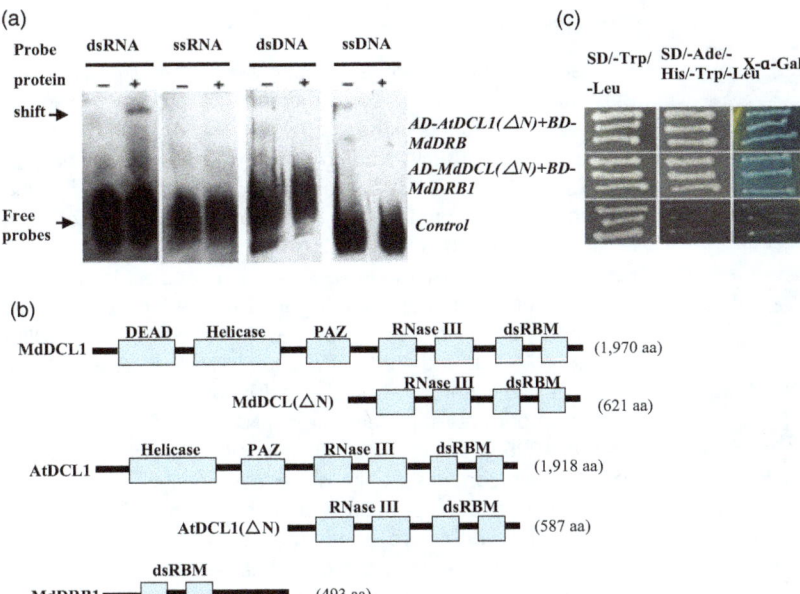

Figure 2 MdDRB1 protein binds to dsRNAs and interacts with DCL1s. (a) EMSA assays of the binding activities of MdDRB1 protein to dsRNA, ssRNA, dsDNA and ssDNA. (b) Schematic drawings of MdDCL1, AtDCL21 and MdDRB1 proteins. Abbreviations of domains are as follows: dsRBM, dsRNA-binding motif; RNase III, ribonuclease III domain; NLS, nuclear localization signal; PAZ, Piwi/Argonaute/Zwille domain; Helicase, Helicase-conserved C-terminal domain; DEAD, DEAD/DEAH box. The C-terminal halves of Dicer-like (DCL) proteins containing two RNase III domains and two dsRBMs (designated as Short DCL) were used for yeast two-hybrid (Y2H) assays. (c) Y2H assays for the interaction between MdDRB1 and MdDCL1/AtDCL1 with selective growth combined with an X-α-Gal overlay assay.

in antisense callus, but down-regulated in sense callus, in agreement with the result of miRNA chip assay (Figure 3d).

MbDRB1 promotes adventitious root formation by modulating miRNAs and their target genes in transgenic apple

A wild crabapple *Malus baccata*, which is widely used as rootstock in China, was used as plant material for functional characterization *in planta*. Firstly, the full-length cDNA of *MbDRB1*, the homologue of *MdDRB1* in *M. baccata*, was cloned. The sequence analysis showed that *MbDRB1* and *MdDRB1* was the same gene. To further characterize its function *in planta*, *MbDRB1* gene driven by a CaMV 35S promoter was genetically transformed into *M. baccata*. As a result, *in vitro* cultures of transgenic lines were obtained. Three transgenic lines L1, L2 and L3 were chosen for phenotypical observation and further investigation. Firstly, semi-quantitative RT-PCRs were conducted to check the transcript levels of *MbDRB1*. The result showed that all three lines produced more *MbDRB1* transcripts than the nontransgenic control, indicating that *MbDRB1* gene was overexpressed in three transgenic lines. Among them, line L3 produced the highest level of *MbDRB1* transcripts, while L1 generated the lowest level (Figure 4c).

Subsequently, *in vitro* microcuttings of three transgenic lines were used for rooting test. When the microcuttings were transferred onto 1/2MS medium without hormones, all three transgenic lines formed adventitious roots, while the control did not at all. On 1/2MS medium plus 0.1 mg/L IBA, all three transgenic lines rooted, while only 25% control rooted. Besides the rooting frequencies, three transgenic lines also generated more adventitious roots per microcuttings than the nontransgenic controls (Figure 4a,b). The rooted plantlets were transferred to pots containing a mixture of soil : perlite (1 : 1) and placed in a greenhouse. Two-month-old self-rooted plantlets were used for the observation on adventitious root growth in soil. The result showed that three transgneic lines formed robust root systems with increased root number and size, compared with the controls (Figure 4c). In addition, the *in vitro* rooting frequencies, root number and size were positively correlated with the transcript levels

of *MbDRB1* genes in transgenic lines, indicating that *MbDRB1* overexpression promoted root formation in transgenic lines.

It is well known that auxin plays a crucial role in adventitious root formation (Smet *et al.*, 2006). To clarify the molecular mechanism by which *MbDRB1* influences adventitious root formation, the accumulation levels of three auxin-associated miRNAs, that is, miR160, miR164 and miR393, were analysed with stem-loop RT-PCRs. The result showed that three transgenic lines accumulated less miR160, miR164 and miR393 than the nontransgenic control and that *MbDRB1* transcript levels were negatively associated with miRNA levels in three transgenic lines, indicating that *MbDRB1* inhibits the accumulation of three miRNAs (Figure 4d).

Furthermore, the expression levels of five genes, including *MbARF10* and *MbARF16* targeted by miR160, *MbNAC1* by miR164, and *MbTIR1* by miR393 (Figure S2), were examined with semi-quantitative RT-PCRs. The result showed that the expression levels of five genes were negatively related to the accumulation levels of the corresponding miRNAs and therefore increased in three transgenic lines compared with the nontransgenic control (Figure 4e).

In addition, the auxin concentrations were determined in three transgenic lines and the WT control. The result showed that three transgenic lines produced less auxin than the WT control, and that the auxin concentrations were negatively correlated with the expression levels of *MbDRB1* (Figure 4f).

MbDRB1 transgenic apple plants produce curly leaf and exhibit columnar-like architecture

Five-month-old self-rooted plants were observed. It was found that the leaves were down-curled in transgenic plants, compared with the WT control (Figure 5a). In *Arabidopsis*, miR165 and its target genes influence leaf curvature (Zhou *et al.*, 2007). Therefore, semi-quantitative stem-loop RT-PCR and semi-quantitative RT-PCR were conducted to analyse the expression levels of miR165 and its candidate apple targets *MbREV*, *MbPHB* and *MbHB15*, respectively (Fig. S2). The result showed that miR165 was up-regulated while target genes *MbREV*, *MbPHB* and *MbHB15* down-regulated, indicating that *MbDRB1* inhibits the

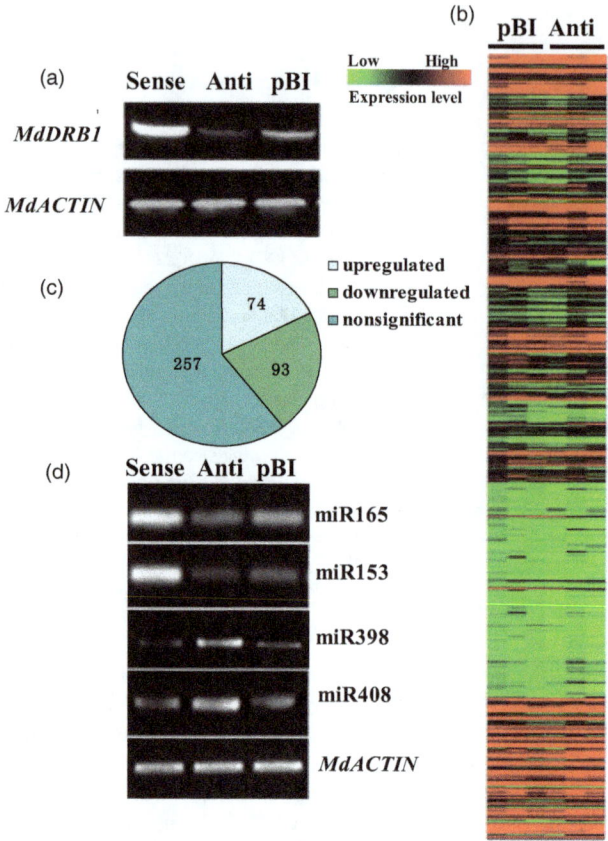

Figure 3 The miRNA microarray assay of MdDRB1 transgenic apple calluses. (a) Semi-quantitative RT-PCR analysis of *MdDRB1* gene in transgenic calluses. 'Sense' and 'anti' mean that transgenic callus ectopically expressing *MdDRB1* in sense and antisense orientations, respectively. Empty pBI vector was used as control. (b) Heat map of 426 miRNAs affected in transgenic apple callus ectopically expressing an antisense *MdDRB1* fragment. (c) Classification of microarray-derived miRNAs into three categories based on the respective fold changes. (d) Stem-loop RT-PCR analysis of mature miR165, miR153, miR398 and miR408. *MdACTIN* was used as loading control.

expressions of three target genes by enhancing miR165 level, finally influencing leaf curvature (Figure 5b).

In addition, 5-month-old transgenic plants also exhibited columnar-like features such as less branches and strong stem compared with the control (Figure 5a). Interestingly, the columnar features became more noticeable in 1-year-old transgenic trees. For example, more than 90% lateral buds sprouted, with most of them becoming spurs <10 cm in length, while only few becoming branches where the secondarily lateral buds sprouted too. In contrast, in 1-year-old nontransgenic tree, only 15% produced branches without secondary spur, with the most lateral buds keeping nonsprouting (Figure 5c,d). Furthermore, 3-year-old nontransgenic and transgenic trees blossomed. Almost all spurs along the main shoot generated floral buds and got blossomed in the transgenic trees, while some buds at the similar position failed to form spur and generate floral buds in the nontransgenic trees (Figure 5e). Finally, semi-quantitative RT-PCRs were conducted to examine the expression levels of *MbLBD1*, *MbLBD2*, *MbLBD3* and *MbARF7*, which may be related to columnar growth habit in apple. The results showed that the transcript levels of all four genes enhanced in three transgenic lines compared with the control (Figure 5f).

Discussion

The dsRNA-binding proteins ubiquitously exist in various organisms and play crucial roles in multiple life processes (Eamens *et al.*, 2009; Saunders and Barber, 2003). All dsRNA-binding proteins contain dsRNA-binding domains or motifs. *Arabidopsis* AtHYL1/AtDRB1 and AtDCL1 are well-dissected dsRNA-binding proteins, which contain two dsRNA-binding (DRB) motifs at N- and C-terminals, respectively (Hiraguri *et al.*, 2005; Lu and Fedoroff, 2000). Just like AtDRB1/AtHYL1, apple MdDRB1 has two DRB motifs at N-terminal (Figure 1a). The highly conserved DRB motifs are specifically binds dsRNA, but not ssRNA, dsDNA and ssDNA (Lu and Fedoroff, 2000; Hiraguri *et al.*, 2005; Figure 2a). The unique structure and binding characteristic are crucially important for pri- and pre-miRNA processing (Krovat and Jantsch, 1996; Yang *et al.*, 2010). In addition, *Arabidopsis* DRB1/ HYL1 protein contains six almost perfect repeats of a 28-amino acid sequence at C-terminal (Lu and Fedoroff, 2000), while MdDRB1 has only two short repeats. Meanwhile, Chinese cabbage BcpLH does not contain repeated sequence, although it shares high identity in two N-terminal dsRNA-binding motifs with AtDRB1/AtHYL1 and MdDRB1, and therefore shows similar function in leaf curvature (Lu and Fedoroff, 2000; Yu, 2005; Figure 1a).

In *Arabidopsis*, *DRB1/HYL1* cooperates with *DCL1*, *SE* and *CPL2/FRY2* forming a RISC complex to process miRNA biogenesis (Hiraguri *et al.*, 2005; Kurihara *et al.*, 2006; Manavella *et al.*, 2012; Yang *et al.*, 2006). Its N-terminal DRB domains are sufficient for pre-miRNA processing (Wu *et al.*, 2007), which explained why the C-terminal difference exerts less impact on the miRNA biogenesis among different plant species (Figure 1a; Yu, 2005). The mutant *hyl1* exhibits altered miRNA accumulation (Han *et al.*, 2004). Similarly, MdDRB1 interacts with RISC component DCL1s and involves in the regulation of miRNA biogenesis (Figure 2c).

As a key component of the protein complex RISC, dsRNA-binding protein plays important roles in plant growth and development by modulating miRNA accumulation. HYL1 and its homologue such as BcpLH are characterized with functions in the maintenance of venation and polarity of growing leaves by altering the level of miR165 that targets and cuts HD-ZIP III genes such as *REV* and *PHB* (Lu and Fedoroff, 2000; McConnell and Barton, 1998; Yu *et al.*, 2005). Similarly, *MbDRB1* (*MdDRB1*) overexpression enhanced the accumulation of miR165 and therefore decreased the transcript levels of its target genes *MbREV*, *MbPHB* and *MbHB15*. Usually, overexpression of *REV* or *PHB* leads to upward curvature of leaves, whereas underexpression sometimes causes downward curvature (Liu *et al.*, 2011). Therefore, *MbDRB1* transgenic apple plants produced downwardly growing leaves (Figure 5a).

In addition, the expression levels of *MdDRB1* and its homologues are remarkably induced by auxin and cytokinin (Figure 1c; Lu and Fedoroff, 2000; Yu *et al.*, 2000). Interestingly, *MdDRB1* transgenic plants enhanced the capacity to generate adventitious root and exhibited columnar-like growth habit, which also reminds of phenotypes associated with auxin and cytokinin. Auxin and cytokinin are two crucial phytohormones that are involved in the modulation of root growth and growth habit in plants (Lu and Fedoroff, 2000; Muller and Sheen, 2007). In this study, *MbDRB1* (*MdDRB1*) transgenic apple accumulated less miR160, miR164 and miR393 and more transcripts of their target genes, which enhances the capacity to regenerate adventitious

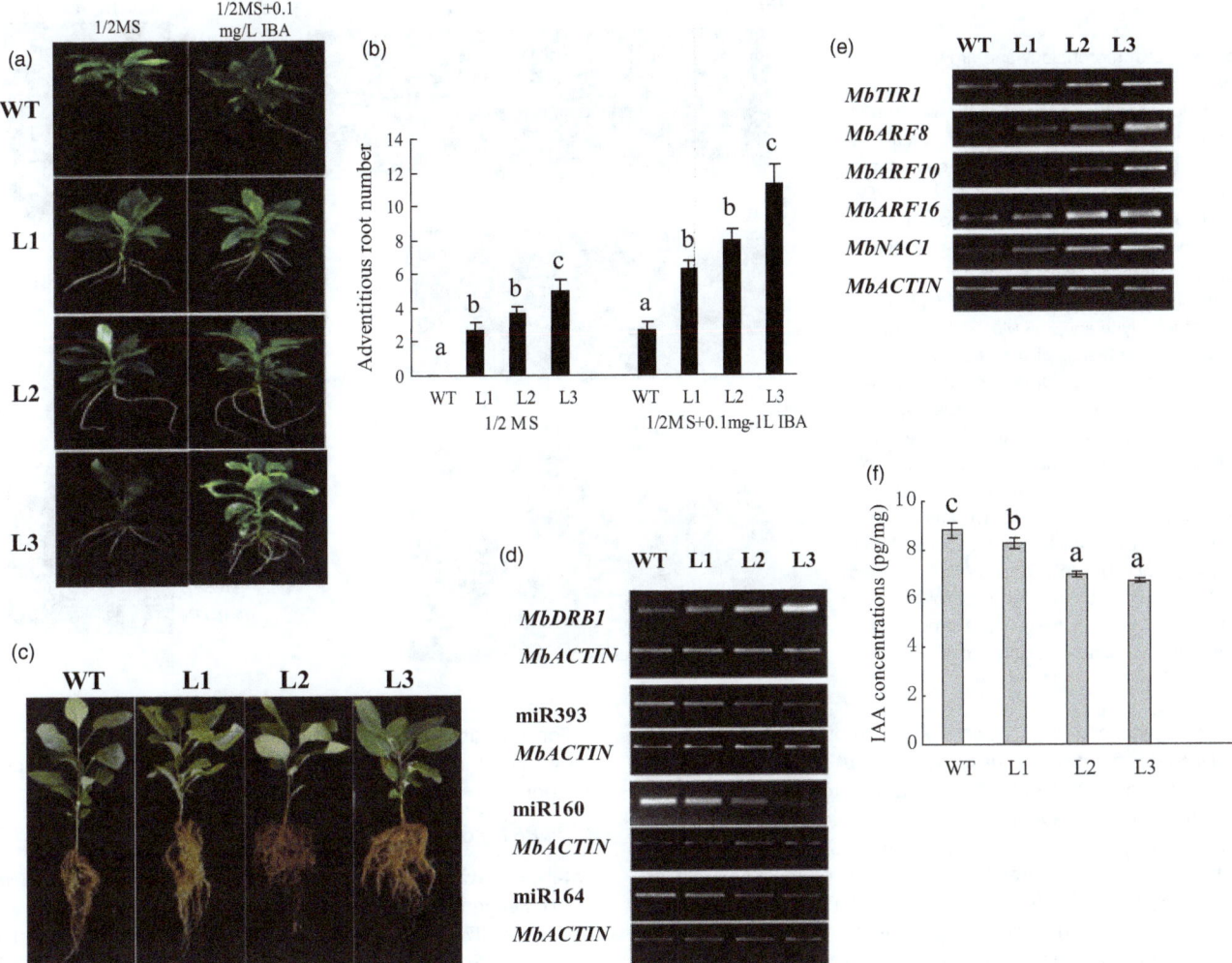

Figure 4 Adventitious root formation in *MbDRB1* transgenic apples. (a) Ectopic expression of *MbDRB1* gene promoted the capacity to regenerate adventitious roots from the microcuttings of transgenic lines L1, L2 and L3. (b) The adventitious root number of wild-type and *35S:MbDRB1* transgenic plants L1, L2 and L3. Data are means of three different independent replicates. Different letters indicate a significant difference ($P < 0.05$), respectively, according to the Duncan's Multiple Range Test. (c), *MbDRB1* transgenic plantlets exhibited powerful root systems compared with the WT control. (d) Stem-loop RT-PCR analysis of mature miRNAs, that is, miR393, miR160, miR167 and miR164, which are associated with auxin response and adventitious rooting. (e) Semi-quantitative RT-PCR analysis of genes that are targeted by those miRNAs. The expression of *MbACTIN* was used as loading control. (f)Free IAA measurement of wild-type and three independent transgenic lines L1, L2 and L3. Data are means of three different independent replicates. Different letters indicate a significant difference ($P < 0.05$), respectively, according to the Duncan's multiple range test.

roots (Figure 4a,b). In *Arabidopsis*, these miRNAs and their target genes associated with auxin and cytokinin signals were involved in the regulation of root formation and development. For example, ARF10 and ARF16, targeted by miR160, control root tip growth and gravity sensing (Mallory *et al.*, 2005), while NAC1, targeted by miR164, transmits auxin signals for lateral root development (Guo *et al.*, 2005). Furthermore, the regulatory pathway miR393-TIR1-NAC1, which likely interacts with the miR164-NAC1 pathway, may involve in lateral root development (Xie *et al.*, 2000). Therefore, auxin and cytokinin control root growth partially by regulating the expression level of MbDRB1 and the accumulation of few miRNAs. In addition, MbDRB1 positively regulates the expression level of *MbTIR1* which is an auxin receptor, indicating MbDRB1 transgenic lines are more sensitive to auxin signal. As a result, MbDRB1 overexpression decreases the auxin concentrations in transgenic lines probably by a feedback regulation pathway.

More interestingly, *MbDRB1* (*MdDRB1*) transgenic apple plants showed columnar-like features (Figure 5c). The first apple cultivar with columnar growth habit is 'Wijcik' which is a spontaneous sport of 'McIntosh' and controlled by the dominant *Co* locus (Lapins, 1969, 1976; Tobutt, 1985). From then on, numerous attempts have been made to identify *Co* locus. Most recently, high throughout expression analyses found that *Co* locus alters the expression levels of numerous genes, especially hormone-related ones (Krost *et al.*, 2013; Zhang *et al.*, 2012). Meanwhile, Bai *et al.* (2012) propose that *MbLBD1*, *MbLBD2* and *MbLBD3* in chromosome 10 are strong candidates for *Co* locus. Three *MbLBDs* are closely related to *ASL16/LBD29* and *ASL18/LBD16*, which regulate lateral organ development in Arabidopsis and other plant species (Bai *et al.*, 2012). In addition, *ARF7/19* acts upstream *ASL16/LBD29* and *ASL18/LBD16* (Okushima *et al.*, 2007). In *MbDRB1* (*MdDRB1*) transgenic apple, *MbARF7*, *MbLBD1*, *MbLBD2* and *MbLBD3* genes were up-regulated (Fig-

Figure 5 Leaf morphology and growth habit of *MbDRB1* transgenic apple plants. (a) Leaf morphology and growth habit of the 5-month-old transgenic plantlets. (b) Stem-loop RT-PCR analysis of mature miRNA miR165 and semi-quantitative RT analysis of its target HD-ZIP III genes *MbREV*, *MbPHB* and *MbHB15*, which are involved in the regulation of leaf curvature. (c) One-year-old *MbDRB1* transgenic apple plants exhibited columnar-like growth habit. (d) Bud germination ratios of wild-type and *35S:MbDRB1* transgenic plants. Data are means of three different independent replicates. Different letters indicate a significant difference (*P* < 0.01), respectively, according to the Duncan's multiple range test. (e) All spurs along the main shoots of 3-year-old *MbDRB1* transgenic apple trees blossomed in a way similar to columnar apple. (f) Semi-quantitative RT-PCR analysis of four genes, that is, *MbARF7*, *MbLBD1*, *MbLBD2* and *MbLBD3*, which may be involved in the regulation of columnar-like growth habit of apple trees. The expression of *MbACTIN* was used as loading control.

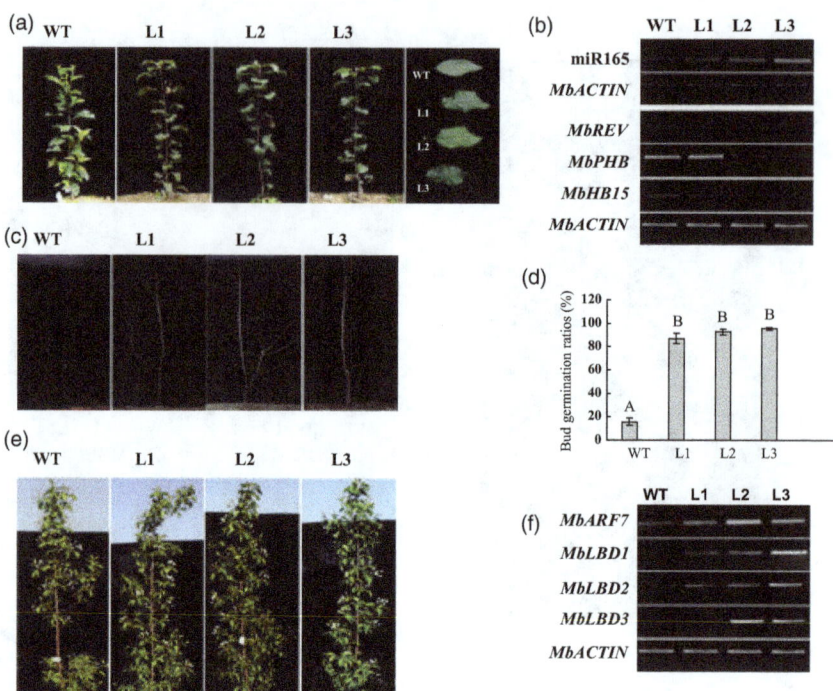

ure 5f). Therefore, it is reasonable to propose that *MbDRB1* (*MdDRB1*) regulates an unknown miRNA that subsequently modulates the MbARFs-MbLBDs pathway to control columnar growth habit in transgenic apple.

In *Arabidopsis*, the mutant *hyl1* exhibits delayed flowering probably due to the altered accumulation of miRNAs such as miR156 and miR172 (Lu and Fedoroff, 2000; Li *et al.*, 2011). However, MbDRB1 transgenic apple lines do not show noticeable alteration in phase transition from juvenile to adult and flower bud differentiation (data not shown).

Although *HYL1* and its homologues have been well functionally characterized in model plants, only few cases are focused on its role in the formation of economically and agronomically important traits in crops. For example, *BcpLH*, a *HYL1* homologue in Chinese cabbage, plays a crucial role in the formation of leafy head (Yu *et al.*, 2000; Yu, 2005). *DRB4* is involved in the defence against Turnip yellow mosaic virus infection (Jakubiec *et al.*, 2012). In this study, *MbDRB1* (*MdDRB1*) transgenic plants exhibit high capacity to regenerate adventitious root and columnar-like growth habit, both of which are desired traits in apple cultivar and rootstock breeding programs. Adventitious root formation is a key step for vegetative propagation, which is the main method for propagation of many horticultural crops, especially woody species (Smolka, 2009). Columnar growth habit features with a compact tree architecture with fruit spurs instead of lateral branches (Krost *et al.*, 2013), therefore being considered as a desired trait suitable for efficient high-density production systems in apple.

Experimental procedures

Plant materials and treatments

The *in vitro* shoot cultures of apple 'Gala' were used for gene cloning, while *M. baccatais* for gene genetic transformation. They were subcultured at a 4-week interval at 25 °C under long-day conditions (16 h light/8 h dark) on MS medium containing 0.5 mg/L 6-BA, 0.2 mg/L NAA and 0.2 mg/L GA.

For hormone treatment, 100 μm IAA and 100 μm 6-BA were sprayed on the leaves, respectively, while distilled water was used as control.

MdDRB1 isolation and sequences analysis

Total RNA was extracted from 'Gala' tissue cultures using the TRizol Reagent (Invitrogen, Carlsbad, CA) and then treated with DNase I at 37 °C for 30 min. First-strand cDNA was synthesized with SuperScriptTM III reverse transcriptase (Invitrogen). The forward primers pMdDRB1-S and pMdDRB1-A, which correspond to the conserved motifs of plant DRB1 proteins, were used to amplify target gene. The PCR product amplified with Ex Taq DNA polymerase (TaKaRa, Dalian, China) was cloned into the T-cloning vector PMD-18T Vector (TaKaRa). Both 3' and 5' end fragments were obtained according to the 3'- and 5'-Full RACE Core Sets (TaKaRa). The full-length MdDRB1 cDNA was amplified using primers MdDRB-S and MdDRB-A. Sequence alignment analysis was performed with *CLUSTALW* (http://www.ebi.ac.uk/Tools/msa/clustalw2) and *BOXSHADE* (http://www.ch.embnet.org/software/BOX_form.html). The dendrogram was produced with MEGA4.1.

Stem-loop RT-PCR and semi-quantitative RT-PCR

Total RNAs were isolated from the roots, shoots, leaves, flowers, fruits and seeds of a mature 'Gala' apple tree. Stem-loop RT-PCRs were performed to determine the accumulation level of mature miRNAs as described by Chen *et al.* (2005). The procedure contained two steps: stem-loop reverse transcription (RT) and PCR. Firstly, the cDNA synthesis was performed using the prime script first-strand cDNA Synthesis Kit (Takara) according to the manufacturer's instructions using a stem-loop RT primer instead of an oligo (dT) primer (Table S1). Subsequently, PCR was performed to detect the accumulation level of mature miRNAs using the miRNAs-specific forward primer and the stem-loop-specific reverse primer (Table S1). Each PCR reaction mixture contained 200 ng cDNA, 2.5 μL 10× Taq buffer, 2 μL dNTPs

(2.5 mM), 1 μL of each primer (10 μM) and 0.5 μL high-fidelity Taq DNA polymerase (TransGen, Beijing, China) in a total volume of 25 μL. The reactions were initially denatured at 94 °C for 5 min followed by 30 cycles at 94 °C for 30 s, 58 °C for 30 s and 72 °C for 30 s, and a final cycle at 72 °C for 5 min.

Semi-quantitative RT-PCR was performed to examine the expression levels of the genes. The final reaction mixture and reactions were the same as those used in the stem-loop RT-PCR. MdACTIN was used to normalize cDNA loading. The primer sequences used were listed in Tables S1 and S2.

5′ RACE analysis of mRNA cleavage

Total RNA was extracted from transgenic plants using TRizol reagent and Poly(A)+ mRNA purified using a PolyA purification kit (Invitrogen). RNA ligase-mediated 5′ RACE was performed using the GeneRacer kit (Invitrogen). The GeneRacer RNA Oligo adapter was directly ligated to mRNA (100 ng) without calf intestinal phosphatase and tobacco acid pyrophosphatase treatment. Initial PCR was performed with the GeneRacer 5′ primer and gene-specific primers. Nested PCR was performed with 1 μL of the initial PCR reaction, the GeneRacer 5′ nested primer and gene-specific internal primer. After the second amplification, PCR products were gel-purified and cloned into PMD-18T (Takara, China) vector. Ten independent clones were sequenced. The primer sequences used were listed in Table S2.

Binding activity assay

The ORF of MdDRB1 was amplified using the primer pair MdDRB1-S and MdDRB1-A (Table S2) and then inserted into vector pGEX-4T-1 (Amersham Biosciences, Chalfont St. Giles, UK) to generate the MdDRB1-GST fusion construct. The construct was transformed into E. coli strain BL-21 cells. After growing to saturation, the trans-formed cells were treated with isopropyl-β-D-thiogalactoside to induce protein expression and then purified using a Pierce GST Spin Purification Kit (Thermo, Brookfield, WI).

EMSAs were performed using biotin-labelled probes and the Light shift Chemiluminescent EMSA kit (Pierce, Rockford, IL) according to the manufacturer's instructions. A volume containing 50 ng of recombinant GST-MdDRB1 was purified for the binding reactions. Synthetic oligonucleotides for the dsRNA, ssRNA, dsDNA and ssDNA were biotinylated and heated at 100 °C for 5 min and then allowed to anneal to the double-stranded oligonucleotides for 12 h at room temperature. The binding reactions were carried out in a total volume of 20 μL of a solution containing 25 mM HEPES-KOH (pH 7.5), 100 mM KCl, 0.1 mM ethylenediaminetetraacetic acid (EDTA), 17% glycerol, 1 mM DTT, 4 μg of poly (dI-dC), 50 ng of purified protein and 1 pmol of labelled probe. The assay mixtures were incubated for 20 min at room temperature. Then, the reaction mixtures were layered on 6% acrylamide gels containing 0.5% TBE buffer and 3.6% glycerol. After a prerun in the 0.5% TBE buffer for 1 h at 100 V at room temperature, the samples were electropho-resed for an additional 2 h under the same conditions. Following native polyacrylamide gel electrophoresis, the DNA was transferred to positively charged nylon membranes (Hybond N+; Amersham, Little Chalfont, Buckinghamshire, UK), and the signal was detected using the chemiluminescent nucleic acid detection method (Pierce).

Yeast two-hybrid (Y2H) assays

Gal4-based two-hybrid system was used as described by the manufacturer (Clontech, Palo Alto, CA). The full-length encoding region of MdDRB1 was amplified with RT-PCR and cloned into the DNA binding domain vector pGBT9 to make the bait plasmid

pGBT9-MdDRB1. The coding sequence, each of AtDCL1 and MdDCL1, was fused to the two-hybrid activation domain vector (pGAD424) as prey. The primers used to create these constructs were listed in Table S2. Each pGAD424 (AtDCL1 and MdDCL1) was separately cotransformed with pGBKT7-MdDRB1 into yeast strain Y2H for a yeast two-hybrid test. Positive colonies were selected on SD/-Trp–Leu–His–Ade medium. α-Galactosidase filter assays were conducted using the substrate 5-bromo-4-chloro-α-D-galactoside (X-α-Gal) as described by Halfter et al. (2000).

MicroRNA microarray

Wild-type and transgenic calluses of the apple cultivar 'Orin' were maintained at 3-week intervals on MS medium containing 0.5 mM 2,4-D and 1.5 mM 6-BA in the dark at 25 °C. RNAs were extracted using the RNA Plant Plus Reagent (TIANGEN, Beijing, China). The Affymetrix GeneChip miRNA array was done by Shanghai Biotechnology Co., Ltd. (Shanghai, China). Plant miRNA microarray included 426 miRNA probes. The data were analysed with software MeV.

Vector construction and genetic transformation

The full-length cDNA of MbDRB1 was amplified using the primer pairs MdDRB-S and MdDRB-A and then cloned into the expression vector pBIN, fusing to GFP-tag under the control of a 35S promoter (Table S2). A 318-bp-specific fragment of MdDRB1 in the 3′-UTR was amplified using the primer 5′-ATATCATGATGTGGCAGCTG-3′ and 5′-ATATATGCTAGGAGAGGACTTG-3′ and cloned into pBIN to construct an MdDRB1 antisense expressing vector (Table S2). Subsequently, the resultant vectors were genetically transformed into the callus of 'Orin' apple and the tissue cultures of M. bac-catais as reported by Li et al. (2002), respectively.

Measurement of free IAA concentration

200 mg tissues of wild-type and 35S:MbDRB1 transgenic plants were harvested, weighed and then frozen in liquid nitrogen. The samples were used for free IAA measurement with a UPLC-MS/MS method as described by Fu et al. (2012).

Adventitious rooting in vitro

Three-week-old apple shoot cultures in vitro were transferred to root-inducing medium, that is, 1/2MS medium plus 0.1 mg/L IBA. Then, in vitro self-rooted plantlets were transferred to pots containing a mixture of soil/perlite (1 : 1) and placed in the greenhouse at 25 °C under long-day conditions (16 h light/8 h dark) for further investigation.

Acknowledgements

We thank Drs. Jinfang Chu, Xiaohong Sun and Cunyu Yan at Institute of Genetics and Developmental Biology, Chinese Acad-emy of Sciences, Beijing, China, for their helps in determining the IAA concentration. This work was supported by grants from NSFC (31171946), Ministry of Agriculture of China (201203075-3) and Ministry of Education of China (IRT1155). The authors have no conflict of interest to declare.

References

Bai, T., Zhu, Y., Fernández-Fernández, F., Keulemans, J., Brown, S. and Xu, K. (2012) Fine genetic mapping of the Co locus controlling columnar growth habit in apple. Mol. Genet. Genomics, 287, 437–450.

Bartel, D.P. (2004) MicroRNAs: genomics, biogenesis, mechanism, and function. Cell, 116, 281–297.

Brodersen, P. and Voinnet, O. (2009) Revisiting the principles of microRNA target recognition and mode of action. *Nature*, **10**, 141–148.

Buhtz, A., Springer, F., Chappell, L., Baulcombe, D.C. and Kehr, J. (2008) Identification and characterization of small RNAs from the phloem of *Biassica napus*. *Plant Journal*, **53**, 739–749.

Chen, C., Ridzon, D.A., Broomer, A.J., Zhou, Z., Lee, D.H., Nguyen, J.T. and Guegler, K.J. (2005) Real-time quantification of microRNAs by stem–loop RT–PCR. *Nucleic Acids Res.* **33**, e179 doi: 10.1093/nar/gni178

Eamens, A.L., Smith, N.A., Curtin, S.J., Wang, M.B. and Waterhouse, P.M. (2009) The *Arabidopsis thaliana* double-stranded RNA binding protein DRB1 directs guide strand selection from microRNA duplexes. *RNA*, **15**, 2219–2235.

Eamens, A.L., Kim, K.W., Curtin, S.J. and Waterhouse, P.M. (2012) DRB2 is required for microRNA biogenesis in *Arabidopsis thaliana*. *PLoS ONE*, **7**, e35933.

Fu, J.H., Chu, J.F., Sun, X.H., Wang, J.D. and Yan, C.Y. (2012) Simple, rapid, and simultaneous assay of multiple carboxyl containing phytohormones in wounded tomatoes by UPLC-MS/MS using single SPE purification and isotope dilution. *Anal. Sci.* **28**, 1081–1087.

Ge, A., Shangguan, L., Zhang, X., Dong, Q., Han, J., Liu, H., Wang, X.C. and Fang, J.G. (2012) Deep sequencing discovery of novel and conserved microRNAs in strawberry (*Fragaria× ananassa*). *Physiol. Plant.* **10**, 1399–3054.

Gleave, A.P., Ampomah-Dwamena, C., Berthold, S., Dejnoprat, S., Karunairetnam, S., Nain, B. and MacDiarmid, R.M. (2008) Identification and characterization of primary microRNAs from apple (*Malus domestica cv. Royal Gala*) expressed sequence tags. *Tree Genet. Genomes*, **4**, 343–358.

Guo, H.S., Xie, Q., Fei, J.F. and Chua, N.H. (2005) MicroRNA directs mRNA cleavage of the transcription factor *NAC1* to downregulate auxin signals for *Arabidopsis* lateral root development. *Plant Cell*, **17**, 1376–1386.

Halfter, U., Ishitani, M. and Zhu, J.K. (2000) The Arabidopsis SOS2 protein kinase physically interacts with and is activated by the calcium-binding protein SOS3. *Proc. Natl Acad. Sci. USA*, **97**, 3735–3740.

Han, M.H., Goud, S., Song, L. and Fedoroff, N. (2004) The *Arabidopsis* double-stranded RNA-binding protein HYL1 plays a role in microRNA-mediated gene regulation. *Proc. Natl Acad. Sci. USA*, **101**, 1093–1098.

Hiraguri, A., Itoh, R., Kondo, N., Nomura, Y., Aizawa, D., Murai, Y., Koiwa, H., Seki, M., Shinozaki, K. and Fukuhara, T. (2005) Specific interactions between Dicer-like proteins and HYL1/DRBfamily dsRNA-binding proteins in *Arabidopsis thaliana*. *Plant Mol. Biol.* **57**, 173–188.

Jacobsen, S.E., Running, M.P. and Meyerowitz, E.M. (1999) Disruption of an RNA helicase/RNaseIII gene in Arabidopsis causes unregulated cell division in floral meristems. *Development*, **126**, 5231–5243.

Jakubiec, A., Yang, S.W. and Chua, N.H. (2012) Arabidopsis DRB4 protein in antiviral defense against turnip yellow mosaic virus infection. *Plant Journal*, **69**, 14–25.

Jones-Rhoades, M.W., Bartel, D.P. and Bartel, B. (2006) MicroRNAs and their regulatory roles in plants. *Curr. Opin. Plant Biol.* **57**, 19–53.

Krost, C., Petersen, R., Lokan, S., Brauksiepe, B., Braun, P. and Schmidt, E.R. (2013) Evaluation of the hormonal state of columnar apple trees (*Malus x domestica*) based on high throughput gene expression studies. *Plant Mol. Biol.* **81**, 211–220.

Krovat, B.C. and Jantsch, M.F. (1996) Comparative mutational analysis of the double-stranded RNA binding domains of *Xenopus laevis* RNA-binding protein A. *J. Biol. Chem.* **271**, 28112–28119.

Kurihara, Y., Takashi, Y. and Watanabe, Y. (2006) The interaction between DCL1 and HYL1 is important for efficient and precise processing of pri-miRNA in plant microRNA biogenesis. *RNA*, **12**, 206–212.

Lapins, K.O. (1969) Segregation of compact growth types in certain apple seedlings progenies. *Can. J. Plant Sci.* **49**, 765–768.

Lapins, K.O. (1976) Inheritance of compact growth type in apple. *J. Am. Soc. Hort. Sci.* **101**, 133–135.

Li, D.D., Shi, W. and Deng, X.X. (2002) Agrobacterium-mediated transformation of embryogenic calluses of Ponkan mandarin and the regeneration of plants containing the chimeric ribonuclease gene. *Plant Cell Rep.* **21**, 153–156.

Li, S., Yang, X., Wu, F. and He, Y. (2012) HYL1 controls the miR156-mediated juvenile phase of vegetative growth. *J. Exp. Bot.*, **63**, 2787–2798.

Liu, Z., Jia, L., Wang, H. and He, Y. (2011) HYL1 regulates the balance between adaxial and abaxial identity for leaf flattening via miRNA-mediated pathways. *J. Exp. Bot.* **62**, 4367–4381.

Lu, C. and Fedoroff, N. (2000) A mutation in the *Arabidopsis HYL1* gene encoding a dsRNA binding protein affects responses to abscisic acid, auxin, and cytokinin. *Plant Cell*, **12**, 2351–2365.

Mallory, A.C., Bartel, D.P. and Bartel, B. (2005) MicroRNA-directed regulation of *Arabidopsis AUXIN RESPONSE FACTOR17* is essential for proper development and modulates expression of early auxin response genes. *Plant Cell*, **17**, 1360–1375.

Manavella, P.A., Hagmann, J., Ott, F., Laubinger, S., Franz, M., Macek, B. and Weigel, D. (2012) Fast-forward genetics identifies plant CPL phosphatases as regulators of miRNA processing factor HYL1. *Cell*, **151**, 859–870.

Marrocco, K., Criqui, M.C., Zervudacki, J., Schott, G., Eisler, H., Parnet, A., Dunoyer, P. and Genschik, P. (2012) APC/C-mediated degradation of dsRNA-binding protein 4 (DRB4) involved in RNA silencing. *PLoS ONE*, **7**, e35173.

Mathieu, J., Yant, L.J., Mürdter, F., Küttner, F. and Schmid, M. (2009) Repression of flowering by the miR172 target *SMZ*. *PLoS Biol.* **7**, e1000148.

McConnell, J.R. and Barton, M.K. (1998) Leaf polarity and meristem formation in Arabidopsis. *Development*, **125**, 2935–2942.

Muller, B. and Sheen, J. (2007) Arabidopsis cytokinin signaling pathway. *Science Signal.* **407**, cm4.

Okushima, Y., Fukaki, H., Onoda, M., Theologis, A. and Tasaka, M. (2007) ARF7 and ARF19 regulate lateral root formation via direct activation of LBD/ASL genes in *Arabidopsis*. *Plant Cell*, **19**, 118–130.

Pantaleo, V., Szittya, G., Moxon, S., Miozzi, L., Moulton, V., Dalmay, T. and Burgyan, J. (2010) Identification of grapevine microRNAs and their targets using high-throughput sequencing and degradome analysis. *Plant J.* **62**, 960–976.

Park, W., Li, J., Song, R., Messing, J. and Chen, X. (2002) CARPEL FACTORY, a dicer homolog, and HEN1, a novel protein, act in microRNA metabolism in *Arabidopsis thaliana*. *Curr. Biol.*, **12**, 1484–1495.

Pélissier, T., Clavel, M., Chaparro, C., Pouch-Pélissier, M.N., Vaucheret, H. and Deragon, J.M. (2011) Double-stranded RNA binding proteins DRB2 and DRB4 have an antagonistic impact on polymerase IV-dependent siRNA levels in *Arabidopsis*. *RNA*, **17**, 1502–1510.

Reinhart, B.J., Weinstein, E.G., Rhoades, M.W., Bartel, B. and Bartel, D.P. (2002) MicroRNAs in plants. *Genes Dev.* **16**, 1616–1626.

Saunders, L.R. and Barber, G.N. (2003) The dsRNA binding protein family: critical roles, diverse cellular functions. *FASEB J.* **17**, 961–983.

Schauer, S.E., Jacobsen, S.E., Meinke, D.W. and Ray, A. (2002) DICER-LIKE1: blind men and elephants in *Arabidopsis* development. *Trends Plant Sci.* **7**, 487–491.

Smet, I.D., Vanneste, S., Inze, D. and Beeckman, T. (2006) Lateral root initiation or the birth of a new meristem. *Plant Mol. Biol.* **60**, 871–887.

Smolka, A. (2009) *Understanding of molecular mechanisms and improvement of adventitious root formation in apple*, PhD thesis, Swedish University of Agricultural Sciences.

Tang, X.R., Wu, H., Jia, M., Yu, X.H. and He, Y.K. (2002) Isolation and expressional analysis of cDNA encoding a dsRNA binding protein homologue OsRBP of rice. *J. Plant Physiol. Mol. Biol.* **28**, 41–45.

Tobutt, K.R. (1985) Breeding columnar apples at East Malling. *Acta Hort.* **159**, 63–68.

Undan, J.R., Tamiru, M., Abe, A., Yoshida, K., Kosugi, S., Takagi, H., *et al.* (2012) Mutation in *OsLMS*, a gene encoding a protein with two double-stranded RNA binding motifs, causes lesion mimic phenotype and early senescence in rice (*Oryza sativa L.*). *Genes Genet. Syst.* **87**, 169–179.

Wu, F., Yu, L., Cao, W., Mao, Y., Liu, Z. and He, Y. (2007) The N-terminal double-stranded RNA binding domains of *Arabidopsis* HYPONASTIC LEAVES1 are sufficient for pre-microRNA processing. *Plant Cell*, **19**, 914–925.

Xie, Q., Frugis, G., Colgan, D. and Chua, N.H. (2000) *Arabidopsis* NAC1 transduces auxin signal downstream of TIR1 to promote lateral root development. *Genes Dev.* **14**, 3024–3036.

Xu, Q., Liu, Y., Zhu, A., Wu, X., Ye, J., Yu, K., Guo, W. and Deng, X. (2010) Discovery and comparative profiling of microRNAs in a sweet orange red-flesh mutant and its wild type. *BMC Genomics*, **11**, 246.

Yang, L., Liu, Z., Lu, F., Dong, A. and Huang, H. (2006) SERRATE is a novel nuclear regulator in primary microRNA processing in *Arabidopsis*. *Plant J.* **47**, 841–850.

Yang, J.S., Maurin, T., Robine, N., Rasmussen, K.D., Jeffrey, K.L., Chandwani, R., Papapetrou, E.P., Sadelain, M., O'Carroll, D. and Lai, E.C. (2010) Conserved vertebrate mir-451 provides a platform for Dicer-independent, Ago2-mediated microRNA biogenesis. *Proc. Natl. Acad. Sci. USA*, **107**, 15163–15168.

Yu, X., Peng, J., Feng, X., Yang, S., Zheng, Z., Tang, X., Shen, R., Liu, P. and He, Y. (2000) Cloning and structural and expressional characterization of BcpLH gene preferentially expressed in folding leaf of Chinese cabbage. *Sci. China*, **43**, 321–329.

Yu, L. (2005) *Molecular mechanism involved in gene regulation of leaf curvature*. PhD thesis. Shanghai, China: Chinese Academy of Sciences.

Yu, L., Yu, X., Shen, R. and He, Y. (2005) *HYL1* gene maintains venation and polarity of leaves. *Planta*, **221**, 231–242.

Zhang, Y., Zhu, J. and Dai, H. (2012) Characterization of transcriptional differences between columnar and standard apple trees using RNA-Seq. *Plant Mol. Biol. Rep.* **30**, 957–965.

Zhou, G.K., Kubo, M., Zhong, R., Demura, T. and Ye, Z.H. (2007) Overexpression of miR165 affects apical meristem formation, organ polarity establishment and vascular development in Arabidopsis. *Plant Cell Physiol.* **48**, 391–404.

Zhou, X., Wang, G., Sutoh, K., Zhu, J.K. and Zhang, W. (2008) Identification of cold-inducible miroRNAs in plants by transcriptome analysis. *Biochim. Biophys. Acta*, **1779**, 780–788.

17

Inducible expression of a fusion gene encoding two proteinase inhibitors leads to insect and pathogen resistance in transgenic rice

Jordi Quilis[1,†], Belén López-García[1,†], Donaldo Meynard[2], Emmanuel Guiderdoni[2] and Blanca San Segundo[1,*]

[1]Centre for Research in Agricultural Genomics (CRAG), CSIC-IRTA-UAB-UB, Edifici CRAG, Barcelona, Spain
[2]CIRAD, UMR AGAP, Montpellier Cedex 5, France

*Correspondence

email blanca.sansegundo@cragenomica.es
†These authors have contributed equally to this work.

Keywords: *Chilo suppressalis*, Fusion protein, *Magnaporthe oryzae*, Maize proteinase inhibitor, *Oryza sativa*, Potato carboxypeptidase inhibitor.

Summary

Plant proteinase inhibitors (PIs) are considered as candidates for increased insect resistance in transgenic plants. Insect adaptation to PI ingestion might, however, compromise the benefits received by transgenic expression of PIs. In this study, the maize proteinase inhibitor (MPI), an inhibitor of insect serine proteinases, and the potato carboxypeptidase inhibitor (PCI) were fused into a single open reading frame and introduced into rice plants. The two PIs were linked using either the processing site of the *Bacillus thuringiensis* Cry1B precursor protein or the 2A sequence from the foot-and-mouth disease virus (FMDV). Expression of each fusion gene was driven by the wound- and pathogen-inducible *mpi* promoter. The *mpi-pci* fusion gene was stably inherited for at least three generations with no penalty on plant phenotype. An important reduction in larval weight of *Chilo suppressalis* fed on *mpi-pci* rice, compared with larvae fed on wild-type plants, was observed. Expression of the *mpi-pci* fusion gene confers resistance to *C. suppressalis* (striped stem borer), one of the most important insect pest of rice. The *mpi-pci* expression systems described may represent a suitable strategy for insect pest control, better than strategies based on the use of single PI genes, by preventing insect adaptive responses. The rice plants expressing the *mpi-pci* fusion gene also showed enhanced resistance to infection by the fungus *Magnaporthe oryzae*, the causal agent of the rice blast disease. Our results illustrate the usefulness of the inducible expression of the *mpi-pci* fusion gene for dual resistance against insects and pathogens in rice plants.

Introduction

Rice (*Oryza sativa* L.) is one of the most important cereal crops in the world and a source of food for more than half of the world's population. Rice yields can be severely compromised by the lepidopteran insect *Chilo suppressalis* (striped stem borer) (Dale, 1994). Strategies for controlling this pest based on the use of chemical insecticides are not effective for the control of *C. suppressalis* because larvae enter shortly after hatching into the plant and are protected from the effect of insecticide treatment. Moreover, the repeated use of agrochemicals for the control of *C. suppressalis* has several drawbacks, including the lack of specificity, the possible effects on beneficial organisms, the incidence of resistance development upon prolonged application and the adverse impact on human health and the environment.

An alternative approach for the control of *C. suppressalis* is the development of genetically modified rice plants expressing insecticidal genes. Along with this, many crop plants expressing *Bt* genes from the soil bacterium *Bacillus thuringiensis* (encoding δ-endotoxins, also known as Cry proteins) that show insect resistance have been developed (Sanahuja *et al.*, 2011; Stevens *et al.*, 2012). Knowing the ability of insects to develop resistance against insecticidal compounds and that insect control products based on topically applied Bt toxins are widely used, it is increasingly clear that alternative or complementary control strategies must by developed to assure that we do not compromise the benefits provided by *Bt*-based insect pest control strategies.

Plant proteinase inhibitors (PIs) have been also identified as candidates for the development of insect-resistant transgenic crops. Plant PIs are usually present in reproductive tissues (seeds, tubers) where they are synthesized and stored (Koiwa *et al.*, 1997; Ryan, 1990). Furthermore, PI gene expression is induced in vegetative tissues by chewing insects and wounding. The usefulness of plant PIs to reduce insect attack in controlled greenhouse conditions and in the field is also documented (Duan *et al.*, 1996; Dunse *et al.*, 2010; Haq *et al.*, 2004; Hilder *et al.*, 1987; Johnson *et al.*, 1989; Stevens *et al.*, 2012; Vila *et al.*, 2005). Moreover, the use of PIs has been regarded as a suitable alternative to Bt in pest control, as milder toxicity of these genes is expected to exert a lower selection pressure to the insect.

Plant PIs function in the plant defence response against herbivorous insects via the inhibition of insect digestive proteinases (Broadway, 1996; Johnson *et al.*, 1989; Srinivasan *et al.*, 2005). When plant PIs bind to the target proteinases, they block the digestion of proteins, leading to growth and developmental delays, and increased mortality. However, insects are able to adapt to the presence of inhibitors in the diet by either overproducing PI-sensitive proteinases, replacing the inhibited enzymes by PI-insensitive proteinases, or inducing the production of PI-degrading enzymes (Broadway, 1997; Gatehouse, 2002; Girard *et al.*, 1998a; Giri *et al.*, 1998; Jongsma *et al.*, 1995). A phenomenon of overcompensation to dietary proteinase inhibitors produced in transgenic plants resulting in increased weight gain of the target pest has also been described (Cloutier *et al.*, 2000; De Leo *et al.*, 1998; Girard *et al.*, 1998b). Most of these problems can be associated with the use of a single PI transgene targeting only one class of proteases in the insect midgut. In this context, the simultaneous expression of PIs with different modes of action in transgenic plants not only might confer increased level of protection against insect attack but also would hinder insect adaptation to PI ingestion.

We previously reported that wounding and insect feeding induce the expression of the *mpi* (*maize proteinase inhibitor*) gene in maize plants (Cordero *et al.*, 1994). MPI is a bifunctional inhibitor that effectively inhibits digestive serine proteinases, namely elastase and chymotrypsin proteinases from the lepidopteran insects *Spodoptera littoralis* and *Chilo suppressalis* (Tamayo *et al.*, 2000; Vila *et al.*, 2005). In addition to wounding, *mpi* expression is induced by fungal infection in maize plants (Cordero *et al.*, 1994). In this respect, several proteinase inhibitors have been reported to inhibit the *in vitro* growth of phytopathogenic fungi, and, for some of them, a defence role in resistance to fungal pathogens has been proposed (Haq *et al.*, 2004; Kim *et al.*, 2009; Laluk and Mengiste, 2011). However, despite the body of information currently available on plant proteinase inhibitors as insecticidal agents, their potential role as antimicrobial agents remains less explored.

On the other hand, the potato carboxypeptidase inhibitor (PCI) accumulates in potato tubers as well as in wounded leaves of potato plants (Graham and Ryan, 1981; Villanueva *et al.*, 1998). PCI inhibits digestive carboxypeptidase activities from the lepidopteran insect *Helicoverpa armigera* (Bayés *et al.*, 2003). However, the constitutive expression of *pci* in rice has an apparent stimulatory effect on the growth of *C. suppressalis* larvae fed on the *pci* transgenic plants, suggesting an adaptive mechanism in the larval gut to compensate for the effect of PCI ingestion (Quilis *et al.*, 2007).

In this study, we examined the effect of expression of a fusion gene consisting of two plant proteinase inhibitors, *mpi* and *pci*, in transgenic rice focusing on resistance against insect attack and pathogen infection. Two different strategies were assayed for obtaining the fusion gene depending on the sequence used to fuse the two individual *PI* genes in the form of a single transcriptional unit. In both cases, the expression of the *mpi-pci* fusion gene is driven by the wound- and pathogen-inducible *mpi* promoter. We report that the inducible expression of the *mpi-pci* fusion gene confers resistance to the lepidopteran insect *C. suppressalis* as well as to the fungal pathogen *Magnaporthe oryzae*, the causal agent of the rice blast disease.

Results

Production of transgenic rice plants expressing a *mpi-pci* fusion gene

Two different constructs containing a *mpi-pci* hybrid gene that differed on the linker sequence used to fuse the two genes were prepared (Figure 1a; Methods S1). The first strategy was selected based on the potential of the FMDV2A (foot-and-mouth disease virus 2A) sequence to mediate self-processing of a polyprotein (Halpin *et al.*, 1999). In this way, the MPI and PCI polypeptides were fused into a single open reading frame using the FMVD 2A sequence (hereafter referred to as MPI-2A-PCI protein). The second strategy consisted the use of the nucleotide sequence encoding the processing site of the cry1B precursor protein to fuse the *mpi-* and *pci-*coding sequences (hereafter referred to as MPI-C-PCI protein). In this respect, it is well known that the Cry proteins are synthesised as longer inactive precursor proteins which are proteolytically processed by trypsin-like proteinases in

the insect gut to become an active toxin (Miranda *et al.*, 2001). Knowing the functional importance of the C-terminal tail of PCI for inhibition of carboxypeptidase activities (Marino-Buslje *et al.*, 2000), the *pci* gene was fused in frame and downstream of the *mpi* gene in the two strategies.

The two *mpi-pci* fusion genes were expressed in rice under the control of the *mpi* promoter (−1872/+197 promoter region). This promoter has been shown to be activated in vegetative tissues of transgenic rice in response to mechanical wounding (a plant response to insect feeding) and fungal infection (Breitler *et al.*, 2001; Moreno *et al.*, 2005). The components of the two plant expression vectors used for wound-inducible expression of each *mpi-pci* fusion gene in rice plants are shown in Figure 1a.

Transgenic rice (*O. sativa* elite *japonica* cultivar Ariete) plants were generated by *Agrobacterium*-mediated transformation. Transgene integration in the rice genome was assessed by Southern blot hybridization, most of the plants showing 1 or 2 integration events (Figure S1). The wound-inducible expression of the *mpi-pci* gene in leaves of transgenic rice plants was confirmed through successive generations (T0 to T3 progeny plants) by Northern blot analysis (Figure 1b shows the results obtained on the analysis of T2 homozygous plants). No accumulation of *mpi-pci* transcripts occurred in either wounded leaves from untransformed plants (Figure 1b, WT; see Methods S1 for wounding of plant leaves) or unwounded leaves of transgenic plants (results not shown). Finally, no visible pleiotropic effects on plant growth and development due to the expression of the *mpi-pci* fusion gene were observed (results not shown).

Effect of transgenic expression of the *mpi-pci* fusion gene on insect growth

To determine the effect of transgenic expression of a *mpi-pci* gene on the growth of *C. suppressalis* larvae, a series of feeding trials were carried out. In these experiments, five independent homozygous lines for each fusion gene (*mpi-C-pci* and *mpi-2A-pci* lines) and wild-type plants were assayed (Methods S1). These studies revealed a significant reduction in weight gain in the larvae fed on rice lines expressing one or another *mpi-pci* fusion gene when compared with larvae fed on wild-type plants (Figure 2a). After feeding for 14 days, the average weight of larvae fed on control plants was 38.8 ± 3.1 mg. The weight of larvae fed on *mpi-C-pci* lines was, however, significantly lower than that of larvae fed on control plants, their mean weight ranging from 13.7 to 23.4 mg depending on the line (percentage of weight reduction, 64.6%–39.6%) (Figure 2a). Similarly, larvae fed on *mpi-2A-pci* plants grew significantly slower that those fed on control plants, these larvae reaching a mean weight of 22.1 to 32.5 mg (percentage of weight reduction, 42.9%–16.2%) (Figure 2a). Representative images of larvae recovered at the completion of the 14-day experimental period are shown in Figure 2b. On the basis of the weight of larvae recovered in these feeding experiments, the larval population was distributed into different groups. Whereas larvae recovered from wild-type plants distributed along the various weight classes and reached the higher weight classes (>70 mg), most of the larvae recovered from *mpi-pci* lines remained in the lower weight classes, and none of the larvae recovered from these plants reached the 70 mg weight mark (Figure S2). Together, feeding trials established that larvae fed on rice plants expressing a *mpi-pci* gene under a wound-inducible regime showed an important reduction in their weight when compared with that of larvae fed on control plants.

(a) **Vector 1:** *pC1300::mpi prom::mpi-2A-pci::mpi ter*

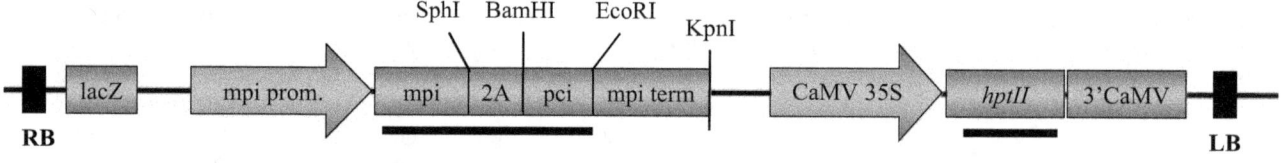

(b) **Vector 2:** *pC1300::mpi prom::mpi-C-pci::mpi ter*

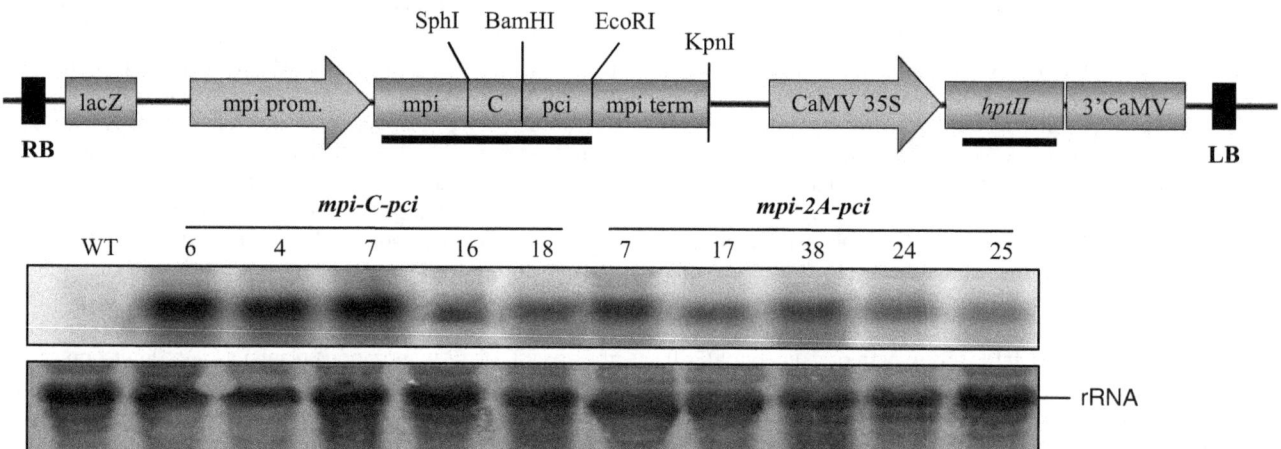

Figure 1 Expression of the *mpi-pci* fusion gene in transgenic rice plants. (a) Schematic representation of vectors used for rice transformation. Vector 1 contained the *mpi-2A-pci* fusion gene, whereas vector 2 contained the *mpi-C-pci* fusion gene. The wound-inducible *mpi* promoter was used to drive the expression of the *mpi-pci* fusion gene. The *hptII* gene encoding resistance to hygromycin served as the selectable marker in rice transformation. The regions used as probe for genomic Southern blot analyses (Figure S1) are indicated with solid lines. (b) Northern blot analysis of total RNA isolated from wounded leaves of transgenic and control untransformed (WT, Ariete) rice plants. Representative lines transformed with the *mpi-2A-pci* or the *mpi-C-pci* fusion gene (T2 generation) are shown. Hybridization to the 18S ribosomal RNA is shown in the lower panel as a control for loading.

Wound-inducible expression of the *mpi-pci* fusion gene in rice confers insect resistance

Knowing that rice plants expressing a *mpi-pci* fusion gene had a negative impact on larval growth, we assessed the performance of rice plants during infestation with *C. suppressalis* larvae under controlled greenhouse conditions. Three weeks after infestation, transgenic plants, either the *mpi-C-pci* or the *mpi-2A-pci* lines, showed normal development, whereas control plants were severely damaged or dead (Figure 3a). Furthermore, Northern blot analysis revealed important levels of *mpi-pci* transcript accumulation in leaves of all the transgenic lines used in these bioassays, this observation indicating that the *mpi* promoter directs high level of transgene expression in *C. suppressalis*-infested rice plants (Figure 3b). This is in agreement with the results previously reported on the *mpi* promoter activity in transgenic rice (Breitler *et al.*, 2001, 2004; Vila *et al.*, 2005).

It is well known that infestation by *C. suppressalis* leads to the formation of empty panicles, a condition commonly called 'white head'. Accordingly, we determined the percentage of white panicles in rice plants that have been infested with *C. suppressalis* larvae. For this, a subset of noninfested and *C. suppressalis*-infested transgenic and wild-type plants were allowed to continue growth and further analysed for production of white panicles (Figure 3c). Whereas the percentage of white panicles in wild-type plants was 20.5%, the transgenic lines produced a percentage of white panicles ranging from 5.9% to 12.5% in the *mpi-C-pci lines* or from 7.7% to 13.6% in the *mpi-2A-pci* lines.

Collectively, results obtained in insect bioassays revealed that the expression of the *mpi-pci* fusion gene in rice plants confers protection against the lepidopteran insect *C. suppressalis*. In addition to prevent plant damage caused by *C. suppressalis*, rice plants expressing a *mpi-pci* gene produced less white panicles than wild-type plants.

Analysis of transgene-derived protein products

Immunoblot experiments were carried out to examine the accumulation of the transgene products in rice tissues in each strategy assayed in this work. For this, protein extracts were prepared from wounded leaves of wild-type and transgenic plants, both *mpi-C-pci* and *mpi-2A-pci* plants (homozygous T2 generation), and probed with an anti-MPI antibody raised against the pure MPI protein (Tamayo *et al.*, 2000) (for details on Western blot analysis see Methods S1). As a control, protein extracts from wounded leaves of rice plants expressing the *mpi* gene under the control of its own promoter (Vila *et al.*, 2005) were analysed. As expected, the MPI polypeptide accumulated in *mpi* rice plants (Figure 4a). Immunoblot analysis of protein extracts from *mpi-C-pci* lines also confirmed the accumulation of the fusion protein, which was absent in wild-type plants (Figure 4a). No immunological reactions occurred with the protein extracts from rice plants when the pre-immune serum was used (results not shown).

The expression strategy using the 2A sequence as the linker peptide aimed to produce the two individual proteinase inhibitor proteins, MPI and PCI, from the hybrid protein in rice tissues. Immunoblot analysis of protein extracts obtained

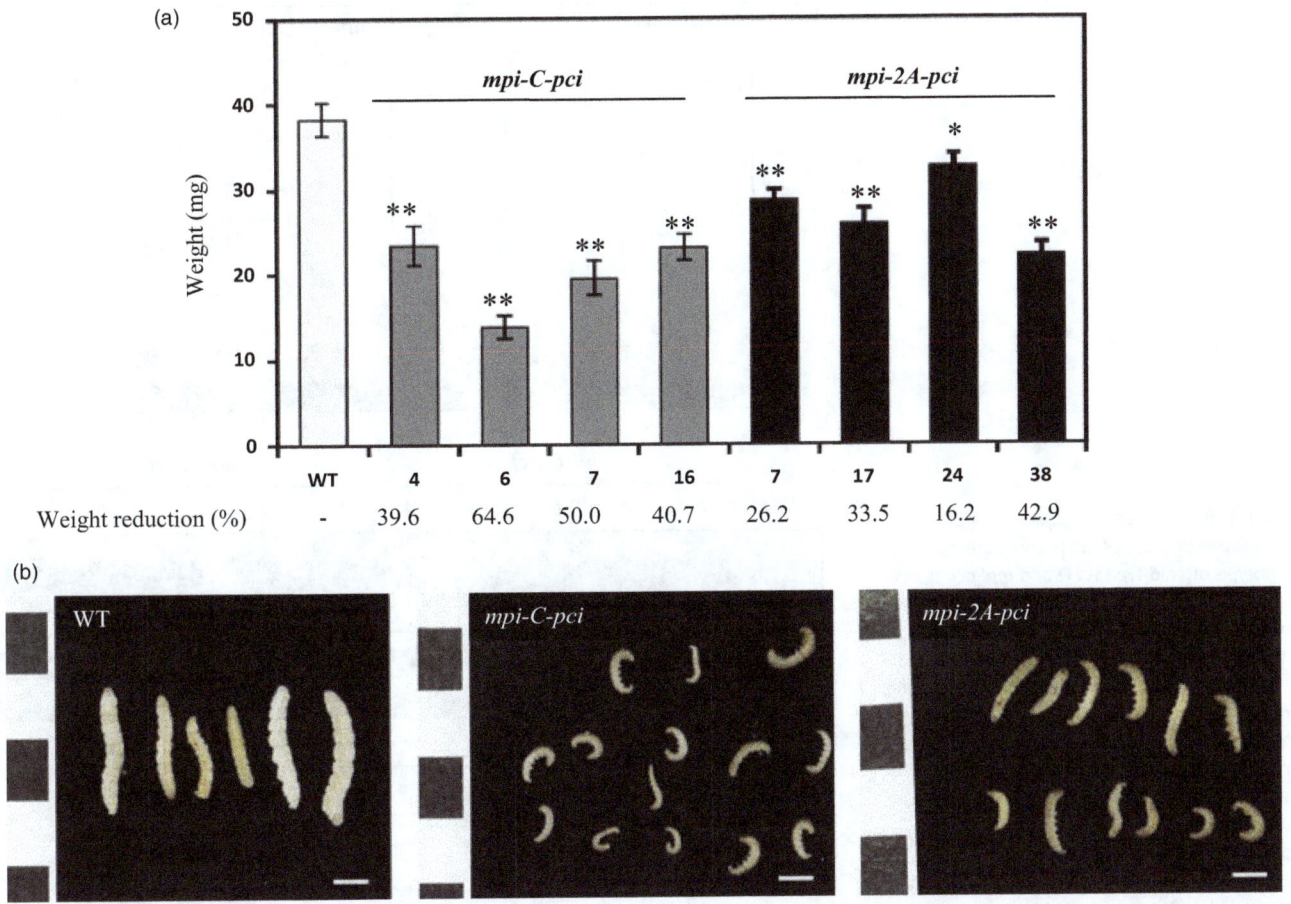

Figure 2 Insect feeding bioassays. (a) Growth of *Chilo suppressalis* fed on control wild-type (WT, white bar), *mpi-C-pci* (grey bars) and *mpi-2A-pci* (black bars) plants (homozygous T2 lines). Fourteen days after infestation with L2 *C. suppressalis* larvae, plants were dissected and larvae were recovered and individually weighted. The percentage of weight reduction in larvae fed on each transgenic line relative to larvae recovered from control plants is indicated at the bottom. Data sets were analysed for significant differences using Student's t-test (**$P \leq 0.001$; *$P \leq 0.05$). Error bars indicate the standard deviation of the error (SE). The distribution of the larval population according to their weight for each line is presented in Figure S2. (b) Representative *C. suppressalis* larvae recovered from wild-type (WT), *mpi-C-pci* (line 6) or *mpi-2A-pci* (line 38) rice plants. Bar = 0.5 cm

from the *mpi-2A-pci* lines revealed accumulation of both, the fusion uncleaved MPI-2A-PCI protein and the MPI protein, indicating that the fusion protein was partially processed in the rice leaves (Figure 4a). The cleavage efficiency varied among the different transgenic lines. By comparing the band intensities with those of known amounts of MPI protein, the MPI content in *mpi-2A-pci* lines was estimated to be 0.10–0.25% of total soluble proteins (% of MPI in total soluble protein) depending on the transgenic line. Longer exposure times of immunoblots containing protein extracts from *mpi-2A-pci* plants indicated that the MPI-2A-PCI polyprotein was properly cleaved as revealed by the absence of MPI-containing truncated protein signals (Figure S3).

During the course of this work, the pure PCI protein was used to raise polyclonal antibodies in either rabbits or egg yolk which were then used for immunoblot analysis of protein extracts from rice leaves. Unfortunately, these analyses did not allowed us to detect the PCI polypeptide in the same protein extracts in which MPI was detected. Detection problems arising from inefficient extraction of PCI from plant tissues, or low antigenicity of the PCI polypeptide, might explain these results. In favour of the low antigenicity of PCI, pure PCI polypeptide was undetectable by Western blot analysis (data not shown).

Failure to detect PCI in rice protein extracts prompted us to examine PCI activity in leaves of *mpi-pci* rice plants. For this, total protein extracts were prepared from wounded leaves of either *mpi-C-pci* or *mpi-2A-pci* plants which were then subjected to affinity chromatography on immobilized carboxypeptidase A (CPA). As controls, pure PCI protein as well as protein extracts from leaves of rice plants constitutively expressing *pci* (Quilis *et al.*, 2007) was subjected to affinity chromatography. In all the cases, the eluted fractions were assayed for their ability to inhibit bovine CPA activity. When pure PCI protein or protein extracts from *pci* rice plants were analysed, the inhibitory activity of PCI against CPA was recorded in fractions eluted at pH 11.0, but not in protein fractions eluted at pH 9.0 (Figure 4b, upper panels). Equally, when protein extracts from *mpi-C-pci* or *mpi-2A-pci* plants were subjected to affinity chromatography, the protein fractions eluted at pH 11.0 showed inhibitory activity against CPA (Figure 4b, lower panels). A certain level of CPA inhibition in the first fraction eluted at pH 9.0 (*mpi-C-pci* plants) was also detected, pointing to a weaker interaction of the fusion protein with CPA during affinity chromatography relative to that of the PCI alone.

Together, these results suggest that the two *mpi-pci* fusion genes are efficiently transcribed and translated to produce a MPI-

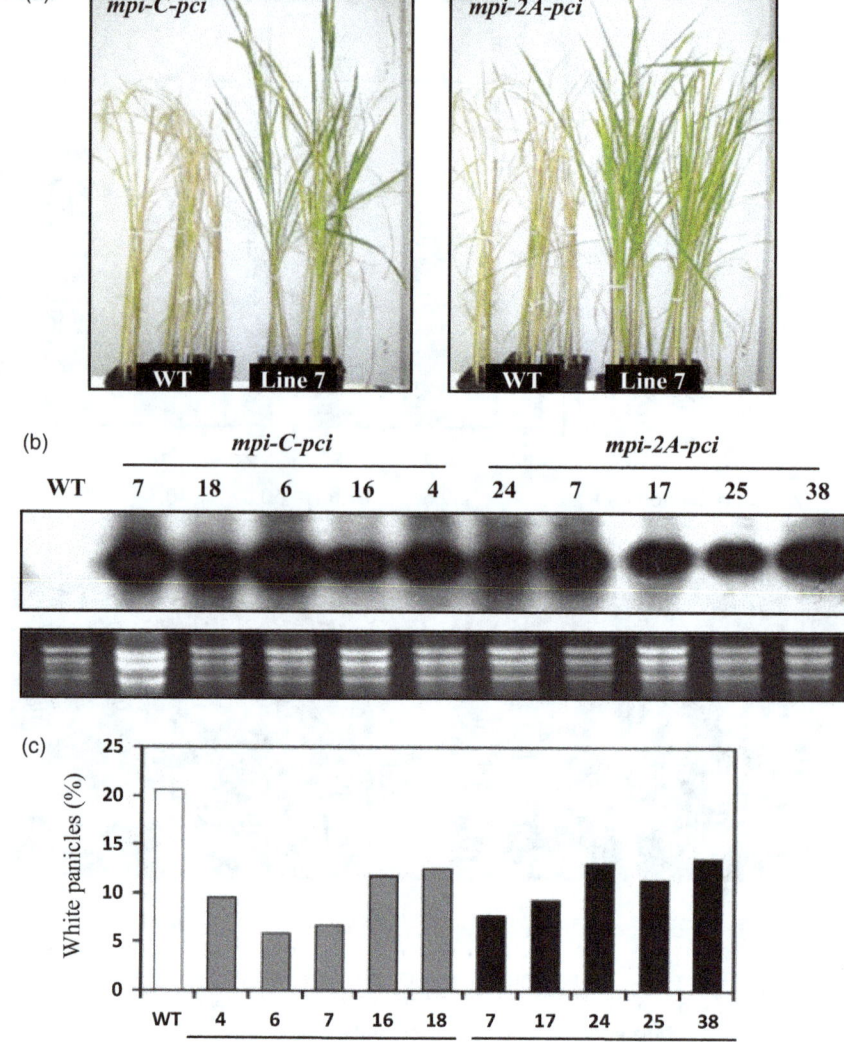

Figure 3 Resistance to *C. suppressalis* of rice plants expressing a *mpi-pci* fusion gene. (a) Phenotype of wild-type (WT) and *mpi-pci* plants that had been infested with *C. suppressalis* larvae. Bioassays for insect resistance were performed with *mpi-C-pci* (lines 4, 6, 7, 16 and 18) and *mpi-2A-pci* (lines 7, 17, 24, 25 and 38) plants. Results for the *mpi-C-pci* line 7 and the *mpi-2A-pci* line 7 are shown (similar results were obtained for the other lines here assayed for each gene). (b) Northern blot analysis of RNAs obtained from *mpi-C-pci* or *mpi-2A-pci* rice plants used in bioassay experiments. Leaves from at least 6 plants for each line were pooled and analysed by Northern blot (20 µg of total RNA each sample). Ethidium bromide staining served as loading controls. (c) Percentage of white panicles in rice plants infested with *C. suppressalis* larvae relative to the total panicles obtained for each line. Data shown are from one of two experiments that produced similar results.

PCI fusion protein when fused by either the FMDV 2A sequence or the processing sequence of the Cry1B precursor protein. Concerning the *mpi-2A-pci* rice plants, processing of the fusion protein and accumulation of the MPI polypeptide in rice tissues was demonstrated. Limitations for immunodetection of PCI by Western blot analysis hampered detection of the PCI polypeptide produced by cleavage of the MPI-2A-PCI fusion protein in rice tissues. Even though the accumulation of the PCI polypeptide plants could not be clarified in this work, we have shown that protein extracts from *mpi-C-pci* plants exhibited inhibitory activity against CPA, supporting that the PCI domain of the fusion protein is active and effective for inhibition of carboxypeptidase activity.

Resistance of *mpi-pci* rice plants to infection by the rice blast fungus *Magnaporthe oryzae*

In this work, the functional relevance of transgenic expression of the *mpi-pci* fusion gene in terms of resistance to infection by *M. oryzae* was examined. Results previously reported by our group indicated that rice plants constitutively expressing the *pci* gene exhibit enhanced resistance to infection by this fungus (Quilis *et al.*, 2007). Accordingly, resistance to fungal infection of *mpi-pci* rice plants was determined and compared to that of *pci* rice plants (Methods S1). For this, leaves of the transgenic plants were locally

inoculated with spores from a *gfp*-expressing *M. oryzae* isolate as previously described (Campos-Soriano and San Segundo, 2009). Differences in the degree of disease symptoms caused by *M. oryzae* between *mpi-pci* lines and wild-type plants were clearly observed. At 3 days after inoculation, small lesions were generally observed in leaves of the *mpi-pci* and *pci* lines, whereas the leaves from control plants were visibly damaged under the same experimental conditions (Figure 5a). Fluorescence microscopy revealed the presence of a fluorescent hyphae growing on the leaf surface of wild-type plants (Figure 5a).

Finally, image analysis was used to determine the percentage of leaf area affected by blast lesions in *mpi-pci* and *pci* plants at 6 days postinoculation. In agreement with the visual inspection, the inoculated leaves from *mpi-pci* and *pci* transgenic lines exhibited a lower percentage of diseased area relative to inoculated leaves from wild-type plants (Figure 5b). Generally, the rice lines expressing the *mpi-C-pci* gene exhibited relatively reduced blast disease symptoms compared to the transgenic lines expressing either the *mpi-2A-pci* gene or the *pci* gene alone.

Together, disease resistance assays demonstrated that expression of the *mpi-pci* fusion gene in rice, either the *mpi-2A-pci* or the *mpi-C-pci* gene, confers enhanced resistance against the rice blast fungus *M. oryzae*. Furthermore, these results indicated that the

(a)

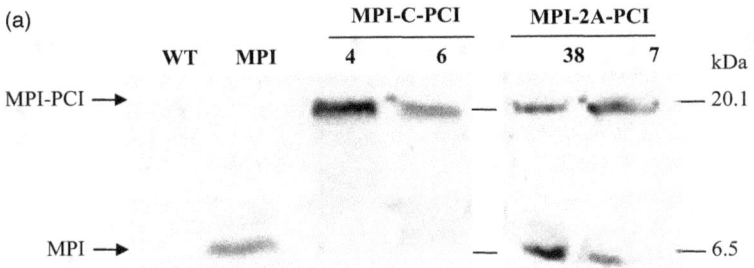

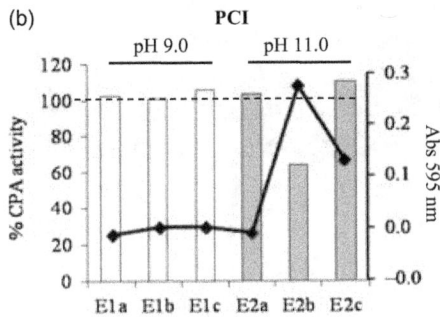

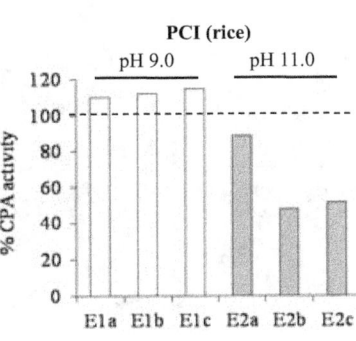

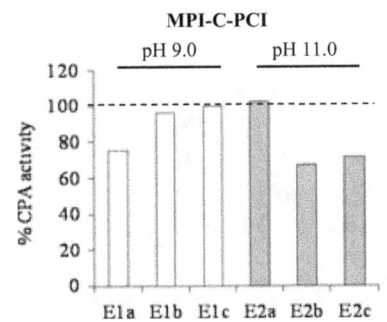

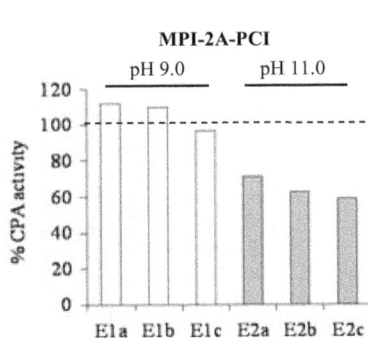

Figure 4 Protein analysis of transgenic plants expressing a *mpi-pci* fusion gene. (a) Immunodetection of MPI in total protein extracts obtained from leaves of wild-type (WT) plants and plants harbouring the *mpi* gene (Vila *et al.*, 2005), the *mpi-C-pci* (lines 4 and 6) or the *mpi-2A-pci* (lines 38 and 7). In all the cases, transgene expression was driven by the wound-inducible *mpi* promoter. Leaves were mechanically wounded and harvested at 16 h after wounding. Protein extracts (40 µg each extract) were separated by 15% SDS-PAGE and probed with an anti-MPI antibody (Tamayo *et al.*, 2000). (b) Affinity chromatography on immobilized CPA of pure PCI (Sigma) and protein extracts from rice plants constitutively expressing *pci* (line 7.34), *mpi-C-pci* plants (line 7) and *mpi-2A-pci* (line 7) plants. Eluted fractions were analysed for inhibition of bovine CPA activity using AAFP as the substrate (Quilis *et al.*, 2007). Lanes E1a to E1c, fractions eluted using buffer at pH 9.0. Lanes E2a to E2c, fractions eluted using buffer at pH 11.0. In each sample, CPA activity is referred to the activity of the enzyme in the corresponding buffer (considered as 100%).

MPI-C-PCI fusion protein might be effective for inhibition of fungal proteinases (as it was previously demonstrated for PCI; Quilis *et al.*, 2007). Knowing that the fusion protein is partially processed in rice leaves of *mpi-2A-pci* (see Figure 4a), the contribution to the phenotype of disease resistance that is observed in these plants might be due to the activity of the MPI-2A-PCI protein and the PCI protein originating from the fusion protein.

Discussion

In this work, we show that the expression of a fusion gene encoding two proteinase inhibitors, namely the MPI and PCI inhibitors, in transgenic rice provides dual resistance against insect attack and pathogen infection. The protective effect of the fusion gene was achieved when using either the FMDV2A linker sequence or the processing region of the Cry1B precursor protein to fuse the two proteinase inhibitor polypeptides. Although evidence exists in the literature on the insecticidal properties of plant PIs, very limited information is currently available about a dual protective effect against insects and pathogens in transgenic crops expressing plant PIs. Only, the simultaneous expression of a trypsin inhibitor (sporamin) gene and a phytocystatin gene in transgenic tobacco plants was reported to be effective for resistance against insect and pathogen attack (Senthilkumar *et al.*, 2010). More recently, the UNUSUAL SERINE PROTEASE INHIBITOR (UPI) gene was reported to function in resistance to necrotrophic fungi and insect herbivory in Arabidopsis plants (Laluk and Mengiste, 2011).

The transgenic rice expressing a *mpi-pci* fusion gene under the control of the *mpi* promoter showed normal growth and development. In other studies, the constitutive expression of proteinase inhibitor genes in transgenic plants has been shown to have a negative impact on plant development (Zavala *et al.*, 2004). Clearly, the use of a wound- and pathogen-inducible *mpi* promoter (instead of a constitutive promoter) minimizes the possibility of alterations in fitness of the transgenic plants while driving levels of transgene expression sufficient to confer protection to the plant. An additional advantage of using the *mpi* promoter is that this promoter is not active in pollen, seed embryo or endosperm of rice plants, thus reducing food and environmental concerns (Breitler *et al.*, 2001, 2004). Together, the results here presented suggest that the *mpi* promoter can be considered a good candidate to drive conditional expression of insecticidal and antifungal genes in transgenic rice.

Insect feeding experiments revealed that the weight gain of *C. suppressalis* larvae fed on *mpi-pci* plants was reduced compared to that of larvae fed on wild-type plants. Based on the inhibitory properties of MPI and PCI (MPI inhibits insect elastase and chymotrypsin activities, whereas PCI inhibits insect carboxypeptidases), a broad spectrum of insect digestive proteinases is expected to be affected by the coordinated and complementary activity of the two inhibitors. This fact might well explain

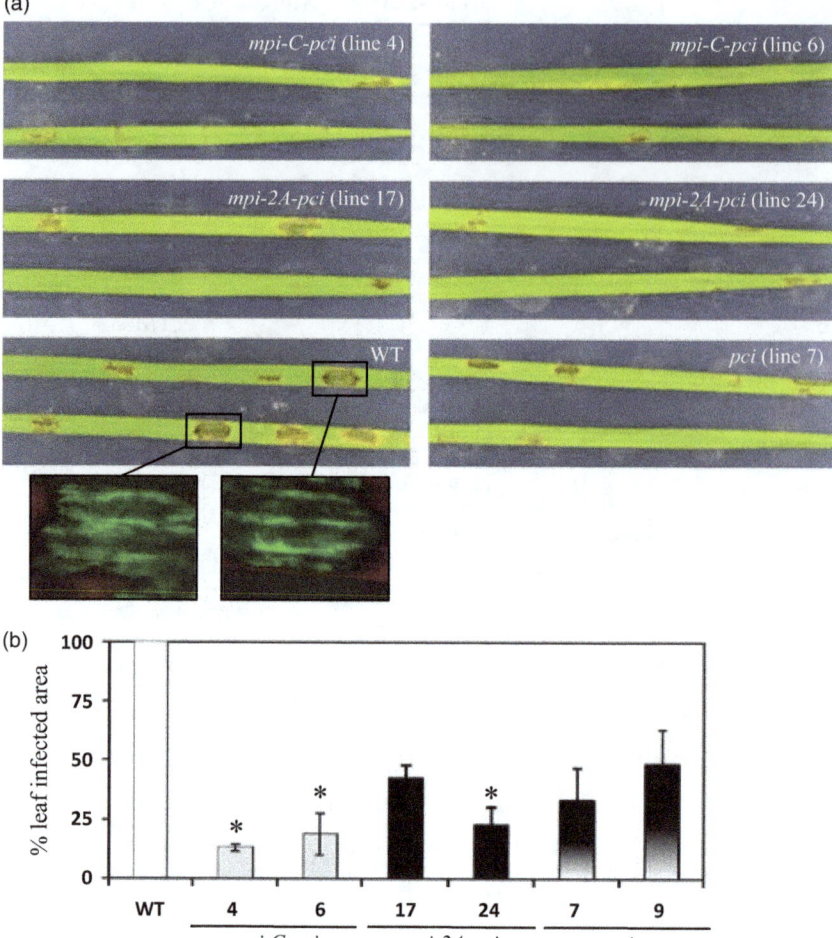

Figure 5 Resistance of rice plants expressing the *mpi-pci* gene to *M. oryzae*. (a) Leaves from transgenic rice lines harbouring the *mpi-C-pci*, *mpi-2A-pci*, *pci* or wild-type (WT) plants were inoculated with spores of the *gfp*-expressing *M. oryzae* isolate (10^6 spores/mL) as previously described (Campos-Soriano *et al.*, 2013). Disease symptoms were observed by light and fluorescent microscopy at 3 days after inoculation. Representative results from one of three experiments that produced similar results are shown. (b) Percentage of leaf area affected by blast lesions was determined using the image analysis software Assess v. 2.0 for plant disease quantification at 6 days after inoculation with *M. oryzae* spores. Data sets were analysed for significant differences using Student's *t*-test (*$P \leq 0.1$). Results are shown as mean values ±SE.

the observed effectiveness of this strategy in conferring insect resistance. The mean weight reduction in larvae fed on rice plants expressing a fusion gene was 39.6%–64.6% and 16.2%–42.9% for the *mpi-C-pci* and *mpi-2A-pci* lines, respectively. In this respect, we previously reported that transgenic expression of the *mpi* gene alone (also under the control of its own promoter) results in a mean weight reduction in *C. suppressalis* larvae of 23.0%–28.6% depending on the line (Vila *et al.*, 2005). Further studies are, however, needed to conclusively determine whether transgenic expression of the *mpi-pci* fusion gene is more effective for inhibition of larval growth than the expression of the *mpi* gene alone.

On the other hand, results previously reported by our group indicated that production of the PCI inhibitor alone in the rice plant might be counterproductive in terms of insect resistance, as the weight gain of *C. suppressalis* larvae fed on rice plants constitutively expressing the *pci* gene was significantly larger than that of larvae fed on wild-type plants (Quilis *et al.*, 2007). It can then be hypothesized that the production of the two plant inhibitors, MPI and PCI, compensates the negative effects that occur in *C. suppressalis* larvae fed on rice plants expressing the *pci* gene alone. If so, the simultaneous production of the PCI and MPI inhibitors in rice would counteract adaptive mechanisms in the complement of digestive proteinases and minimize the compensatory response of *C. suppressalis* larvae after ingestion of the PCI alone.

Concerning the mechanisms underlying insect resistance in *mpi-pci* rice plants, two scenarios can be envisaged. Firstly, there

is the possibility that the MPI-PCI fusion protein acts as a multifunctional proteinase inhibitor. In this respect, efforts have been made to obtain multifunctional proteinase inhibitors exhibiting activity against multiple target proteinases (Schluter *et al.*, 2010). Such approaches involve chimeric proteins that integrate complete or partial functionally significant regions of the inhibitor sequences with a broader inhibitory range against target proteinases (Brunelle *et al.*, 2005; Inanaga *et al.*, 2001; Outchkourov *et al.*, 2004). Most of these studies, however, have been approached by *in vitro* inhibition assays on insect gut protein extracts or by insect feeding assays using artificial diets supplemented with isolated PIs. Secondly, the activity of the two partners originating from the fusion protein either in the rice tissue (*mpi-2A-pci* plants) or in the gut of *C. suppressalis* larvae (*mpi-C-pci* plants) would be responsible for the observed phenotype of resistance. In this respect, whereas the *in vivo* processing of the MPI-2A-PCI fusion protein in rice leaves has been confirmed in this work, the proteolytic processing of the fusion protein in the insect gut has not been demonstrated yet. Further analysis will aid in determining whether insect resistance in *mpi-C-pci* and *mpi-2A-pci* rice plants is mediated by the MPI-PCI fusion protein acting as a multifunctional proteinase inhibitor or by the activity of the two independent inhibitors.

As for the PCI inhibitor, it is well known that the C-tail residues of the PCI polypeptide are essential for enzyme inhibition (Marino-Buslje *et al.*, 2000). There is then a possibility that the PCI protein, either alone or in the form of uncleaved MPI-PCI fusion protein, exerts its inhibitory activity against carboxypeptidase activities. The observation that protein extracts from *mpi-C-pci* plants exhibit inhibitory properties against carboxypeptidase activity, together with the observed phenotype of blast disease resistance in these plants, favours the possibility that the PCI domain of the MPI-C-PCI fusion protein is functional and active for inhibition of target carboxypeptidases. Then, fusing the PCI polypeptide to the C-terminus region of proteins of interest in the form of polyproteins might then be a suitable strategy to obtain insect and/or pathogen resistance in transgenic plants.

Moreover, we show that expression of a *mpi-pci* fusion gene in rice does not confer levels of protection against *M. oryzae* significantly higher than those occurring in the rice plants expressing the *pci* gene alone. Contrary to the *pci* rice plants, the rice plants expressing the *mpi* gene alone do not show resistance to infection by the rice blast fungus *M. oryzae* (results not shown). From these observations, it can be reasoned that the protective effect that is observed in *mpi-C-pci* and *mpi-2A-pci* plants results from the antifungal properties of PCI (Quilis *et al.*, 2007). If so, the two inhibitors would not act in a synergistic manner in conferring resistance to *M. oryzae* infection in transgenic rice.

Concerning the strategy based on the use of the FMVD2A sequence, it should be here mentioned that this strategy has been used for co-expression of different genes in transgenic plants. Some examples are the expression of reporter genes in tobacco (de Felipe *et al.*, 2006; Halpin *et al.*, 1999; Ma and Mitra, 2002), the production of lignocelluloses degradation enzymes or self-cleaving antimicrobial polyproteins in tobacco or Arabidopsis plants (François *et al.*, 2004; Lee *et al.*, 2012), or coexpression of carotenoid or tyramine derivatives biosynthetic enzymes in the rice endosperm (Park *et al.*, 2009; Ralley *et al.*, 2004). However, when using the FMDV2A sequence to fuse two antimicrobial proteins (snaking and defensin) in the form of a hybrid protein, no cleavage products were detected in tobacco and potato plants (Kovalskaya *et al.*, 2011). Our *in vivo* experiments demonstrated a clear phenotype of resistance to insect attack and pathogen infection in the *mpi-2A-pci* rice plants, despite the lack of complete processing of the *in planta*-produced fusion protein. At present, the contribution of the fusion MPI-2A-PCI protein and the individual proteins derived from the fusion protein (MPI and PCI) to the observed phenotype of resistance against *C. suppressalis* remains unknown. Altogether, these findings support the need of addressing on a case-by-case basis the evaluation of the FMVD2A sequence as a linker peptide to fuse proteins of interest in a particular plant species.

An important advantage of the strategy assayed in this work is that a single transformation event can lead to the production of plant PIs with different modes of action using only a single promoter, thus avoiding the risk of homology-based silencing, transgene segregation along successive generation or the need of different promoters or selectable markers (as it occurs in other strategies for co-expression of multiple transgenes). Clearly, different approaches can be used for the simultaneous expression of insecticidal and/or antifungal genes in plants, each one having its own advantages and drawbacks to be adopted in crop biotechnology. For instance, the integration of multiple transgenes in a single plant can be achieved by either crossing individual transformants, by sequential transformation, and by co-transformation either with a single plasmid equipped with multicassettes or with multiple plasmids can be approached. Such strategies are, however, time-consuming and require the use of different selectable marker genes. For single-plasmid co-transformation,

the main technical limitation is the difficulty to assemble complex plasmids with multiple gene cassettes, whereas the success of co-transformation with multiple plasmids depends on the frequency by which independent transgenes are integrated into the host genome (transgenes can also segregate in successive generations). Expression of multiple genes in a single transgenic plant also requires the use of different promoters to avoid homology-dependent transgene silencing. Here, it is worthy to mention that there are few reports on insect resistance based on the co-expression of multiple proteinase inhibitors in transgenic plants. Gene pyramiding for insect resistance using plant PIs has been reported by crossing transgenic lines expressing individual genes or using a plant transformation vector that harbours a double expression cassette (Abdeen *et al.*, 2005; Dunse *et al.*, 2010).

Due to the focus on agricultural importance, expression of genes encoding multiple insecticidal factors appears to be critical for long-term control of insect pests. Knowing that the two components of the MPI-PCI fusion protein are effective for inhibition of insect proteinases with different mechanisms of action (quimotrypsin and elastase by MPI, carboxypeptidases by PCI), the simultaneous expression of these plant proteinase inhibitors is expected to maintain its insecticidal properties longer than inhibitors acting on proteinases with a single mechanistic mode of action by hindering insect adaptation to proteinase inhibitor ingestion. The use of *mpi-pci* fusion genes, either alone or adjunct to Bt, can thus be considered as a promising strategy for the long-term protection of rice plants against the striped stem borer *C. suppressalis*. As an additional advantage, the transgenic expression of the *mpi-pci* fusion gene in rice confers protection against the rice blast fungus *M. oryzae*. This fungus is the causal agent of the rice blast disease, one the most devastating fungal disease of cultivated rice worldwide (Talbot, 2003). Currently, the control of the rice blast disease depends on the use of chemically synthesized fungicides. Along with this, the *mpi-pci* rice plants represent an alternative to the economically costly and environmentally undesirable chemical control of the rice blast disease. The ultimate goal of insect and pathogen resistance of the rice plants expressing a *mpi-pci* fusion gene needs now to be confirmed by further evaluation in the field.

Experimental procedures

Construction of plant expression vectors and rice transformation

Two vectors containing a *mpi-pci* fusion gene were prepared in this work and used for rice transformation (Figure 1a, Table S1). In both cases, the coding sequence of the *pci* gene was fused in frame and downstream of the *mpi* gene through either the FMVD2A linker sequence (*mpi-2A-pci* gene) or the cleavage site of the Cry1B precursor protein (*mpi-C-pci* gene). Expression of each fusion gene is driven by the *mpi* promoter and terminator regions (Moreno *et al.*, 2005). Transformation was carried out using the Mediterranean elite *japonica* rice (*Oryza sativa* L.) cultivar Ariete. Details on preparation of expression vectors and production of transgenic rice plants by *Agrobacterium*-mediated transformation of embryogenic calli derived from mature embryos are indicated in Supporting Information. T_0 plants were transferred to a containment glasshouse and examined for transgene expression and integration. Homozygous transgenic lines were identified in the T_2 generation. All rice plants were grown at 27 ± 2 °C under 18-h/6-h light/dark photoperiod.

RNA and DNA analyses

Total RNA was isolated from the leaves of transgenic and control plants using Trizol reagent according to the manufacturer's instructions (Invitrogen, Carlsbad, CA, USA). For Northern blot analysis, RNAs were subjected to formaldehyde-containing agarose gel electrophoresis and transferred to Hybond-N membranes (Amersham Biosciences, Piscataway, NJ). A DNA fragment coding for the specific mpi-pci fusion gene (mpi-2A-pci or mpi-C-pci) was used as a probe. All probes were gel-purified and radioactively labelled by random priming, according to the manufacturer's protocol (Roche Diagnostics, Mannheim, Germany). Hybridizations were conducted at 42 °C in 40% formamide in the hybridization solution.

Genomic DNA was extracted using MATAB (mixed alkyltrimethylammonium bromide) as the extraction buffer (Coca et al., 2004). For Southern blot analysis, genomic DNA was digested with KpnI restriction enzyme, electrophoresed on 0.8% agarose gels, transferred onto nylon membranes (Hybond-N, Amersham, UK) and hybridized to DNA probes ^{32}P-labelled by random priming (Roche Diagnostics GmbH, Mannheim, Germany). Hybridizations were carried out at 65 °C (Coca et al., 2004).

Preparation of protein extracts and immunoblotting

Protein extracts were prepared from wounded leaves of transgenic and nontransformed plants. Plant material was ground with liquid nitrogen, and total proteins were extracted using 50 mM Tris–HCl, pH 8, 1 mM EDTA, 5% glycerol, 1 mM dithiothreitol and 0.1% (v/v) Triton X-100 as the extraction buffer. Extractions were carried out at 4 °C for 2 h with continuous slow stirring. Protein quantification was performed using Bio-Rad protein assay reagents (Bio-Rad, Madrid, Spain). Immunoblots were prepared as indicated in Supporting Information. To determine the level of MPI accumulation in total protein extracts from leaves of mpi-2A-pci plants, the MPI protein was expressed in Escherichia coli (Tamayo et al., 2000), and different dilutions of MPI protein were applied to each blot as standards. The MPI levels were measured from digitalized images of the blots (Quantity One Program from Bio-Rad).

Carboxypeptidase A affinity chromatography and activity assays

Total protein extracts from transgenic plants were fractionated by affinity chromatography using a column of immobilized bovine CPA (CPA-agarose, Sigma, St Louis, MO, USA). The column was washed successively with 2.5 mL of 10 mM Tris–HCl, pH 7.5, 0.15 M NaCl; 50 mM Tris–HCl, pH 9.0; and 50 mM Na$_2$HPO$_4$, pH 11.0. Eluted fractions (0.5 mL) were collected with each buffer and neutralized to pH 7.5.

To determine the inhibitory activity of eluted fractions, 95 μL of each fraction was pre-incubated with 5 μL of 0.5 μM bovine CPA (Sigma) for 10 min on ice. Then, CPA activity was assayed using N-(4-methoxyphenyl-azoformyl)-L-phenylalanine (AAFP, Bachem, Bubendorf, Switzerland) at a final concentration of 0.05 mM (Quilis et al., 2007). The final volume of 200 μL was added to a 96-well plates, and CPA activity was determined by decreasing absorbance at 350 nm over time. As positive control of inhibition, CPA activity was determined using different concentrations of commercial PCI (Sigma). For each fraction, the relative CPA activity was calculated as comparison of the CPA activity after addition of the corresponding elution buffer.

Insect bioassays of transgenic plants

Larvae of C. suppressalis were obtained from rice fields in the Camargue (France) and Delta del Ebro (Spain). Feeding assays with C. suppressalis larvae were performed as described by Vila and co-workers (Vila et al., 2005). Further experimental details can be found in Supporting Information. Resistance of transgenic rice plants to C. suppressalis was performed as described above, and plant symptoms (white panicles) were recorded after 21 days postinfestation.

Blast resistance assays

Resistance of mpi-pci and pci rice plants to infection by the rice blast fungus M. oryzae was carried out using the detached leaf assay (Coca et al., 2004) using a gfp-expressing M. oryzae isolate (gfp-PR9) (Campos-Soriano and San Segundo, 2009; Campos-Soriano et al., 2013). Further experimental details can be found in Supporting Information.

Acknowledgements

BLG was a researcher from the Ministerio de Ciencia e Innovación (MICINN), Spain ('Ramon y Cajal'). We are grateful to M. Coca (CRAG, Spain) for her collaboration in this work and M. Royer (CIRAD, BGPI unit, France) for providing us with the C region of the Cry1B sequence synthesized in the framework of an UNDP-funded project. We are also grateful to C. Halpin, Dundee University, UK) for providing us with the pGUS2AGFP plasmid). This work was supported by grants BIO2003-04936-C02-01/BIO2012-32838 to B.S.S and AGL2010-16847 to B.L-G from Ministerio de Economía y Competitividad (MINECO). We also thank the Consolider-Ingenio CSD2007-00036 to CRAG and the Generalitat de Catalunya (Xarxa de Referencia en Biotecnología and SGR 09626) for substantial support.

References

Abdeen, A., Virgós, A., Olivella, E., Villanueva, J., Avilés, X., Gabarra, R. and Prat, S. (2005) Multiple insect resistance in transgenic tomato plants over-expressing two families of plant proteinase inhibitors. Plant Mol. Biol. 57, 189–202.

Bayés, A., Sonnenschein, A., Daura, X., Vendrell, J. and Avilés, F.X. (2003) Procarboxypeptidase A from the insect pest Helicoverpa armigera and its derived enzyme. Eur. J. Biochem. 270, 3026–3035.

Breitler, J.C., Cordero, M.J., Royer, M., Meynard, D., San Segundo, B. and Guiderdoni, E. (2001) The −689/+197 region of the maize protease inhibitor gene directs high level, wound-inducible expression of the cry1B gene which protects transgenic rice plants from stemborer attack. Mol. Breeding 7, 259–274.

Breitler, J.C., Vassal, J.M., Catala, M.M., Meynard, D., Marfà, V., Melé, E., Royer, M., Murillo, I., San Segundo, B., Guiderdoni, E. and Messeguer, J. (2004) Bt rice harbouring cry genes controlled by a constitutive or wound-inducible promoter: protection and transgene expression under Mediterranean field conditions. Plant Biotechnol. J. 2, 417–430.

Broadway, R.M. (1996) Dietary proteinase inhibitors alter complement of midgut proteases. Arch. Insect Biochem. Physiol. 32, 39–53.

Broadway, R.M. (1997) Dietary regulation of serine proteinases that are resistant to serine proteinase inhibitors. J. Insect Physiol. 43, 855–874.

Brunelle, F., Girard, C., Cloutier, C. and Michaud, D. (2005) A hybrid, broad-spectrum inhibitor of Colorado potato beetle aspartate and cysteine digestive proteinases. Arch. Insect Biochem. Physiol. 60, 20–31.

Campos-Soriano, L. and San Segundo, B. (2009) Assessment of blast disease resistance in transgenic PRms rice using a gfp-expressing Magnaporthe oryzae strain. Plant. Pathol. 58, 677–689.

Campos-Soriano, L., Valè, G., Lupotto, E. and San Segundo, B. (2013) Investigation of rice blast development in susceptible and resistant rice cultivars using a gfp-expressing Magnaporthe oryzae isolate. Plant. Pathol. 62, 1030–1037.

Cloutier, C., Jean, C., Fournier, M., Yelle, S. and Michaud, D. (2000) Adult Colorado potato beetles, Leptinotarsa decemlineata compensate for nutritional stress on oryzacystatin I-transgenic potato plants by hypertrophic behavior and over-production of insensitive proteases. Arch. Insect Biochem. Physiol. 44, 69–81.

Coca, M., Bortolotti, C., Rufat, M., Peñas, G., Eritja, R., Tharreau, D., Martínez del Pozo, A., Messeguer, J. and San Segundo, B. (2004) Transgenic rice plants expressing the antifungal AFP protein from Aspergillus giganteus show enhanced resistance to the rice blast fungus Magnaporthe grisea. Plant Mol. Biol. 54, 245–259.

Cordero, M.J., Raventós, D. and San Segundo, B. (1994) Expression of a maize proteinase-inhibitor gene is induced in response to wounding and fungal infection: systemic wound-response of a monocot gene. Plant J. 6, 141–150.

Dale, D. (1994) Insect pests of the rice plant - Their biology and ecology. In Biology and Management of Rice Insects (Heinrichs, E.A., ed), pp. 363–485. Los Baños, Philippines: IRRI.

De Leo, F., Bonadé-Bottino, M.A., Ceci, L.R., Gallerani, R. and Jouanin, L. (1998) Opposite effects on Spodoptera littoralis larvae of high expression level of a trypsin proteinase inhibitor in transgenic plants. Plant Physiol. 118, 997–1004.

Duan, X., Li, X., Xue, Q., Abo-El-Saad, M., Xu, D. and Wu, R. (1996) Transgenic rice plants harboring an introduced potato proteinase inhibitor II gene are insect resistant. Nat. Biotechnol. 14, 494–498.

Dunse, K.M., Stevens, J.A., Lay, F.T., Gaspar, Y.M., Heath, R.L. and Anderson, M.A. (2010) Coexpression of potato type I and II proteinase inhibitors gives cotton plants protection against insect damage in the field. Proc. Natl Acad. Sci. USA 107, 15011–15015.

de Felipe, P., Luke, G.A., Hughes, L.E., Gani, D., Halpin, C. and Ryan, M.D. (2006) E unum pluribus: multiple proteins from a self-processing polyprotein. Trends Biotechnol. 24, 68–75.

François, I.E.J.A., Van Hemelrijck, W., Aerts, A.M., Wouters, P.F.J., Proost, P., Broekaert, W.F. and Cammue, B.P.A. (2004) Processing in Arabidopsis thaliana of a heterologous polyprotein resulting in differential targeting of the individual plant defensins. Plant Sci. 166, 113–121.

Gatehouse, J.A. (2002) Plant resistance towards insect herbivores: a dynamic interaction. New Phytol. 156, 145–169.

Girard, C., Le Métayer, M., Bonadé-Bottino, M., Pham-Delegue, M.H. and Jouanin, L. (1998a) High level of resistance to proteinase inhibitors may be conferred by proteolytic cleavage in beetle larvae. Insect Biochem. Mol. Biol. 28, 229–237.

Girard, C., Le Métayer, M., Zaccomer, B., Bartlet, E., Williams, I., Bonadé-Bottino, M., Pham-Delegue, M.-H. and Jouanin, L. (1998b) Growth stimulation of beetle larvae reared on a transgenic oilseed rape expressing a cysteine proteinase inhibitor. J. Insect Physiol. 44, 263–270.

Giri, A.P., Harsulkar, A.M., Deshpande, V.V., Sainani, M.N., Gupta, V.S. and Ranjekar, P.K. (1998) Chickpea defensive proteinase inhibitors can be inactivated by podborer gut proteinases. Plant Physiol. 116, 393–401.

Graham, J.S. and Ryan, C.A. (1981) Accumulation of a metallo-carboxypeptidase inhibitor in leaves of wounded potato plants. Biochem. Biophys. Res. Commun. 101, 1164–1170.

Halpin, C., Cooke, S.E., Barakate, A., El Amrani, A. and Ryan, M.D. (1999) Self-processing 2A-polyproteins - a system for co-ordinate expression of multiple proteins in transgenic plants. Plant J. 17, 453–459.

Haq, S.K., Atif, S.M. and Khan, R.H. (2004) Protein proteinase inhibitor genes in combat against insects, pests, and pathogens: natural and engineered phytoprotection. Arch. Biochem. Biophys. 431, 145–159.

Hilder, V.A., Gatehouse, A.M.R., Sheerman, S.E., Barker, R.F. and Boulter, D. (1987) A novel mechanism of insect resistance engineered into tobacco. Nature 330, 160–163.

Inanaga, H., Kobayasi, D., Kouzuma, Y., Aoki-Yasunaga, C., Iiyama, K. and Kimura, M. (2001) Protein engineering of novel proteinase inhibitors and their effects on the growth of Spodoptera exigua larvae. Biosci. Biotechnol. Biochem. 65, 2259–2264.

Johnson, R., Narvaez, J., An, G.H. and Ryan, C. (1989) Expression of proteinase inhibitors I and II in transgenic tobacco plants: effects on natural defense against Manduca sexta larvae. Proc. Natl Acad. Sci. USA 86, 9871–9875.

Jongsma, M.A., Bakker, P.L., Peters, J., Bosch, D. and Stiekema, W.J. (1995) Adaptation of Spodoptera exigua larvae to plant proteinase inhibitors by induction of gut proteinase activity insensitive to inhibition. Proc. Natl Acad. Sci. USA 92, 8041–8045.

Kim, J.Y., Park, S.C., Hwang, I., Cheong, H., Nah, J.W., Hahm, K.S. and Park, Y. (2009) Protease inhibitors from plants with antimicrobial activity. Int. J. Mol. Sci. 10, 2860–2872.

Koiwa, H., Bressan, R.A. and Hasegawa, P.M. (1997) Regulation of protease inhibitors and plant defense. Trends Plant Sci. 2, 379–384.

Kovalskaya, N., Zhao, Y. and Hammond, R.W. (2011) Antibacterial and antifungal activity of a snakin-defensin hybrid protein expressed in tobacco and potato plants. Open Plant Sci. J. 5, 29–42.

Laluk, K. and Mengiste, T. (2011) The Arabidopsis extracellular UNUSUAL SERINE PROTEASE INHIBITOR functions in resistance to necrotrophic fungi and insect herbivory. Plant J. 68, 480–494.

Lee, D.-S., Lee, K.-H., Jung, S., Jo, E.-J., Han, K.-H. and Bae, H.-J. (2012) Synergistic effects of 2A-mediated polyproteins on the production of lignocellulose degradation enzymes in tobacco plants. J. Exp. Bot. 63, 4797–4810.

Ma, C.L. and Mitra, A. (2002) Expressing multiple genes in a single open reading frame with the 2A region of foot-and-mouth disease virus as a linker. Mol. Breeding 9, 191–199.

Marino-Buslje, C., Venhudová, G., Molina, M.A., Oliva, B., Jorba, X., Canals, F., Avilés, F.X. and Querol, E. (2000) Contribution of C-tail residues of potato carboxypeptidase inhibitor to the binding to carboxypeptidase A. Eur. J. Biochem. 267, 1502–1509.

Miranda, R., Zamudio, F.Z. and Bravo, A. (2001) Processing of Cry1Ab delta-endotoxin from Bacillus thuringiensis by Manduca sexta and Spodoptera frugiperda midgut proteases: role in protoxin activation and toxin inactivation. Insect Biochem. Mol. Biol. 31, 1155–1163.

Moreno, A.B., Peñas, G., Rufat, M., Bravo, J.M., Estopà, M., Messeguer, J. and San Segundo, B. (2005) Pathogen-induced production of the antifungal AFP protein from Aspergillus giganteus confers resistance to the blast fungus Magnaporthe grisea in transgenic rice. Mol. Plant-Microbe Interact. 18, 960–972.

Outchkourov, N.S., de Kogel, W.J., Wiegers, G.L., Abrahamson, M. and Jongsma, M.A. (2004) Engineered multidomain cysteine protease inhibitors yield resistance against western flower thrips (Franklinielia occidentalis) in greenhouse trials. Plant Biotechnol. J. 2, 449–458.

Park, S., Kang, K., Kim, Y.S. and Back, K. (2009) Endosperm-specific expression of tyramine N-hydroxycinnamoyltransferase and tyrosine decarboxylase from a single self-processing polypeptide produces high levels of tyramine derivatives in rice seeds. Biotechnol. Lett. 31, 911–915.

Quilis, J., Meynard, D., Vila, L., Avilés, F.X., Guiderdoni, E. and San Segundo, B. (2007) A potato carboxypeptidase inhibitor gene provides pathogen resistance in transgenic rice. Plant Biotechnol. J. 5, 537–553.

Ralley, L., Enfissi, E.M.A., Misawa, N., Schuch, W., Bramley, P.M. and Fraser, P.D. (2004) Metabolic engineering of ketocarotenoid formation in higher plants. Plant J. 39, 477–486.

Ryan, C.A. (1990) Protease Inhibitors in Plants: Genes for Improving Defenses Against Insects and Pathogens. Annu. Rev. Phytopathol. 28, 425–449.

Sanahuja, G., Banakar, R., Twyman, R.M., Capell, T. and Christou, P. (2011) Bacillus thuringiensis: a century of research, development and commercial applications. Plant Biotechnol. J. 9, 283–300.

Schluter, U., Benchabane, M., Munger, A., Kiggundu, A., Vorster, J., Goulet, M.C., Cloutier, C. and Michaud, D. (2010) Recombinant protease inhibitors for herbivore pest control: a multitrophic perspective. J. Exp. Bot. 61, 4169–4183.

Senthilkumar, R., Cheng, C.P. and Yeh, K.W. (2010) Genetically pyramiding protease-inhibitor genes for dual broad-spectrum resistance against insect and phytopathogens in transgenic tobacco. Plant Biotechnol. J. 8, 65–75.

Srinivasan, A., Giri, A.P., Harsulkar, A.M., Gatehouse, J.A. and Gupta, V.S. (2005) A Kunitz trypsin inhibitor from chickpea (Cicer arietinum L.) that exerts anti-metabolic effect on podborer (Helicoverpa armigera) larvae. Plant Mol. Biol. 57, 359–374.

Stevens, J., Dunse, K., Fox, J., Evans, S. and Anderson, M. (2012) Biotechnological Approaches for the Control of Insects Pests in Crop Plants. In: *Pesticides - Advances in Chemical and Botanical Pesticides* (Soundararajan, R.P., ed.) pp. 269–308. Rijeka, Croatia: InTech.

Talbot, N.J. (2003) On the trail of a cereal killer: exploring the biology of *Magnaporthe grisea*. *Annu. Rev. Microbiol.* **57**, 177–202.

Tamayo, M.C., Rufat, M., Bravo, J.M. and San Segundo, B. (2000) Accumulation of a maize proteinase inhibitor in response to wounding and insect feeding, and characterization of its activity toward digestive proteinases of *Spodoptera littoralis* larvae. *Planta* **211**, 62–71.

Vila, L., Quilis, J., Meynard, D., Breitler, J.C., Marfà, V., Murillo, I., Vassal, J.M., Messeguer, J., Guiderdoni, E. and San Segundo, B. (2005) Expression of the maize proteinase inhibitor (*mpi*) gene in rice plants enhances resistance against the striped stem borer (*Chilo suppressalis*): effects on larval growth and insect gut proteinases. *Plant Biotechnol. J.* **3**, 187–202.

Villanueva, J., Canals, F., Prat, S., Ludevid, D., Querol, E. and Avilés, F.X. (1998) Characterization of the wound-induced metallocarboxypeptidase inhibitor from potato - cDNA sequence, induction of gene expression, subcellular immunolocalization and potential roles of the C-terminal propeptide. *FEBS Lett.* **440**, 175–182.

Zavala, J.A., Patankar, A.G., Gase, K. and Baldwin, I.T. (2004) Constitutive and inducible trypsin proteinase inhibitor production incurs large fitness costs in *Nicotiana attenuata*. *Proc. Natl Acad. Sci. USA* **101**, 1607–1612.

18

Identification of the factors that control synthesis and accumulation of a therapeutic protein, human immune-regulatory interleukin-10, in *Arabidopsis thaliana*

Ling Chen, Brian R. Dempsey[†], Laszlo Gyenis[†], Rima Menassa, Jim E. Brandle[‡] and Sangeeta Dhaubhadel*

Southern Crop Protection and Food Research Centre, Agriculture and Agri-Food Canada, London, ON, Canada

*Correspondence

email sangeeta.dhaubhadel@agr.gc.ca

[†]Present address: Department of Biochemistry, Western University, 1151 Richmond St., London, ON, Canada N6A 5C1.

[‡]Present address: Vineland Research and Innovation Centre, 4890 Victoria Avenue North, Vineland Station, ON, Canada L0R 2E0.

Keywords: recombinant protein, trans-gene expression, IL-10, molecular farming, gene regulation.

Summary

Plants are one of the most economical platforms for large-scale production of recombinant proteins for biopharmaceutical and industrial uses. A large number of human recombinant proteins of therapeutic value have been successfully produced in plant systems. One of the main technical challenges of producing recombinant proteins in plants is to obtain sufficient level of protein. This research aims to identify the factors that control synthesis and accumulation of recombinant proteins in stable transgenic plants. A stepwise dissection of human immune-regulatory interleukin-10 (IL-10) protein production was carried out using *Arabidopsis thaliana* as a model system. EMS-mutagenized transgenic *Arabidopsis* IL-10 lines, at2762 and at3262, produced significantly higher amount of IL-10 protein than the non-mutagenized IL-10 line (WT-IL-10). The fates of trans-gene in these sets of plants were compared in detail by measuring synthesis and accumulation of *IL-10* transcript, transcript stability, protein synthesis and IL-10 protein accumulation. The *IL-10* transcripts were more stable in at2762 and at3262 lines than WT-IL-10, which may contribute to higher protein synthesis in these lines. To evaluate whether translational regulation of IL-10 controls its synthesis in non-mutagenized WT-IL-10 and higher IL-10 accumulating mutant lines, we measured the efficiency of the translational machinery. Our results indicate that mutant lines with higher trans-gene expression contain more robust and efficient translational machinery compared with the control line.

Introduction

Plants provide one of the most economical production platforms for the large-scale production of recombinant proteins and specialized metabolites for biopharmaceutical and industrial uses. Even though these high-value products can be produced in microbial- or animal cell-based systems, plant system offers additional benefits. Plant-based systems possess a minimal possibility of product contamination with endotoxins or animal and human pathogens, and plants do not require complex bioreactors or expensive culture media for growth (Howard, 2005). With a growing demand for protein-based therapeutics and diagnostics, the current capacity to meet those requirements is insufficient, and plants provide a promising high-volume production platform. To date, a large number of human recombinant proteins of therapeutic value, including antibodies, vaccines, enzymes, hormones cytokines and growth factors, have been successfully produced in transgenic plants (Kermode, 2006; Martínez *et al.*, 2012; Matić *et al.*, 2012; Streatfield, 2007). However, accumulation levels of most proteins fall below the 1% of total soluble protein (TSP) critical for economic production (Lössl and Waheed, 2011).

Production of recombinant proteins using transient expression systems allows faster timelines (1–2 weeks) and higher expression levels (up to 50% of TSP) than stable transgenic plants (Conley *et al.*, 2011; Gleba *et al.*, 2005; Livingstone *et al.*, 2010;

Pogue *et al.*, 2010). However, transient expression is more labour intensive and can only be used at industrial levels for high-value proteins that are needed in relatively small amounts. A major advantage of stable transgenic plants is that the heterologous protein production trait is heritable, resulting in a permanent resource, allowing for simple and rapid scale-up and almost unlimited and sustainable production capacity only requiring planting of seeds in a large area and harvesting it (Wang *et al.*, 1996). However, for production of recombinant proteins in stable transgenic plants to be economically feasible, the key issue that must be addressed is improving protein yields (Conley *et al.*, 2011; Fischer *et al.*, 2004; Kermode, 2006). Protein production is a multi-step process that starts with transcription and is regulated at transcriptional, post-transcriptional, translational and post-translational levels. Several elements have been used for optimizing trans-gene expression, including strong promoters to boost transcription, 3' untranslated regions for stabilization of the transcripts and proper processing, and translational enhancers to boost translation (reviewed in Streatfield, 2007). Additionally, positional effects on trans-gene integration site, trans-gene copy number effect (Zhong *et al.*, 1999), subcellular targeting (Streatfield *et al.*, 2003) and protection from proteolytic degradation (Benchabane *et al.*, 2008; Doran, 2006) are critical parameters that affect accumulation of recombinant proteins. Despite a two and half decade long search for methods to boost trans-gene expression, it is still very difficult to produce large amounts of therapeutic proteins in stable transgenic plants.

The goal of this study was to identify the native plant factor(s) that limit the synthesis and accumulation of a recombinant protein, human immune-regulatory protein interleukin-10 (IL-10), in transgenic plants. For this, we generated a transgenic *Arabidopsis thaliana* line homozygous for *IL-10* gene. Ethyl methane sulphonate (EMS) mutagenesis of the IL-10 transgenic line was performed for generating mutant lines accumulating higher levels of IL-10, and a stepwise dissection of *IL-10* expression was carried out using the wild type and the mutant lines. Overall, the results indicate that in addition to increased *IL-10* transcript stability, the mutant lines possess a more robust and efficient translational machinery compared with the control line.

Results and discussion

Identification and characterization of mutants expressing high levels of IL-10 protein

To identify the factor(s) responsible for low yield of recombinant proteins in plants, we used a transgenic Arabidopsis line homozygous for human *IL-10* gene (WT-IL-10) and employed random chemical mutagenesis. This unbiased approach allowed us the random screening of genes that affected protein expression and accumulation. The expression of *IL-10* was driven by 35S promoter that directs constitutive expression in leaves and other vegetative tissues. Previously, *IL-10* has been overexpressed in a tobacco system that produced inherently low yields (Menassa *et al.*, 2001, 2004). In feeding trials using a mouse model system of inflammatory bowel disease, IL-10 null mice displaying an aggressive form of colitis required a diet supplemented with 30% IL-10 expressing tobacco to achieve a relatively modest reduction in the symptoms (Menassa *et al.*, 2007). It is not feasible to supplement a human diet with such a high dose of tobacco for a significant reduction in disease state. Thus, *IL-10* expression in plant systems exemplifies the problems intrinsic to the plant-made pharmaceutical field and makes for an excellent model system.

The Arabidopsis expression system chosen for the present study utilizes targeting of the IL-10 protein to the endoplasmic reticulum to increase protein accumulation to a moderate level when compared to targeting other subcellular locations (Menassa *et al.*, 2001). The attempt to produce IL-10 in tobacco leaves yielded 0.0055% IL-10/TSP in the highest IL-10 producing transgenic tobacco plant (Menassa *et al.*, 2001). When we expressed the same construct in Arabidopsis and measured the level of IL-10 accumulation in several *IL-10* overexpressing Arabidopsis plants, we obtained an average of 0.007 ± 0.0002% IL-10/TSP. The lower level of IL-10 accumulation in tobacco may be due to positional effects of the trans-gene in the genome or differences in organism physiology. Nevertheless, the modest level of *IL-10* expression in Arabidopsis is ideal for the purposes of this study, allowing for observation of an increase in protein accumulation.

To determine what factors are critical for recombinant protein synthesis and accumulation, we performed mutagenesis with varying concentrations of EMS. Treatment of the WT-IL-10 seeds with 0.2%, 0.3% and 0.4% EMS caused reduction in the seed germination rate by 16%, 24% and 25%, respectively. This falls into the suggested range of 10%–25% reduction in germination (Lightner and Caspar, 1998), indicating that an optimum rate of mutation was achieved. A total of 4500 M1 plants were screened for changes in the level of IL-10 protein accumulation compared with WT-IL-10. A schematic diagram of the mutant generation

and screening is shown in Figure 1. This screening process identified four mutant lines with a significant increase in percentage of IL-10/TSP (Figure 2a). The lines with higher IL-10 levels were further advanced to obtain homozygous lines. While the absolute value of percentage of IL-10/TSP for these mutants would not normally be considered high, the fact that it was possible to increase IL-10 accumulation by threefold in magnitude through chemical point mutation was a significant achievement.

During the production of the WT-IL-10 and mutant lines, segregation on selection media was used to ensure that the transgenic plants were homozygous. To confirm the single copy *IL-10* insertion in Arabidopsis genome, genomic DNA from WT-IL-10 and mutant lines were extracted, and the DNA was completely digested with the restriction enzyme EcoRV. The enzyme EcoRV was selected because it digests the DNA with reasonable frequency, but does not cut within the *IL-10* gene (Figure 1). Southern blot analysis was performed using *IL-10* probe (Figure 2b). The result showed only one copy of *IL-10* gene in the transgenic lines while wild-type Arabidopsis Col-0 did not show presence of *IL-10* in it. The blot was reprobed using a *nptII* probe against the kanamycin resistance gene. This *nptII* blot confirmed the results of the *IL-10* Southern blot (data not shown), indicating the presence of a single copy of the *IL-10* gene cassette in the WT-IL-10 and high IL-10 accumulating mutants.

To evaluate whether improved IL-10 protein accumulation observed in EMS mutant lines was a result of increased levels of *IL-10* gene transcript, we compared amounts of *IL-10* mRNA accumulation in WT-IL-10 and the mutant lines. Total RNA was isolated from mature rosette leaf tissues and subjected to semi-quantitative RT-PCR. As shown in Figure 2c, no significant difference in the *IL-10* transcript accumulation was observed between WT-IL-10 and mutant lines based on student's *t*-test ($P < 0.05$). These results, together with the complete sequencing of IL-10 gene in WT-IL-10 and mutant lines, suggested that increases in *IL-10* protein level in mutant lines are not due to alteration of the *IL-10* gene or steady-state transcript levels but possibly as a result of the differences at the post-transcriptional processes.

Arabidopsis lines with higher IL-10 accumulation show similar IL-10 nascent mRNA synthesis but differ in IL-10 transcript stability compared with WT-IL-10

To identify the mechanism of increased IL-10 production in EMS-mutagenized Arabidopsis lines, we selected the two highest IL-10 accumulating mutant lines *at2762* and *at3262* for detailed characterization. The amount of nascent *IL-10* RNA produced and *IL-10* transcript stability were measured in the mutant lines and compared with WT-IL-10. To analyze potential differences in the transcript initiation between WT-IL-10 and mutants, a nuclear run-on assay was performed in nuclei isolated from leaves of *at2762*, *at3262* and WT-IL-10 lines. The nuclear run-on assay provides a measure of the frequency of transcription initiation and is not dependent on the effects of transcript stability (Smale, 2009). Isolated nuclei were incubated with radio-labelled uridine 5′-triphosphate and other nucleoside triphosphates. During this incubation process, new transcripts are not initiated, but radio-labelled nucleotides become incorporated into transcripts that were in the process of being synthesized when the nuclei were isolated. The number of nascent transcripts for a gene is proportional to the transcriptional initiation. Comparison of nascent *IL-10* transcripts in *at2762*, *at3262* and WT-IL-10 lines showed no significant difference in transcript levels suggesting that higher levels of IL-10 protein accumulation in *at2762* and *at3262*

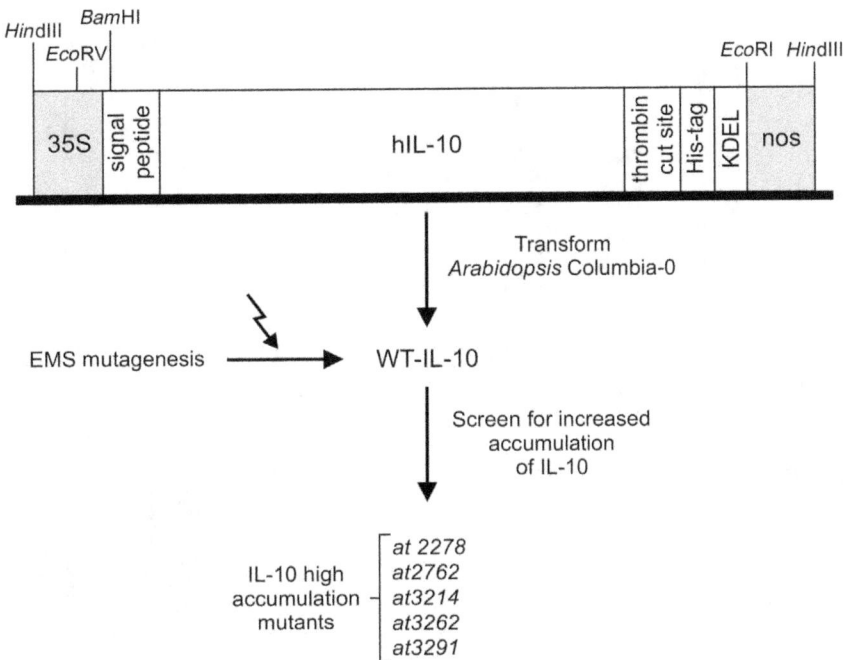

Figure 1 Schematic diagram of *IL-10* construct and experimental strategy. The human *IL-10* gene (*hIL-10*) was placed under control of the enhanced 35S cauliflower mosaic virus promoter and nos terminator. The hIL-10 protein contains a C-terminal, thrombin cleavable, His-tag and KDEL endoplasmic reticulum (ER) retention sequence as well as an N-terminal signal sequence for ER targeting. The WT-IL-10 line was created by transforming the construct into the Arabidopsis Col-0. The WT-IL-10 line was chemically mutagenized with ethyl methanesulphonate EMS, and the mutant plants were screened and identified by cytokine ELISA for increased accumulation of IL-10.

lines were not due to increased transcript initiation but possibly due to post-transcriptional regulation of the *IL-10* gene (Figure 3).

In eukaryotic cells, rates of mRNA degradation are highly variable, and these differences in mRNA stability allow for precise control of gene expression (Gutiérrez *et al.*, 1999). To determine whether *IL-10* mRNAs are more stable in *at2762* and *at3262* lines compared with WT-IL-10, we measured *IL-10* mRNA half-lives in these three lines. Two-week-old *at2762*, *at3262* and WT-IL-10 Arabidopsis seedlings were treated with the transcription inhibitor cordycepin. *IL-10* mRNA level was examined by conventional Northern blot hybridization using an *IL-10* cDNA probe. *IL-10* transcripts were relatively more stable in the lines *at2762* and *at3262* compared with WT-IL-10. The experiment was repeated three times with three independent biological replicates. The results were reproducible and similar pattern of *IL-10* transcript was observed in all three replicates. A representative example of the *IL-10* mRNA stability study is shown in Figure 4. This observation supports our analysis of steady-state level of *IL-10* mRNA and that the subtle increase in the stability of the transcripts in the mutant lines (Figure 4) was not reflected in the transcript accumulation measured by the semi-quantitative RT-PCR (Figure 2c). Our results suggest that the slightly altered mRNA turnover in the mutant plant lines possibly causes an increase in IL-10 protein accumulation. EMS mutagenesis causes random point mutations in the genome; however, the frequency of point mutation cannot be controlled; therefore, the exact causes for increased *IL-10* transcript or decreased *IL-10* mRNA turnover in mutant lines are not yet known. The molecular mechanism of mRNA stability is often controlled at several levels that include a basal decay process, a gene- or sequence-specific decay process and an external stimulus-induced decay process (Gutiérrez *et al.*, 1999). It is

possible that increased stability of *IL-10* transcripts in *at2762* and *at3262* lines compared with WT-IL-10 may be due to a sequence-specific process. Indeed, because all three Arabidopsis lines were exposed to a similar environment and there was no significant difference in the stability of *ubiquitin* control transcript in the three lines, the mutation may have occurred within the *IL-10* gene cassette, resulting in increased stability of the transcript in the mutants versus WT-IL-10. Several studies have demonstrated the significance of a downstream element, AUUUA repeats and non-sense codons in controlling mRNA stability (Reviewed in Gutiérrez *et al.*, 1999). Two very recent discoveries have shown that the promoter elements from which expression of the transcript was driven also are capable of regulating mRNA decay (Bellofatto and Wilusz, 2011; Bregman *et al.*, 2011; Burgess, 2012; Trcek *et al.*, 2011); therefore, a mutation within or close to the promoter may be responsible for reduced IL-10 mRNA turnover in our mutant lines.

Protein synthesis machinery is altered in high IL-10-producing mutant lines

The abundance of cytosolic transcripts is not always directly proportional to the protein product. Translational and post-translational processes are important events that are involved in the control of gene expression in higher plants (Kawaguchi and Bailey-Serres, 2002; Kawaguchi *et al.*, 2004). Ribosome whose occupancy determines the fraction of transcripts engaged in translation and ribosome density, which affects the number of ribosomes per mRNA transcript, are the two rate-limiting steps in translation (Arava *et al.*, 2003). These characteristics delineate the translational efficiency of a transcript and are affected by exposure to differential environmental stimuli or mutations.

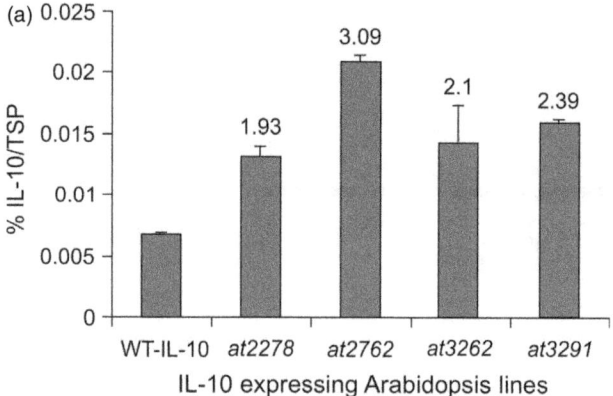

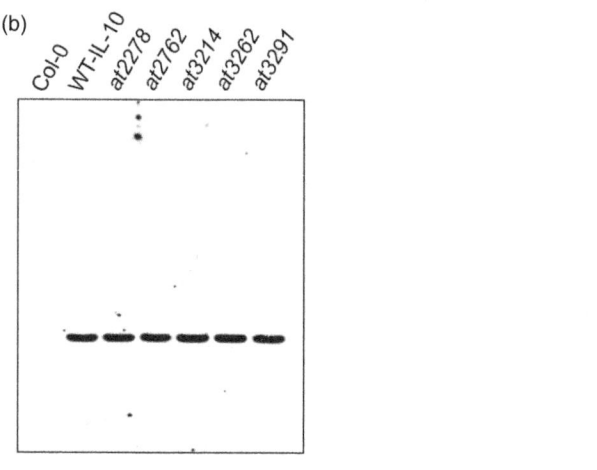

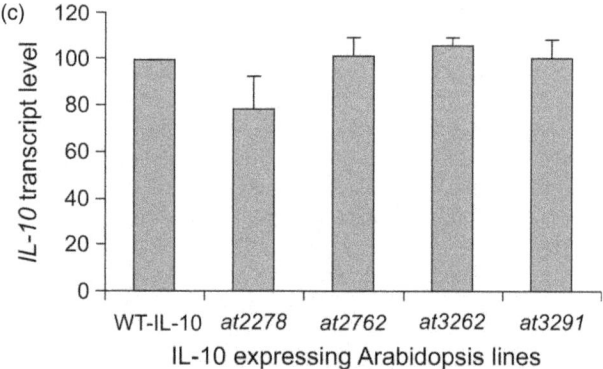

Figure 2 Accumulation of IL-10 protein and transcripts in WT-IL-10 and EMS mutant lines. Four-week-old Arabidopsis leaf tissues were used for RNA and protein analysis. (a) Total proteins were isolated, and protein concentrations were determined by Bradford method. The amount of IL-10 accumulation in each line was measured by ELISA. The value on the top of each bar indicates the fold increase in IL-10 accumulation relative to the WT-IL-10. (b) Southern blot analysis to investigate the copy of *IL-10* gene insertion in wild-type Arabidopsis Col-0, WT-IL-10 and the mutant lines. (c) Total RNA (2 μg) was used as template for semi-quantitative RT-PCR analysis. The *IL-10* mRNA level in each line was measured by using gene-specific primers for *IL-10*. The data were normalized by using Arabidopsis *actin* gene as loading control. For (a) and (c), data shown are mean values from three independent experiments. Bars indicate the standard errors of means.

To investigate whether translational regulation of IL-10 controls its synthesis in WT-IL-10, *at2762* and *at3262* lines, we performed polysomal RNA analysis and measured the efficiency

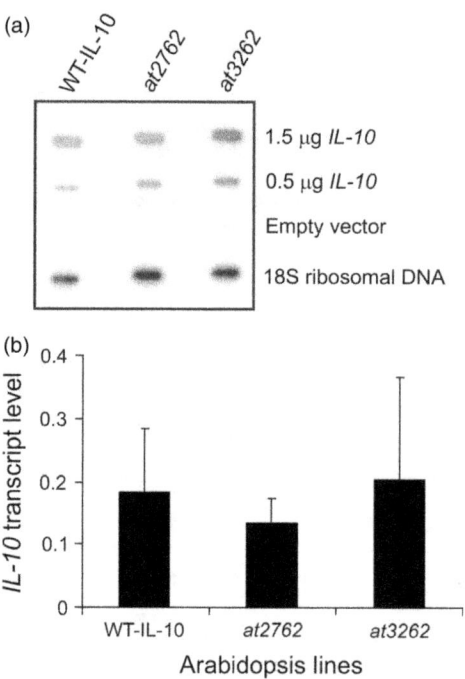

Figure 3 Nuclear run-on assay of *IL-10* transcripts. (a) Slot-blot filter containing gene-specific single stranded DNAs as indicated were hybridized to ^{32}P-labelled nascent transcripts from WT-IL-10, *at3262* and *at2762* lines which were synthesized by run-on transcription. Empty vectors (1 μg) and 18S ribosomal DNA (0.15 μg) were used as controls. (b) Quantification of the amount of *IL-10* transcript corresponding to 1.5 μg of *IL-10* in Figure a. Data are representative of three independent experiments.

of the translational machinery. The measurement of the proportion of *IL-10* mRNA species in polysomal complexes in these lines reflects the efficiency of initiation and re-initiation, as well as the rate of elongation and termination of IL-10 translation. Although there is evidence that not all polysome-bound mRNAs are translated into protein and that some fraction may arrest without translation (Dhaubhadel *et al.*, 2002), the distribution of an mRNA within the polysome gradient reflects its translational efficiency.

Polysomes were isolated and fractionated over sucrose density gradients, and the distribution of the *IL-10* mRNA on polysomes (ribosomes) was determined in all 13 fractions by Northern blot analysis (Figure 5a). Fraction 2 to 7 contain non-polysomal RNAs, 8 and 9 contain small polysomes (two to five ribosomes per mRNA) and fractions 10 to 13 contain large polysomes (Kawaguchi *et al.*, 2003). The majority of the *IL-10* mRNA was found in polysomal fractions (Fractions 8 to 13) in all three lines indicating that *IL-10* mRNAs were actively engaged in translation (Figure 5b). The experiment was repeated at least three times, and the results were reproducible. *In vitro*-transcribed *GFP* RNA was used as a spike-in control. The spike-in control is necessary for correcting variations between the samples during sample processing (Melamed *et al.*, 2009). Normalization of *IL-10* accumulation signals with *GFP* control showed higher levels of *IL-10* mRNA loading in small polysomal fractions in *at2762* and *at3262* compared with WT-IL-10 (Figure 5b). Our results suggest that there is a relatively higher level of small polysome recruitment in *at2762* and *at3262* lines compared with WT-IL-10 and that

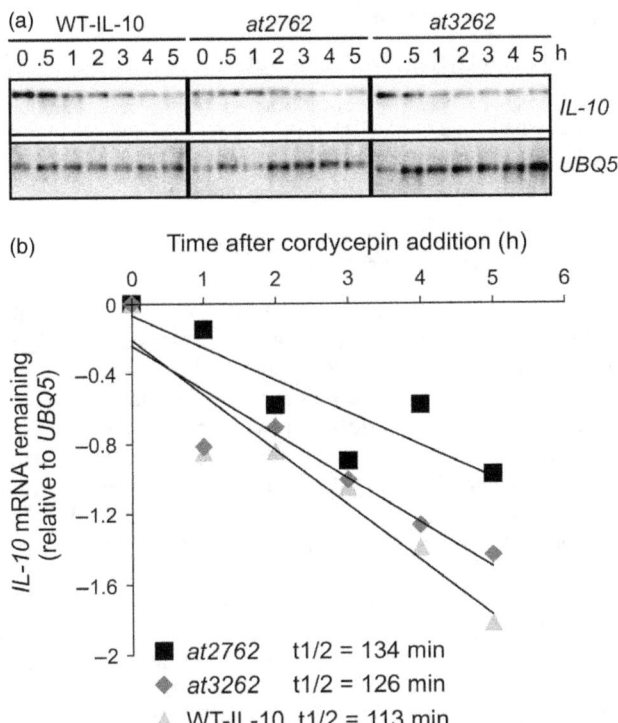

Figure 4 Kinetics of *IL-10* mRNA decay in WT-IL-10 and mutant lines *at3262* and *at2762*. (a) Two-week-old Arabidopsis plants were treated with 0.6 mM cordycepin, and tissues were collected at different time points as indicated. Total RNA (3 μg) was separated by denaturing formaldehyde gel electrophoresis, transferred into a nylon membrane and hybridized with a ³²P-labelled *IL-10* gene fragment. The membrane was stripped and rehybridized with *ubiquitin5* (*UBQ5*). Representative Northern blot analysis of half-life experiments in WT-IL-10, *at2762* and *at3262* lines are shown. (b) Quantification of decrease in *IL-10* transcript abundance and half-life estimation for the representative experiment are shown in the Figure a. The data were normalized to the amount of *UBQ* transcript that does not change significantly during the time course.

higher *IL-10* mRNA distribution in the polysome may have resulted into higher IL-10 protein accumulation in *at2762* and *at3262* than in WT-IL-10.

To determine whether the levels of translation factors in high IL-10 accumulating mutant lines were different from those in WT-IL-10 and, if so, to assess whether they could be correlated with the levels of IL-10 synthesis in those mutant lines, we analyzed the accumulation of translation initiation and elongation factors in *at2762*, *at3262* and WT-IL-10 lines. Previously, antibodies raised against wheat and Arabidopsis translation initiation factors (eIFs) and elongation factors (eEFs) have been used to determine the levels of these translation factors by Western blotting (Browning, 1996; Browning *et al.*, 1990; Mayberry *et al.*, 2007). Antibodies to wheat factors that cross-reacted with Arabidopsis proteins of the sizes consistent with wheat proteins were used.

Of several eIFs used in the experiment, antisera raised against wheat eIF4A and Arabidopsis eIFiso4G recognized a similar size protein in Arabidopsis were expressed at higher levels in *at2762* and *at3262* mutant lines compared with WT-IL-10 (Figure 6). eIF4A works in complex with eIF4F/eIFiso4F and eIF4B in the ATP-dependent unwinding of the 5′-untranslated region of mRNA to remove secondary structures prior to binding of the

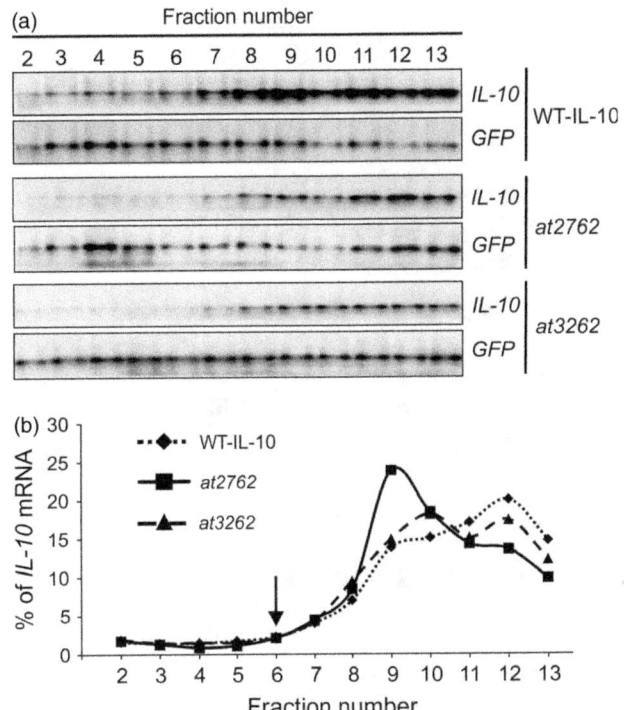

Figure 5 Polysomal RNA analysis. (a) Polysomal RNA was isolated from WT-IL-10, *at2762* and *at3262* lines, subjected to 15–60% sucrose gradient centrifugation and fractionated into 13 fractions (fraction 1 at top of the gradient). Each fraction was spiked-in with *in vitro*-transcribed *GFP* and RNA was purified, separated, blotted onto nylon membrane and hybridized with ³²P-labelled *IL-10* probe. Blots were stripped and reprobed with *GFP* for data normalization. A representative data of three independent experiments are shown. (b) The blot in (a) was quantified by densitometric analysis. The normalized amount of *IL-10* in each fraction was calculated using *GFP* control.

40S ribosomal unit (Balasta *et al.*, 1993). The function of eIFiso4G (as part of eIFiso4F) is largely to support translation from capped mRNAs with unstructured 5′-leader sequence (Gallie and Browning, 2001).

Among the eEFs, antisera raised against wheat eEF1A and eEF2 recognized similar size proteins in Arabidopsis were accumulated at much higher levels in *at2762 and at3262* mutant lines compared with WT-IL-10 (Figure 6). The major role of eEF1A is to bind GTP and aminoacyl-tRNA and bring them to elongating 80S ribosome during the translation elongation phase, whereas eEF2 functions to recycle the eEF1AGDP complex and catalyze the GTP-dependent translocation of peptidyl-tRNA from the A site to P site on the ribosome (Browning, 1996). The amount of these translational factors in wheat seeds correlates with the period of greatest protein synthetic activity (Gallie *et al.*, 1998a,b). From the observation of past studies and our own results, we extrapolate that the EMS mutation may have created a more robust and efficient translational apparatus in the mutant lines *at2762* and *at3262*. Even though the frequency of EMS mutation depends on the position of the gene in the genome and the treatment condition (Kovalchuk *et al.*, 2000), it generally results in high point mutational densities of 1 mutation/300 kb in Arabidopsis (Greene *et al.*, 2003). The precise mutations in the mutant lines that trigger the IL-10 accumulation are yet to be determined.

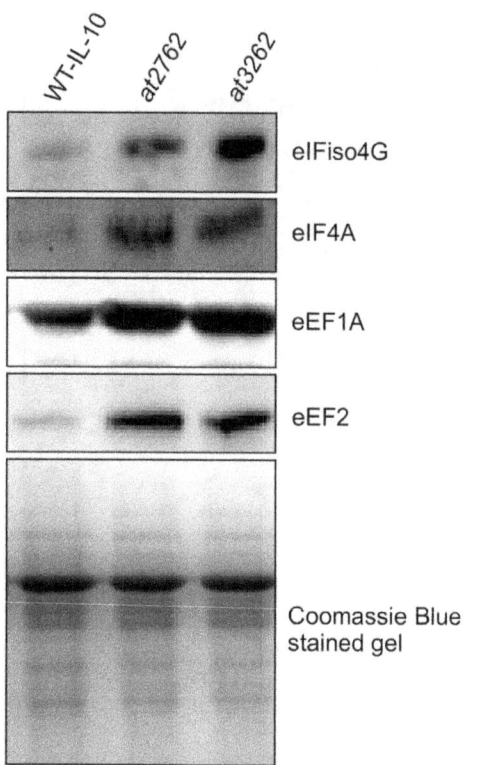

Figure 6 Accumulation of translation initiation and elongation factors in WT-IL-10, *at2762* and *at3262* lines. Four-week-old Arabidopsis leaves were used for protein extraction. Total soluble protein (15 μg) was separated on a SDS-PAGE and transferred onto PVDF membrane by electroblotting. The eIFs and eEFs were detected by sequential incubation of the membrane with anti-eIFiso4G, eIF4A, eEF1A and eEF2 antibodies and HRP-conjugated anti-rabbit IgG followed by the chemiluminescent reaction. A Coomassie Brilliant Blue stained gel is shown as a loading control.

In conclusion, the mutant lines *at3262* and *at2762* accumulate significantly higher level of IL-10 protein compared with WT-IL-10. The transcript accumulation data demonstrated that there were no significant difference in both the nascent and steady-state *IL-10* transcript level in WT-IL-10 and the mutant lines. However, stability of *IL-10* transcripts was slightly higher and recruited more small polysomes onto *IL-10* mRNAs in *at3262* and *at2762* lines compared with WT-IL-10 possibly leading to a higher IL-10 accumulation in *at3262* and *at2762* than WT-IL-10. Most importantly, four translational factors, eIF4A, eIFiso4G, eEF1A and eEF2, accumulate to higher levels in the mutant lines compared with the WT-IL-10 lines.

Overall, our results suggest that the high IL-10-accumulating lines possess a more efficient and robust translational machinery. The advantage of the random mutagenesis approach is that one can screen for desired phenotypic effects, then identify the gene (s) responsible for these effects. Subsequently, this gene(s) could be down-regulated by RNAi to test whether a similar effect is encountered. In our case, if manipulating the expression of the identified gene(s) result in more robust translational machinery, such a system could be adapted to tobacco or *Nicotiana benthamiana* and might become useful for the production of recombinant proteins in transient systems as well as in stable transgenic plants.

Experimental procedures

Plant material and growth condition

Arabidopsis Col-0 seeds were obtained from the Arabidopsis Biological Resource Center (ABRC). Plants were grown in pots under the condition of a 16 h light cycle at 25 °C/8 h dark cycle at 20 °C with 70%–80% relative humidity.

Generation and analysis of transgenic Arabidopsis lines overexpressing IL-10 proteins

To generate human IL-10 expressing Arabidopsis plants, 6-weeks-old Arabidopsis plants were transformed with *Agrobacterium tumefaciens* LBA4404 containing phIL-10C (Menassa *et al.*, 2001) using floral dip method (Clough and Bent, 1998). The T_0 seeds were sterilized and planted on in Murashige and Skoog basal medium (Murashige and Skoog, 1962) containing 50 μg/mL kanamycin and 0.8% agar. Seeds were vernalized for 48 hours at 4 °C in the dark, then transferred to 24-h illumination at 40–55 μmol/m²s at 22 °C. Four-leaf stage kanamycin resistant seedlings were transferred to soil and grown at the above-described condition. Four of 5 mm leaf disc samples from fully grown rosette leaves were collected, frozen immediately in liquid nitrogen and stored at −80 °C until needed for analysis.

The leaf tissue was ground with zirconia/silica beads in a tissue mill in protein extraction buffer phosphate-buffered saline (PBS) pH 7.4 containing 0.1% Tween 20, 2% (w/v) polyvinylpolypyrrolidone, 1 mM EDTA, 1 mM phenylmethylsulphonyl fluoride, 1 μg/mL leupeptin and 100 mM ascorbate. Extracts were cleared by centrifugation at 2300 *g* for 15 min at 4 °C. Each transgenic plant was tested for the IL-10 expression by double antibody sandwich ELISA using anti-human IL-10 monoclonal antibodies following manufacturer's instructions (BD Pharmingen). Purified human IL-10 was used as standards in the ELISA (BD Pharmingen, Mississauga, ON, Canada). The total soluble protein level of each transgenic Arabidopsis plant was measured by Bradford assay (Bio-Rad Ltd, Mississauga, ON, Canada) using BSA as a standard. The highest IL-10 expressing lines that showed Mendelian segregation of a single gene were further advanced to the next transgenic generation.

Generation of mutant Arabidopsis IL-10 lines

The T_3 generation of highest IL-10 expressing homozygous line (WT-IL-10) was subjected to EMS mutagenesis following a protocol by Redei *et al.* (1984). The pooling seed mutagenesis method (Lightner and Caspar, 1998) was followed where 40 mg of WT-IL-10 seeds was treated for 16 h with different concentration of EMS at 0.2%, 0.3% and 0.4% or water only (control). After EMS treatment, the seeds were neutralized with 100 mM $Na_2S_2O_3$ for 15 min two times followed by several rinses with distilled water. The mutagenized seeds were planted in moist sand in trays in the fume hood. The trays were kept in the fume hood for 2 days to allow volatilization of remaining EMS. The seeds were then vernalized for 2 days at 4 °C and planted into containers establishing 12 pools of seeds for each EMS level. Seed were harvested from each pool and maintained as a unit. Equal amount of seeds were planted from each pool, and IL-10 level was determined by double antibody ELISA. The mutant lines that maintained increased IL-10 production (at least two standard deviations above the wild-type level) after the third mutant generation level in two experiments were subjected to genetic and phenotypic characterization.

Southern blot analysis

For genomic DNA isolation, leaf tissues were collected and pooled from 15 plants for each line. The genomic DNA was purified using a CTAB-based purification protocol (Lukowitz et al., 2000; Murray and Thompson, 1980). A total of 15 μg of genomic DNA was digested using the EcoRV restriction enzyme, and Southern blot was performed according to the protocol of Sambrook and Russell (Russell, 2001). Creation of a digoxigenin-(DIG) labelled IL-10 probe and development of the Southern blot were carried out according to the methods outlined in the DIG DNA Labelling and Detection Starter Kit II (Roche, Mannheim, Germany). The template for the IL-10 probe was a PCR product produced from primers located at the ends of the IL-10 gene sequence. After development, the blot was stripped and reprobed using a DIG labelled probe against the nptII gene.

RNA isolation and RT-PCR

Total RNA was isolated from mature rosette leaf tissues collected from 4-week-old Arabidopsis plants using RNeasy Plant Mini Kit (Qiagen, CA, USA). RNA was quantified using nanodrop at 260 nm. Reverse transcription reactions were performed with total RNA (1 μg) using Thermoscript™ RT-PCR system (Invitrogen, Carsbad, CA, USA) according to the manufacturer's instructions. PCR reactions were performed using IL-10 primers. A 634 bp IL-10 amplicon was obtained using the primers IL-10BamHI 5'-GGGAT CCACAAGACAGACTTGCAAAAGAAGG-3' and IL-10SacII 5'-TCC CCGCGGTCAGAGCTCGTCCTTGTG-3'. Actin was used as a reference gene for data normalization. Primer sequences for actin PCR were as follows: actinF 5'- ATGGCAGACGGTGAGGATATT CA-3' and actinR 5'-GCCTTTGCAATCCACATCTGTTG-3'. This primer set generated an amplicon size of 1085 bp. Three biological replicates and two technical replicates for each biological replicate were used. The densitometric analysis was performed using Quantity One software (Bio-Rad Ltd, Mississauga, ON, Canada).

Nuclei isolation and nuclear run-on assay

Nuclei were isolated from 4-week-old Arabidopsis leaves (6 g) according to Folta and Kaufman (2006). Nuclei from different IL-10 Arabidopsis lines were isolated at the same time and processed simultaneously for nuclear run-on assay. The nuclear run-on assay was performed as described before (Folta and Kaufman, 2006). In vitro synthesis of transcripts was performed by using MEGAscript® kit (Ambion Inc., Burlington, ON, Canada) and used as probes in slot-blot hybridization. Two concentrations of IL-10 DNA and control DNAs (empty vector and ribosomal DNA) were blotted into Hybond-N+ nylon membrane (Amersham, GE Healthcare Ltd.) using slot-blot apparatus and hybridized with [α-^{32} P] UTP-labelled in vitro-synthesized transcripts. Following hybridization, the blots were washed three times with 0.5× SSC and 0.1% SDS at 65 °C for 15 min followed by one wash with 0.5× SSC and 0.1% SDS at room temperature for 15 min. The membrane was exposed to phosphor imaging screen (Bio-Rad Laboratories, Hercules, CA, USA). The hybridization signals were detected using a Molecular Imager Fx phosphorimager (Bio-Rad Laboratories, Hercules, CA, USA) and quantified using Quantity One software (Bio-Rad Laboratories, Inc.).

mRNA half-life measurements and Northern blot analysis

For the half-life measurement, two-week-old Arabidopsis seedlings grown on seed selection media were incubated in leaf incubation buffer for 30 min at room temperature followed by cordycepin treatment to a final concentration of 0.6 mM as described by Seeley et al.(1992). Tissue samples were collected at regular intervals after cordycepin treatment, frozen immediately in liquid nitrogen and stored at -80^0 C. Total RNA was isolated using RNeasy Plant Mini Kit (Qiagen, CA, USA). Northern blot hybridization was carried out as described previously (Dhaubhadel et al., 1999) using ^{32}P-labelled IL-10 probe. For a loading control in Northern blot hybridization, Arabidopsis thaliana ubiquitin 5 (UBQ5) (accession no. NM_116090) was amplified using the primer set 5'-GTGGTGCTAAGAAGAGGAAGA-3' and 5'-TCAAGCTTCAACTCCTTCTTT-3' and was used as a probe. Following hybridization with IL-10 probe, the blots were washed, and signals were measured as described earlier. The blot was stripped and hybridized with UBQ5. The signals were quantified by densitometric analysis, and mRNA half-life was determined using the following equation: $t_{1/2}=\ln(2)/k_d$, where k_d is a decay constant and $t_{1/2}$ is half-life.

In vitro transcription

The smGFP gene was amplified with primers GFP-F: 5'-GGGG CGGGATCCATGAGTAAAGGAGAAG-3' and GFP-R: 5'-GCGGG GGCGGAATTCTTATTTGTATAGTTCATC-3', cloned into pGEM-T vector (Promega, Madison, USA), sequence verified, followed by in vitro transcription using MEGAscript® kit using manufacturer's instruction (Ambion Inc.).

Polysome isolation and polysomal RNA analysis

Mature rosette leaves (6 g) from four-week-old Arabidopsis plants were used for polysomal RNA analysis. Polysomes were isolated as described by Kawaguchi et al. (2004). Polysomal RNA from Arabidopsis lines at2762, at3262 and WT-IL-10 were measured at 260 nm, and equal amounts (OD$_{260}$ = 5) were separated on a 15%–60% (v/v) sucrose density gradient by ultracentrifugation. Following the centrifugation, the density gradient was fractionated into 300-μL fractions (total of 14 fractions) using a density gradient fractionation system connected to an UA-6 detector (ISCO, Lincoln., NE, USA). Each fraction was spiked-in with 80 ng of in vitro-transcribed GFP mRNA before RNA purification. The spike-in control was used to correct for variations between the collected fractions in all subsequent steps. RNA from each fraction was extracted twice with phenol/chloroform (3:1), then precipitated with 3 M sodium acetate (pH 5.2) and absolute ethanol at -20°C. The RNA pellet was recovered by centrifugation at 16,000 g for 20 min at 4°C. Northern blotting was performed as described previously. The membrane was hybridized with ^{32}P-labelled IL-10 probe, washed and exposed to a phosphor imaging screen (Bio-Rad Laboratories, Hercules, CA, USA). The phosphor imaging screen was scanned using a Molecular Imager Fx phosphorimager (Bio-Rad, Hercules, USA). The membrane was stripped and used for hybridization with ^{32}P-labelled GFP probe.

Western blotting analysis

Protein extraction and quantification was performed as described previously (Dhaubhadel et al., 2005). For Western blot analysis, 15 μg protein from each sample was separated on a SDS-PAGE according to the method of Laemmli (1970) and transferred onto PVDF membrane by electroblotting using a semi-dry electrophoretic transfer system (Bio-Rad Laboratories, Mississauga, ON, Canada). Antibodies against various translational initiation and elongation factors were obtained from Dr. Karen Browning (University of Texas, USA), and detection of corresponding protein accumulation in the samples was performed by sequential incubation of the blot with primary antibody (against eIFs and eEFs) and HRP-conjugated anti-rabbit IgG at the dilution of 1 : 1000 and 1 : 20 000, respectively. The proteins were detected by the enhanced chemiluminescent detection system using Supersignal® West Femto Maximum Sensitivity Substrate (Pierce Biotechnology, Rockford, IL).

Acknowledgements

We are grateful to Dr Karen Browning (University of Texas, USA) for the translational factor antibodies, Dr. Julia Bailey-Serres (University of California, USA) for advice on polysomal RNA analysis, Dr Fred Souret and Dr Pamela Green (University of Delaware, USA) for information on RNA stability and half-life calculation and Alex Molnar (Agriculture Canada, London) for help with the figures. This work was supported by Agriculture and Agri-Food Canada's Abase grant awarded to S.D.

References

Arava, Y., Wang, Y., Storey, J.D., Liu, C.L., Brown, P.O. and Herschlag, D. (2003) Genome-wide analysis of mRNA translation profiles in *Saccharomyces cerevisiae*. *Proc. Natl Acad. Sci.* **100**, 3889–3894.

Balasta, M.L., Carberry, S.E., Friedland, D.E., Perez, R.A. and Goss, D.J. (1993) Characterization of the ATP-dependent binding of wheat germ protein synthesis initiation factors eIF-(iso)4F and eIF-4A to mRNA. *J. Biol. Chem.* **268**, 18599–18603.

Bellofatto, V. and Wilusz, J. (2011) Transcription and mRNA stability: parental guidance suggested. *Cell* **147**, 1438–1439.

Benchabane, M., Goulet, C., Rivard, D., Faye, L., Gomord, V. and Michaud, D. (2008) Preventing unintended proteolysis in plant protein biofactories. *Plant Biotechnol. J.* **6**, 633–648.

Bregman, A., Avraham-Kelbert, M., Barkai, O., Duek, L., Guterman, A. and Choder, M. (2011) Promoter elements regulate cytoplasmic mRNA decay. *Cell* **147**, 1473–1483.

Browning, K.S. (1996) The plant translational apparatus. *Plant Mol. Biol.* **32**, 107–144.

Browning, K.S., Humphreys, J., Hobbs, W., Smith, G.B. and Ravel, J.M. (1990) Determination of the amounts of the protein synthesis initiation and elongation factors in wheat germ. *J. Biol. Chem.* **265**, 17967–17973.

Burgess, D.J. (2012) RNA stability: remember your driver. *Nat. Rev. Genet.* **13**, 72–73.

Clough, S.J. and Bent, A.F. (1998) Floral dip: a simplified method for *Agrobacterium*-mediated transformation of *Arabidopsis thaliana*. *Plant J.* **16**, 735–743.

Conley, A.J., Joensuu, J.J., Richman, A. and Menassa, R. (2011) Protein body-inducing fusions for high-level production and purification of recombinant proteins in plants. *Plant Biotechnol. J.* **9**, 419–433.

Dhaubhadel, S., Chaudhary, S., Dobinson, K.F. and Krishna, P. (1999) Treatment with 24-epibrassinolide, a brassinosteroid, increases the basic thermotolerance of Brassica napus and tomato seedlings. *Plant Mol. Biol.* **40**, 333–342.

Dhaubhadel, S., Browning, K.S., Gallie, D.R. and Krishna, P. (2002) Brassinosteroid functions to protect the translational machinery and heat-shock protein synthesis following thermal stress. *Plant J.* **29**, 681–691.

Dhaubhadel, S., Kuflu, K., Romero, M.C. and Gijzen, M. (2005) A soybean seed protein with carboxylate-binding activity. *J. Exp. Bot.* **56**, 2335–2344.

Doran, P.M. (2006) Foreign protein degradation and instability in plants and plant tissue cultures. *Trends Biotechnol.* **24**, 426–432.

Fischer, R., Stoger, E., Schillberg, S., Christou, P. and Twyman, R.M. (2004) Plant-based production of biopharmaceuticals. *Curr. Opin. Plant Biol.* **7**, 152–158.

Folta, K.M. and Kaufman, L.S. (2006) Isolation of Arabidopsis nuclei and measurement of gene transcription rates using nuclear run-on assays. *Nat. Protoc.* **1**, 3094–3100.

Gallie, D.R. and Browning, K.S. (2001) eIF4G functionally differs from eIFiso4G in promoting internal initiation, cap-independent translation, and translation of structured mRNAs. *J. Biol. Chem.* **276**, 36951–36960.

Gallie, D.R., Le, H., Caldwell, C. and Browning, K.S. (1998a) Analysis of translation elongation factors from wheat during development and following heat shock. *Biochem. Biophys. Res. Commun.* **245**, 295–300.

Gallie, D.R., Le, H., Tanguay, R.L. and Browning, K.S. (1998b) Translation initiation factors are differentially regulated in cereals during development and following heat shock. *Plant J.* **14**, 715–722.

Gleba, Y., Klimyuk, V. and Marillonnet, S. (2005) Magnifection—a new platform for expressing recombinant vaccines in plants. *Vaccine*, **23**, 2042–2048.

Greene, E.A., Codomo, C.A., Taylor, N.E., Henikoff, J.G., Till, B.J., Reynolds, S.H., Enns, L.C., Burtner, C., Johnson, J.E., Odden, A.R., Comai, L. and Henikoff, S. (2003) Spectrum of chemically induced mutations from a large-scale reverse-genetic screen in Arabidopsis. *Genetics*, **164**, 731–740.

Gutiérrez, R.A., MacIntosh, G.C. and Green, P.J. (1999) Current perspectives on mRNA stability in plants: multiple levels and mechanisms of control. *Trends Plant Sci.* **4**, 429–438.

Howard, J.A. (2005) Commercialization of biopharmaceutical and bioindustrial proteins from plants. *Crop Sci.* **45**, 468–472.

Kawaguchi, R. and Bailey-Serres, J. (2002) Regulation of translational initiation in plants. *Curr. Opin. Plant Biol.* **5**, 460–465.

Kawaguchi, R., Williams, A.J., Bray, E.A. and Bailey-Serres, J. (2003) Water-deficit-induced translational control in *Nicotiana tabacum*. *Plant, Cell Environ.* **26**, 221–229.

Kawaguchi, R., Girke, T., Bray, E.A. and Bailey-Serres, J. (2004) Differential mRNA translation contributes to gene regulation under non-stress and dehydration stress conditions in *Arabidopsis thaliana*. *Plant J.* **38**, 823–839.

Kermode, A.R. (2006) Plants as factories for production of biopharmaceutical and bioindustrial proteins: lessons from cell biology. *Can. J. Bot.* **84**, 679–694.

Kovalchuk, I., Kovalchuk, O. and Hohn, B. (2000) Genome-wide variation of the somatic mutation frequency in transgenic plants. *EMBO J.* **19**, 4431–4438.

Laemmli, U.K. (1970) Cleavage of structural proteins during assembly of the head of bacteriophage T4. *Nature*, **227**, 680–685.

Lightner, J. and Caspar, T. (1998) Seed mutagenesis of Arabidopsis. In *Methods in Molecular Biology* (Martine-Zapater, J.M. and Salinas, J. eds.), pp. 91–103. Totowa, NJ: Human Press Inc.

Livingstone, J.M., Seguin, P. and Strömvik, M.V. (2010) An in silico study of the genes for the isoflavonoid pathway enzymes in soybean reveals novel expressed homologues. *Can. J. Plant Sci.* **90**, 453–469.

Lössl, A.G. and Waheed, M.T. (2011) Chloroplast-derived vaccines against human diseases: achievements, challenges and scopes. *Plant Biotechnol. J.* **9**, 527–539.

Lukowitz, W.C., Gillmor, S. and Scheible, W.R. (2000) Positional cloning in Arabidopsis. Why it feels good to have a genome initiative working for you. *Plant Physiol.* **123**, 795–805.

Martínez, C.A., Giulietti, A.M. and Rodríguez Talou, J. (2012) Research advances in plant-made flavivirus antigens. *Biotechnol. Adv.* **30**, 1493–1505.

Matić, S., Masenga, V., Poli, A., Rinaldi, R., Milne, R.G., Vecchiati, M. and Noris, E. (2012) Comparative analysis of recombinant Human Papillomavirus 8 L1 production in plants by a variety of expression systems and purification methods. *Plant Biotechnol. J.* **10**, 410–421.

Mayberry, L.K., Dennis, M.D., Leah Allen, M., Ruud Nitka, K., Murphy, P.A., Campbell, L. and Browning, K.S. (2007) Expression and purification of recombinant wheat translation initiation factors eIF1, eIF1A, eIF4A, eIF4B, eIF4F, eIF(iso)4F, and eIF5. In *Methods in Enzymology*, Vol. **430** (Lorsch, J. ed) pp 397–408. San Diego, CL: Academic Press.

Melamed, D., Eliyahu, E. and Arava, Y. (2009) Exploring translation regulation by global analysis of ribosomal association. *Methods*, **48**, 301–305.

Menassa, R., Nguyen, V., Jevnikar, A. and Brandle, J. (2001) A self-contained system for the field production of plant recombinant interleukin-10. *Mol. Breeding*, **8**, 177–185.

Menassa, R., Kennette, W., Nguyen, V., Rymerson, R., Jevnikar, A. and Brandle, J. (2004) Subcellular targeting of human interleukin-10 in plants. *J. Biotechnol.* **108**, 179–183.

Menassa, R., Du, C., Yin, Z.Q., Ma, S., Poussier, P., Brandle, J. and Jevnikar, A.M. (2007) Therapeutic effectiveness of orally administered transgenic low-alkaloid tobacco expressing human interleukin-10 in a mouse model of colitis. *Plant Biotechnol. J.* **5**, 50–59.

Murashige, T. and Skoog, F. (1962) A revised medium for rapid growth and bioassays with tobacco tissue culture. *Physiol. Plant.* **15**, 473–497.

Murray, M.G. and Thompson, W.F. (1980) Rapid isolation of high molecular weight plant DNA. *Nucleic Acids Res.* **8**, 4321–4325.

Pogue, G.P., Vojdani, F., Palmer, K.E., Hiatt, E., Hume, S., Phelps, J., Long, L., Bohorova, N., Kim, D., Pauly, M., Velasco, J., Whaley, K., Zeitlin, L., Garger, S.J., White, E., Bai, Y., Haydon, H. and Bratcher, B. (2010) Production of pharmaceutical-grade recombinant aprotinin and a monoclonal antibody product using plant-based transient expression systems. *Plant Biotechnol. J.* **8**, 638–654.

Redei, G.P., Acedo, G.N. and Sandhu, S.S. (1984) Mutation induction and detection in Arabidopsis. In *Mutation, Cancer and Malformation* (Chu, E.H.Y. and Generoso, W.M., eds), pp. 285–313. New York: Plenum.

Russell, J.S. (2001) *Molecular Cloning.* A Laboratory Manual. Cold Spring Harbor: Cold Spring Harbor Laboratory Press.

Seeley, K.A., Byrne, D.H. and Colbert, J.T. (1992) Red light-independent instability of oat phytochrome mRNA in vivo. *Plant Cell,* **4**, 29–38.

Smale, S.T. (2009) Nuclear Run-On Assay. *Cold Spring Harbor Protocols 2009:* pdb.prot5329.

Streatfield, S.J. (2007) Approaches to achieve high-level heterologous protein production in plants. *Plant Biotechnol. J.* **5**, 2–15.

Streatfield, S.J., Lane, J.R., Brooks, C.A., Barker, D.K., Poage, M.L., Mayor, J.M., Lamphear, B.J., Drees, C.F., Jilka, J.M., Hood, E.E. and Howard, J.A. (2003) Corn as a production system for human and animal vaccines. *Vaccine,* **21**, 812–815.

Trcek, T., Larson Daniel, R., Moldón, A., Query Charles, C. and Singer Robert, H. (2011) Single-molecule mRNA decay measurements reveal promoter-regulated mRNA stability in yeast. *Cell,* **147**, 1484–1497.

Wang, H., Kohalmi, S.E. and Cutler, A.J. (1996) An improved method for polymerase chain reaction using whole yeast cells. *Anal. Biochem.* **237**, 145–146.

Zhong, G.-Y., Peterson, D., Delaney, D.E., Bailey, M., Witcher, D.R., Register Iii, J.C., Bond, D., Li, C.-P., Marshall, L., Kulisek, E., Ritland, D., Meyer, T., Hood, E.E. and Howard, J.A. (1999) Commercial production of aprotinin in transgenic maize seeds. *Mol. Breeding,* **5**, 345–356.

19

Flocculation increases the efficacy of depth filtration during the downstream processing of recombinant pharmaceutical proteins produced in tobacco

Johannes F. Buyel[1],* and Rainer Fischer[1,2]

[1]Institute for Molecular Biotechnology, RWTH Aachen University, Aachen, Germany
[2]Fraunhofer Institute for Molecular Biology and Applied Ecology, Aachen, Germany

*Correspondence

email johannes.buyel@rwth-aachen.de

Keywords: plant-derived biopharmaceuticals, downstream processing, flocculation, design of experiments, process development, monoclonal antibody.

Summary

Flocculation is a cost-effective method that is used to improve the efficiency of clarification by causing dispersed particles to clump together, allowing their removal by sedimentation, centrifugation or filtration. The efficacy of flocculation for any given process depends on the nature and concentration of the particulates in the feed stream, the concentration, charge density and length of the flocculant polymer, the shear rate, the properties of the feed stream (e.g. pH and ionic strength) and the properties of the target products. We tested a range of flocculants and process conditions using a design of experiments approach to identify the most suitable polymers for the clarification step during the production of a HIV-neutralizing monoclonal antibody (2G12) and a fluorescent marker protein (DsRed) expressed in transgenic tobacco leaves. Among the 23 different flocculants we tested, the greatest reduction in turbidity was achieved with Polymin P, a branched, cationic polyethylenimine with a charge density of 13.0 meq/g. This flocculant reduced turbidity by more than 90% under a wide range of process conditions. We developed a model that predicted its performance under different process conditions, and this enabled us to increase the depth filter capacity three–sevenfold depending on the process scale, depth filter type and plant species. The costs of filter consumables were reduced by more than 50% compared with a process without flocculant, and there was no loss of recovery for either 2G12 or DsRed.

Introduction

The production of biopharmaceutical proteins in plants is less expensive and more scalable than fermenter-based systems (Fischer et al., 2013). Additionally, plants carry out complex post-translational modifications but do not support the replication of human pathogens (Commandeur et al., 2003; Fischer and Emans, 2000). To develop plants into a commercially viable production platform, it is necessary to consider the economics of the entire process and particularly the costs of downstream processing (DSP). These can account for up to 80% of the total costs, and the most expensive steps are often product specific, based on the chromatography media needed to purify specific proteins (Fischer et al., 2012; Wilken and Nikolov, 2012). Significant cost savings can be achieved by matching the early unit operations to the properties of the host, taking into account the unique components and contaminants found in particular plants, for example the removal of fibres, oils and metabolic by-products during initial clarification. The optimization of upstream production can also improve the safety profile of an entire process based on plant biomass, for example, by introducing rockwool as a 'low-bioburden' growth substrate in contrast to soil (Fischer et al., 2012).

Clarification methods exploit differences between the target protein and impurities in terms of size, solubility and/or charge density. Process-scale clarification may involve centrifugation and/or microfiltration, but depth filtration is emerging as the key clarification technology because disposable filter cassettes obviate equipment cleaning and reduce the risk of cross-contamination between campaigns (Laukel et al., 2011; Whitford, 2010). Filter media with progressively smaller pore sizes can be chosen to improve product recovery by simultaneously removing particulates and soluble contaminants (O'Brien et al., 2012; Pegel et al., 2011; Yigzaw et al., 2006). If the flow rate is constant, large particles retained on an initial coarse filter build-up a filter cake over time, which progressively improves filter efficiency although the backpressure also increases. Bag-shaped filters made from needle felt or nylon filaments are often used for this step. Small particles with sizes in the range 0.1–10 μm are typically removed by subsequent depth filtration (Gottschalk, 2009; Pegel et al., 2011). Small-scale experiments are often used to identify the most effective filter combinations but the set-up may require further validation at the medium or pilot scales to generate reliable quantitative capacity data for process scale-up (Jornitz, 2006; Kandula et al., 2009).

It is easier to remove larger than smaller particles from the feed stream, so any method that increases the mean particle size makes the process of clarification more efficient. Flocculation involves the addition of substances that cause suspended particles to (i) clump together by increasing the attractive forces between them until they overcome repulsive forces caused by like surface charges, or (ii) precipitate because their surface charges are neutralized (Barany and Szepesszentgyorgyi, 2004; Gregory and Barany, 2011; Zhou and Franks, 2006). Flocculants are inexpensive and form part of the structure of the particles they generate, so they offer a cost-effective strategy to increase the efficiency of clarification without contaminating the feed stream. Flocculation can remove whole cells, debris and precipitated proteins, reducing the passage of small particles through the filter train (Aspelund *et al.*, 2008; Peram *et al.*, 2010; Westoby *et al.*, 2011).

Flocculation can be induced by polymers with either low or high charge densities (Gregory and Barany, 2011; Roush and Lu, 2008). The former often work by bridging dispersed particles, and the efficacy of flocculation therefore depends on the ability of the polymer to extend beyond the diffuse double layer surrounding each particle and attach to a second particle (Russel *et al.*, 1989). If the flocculant concentration exceeds an optimal value, then the dispersed particles can become stable and the efficacy of clarification is reduced. Polymers with high charge densities often work by charge neutralization, that is, binding to and shielding charged surface patches so that dispersed particles become insoluble (Gregory and Barany, 2011; Runkana *et al.*, 2004). Here, the efficiency of flocculation is independent of the polymer concentration but complete aggregation requires quantitative amounts of charged flocculant so that all particle charges are neutralized. The efficiency of flocculation also depends on the adsorption rate, which is related to shear rates, although excessive shear can also break up the flocks and reduce particle aggregation, so optimal flocculation occurs when these two effects are in equilibrium (Gregory, 1988; Zhou and Franks, 2006).

The optimization of flocculation for any given process depends on the nature and concentration of the particulates, the concentration, charge density and length of the polymer, the shear rate, the properties of the feed stream (e.g. pH and ionic strength) and the properties of the target products (Barany and Szepesszentgyorgyi, 2004; Gregory and Barany, 2011; Pearson *et al.*, 2004; Zhou and Franks, 2006). High salt concentrations can affect flocculation by screening particle surface charges, reducing the thickness of the diffuse double layer surrounding the charged particle, reducing the flatness of the adsorbed polymer, reducing polymer binding affinity and reducing the polymer effective charge (Gregory and Barany, 2011; Zhou and Franks, 2006). Higher salt concentrations therefore increase the bridging efficiency of low-charge-density flocculants, but reduce the neutralization efficiency of high-charge-density flocculants (Zhou and Franks, 2006). Increasing the salt concentration and modifying other process conditions such as pH can also affect the

solubility of the target protein, so the optimal flocculation strategy is product dependent (Fisher and Glatz, 1988; Holler *et al.*, 2007).

The efficient recovery and purification of biopharmaceutical proteins must take priority over the selection of suitable flocculation conditions. Polymers should therefore be selected to induce flocculation under the standardized conditions typically used for protein extraction. For example, this can be achieved by investigating the ζ-potential to streamline process development and favour regulatory compliance (Pearson *et al.*, 2004; Rasteiro *et al.*, 2010).

We therefore tested a range of flocculants and process conditions using a design of experiments (DoE) approach to develop an optimized clarification step for the production of a HIV-neutralizing monoclonal antibody (2G12) and a fluorescent marker protein (DsRed) expressed in transgenic tobacco leaves. We tested 23 different flocculants in a range of buffers and pH environments, at several temperatures and incubation times, to select a polymer that would maximize depth filter capacity and product recovery under typical process conditions. We also confirmed the suitability of the best-performing polymer for process applications using three different depth filters at the laboratory and pilot scales, as well as three different *Nicotiana* species.

Results

Process layout

We scaled down a 200-kg (pilot-scale) GMP process for the production of the HIV-neutralizing monoclonal antibody 2G12 and the fluorescent protein DsRed to a 500-g laboratory-scale process that was used to identify alternatives to the initial three-step depth filter train (Figure 1a). The capacities of the individual depth filters PDH4 and PDF4 were similar to the initial PB2 + PC2 combination when the feed stream contained extracts from tobacco plants grown in soil (Figure 1b). However, the capacities of all filters were reduced to 25%–30% when rockwool was used instead of soil to improve the safety profile of the overall process. Therefore, we set out to identify flocculants that could reduce extract turbidity, thus prolonging filter capacity and ultimately reducing the overall DSP costs for plant-derived biopharmaceuticals.

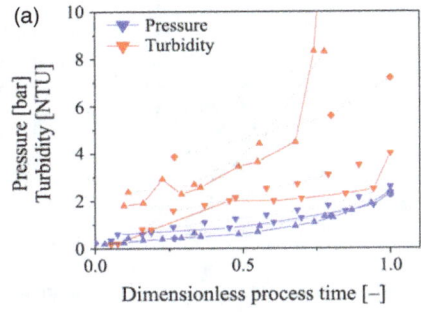 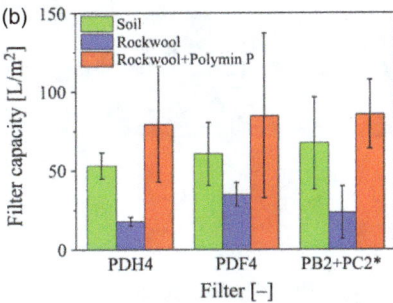

Figure 1 Process scale-down and depth filter comparison. (a) The laboratory-scale (dotted lines) turbidity breakthrough and pressure curves (red and blue, respectively) matched well with those for process-scale experiments (solid lines) for the PB2 + PC2 depth filter combination (▼) as well as the PDF4 (▲) and PDH4 (♦) individual filters. (b) The capacity of all filters loaded with extract from plants grown on rockwool was only 25%–30% of that determined for extracts from soil-grown plants. The addition of Polymin P at concentrations ~2 g/L not only restored filter capacity but resulted in an additional 25% capacity increase compared with the process using soil-grown plants. Error bars indicate standard deviations (*n* ≥ 3), and * indicates a combination of two equal-sized depth filters.

Selecting and screening the initial panel of flocculants

There is little information, even from flocculant manufacturers, about the types of polymers that are suitable for processes involving homogenized plant tissues. We therefore selected polymers with a broad range of properties affecting flocculation efficacy, including charge (anionic, cationic), charge density (1.0–19.0 meq/g), size (50–5000 kDa), structure (linear, branched) and backbone chemistry (Tables S1 and S2) (Rasteiro et al., 2010). The shear rate was kept constant, and the particle concentration in the feed stream, measured as the extract turbidity, varied between batches with an average value of 12 000 ± 5600 NTU (n = 47). We used a DoE approach to evaluate the flocculants under different buffer pH and conductivity conditions.

An initial screen was carried out by maintaining the conductivity of the buffer at 50 mS/cm (the conductivity of our unmodified extraction buffer) and testing the flocculants at pH values of 4.0, 6.0 and 8.0 to identify those effective under conditions commonly reported for the extraction of recombinant proteins from tobacco leaves (Hassan et al., 2008). The flocculant manufacturers recommended concentrations in the range of 0.2–10 g polymer per 1000 g of dispersed solids. We found that 132 ± 2 g (n = 19) of solids was dispersed in the plant homogenate and therefore used polymer concentrations of 0.01, 0.50 and 2.00 g/L corresponding to ~0.08, ~3.79 and ~15.15 g flocculant per 1000 g of dispersed solids. Initially, we observed the onset of flock formation and sedimentation after 15–45 min depending on polymer type and concentration (data not shown). We therefore selected 24-h settling time in the first screen and 1 h for the refinement experiments, the latter representing the maximum duration of a flocculation step that could be readily incorporated into the existing extraction process.

The screening showed that polymer concentrations of 0.01 g/L were insufficient to promote flocculation. The Paestrol flocculants (822 BS, 855 BS, 2350 and 2610), the Sedipur flocculants (CL 950 and CL 951) and Polymin P were able to induce flocculation in the plant extract as indicated by the removal of the grey/brown (pH 6 and 8) or green (pH 4) colour corresponding to the dispersed plant matter at either two or three different pH values, revealing the red (pH 6 and 8) or clear/blue (pH 4) colour of the cleared extract (Figure 2a). The extract appeared red at pH 6 and 8 due to the presence of DsRed, but the colour was not present at pH 4 because DsRed precipitates at ~pH 5. These observations were confirmed by the lower average turbidity following bag filtration (Figure 2b). The Sedipur flocculants had similar physicochemical properties (Table S1) but only Sedipur CL 950 was selected for further testing because its lower charge density compared with Sedipur CL 951 reduced the risk of protein precipitation.

The flocculants were tested at additional concentrations by measuring turbidity directly after bag filtration, indicating that Polymin P reduced the turbidity most efficiently, followed by Sedipur CL 950, then Paestrol 822 BS and 2610 (which performed equally well) and finally Paestrol 855 BS and 2350 (Figure 3).

The efficacy of Polymin P, Sedipur CL 950 and Paestrol 2610 was confirmed at pH 5 and 7. Paestrol 2610 was used instead of Paestrol 822 BS because the stock solution was less viscous, which can simplify dosing during process-scale manufacturing. Polymin P was again the best flocculant, reducing extract turbidity efficiently even when incubated for <1 h. Sedipur CL 950 had a limited impact on turbidity after 1 h and Paestrol 2610 appeared to have no effect at all.

Selection and screening of additional flocculants

The initial screen suggested that branched cationic polymers with high charge densities promoted fast and efficient flocculation (Polymin P, Sedipur CL 950), whereas linear and/or anionic polymers had only a limited effect and required more time for flock formation. We therefore selected additional cationic polymers with high charge densities and the potential to achieve better flocculation (Table S1). Catiofast GM and Lupasol PS

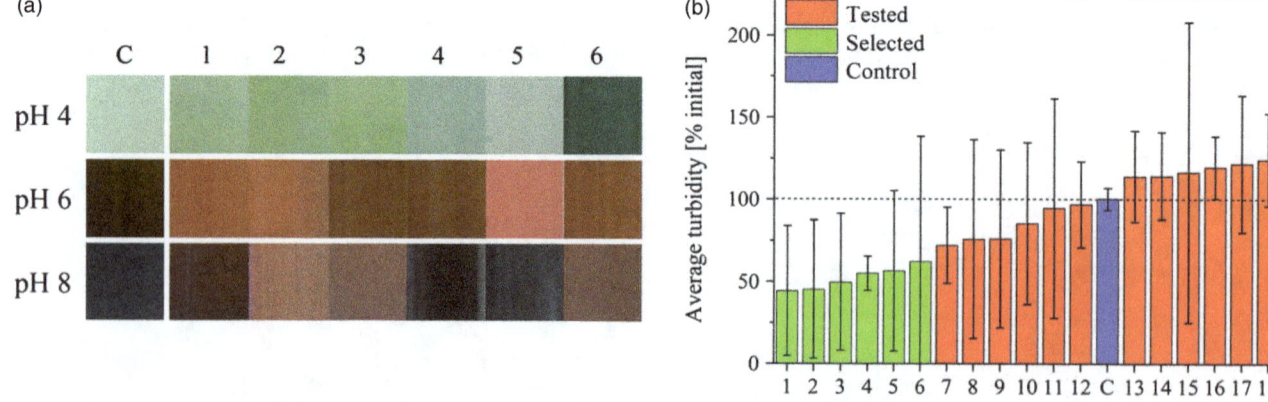

Figure 2 Impact of flocculants on the turbidity of crude tobacco extracts at different pH values. (a) Compared with an untreated extract (c), the flocculants Paestrol 855 BS (1), Paestrol 822 BS (2), Paestrol 2610 (3), Sedipur CL 950 (4), Polymin P (5) and Paestrol 2350 (6), reduced the turbidity of the extract and the green colour at two or all three of the tested pH values after settling for 24 h. The other flocculants had a more limited effect. The treated extract appeared light green at pH 4 because of the residual dispersed particles whereas a red colour was observed at pH values >6 due to DsRed (which precipitates below pH ~5). (b) Turbidities normalized to controls at the corresponding pH values confirmed these findings. Number codes 1–6 as in (a), numbers 7–18 as in Table S1, polymers selected for further testing are highlighted by green columns. Error bars indicate the standard deviation of all measurements made for each polymer at the three pH values under investigation and for concentrations of 0.5 and 2.0 g/L, hence the large values of the bars (n > 3).

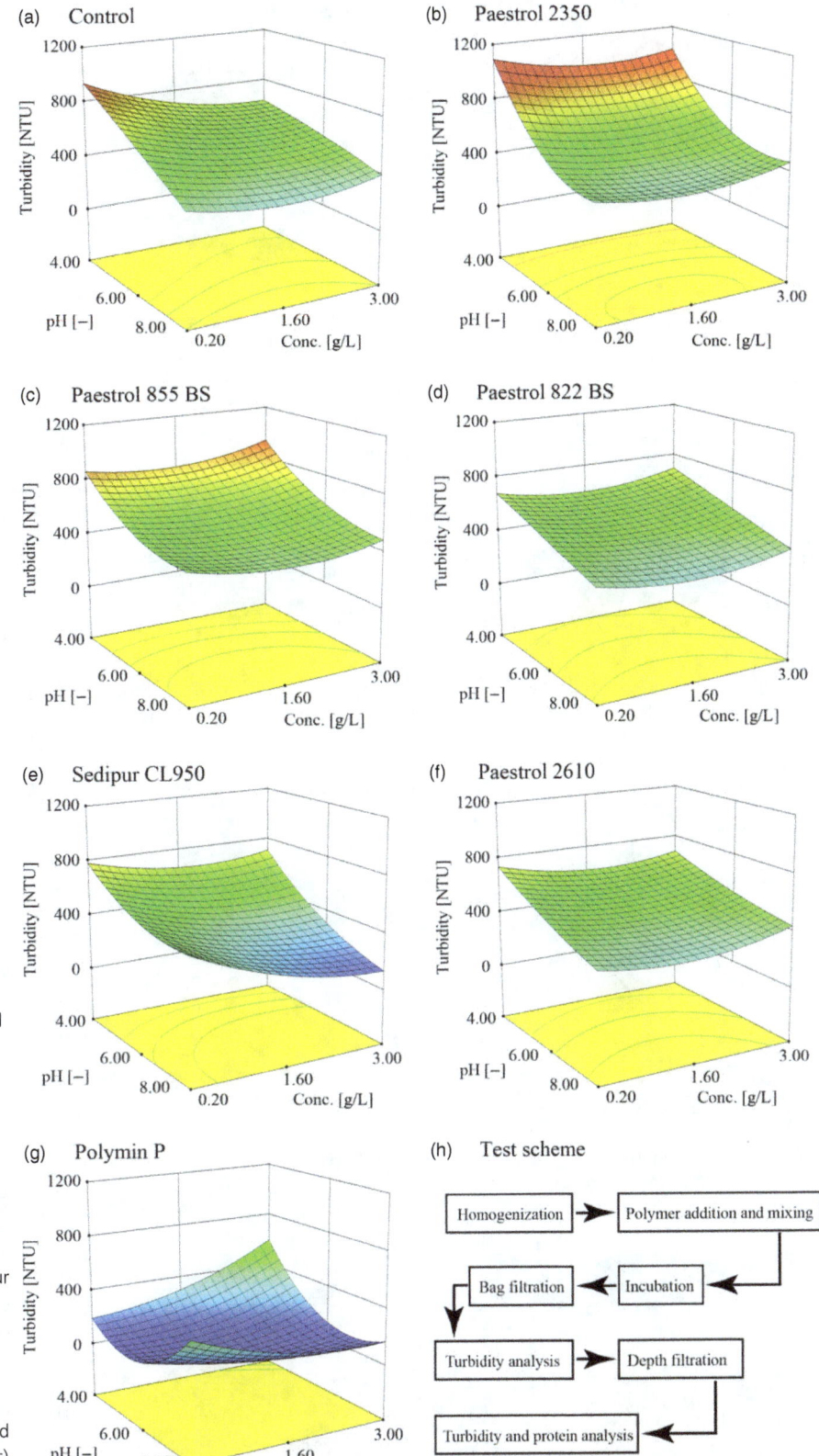

Figure 3 Response surface methodology model for the reduction of tobacco extract turbidity achieved with different flocculants immediately after bag filtration, depending on concentration and pH. Compared with an untreated extract (a), Paestrol 2350 (b) and Paestrol 855 BS (c) increased the turbidity of the extract under low-pH conditions and had no other beneficial effect, whereas Paestrol 822 BS (d) reduced the turbidity slightly. Paestrol 2610 (f) reduced the turbidity slightly under low-pH conditions, Sedipur CL 950 (e) achieved a significant reduction in turbidity under high-pH conditions at high concentrations, and Polymin P (g) reduced turbidity over a broad pH range even at low concentrations. The turbidity values are shown as 1:10 dilutions in buffer. (h) Test scheme applied during flocculant screening (up to turbidity analysis) and refinement (up to turbidity and protein analysis).

achieved efficient flocculation under standard extraction conditions (pH 8.0, conductivity 50 mS/cm) and were chosen for direct comparison with Polymin P at different conductivities and pH values. Catiofast GM reduced the extract turbidity by ~90% at conductivities >35 mS/cm, but only by ~60% at lower conductivities (Figure 4). This complemented the activities of Lupasol PS and Polymin P, which performed best at conductivities <45 mS/cm and achieved similar reductions in turbidity (Figure 4)

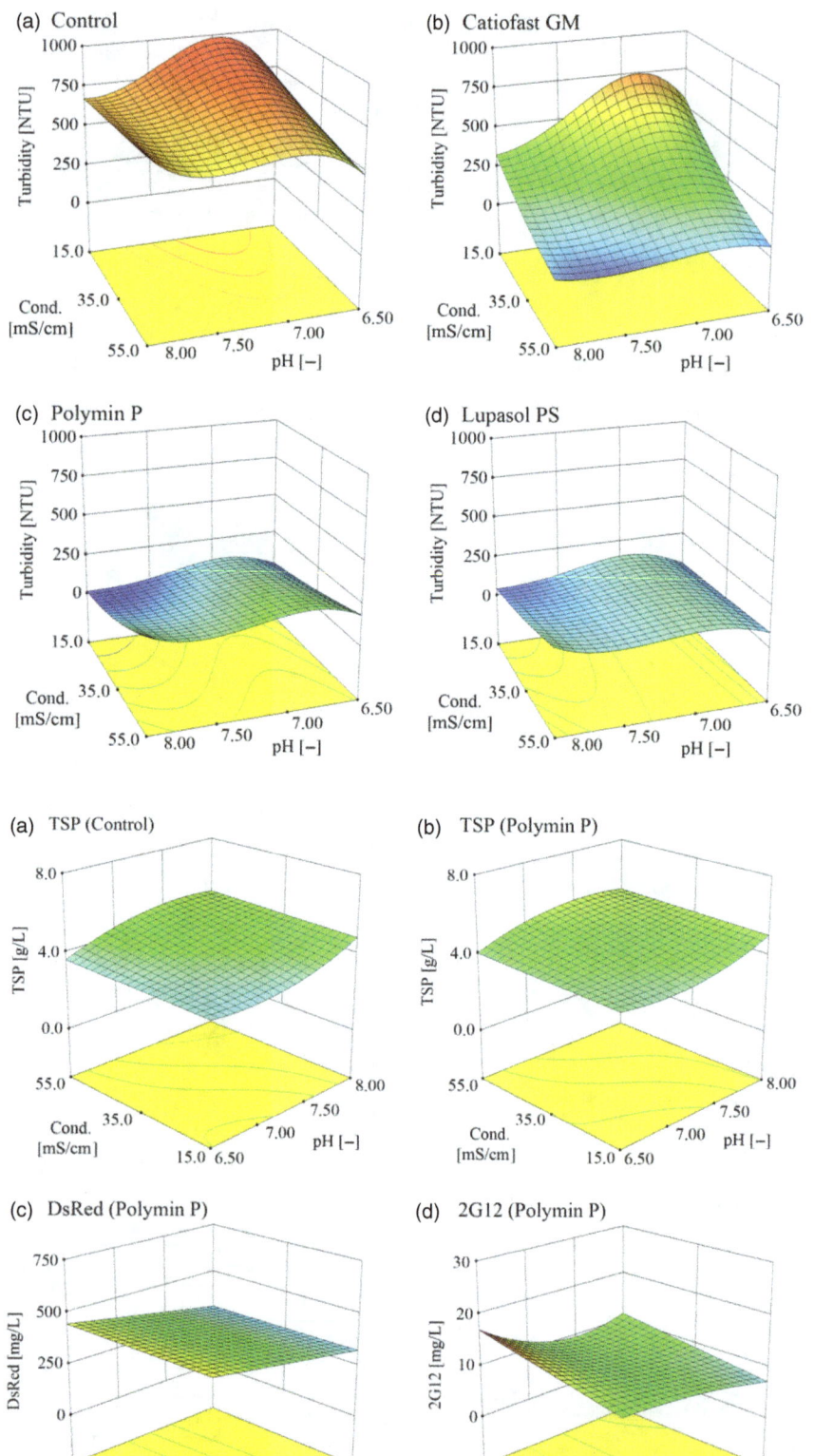

Figure 4 Turbidity reduction achieved with highly charged cationic polymers under different pH and conductivity conditions, compared with an untreated extract (a). Catiofast GM (b) reduced extract turbidity at high conductivities, whereas Polymin P (c) and Lupasol PS (d) reduced extract turbidity at moderate and low conductivities regardless of the pH. All response surfaces were calculated for 4.0 g/L of polymer and an incubation time of 15 min.

Figure 5 Effect of flocculants on total soluble proteins, DsRed and 2G12 concentrations in leaf extracts. None of the polymers affected total soluble protein (TSP), DsRed or 2G12 concentrations significantly compared with controls. The example shown here is Polymin P and TSP (a versus b). More DsRed and 2G12 were extracted at pH 6.5 (c,d respectively), whereas most TSP was extracted at higher pH and conductivity levels (a,d). All response surfaces were calculated for 2.0 g/L of polymer and an incubation time of 15 min.

but flocculation was more rapid with Polymin P and required ~10% less polymer.

The concentrations of total soluble protein (TSP), DsRed and 2G12 varied 5%–10% among the polymers and controls, that is, within the error ranges of the models and quantitation assays (Figure 5). All three polymers therefore had an insignificant impact on protein levels within the tested concentration range (0.5–4.0 g/L) and did not interfere with target protein recovery.

Regardless of the polymer type, more TSP was extracted at higher pH values and conductivities, but a low pH was favourable for the extraction of DsRed and even more so for 2G12, particularly at high conductivities.

Further testing of Polymin P

The superior properties of Polymin P were investigated by converting the constant incubation time of 1 h into a factor included in the DoE (Table 1). The optimal Polymin P concentration was 1.7–3.5 g/L, reducing extract turbidity to below 10% of the initial value, whereas a steep increase in extract turbidity was observed at lower concentrations (Figure 6). The polymer concentrations were calculated based on the Polymin P formulation provided by BASF, a 50% solution in water. Therefore, the effective concentrations are 50% of the values stated in the text. A slight increase in extract turbidity was observed at polymer concentrations >3.5 g/L probably reflecting an oversaturation effect as previously reported (Barany and Szepesszentgyorgyi, 2004). The efficiency of Polymin P declined in the pH range 7.0–7.7 and the extract appeared milky after bag filtration indicating the presence of fine dispersed particles, probably fragments of

 lignified cell walls that are uncharged in this pH range and thus unlikely to interact with the flocculant (Nečesaný, 1971). We observed pH-dependent turbidity reduction in experiments with Catiofast GM and Lupasol PS (Figure 4b,d).

High conductivities improved the performance of Polymin P in low-pH extracts but had the opposite effect at high pH (Figure 6). This may reflect the ability of salt ions to shield positive particle charges at low pH, thereby preventing the repulsion of polycationic flocculants, whereas at high pH, they would compete for binding sites on the particles, thereby preventing interactions with the flocculant. Increasing the incubation time from 0.25 to 1.25 h following the addition of Polymin P improved clarification but the effect was small compared with that of the other factors, indicating rapid kinetics of flock formation in the <15 min range.

A predictive model of turbidity reduction

The experiments described above indicated that Polymin P was the most effective polymer for the flocculation of dispersed particles in plant tissue homogenates at varying pH values and conductivities, with ~2 g/L (effective concentration ~1 g/L) sufficient to induce maximum flock formation in <15 min. Changes in the feed stream conditions can significantly affect the effectiveness of flocculation induced by polymers and thus reduce process robustness, defined as the degree to which reducing extract turbidity is dependent on changes in the process parameters (Pearson et al., 2004). We therefore developed a predictive model for the robustness of turbidity reduction with Polymin P, taking process economics, protein stability and regulatory compliance into account, all of which favour short process times and the minimal use of additives. Polymer concentrations in the range 1.5–2.5 g/L and incubation times of 0.08–0.50 h were therefore combined with a wide range of pH values, conductivities and temperatures potentially required for processing different target

Table 1 Parameter ranges tested during Polymin P optimization by DoE

Parameter	Lower boundary	Upper boundary
Extraction buffer pH (–)	4	8
Extraction buffer conductivity (mS/cm)	15	55
Extract temperature (°C)	4	30
Flocculant concentration (g/L)	0	4
Flocculation time (h)	0.08	1.25

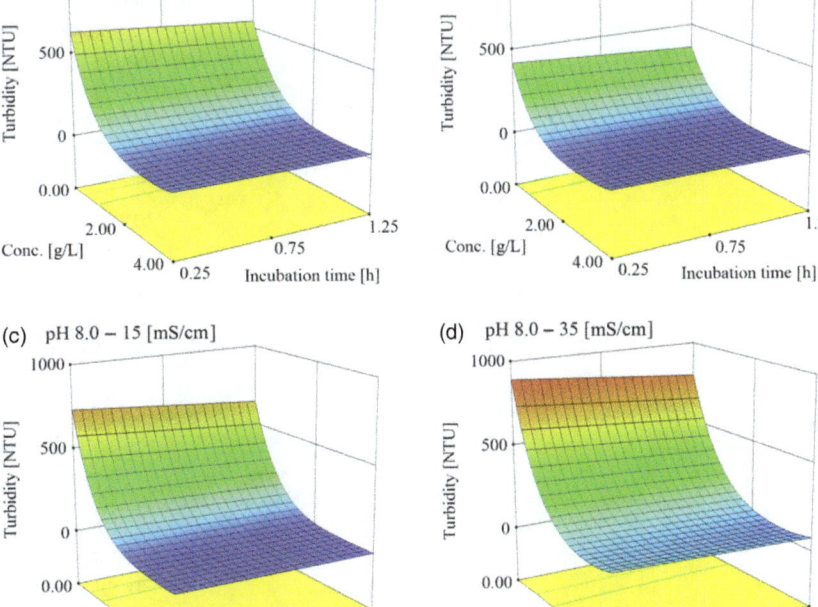

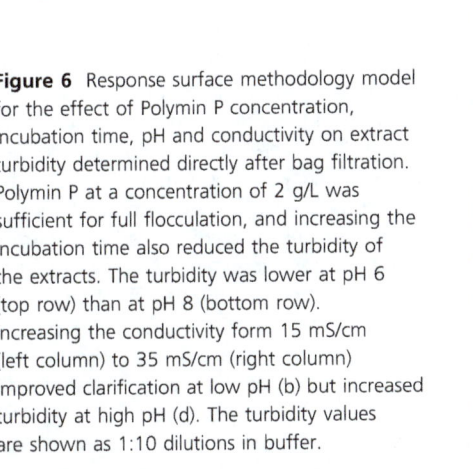

Figure 6 Response surface methodology model for the effect of Polymin P concentration, incubation time, pH and conductivity on extract turbidity determined directly after bag filtration. Polymin P at a concentration of 2 g/L was sufficient for full flocculation, and increasing the incubation time also reduced the turbidity of the extracts. The turbidity was lower at pH 6 (top row) than at pH 8 (bottom row). Increasing the conductivity form 15 mS/cm (left column) to 35 mS/cm (right column) improved clarification at low pH (b) but increased turbidity at high pH (d). The turbidity values are shown as 1:10 dilutions in buffer.

proteins. Based on a DoE approach, it was possible to establish a model that predicted the effects of flocculant concentration, incubation temperature, pH, conductivity and incubation time on extract turbidity (Tables 2 and 3) that was consistent with our previous findings (Figure 7f).

The use of response surfaces to identify factor settings associated with high (undesirable) or low (desirable) extract turbidity was challenging because the flocculation model was complex. Instead, we used the flocculation model to predict

Table 2 Model factors with a significant impact on the turbidity of crude tobacco extracts during flocculation

Source	Sum of squares	Degrees of freedom	F-value	P-value
Model	3205.08	29	39.64	<0.0001
Incubation temperature (°C) (A)	2.93	1	1.05	0.3114
pH (–) (B)	205.50	1	73.72	<0.0001
Conductivity (mS/cm) (C)	98.86	1	35.46	<0.0001
Incubation time (h) (D)	61.29	1	21.99	<0.0001
Concentration (g/L) (E)	3.35	1	1.20	0.2795
AB	42.86	1	15.38	0.0003
AC	2.31	1	0.83	0.3683
AD	0.07	1	0.03	0.8732
AE	13.73	1	4.93	0.0322
BC	2.27	1	0.81	0.3724
BD	17.23	1	6.18	0.0172
BE	149.20	1	53.52	<0.0001
CD	22.42	1	8.04	0.0071
CE	7.27	1	2.61	0.1142
A^2	31.07	1	11.14	0.0018
B^2	319.38	1	114.57	<0.0001
C^2	13.46	1	4.83	0.0338
D^2	23.71	1	8.51	0.0058
E^2	0.15	1	0.05	0.8176
ABC	91.52	1	32.83	<0.0001
ABD	22.12	1	7.94	0.0075
ACD	36.53	1	13.10	0.0008
A^2E	33.46	1	12.00	0.0013
AC^2	17.31	1	6.21	0.0170
B^2C	563.56	1	202.15	<0.0001
BC^2	61.84	1	22.18	<0.0001
BE^2	18.24	1	6.54	0.0144
C^2D	12.00	1	4.31	0.0445
B^3	408.64	1	146.59	<0.0001
Residual	111.51	40	n.a.	n.a.
Lack of fit	100.05	32	2.18	0.1239
Pure error	11.46	8	n.a.	n.a.

Table 3 Parameters used to evaluate the flocculation model

Evaluation parameter	Value
r^2	0.966
Adj r^2	0.942
Pred r^2	0.887
PRESS*	372.509

*Predicted residual sum of squares.

conditions yielding desirable and undesirable extract turbidities allowing the factor levels associated with these conditions to be analysed (Figure 7). This revealed that lower extract turbidities were associated with pH values in the range 4.5–5.5 or ~8.0 as well as incubation times >15 min. No clear trend was observed for incubation temperature or conductivity indicating that changes in these parameters can be compensated by adjusting polymer concentration and incubation time within the predefined limits. Interestingly, extreme Polymin P concentrations were preferred, but this was probably an artefact reflecting the inclusion of the nonsignificant quadratic effect of the Polymin P concentration (E^2 in Table 2). This effect was included to maintain the model hierarchy (Peixoto, 1987, 1990). Polymin P concentrations <2 g/L were associated with high turbidity, as were pH values of ~4.0 or ~7.5 and incubation times <15 min. Again, no clear trend was observed for incubation temperature or conductivity. For certain factors with an ambiguous influence on turbidity, extreme high or low values were over-represented, suggesting that the model becomes increasingly uncertain at the borders of the design space.

Polymin P reduced TSP levels by 20%–25% at all pH values when present at a concentration of 2 g/L in a buffer with a conductivity of 15 mS/cm (Figure 8). However, SDS-PAGE analysis indicated no significant qualitative or quantitative changes, except for DsRed at pH 4.0 (data not shown). Polymin P had no effect on TSP at conductivities >30 mS/cm confirming previous studies, suggesting that Polymin P can be used to precipitate proteins at low conductivities (Holler et al., 2007). The concentration of DsRed was reduced 15%–20% by Polymin P, regardless of the pH or conductivity of the buffer, except at pH 7 in buffers with conductivities <30 mS/cm. The beneficial effect of this combination probably reflected the ability of the charged flocculant to increase the buffer conductivity, thereby promoting the release of DsRed. The presence of Polymin P in leaf extracts did not alter the concentration of 2G12 by more than 10% at pH 6.0 or above, regardless of the conductivity, but the concentration of 2G12 fell by 30% at pH 4.0. The precise quantitation of 2G12 under these conditions was difficult because the polymer generated a high background signal during surface plasmon resonance (SPR) spectroscopy, which was expected due to the opposing charges of Polymin P (positive) and the carboxymethyl matrix (negative) promoting electrostatic interactions especially in samples with low conductivities. However, the results agreed with previous observations that flocculants have a limited impact on proteins under the conditions we tested.

Investigation of scalability and impact on depth filter capacity

The efficacy of Polymin P was confirmed in bench-top and pilot-scale tests. Manual mixing after the addition of flocculant was replaced with automated mixing in the homogenization device. Short mixing times of 10–15 s were optimal, whereas longer mixing times reduced the efficacy of flocculation, probably by disintegrating the flocks through prolonged exposure to shear stress (Figure 8d). In bench-top tests, Polymin P reduced extract turbidity by ~50% (Figure 9a) and increased the capacity of the depth filters (which were downstream of the bag filtration step in the clarification process) by threefold (PB2 + PC2 and PDF4) or sevenfold (PDH4) at concentrations of 1.5–2.0 g/L (Figure 1b). The same effect was observed for N. benthamiana infiltrated with A. tumefaciens 5 days prior to harvest and, to a lesser extent, (~50% capacity increase) for N. tabacum cultivar K326 when the

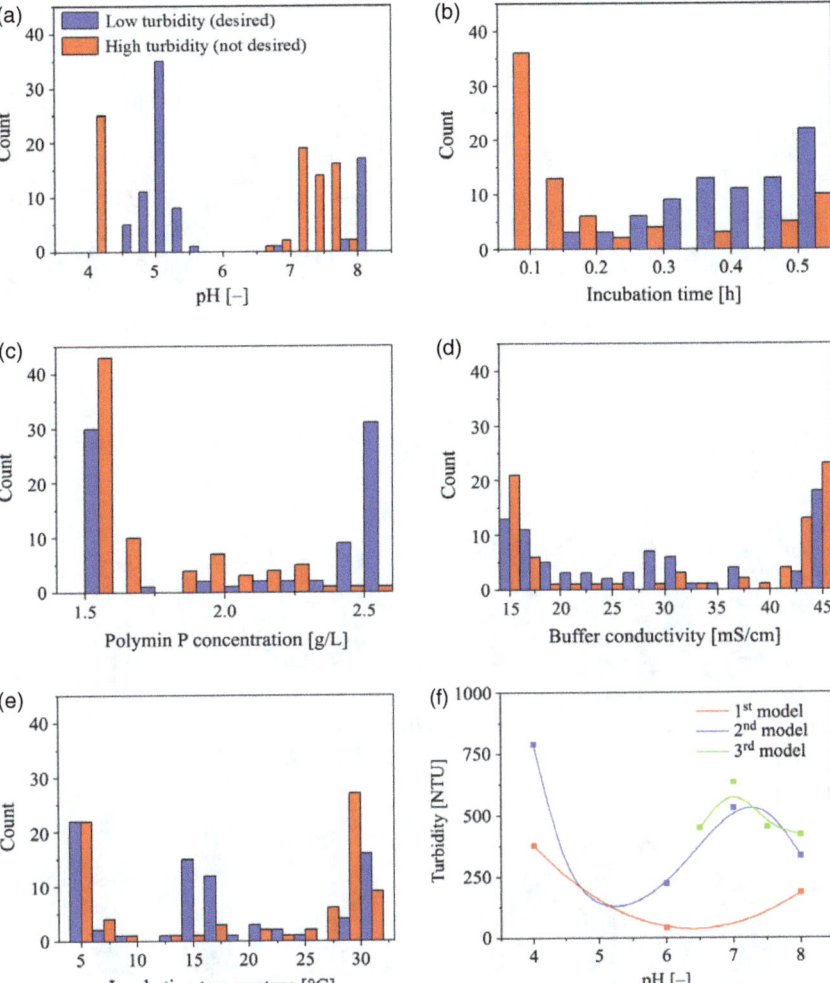

Figure 7 Predicted correlation between process parameters and extract turbidity. The robustness model (Table 2) was used to predict conditions resulting in low (blue, desirable) or high (red, undesirable) extract turbidity. The parameter values for pH (a), incubation time (b), Polymin P concentration (c), conductivity (d) and incubation temperature (e) were analysed separately, revealing their correlation with low or high turbidities. The robustness model (f) was then compared with models built during the second round of flocculant screening and initial optimization, revealing a discrepancy at pH ~7 which indicated that the robustness model had improved over earlier ones due to the inclusion of additional pH values.

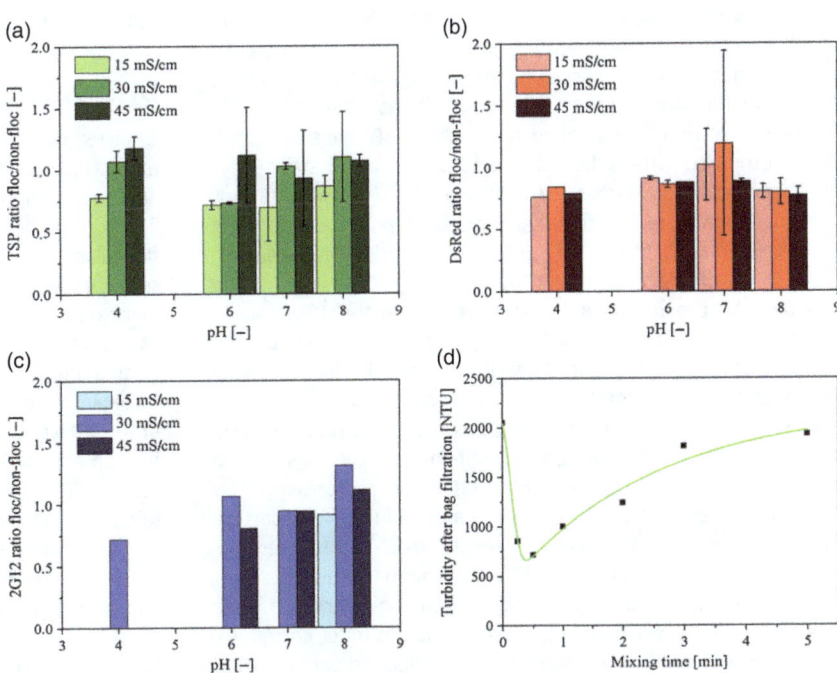

Figure 8 Effect of Polymin P on the concentrations of total soluble proteins, DsRed and 2G12 in tobacco leaf extracts. (a) Under low-pH and low-conductivity conditions, Polymin P reduced total soluble protein levels by ~25%. (b) Regardless of the conductivity, Polymin P slightly increased DsRed concentrations in extracts at pH 7.0 but otherwise reduced them by 15%. (c) The level of 2G12 was reduced by <10% at pH values >6.0, but reduced by 30% at pH 4.0. Error bars indicate standard deviations ($n \geq 3$). (d) The efficacy of flocculation is dependent on the mixing time, with an optimum at short mixing times of ~0.2–0.5 min. [Correction added on 18 June 2014, after first online publication: Data range on x-axis of Figure 8(d) has been amended.]

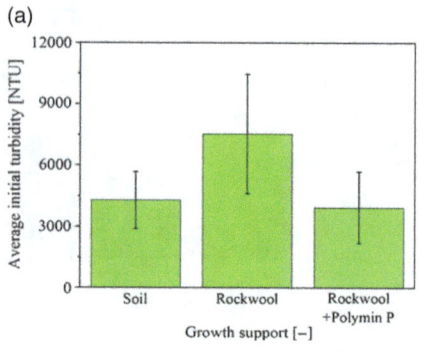

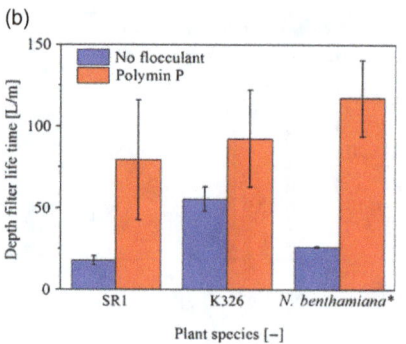

Figure 9 Impact of Polymin P application. (a) The addition of Polymin P reduces predepth filter extract turbidity when processing samples from plants grown on rockwool. The turbidity is reduced to the level observed for soil-grown plants without polymer. (b) Polymin P increases depth filter capacity ~1.5-fivefold when processing extracts from other tobacco varieties and related species. Error bars indicate standard deviations ($n \geq 3$). *Plants had been infiltrated with *A. tumefaciens* 5 days prior to harvest.

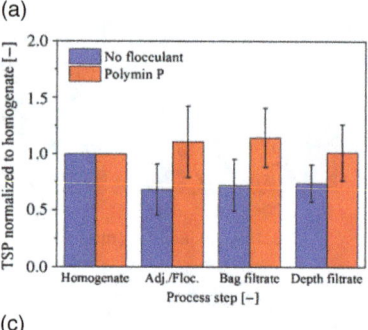

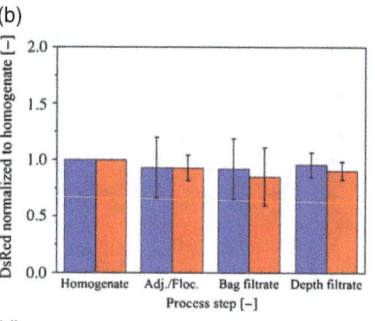

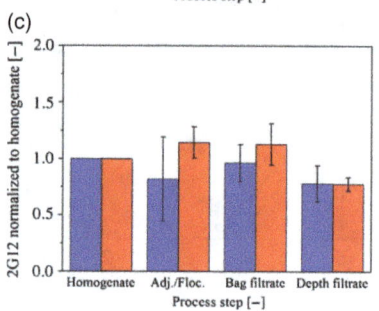

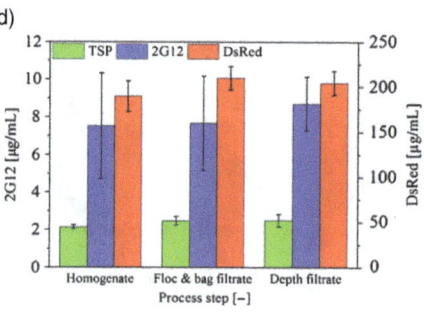

Figure 10 Comparison of process performance with and without Polymin P. (a)–(c). Polymin P did not reduce total soluble protein or target protein levels compared with flocculant-free controls. (d) The same results were obtained in pilot-scale experiments. Error bars indicate standard deviations ($n \geq 3$).

PDH4 depth filter was used (Figure 9b). Similar improvements in depth filter capacity have been reported for mammalian cell cultures, where chitosan was used for flocculation (Riske *et al.*, 2007). The flocculant probably increased the average diameter of particles in the extract (Aspelund *et al.*, 2008), thus favouring filters with larger nominal retention ratings, that is, PDH4 (Table 3). Polymin P was found to increase TSP levels by 20%–25% compared with a flocculant-free set-up at all steps after homogenization, whereas DsRed and 2G12 concentrations remained unaffected at different process steps under standard process conditions, that is, pH 7.5–8.0 with conductivities ranging from 35 to 45 mS/cm (Figure 10).

When the process volume was increased to the 100-L scale, protein concentrations were not significantly affected by the addition of Polymin P (Figure 10d). In contrast, the filter capacity increased from ~80 L/m (Figure 1b) to more than 280 L/m at the pilot-scale. This significant threefold improvement reflected two scale-dependent factors. First, the upstream bag filter was utilized more at the 100-L scale, improving filter performance by promoting the build-up of filter cake and reducing the particle burden to 500 NTU instead of ~5000 NTU in the feed for subsequent depth filtration. Second, the flow pattern at the filter layer surface changed from frontal flow at the bench-top scale (filter disc) to indirect flow in the STAX modules (filter cartridge), which can prevent premature clogging (Giglia and Sciola, 2011;

Giglia and Yavorsky, 2007); Buyel and Fischer, this issue). Process scale-up is therefore likely to reduce the costs further by optimizing the flow pattern and equipment utilization.

Discussion

Charged polymers are frequently used for the clarification of cell extracts and culture supernatants (Aspelund *et al.*, 2008; Barany and Szepesszentgyorgyi, 2004; Pearson *et al.*, 2004; Riske *et al.*, 2007), but there is little information about the compatibility of flocculants with plant tissue homogenates containing >10% (w/v) solids including particles >100 μm in size. Most reports using plant-derived materials focus on the recovery of carbohydrates in the context of biofuels and apply conditions that are not compatible with the purification of biopharmaceutical proteins, for example pH >9.0 or <4.0 and incubation times >2 h (Duarte *et al.*, 2010). Investigations are also typically limited to one or a few polymers (Aspelund *et al.*, 2008; Pearson *et al.*, 2004; Riske *et al.*, 2007) that may not harmonize with the solubility of certain target proteins (Fisher and Glatz, 1988; Holler and Zhang, 2008). We therefore tested 23 polymers with different properties (Table S1) for their ability to flocculate under conditions commonly employed for the extraction of biopharmaceutical proteins from plant leaves. Screening the initial set of flocculants indicated that branched and highly charged (>7 meq/g) cationic polymers in the moderate size range (100–1000 kDa) achieved the fastest and most effective flocculation, but only one of 18 polymers (~5%) could reduce the extract turbidity by more than 50% after bag filtration under various conditions (Polymin P). Using this information, we selected five additional polymers, three of which (60%) reduced the extract turbidity by more than 50% after bag filtration, thus confirming the initial screening results. Lupasol PS induced flocculation effectively under the same conditions as Polymin P, and the two polymers can thus be used as substitutes if one is incompatible with target protein recovery, for example due to precipitation. The optimal conditions for the use of Catiofast GM were complementary to those of Polymin P and Lupasol PS and covered the high conductivity range (>45 mS/cm).

The DoE methodology we developed was useful to (i) identify polymer characteristics important for the flocculation of plant tissue extracts; (ii) provide a small set of highly effective flocculants covering the potential operation range of extraction processes; and (iii) describe suitable windows of operation for these polymers (see below). The effort required amounted to ~3 person-days including analysis.

Polymin P achieved the fastest and most effective flocculation of dispersed cell debris in the initial experiments, probably reflecting electrostatic interactions between this highly charged cationic polymer and negatively charged cell fragments (Gregory and Barany, 2011; Hubbuch et al., 2006; Roush and Lu, 2008; Runkana et al., 2004). Increasing conductivity partially reduced the flock formation induced by Polymin P, probably by screening the particle surface charges (Zhou and Franks, 2006), providing additional evidence for a mode of action based on charge neutralization. Similar effects of increasing salt concentrations have been observed for chitosan (Riske et al., 2007). Sedipur CL 950 probably uses the same mode of action, but its lower charge density may explain the slower flock formation. In contrast, the near-neutral Paestrol 2610 is likely to promote bridging flocculation alone, resulting in slower and less effective flocculation than the others (Gregory and Barany, 2011; Russel et al., 1989). In future studies, we will investigate the precise mechanisms of flocculation for the most effective polymers in more detail.

The optimal concentration of all flocculants was in the range 2–4 g/L, corresponding to 15–30 mg polymer per gram of dispersed solids. This is within the mid-range previously reported for cultures of mammalian cells (0.05–1.00 g/L, optimum 0.2 g/L corresponding to ~10 mg polymer per gram solids (Riske et al., 2007)) and bacteria (5–75 g/L, (Aspelund et al., 2008; Pearson et al., 2004), corresponding to 300 mg polymer per gram dry solids (Aspelund et al., 2008)).

The backbone chemistry also had an important impact on flock formation. The most efficient polymers were those with tertiary amines (e.g. Polymin P). In contrast, primary and secondary amines as well as quaternary ammonium ions (polyacrylamides, polyamines and polyDADMAC, respectively) delayed flocculation and/or reduced the aggregation of cell debris. This could reflect the greater hydrophobicity of tertiary compared with primary and secondary amines due to the larger number of aliphatic substituents, which would increase interactions with cell debris lipids promoting flock formation and would also affect their conformation in solution (Aspelund et al., 2008). The positive charge and small size of two of their substituents (two methyl groups in the case of pDADMAC) could explain why quaternary ammonium ions were less effective than tertiary amines.

Several models have been developed to describe the settling behaviour of flocks (Runkana et al., 2004; Tang et al., 2002).

However, to our knowledge, no reliable model exists today that can predict the efficacy of flock formation by a specific flocculant under specific solution conditions, for example pH. Here, we developed a descriptive model for the best-performing flocculant (Polymin P) that was capable of predicting the turbidity reduction achieved with this polymer over a wide range of operation conditions following the extraction of biopharmaceutical proteins from plants. The model revealed that, within the investigated range, certain pH values may diminish the efficacy of flocculation. Similar observations have been reported for other processes (Barany and Szepesszentgyorgyi, 2004; Riske et al., 2007). Conductivity and temperature were noncritical parameters for the efficacy of Polymin P, but polymer concentrations >1.75 g/L and incubation times >10 min were required to achieve robust clarification. The predictive model can therefore be used to estimate the cost savings that can be achieved by the use of flocculants and compare them with the efforts needed to achieve regulatory compliance. The DoE approach is not limited to clarification and can be transferred easily to other unit operations.

Polymin P was also used successfully for the bench-top scale processing of extracts from transgenic plants producing the monoclonal antibody 2G12 and the fluorescent protein DsRed. Although target protein concentrations were not significantly affected under standard extraction conditions (pH 8.0, 30–45 mS/cm), the filter capacity increased by up to sevenfold (for the filter PDH4) compared with a flocculant-free set-up. This was also the case for different Nicotiana species. However, the addition of flocculants may require a subsequent re-evaluation of the depth filtration step because the properties of the feed stream, for example in terms of particle number and size distribution, can change significantly following the addition of the polymer. Our future studies will therefore focus on the determination of particle size and ζ-potential to understand the flocculation of tobacco leaf extracts by Polymin P, as previously reported for other expression platforms (Pearson et al., 2004; Rasteiro et al., 2010; Serra et al., 2008; Spicer et al., 1996). In this context, we will also investigate how the mode and rate of polymer addition affects flock formation (Barany and Szepesszentgyorgyi, 2004; Weir et al., 1993).

Scale-up to the 100-L process revealed that flocculation can easily be incorporated into the large-scale manufacturing of biopharmaceutical proteins and that the gain in filter capacity can be increased even further due to scale-dependent changes in filter geometry. Furthermore, Polymin P complies with FDA regulation 21 CFR 176.170 ('Components of paper and paperboard in contact with aqueous and fatty foods'), suggesting this polymer would gain regulatory approval in a biopharmaceutical production process.

The shear rate during mixing after adding the polymer may also have a major impact on flock formation and flock size (Gregory, 1988; Gregory and Barany, 2011; Zhou and Franks, 2006). High shear rates often result in smaller flocks due to breakage (Ehrl et al., 2008; Serra et al., 2008), but we found that the shear rate could be neglected as a flocculation parameter because the mixing times following polymer addition were short (10–20 s) compared with the flock formation times observed in our experiments (>2 min) and reported by others (>20 s) (Aspelund et al., 2008; Das and Somasundaran, 2004; Duarte et al., 2010; Rasteiro et al., 2010; Yu et al., 2010; Zhao and Lai, 2010). We also found that only mixing times >30 s had a negative impact on turbidity reduction in our laboratory-scale set-up (Figure 8d).

In conclusion, flocculants were shown to reduce the turbidity of the process stream after bag filtration and to increase depth filter capacity to more than 110 L/m (depending on the scale). The best-performing flocculant was Polymin P. The costs for filter consumables were reduced by more than 50% compared with a process without flocculant, and there were no negative effects on target protein concentration.

Experimental procedures

Biological materials and expression constructs

Agrobacterium tumefaciens strain GV3101/pMP90RK was transformed by electroporation (Main *et al.*, 1995) with plasmid pGFD (Holland *et al.*, 2010), a derivative of pPAM (GenBank AY027531) kindly supplied by Dr Thomas Rademacher, Fraunhofer IME Aachen, Germany. Bacteria were cultivated and prepared for infiltration as previously described (Buyel and Fischer, 2012). Whole *Nicotiana benthamiana* plants were vacuum-infiltrated at ~50 Pa absolute pressure for 15 min, followed by sudden vacuum release to promote transgene expression. Seeds from (i) transgenic tobacco (*Nicotiana tabacum*) Petit Havana SR1 plants carrying the pGFD T-DNA and (ii) wild-type *N. tabacum* K326 plants were also kindly supplied by Dr Thomas Rademacher.

Plant cultivation

Tobacco and *N. benthamiana* seeds were germinated on rockwool blocks (Cultilène, the Netherlands) or in soil and were cultivated in a greenhouse at 25/22 °C day/night temperature with a 16-h photoperiod (180 μmol/s/m^2; λ = 400–700 nm) at 70% relative humidity. The plants were irrigated with a 0.1% solution of Ferty 2 Mega (Kammlott GmbH, Germany). *N. benthamiana* plants were grown for 42 days prior to infiltration with *A. tumefaciens* followed by 5 days postinfiltration incubation under the same conditions prior to harvest. Tobacco plants were harvested 47 days after seeding.

Design of experiments

Flocculants and controls were tested using an IV-optimal response surface methodology (RSM) design of experiments (DoE) comprising 88 conditions covering the pH range 4.0–8.0 and the flocculant concentration range 0.01–2.00 g/L but with a constant conductivity (30 mS/cm) and incubation time (60 min). A refined set of six flocculants was used in an IV-optimal second DoE with 60 different conditions and concentrations up to 3.0 g/L based on the same parameters. We optimized the three best flocculants using an IV-optimal RSM with 30 conditions at additional pH values of 5 and 7. Based on the results, a second set of flocculants was screened using an IV-optimal RSM with 70 conditions at different concentrations (0.5–4.0 g/L), conductivities (15–55 mS/cm), pH values (6.5–8.0) and incubation times (5–30 min). The efficacy of the best-performing flocculant was evaluated in a 57-run IV-optimal DoE with the same parameters. The preferred concentration range of the final flocculant was then evaluated using an IV-optimal RSM design with 70 conditions at different concentrations, conductivities, pH values, incubation times and temperatures (Table 1). Factors with a significant influence on turbidity were preselected from a quadratic or cubic model by automatic backwards selection using a *P*-value threshold of 0.100, and factors with *P*-values >0.050 were removed manually. Exceptions were made to maintain model hierarchy, a requirement for high-quality polynomial regression models (Peixoto, 1987, 1990).

Quantitation of total soluble proteins, 2G12 and DsRed

Samples were taken from the feed stream during extraction and subsequent filtration steps and therefore did not require maceration. Samples were centrifuged twice at 16 000 ***g***, 20 min, 4 °C, and supernatants were stored at −80 °C. The quantity of total soluble protein (TSP) in the supernatants was determined using the Bradford method (Bradford *et al.*, 1976). Briefly, 2.5 or 5.0 μL of supernatant was mixed with 200 μL of Bradford reagent (Thermo Fisher Scientific, Illinois) in 96-well plates and incubated for 10 min at 22 °C before measuring the absorbance at 595 nm using a Synergy HT plate reader (BioTek Instruments, Vermont). Eight dilutions of bovine serum albumin (0–2000 μg/mL) were prepared in triplicate and used to build a standard curve.

DsRed fluorescence in the supernatants was measured using a Synergy HT plate reader fitted with 530/25 nm (excitation) and 590/35 nm (emission) filter sets in 96-well half area plates. Reads were averaged over triplicate samples of 50 μL, and a standard curve was generated with dilutions in the range 0–225 mg/mL.

The quantity and binding specificity of antibody 2G12 was determined by surface plasmon resonance spectroscopy using a Biacore T100 instrument (GE Healthcare, Sweden) to measure the amount of antibody binding to Protein A (Sigma-Aldrich, St. Louis, MO) immobilized on the surface of a CM5 chip by EDC/NHS coupling (Howell *et al.*, 1998; Piliarik *et al.*, 2009). A 585 ng/mL reference solution of 2G12 (Polymun Scientific, Klosterneuburg, Austria) corresponding to ~500 response units (RU) was used for one-point calibration with HBS EP+ as the running buffer, and protein samples were diluted to 2G12 concentrations below the reference solution. A one-point calibration was chosen because calibration curves were found to be linear in a range below 1000 RU (our unpublished data).

Homogenization, flocculation and filtration

The homogenization and bag filtration steps are shown in Figure 3h. Briefly, 500 g of leaf tissue was homogenized in a Polytron PT6100 (Kinematica, Luzern, Switzerland) in the presence of 1.5 L extraction buffer (50 mM phosphate, 500 mM NaCl, pH 8.0). Phosphate was replaced with citric acid to test at pH values <6.0, and the NaCl concentration was adjusted to match the conductivity of the feed stream. After extraction, a water control or one of the 23 different flocculants (Table S1) was added in a pulse to the nonagitated homogenate as 2% or 4% (based on the formulation delivered by the manufacturer) stock solutions in 50 mM phosphate buffer (pH adjusted to that of the extract) and mixed thoroughly for 20 s by manual shaking (screening experiments) or for 10 s at 8200 rpm in the PT6100 (laboratory-scale experiments) to induce flocculation. In screening experiments, flocculation was then allowed to proceed for different times without agitation according to the DoE specifications (Table 1) or for 15 min in bench-top and pilot-scale experiments followed by bag filtration through either a BP-410 filter (Fuhr, Klein-Winternheim, Germany) with a 1-μm nominal retention rating (laboratory-scale experiments) or through two layers of Miracloth (Merck-Millipore, Darmstadt, Germany) for screening experiments. The extract turbidity was determined immediately and 24 h after filtration as a 1:10 dilution in extraction buffer using a Hach 2100P (Hach, Loveland, CO).

Different depth filters were tested alone or in combination to remove fine particles from the bag filtrate (Tables S3 and S4). The volumetric loading flow rate was kept approximately constant over the different scales. The bag and depth filtrates were monitored for turbidity, conductivity, pH, TSP and the concentrations of 2G12 and DsRed.

Acknowledgements

We would like to thank Dr Thomas Rademacher for providing the transgenic SR1 plants and Ibrahim Al Amedi for cultivating the tobacco plants used in our experiments. We would also like to thank Drs Nathalie Sieverling (Ashland, Germany) and Antje Lieske (Fraunhofer IAP, Germany) as well as Mr. Klemens Hauck (BASF, Germany) for kindly providing samples of the different flocculants. Finally, we would like to thank Dr Richard M. Twyman for his assistance with editing the manuscript. This work was in part funded by the European Research Council Advanced Grant 'Future-Pharma', proposal number 269110 and the Fraunhofer Zukunftsstiftung (Future Foundation). The authors have no conflict of interest to declare.

References

Aspelund, M.T., Rozeboom, G., Heng, M. and Glatz, C.E. (2008) Improving permeate flux and product transmission in the microfiltration of a bacterial cell suspension by flocculation with cationic polyelectrolytes. J. Memb. Sci. 324, 198–208.

Barany, S. and Szepesszentgyorgyi, A. (2004) Flocculation of cellular suspensions by polyelectrolytes. Adv. Colloid Interface Sci. 111, 117–129.

Bradford, M.M., Simonian, M.H. and Smith, J.A. (1976) A rapid and sensitive method for the quantitation of microgram quantities of protein utilizing the principle of protein-dye binding. Anal. Biochem. 72, 248–254.

Buyel, J.F. and Fischer, R. (2012) Predictive models for transient protein expression in tobacco (Nicotiana tabacum l.) can optimize process time, yield, and downstream costs. Biotechnol. Bioeng. 109, 2575–2588.

Commandeur, U., Twyman, R.M. and Fischer, R. (2003) The biosafety of molecular farming in plants. AgBiotechNet, 5, 9.

Das, K.K. and Somasundaran, P. (2004) A kinetic investigation of the flocculation of alumina with polyacrylic acid. J. Colloid Interface Sci. 271, 102–109.

Duarte, G.V., Ramarao, B.V. and Amidon, T.E. (2010) Polymer induced flocculation and separation of particulates from extracts of lignocellulosic materials. Bioresour. Technol. 101, 8526–8534.

Ehrl, L., Soos, M. and Morbidelli, M. (2008) Dependence of aggregate strength, structure, and light scattering properties on primary particle size under turbulent conditions in stirred tank. Langmuir, 24, 3070–3081.

Fischer, R. and Emans, N. (2000) Molecular farming of pharmaceutical proteins. Transgenic Res. 9, 279–299. discussion 277.

Fischer, R., Schillberg, S., Buyel, J.F. and Twyman, R.M. (2013) Commercial aspects of pharmaceutical protein production in plants. Curr. Pharm. Des. 19, 5471–5477.

Fischer, R., Schillberg, S., Hellwig, S., Twyman, R.M. and Drossard, J. (2012) Gmp issues for recombinant plant-derived pharmaceutical proteins. Biotechnol. Adv. 30, 434–439.

Fisher, R.R. and Glatz, C.E. (1988) Polyelectrolyte precipitation of proteins: Ii. Models of the particle size distributions. Biotechnol. Bioeng. 32, 786–796.

Giglia, S. and Sciola, L. (2011) Scaling up normal-flow microfiltration processes. Bioprocess Int. 9, 3.

Giglia, S. and Yavorsky, D. (2007) Scaling from discs to pleated devices. PDA J. Pharm. Sci. Tech. 61, 314–323.

Gottschalk, U. (2009) Process Scale Purification of Antibodies. Hoboken, NJ: John Wiley & Sons.

Gregory, J. (1988) Polymer adsorption and flocculation in sheared suspensions. Colloid Surf. 31, 231–253.

Gregory, J. and Barany, S. (2011) Adsorption and flocculation by polymers and polymer mixtures. Adv. Colloid Interface Sci. 169, 1–12.

Hassan, S., van Dolleweerd, C.J., Ioakeimidis, F., Keshavarz-Moore, E. and Ma, J.K. (2008) Considerations for extraction of monoclonal antibodies targeted to different subcellular compartments in transgenic tobacco plants. Plant Biotechnol. J. 6, 733–748.

Holland, T., Sack, M., Rademacher, T., Schmale, K., Altmann, F., Stadlmann, J., Fischer, R. and Hellwig, S. (2010) Optimal nitrogen supply as a key to increased and sustained production of a monoclonal full-size antibody in by-2 suspension culture. Biotechnol. Bioeng. 107, 278–289.

Holler, C. and Zhang, C. (2008) Purification of an acidic recombinant protein from transgenic tobacco. Biotechnol. Bioeng. 99, 902–909.

Holler, C., Vaughan, D. and Zhang, C.M. (2007) Polyethyleneimine precipitation versus anion exchange chromatography in fractionating recombinant beta-glucuronidase from transgenic tobacco extract. J. Chromatogr. A, 1142, 98–105.

Howell, S., Kenmore, M., Kirkland, M. and Badley, R.A. (1998) High-density immobilization of an antibody fragment to a carboxymethylated dextran-linked biosensor surface. J. Mol. Recognit. 11, 200–203.

Hubbuch, J.J., Brixius, P.J., Lin, D.Q., Mollerup, I. and Kula, M.R. (2006) The influence of homogenisation conditions on biomass-adsorbent interactions during ion-exchange expanded bed adsorption. Biotechnol. Bioeng. 94, 543–553.

Jornitz, M.W. (2006) Filter construction and design. Adv. Biochem. Eng. Biotechnol. 98, 105–123.

Kandula, S., Babu, S., Jin, M. and Shukla, A.A. (2009) Design of a filter train for precipitate removal in monoclonal antibody downstream processing. Biotechnol. Appl. Biochem. 54, 149–155.

Laukel, M., Rogge, P. and Dudziak, G. (2011) Disposable downstream processing for clinical manufacturing. BioProcess Int. 9, 14–21.

Main, G.D., Reynolds, S. and Gartland, J.S. (1995) Electroporation protocols for agrobacterium. Methods Mol. Biol. 44, 405–412.

Nečesaný, V. (1971) The isoelectric point of lignified cell walls. Holz als Roh-und Werkstoff, 29, 354–357.

O'Brien, T.P., Brown, L.A., Battersby, D.G., Rudolph, A.S. and Raman, L.P. (2012) Large-scale, single-use depth filtration systems for mammalian cell culture clarification. Bioprocess Int. 10, 50–57.

Pearson, C.R., Heng, M., Gebert, M. and Glatz, C.E. (2004) Zeta potential as a measure of polyelectrolyte flocculation and the effect of polymer dosing conditions on cell removal from fermentation broth. Biotechnol. Bioeng. 87, 54–60.

Pegel, A., Reiser, S., Steurenthaler, M. and Klein, S. (2011) Evaluating disposable depth filtration platforms for mab harvest clarification. Bioprocess Int. 9, 52–56.

Peixoto, J.L. (1987) Hierarchical variable selection in polynomial regression-models. Am. Stat. 41, 311–313.

Peixoto, J.L. (1990) A property of well-formulated polynomial regression-models. Am. Stat. 44, 26–30.

Peram, T., McDonald, P., Carter-Franklin, J. and Fahrner, R. (2010) Monoclonal antibody purification using cationic polyelectrolytes: an alternative to column chromatography. Biotechnol Prog. 26, 1322–1331.

Piliarik, M., Vaisocherova, H. and Homola, J. (2009) Surface plasmon resonance biosensing. Methods Mol. Biol. 503, 65–88.

Rasteiro, M.G., Garcia, F.A.P., Ferreira, P.J., Antunes, E., Hunkeler, D. and Wandrey, C. (2010) Flocculation by cationic polyelectrolytes: relating efficiency with polyelectrolyte characteristics. J. Appl. Polym. Sci. 116, 3603–3612.

Riske, F., Schroeder, J., Belliveau, J., Kang, X.Z., Kutzko, J. and Menon, M.K. (2007) The use of chitosan as a flocculant in mammalian cell culture dramatically improves clarification throughput without adversely impacting monoclonal antibody recovery. J. Biotechnol. 128, 813–823.

Roush, D.J. and Lu, Y.F. (2008) Advances in primary recovery: centrifugation and membrane technology. Biotechnol Prog. 24, 488–495.

Runkana, V., Somasundaran, P. and Kapur, P.C. (2004) Mathematical modeling of polymer-induced flocculation by charge neutralization. J. Colloid Interface Sci. 270, 347–358.

Russel, W.B., Saville, D.A. and Schowalter, W.R. (1989) Colloidal Dispersions. Cambridge, UK: Cambridge University Press.

Serra, T., Colomer, J. and Logan, B.E. (2008) Efficiency of different shear devices on flocculation. Water Res. 42, 1113–1121.

Spicer, P.T., Keller, W. and Pratsinis, S.E. (1996) The effect of impeller type on floc size and structure during shear-induced flocculation. J. Colloid Interface Sci. 184, 112–122.

Tang, P., Greenwood, J. and Raper, J.A. (2002) A model to describe the settling behavior of fractal aggregates. *J. Colloid Interface Sci.* **247**, 210–219.

Weir, S., Ramsden, D.K., Hughes, J. and Thomas, F. (1993) The flocculation of yeast with chitosan in complex fermentation media: the effect of biomass concentration and mode of flocculant addition. *Biotechnol. Tech.* **7**, 199–204.

Westoby, M., Chrostowski, J., de Vilmorin, P., Smelko, J.P. and Romero, J.K. (2011) Effects of solution environment on mammalian cell fermentation broth properties: enhanced impurity removal and clarification performance. *Biotechnol. Bioeng.* **108**, 50–58.

Whitford, W.G. (2010) Single-use systems as principal components in bioproduction. *Bioprocess Int.*, **8**, 34–44.

Wilken, L.R. and Nikolov, Z.L. (2012) Recovery and purification of plant-made recombinant proteins. *Biotechnol. Adv.* **30**, 419–433.

Yigzaw, Y., Piper, R., Tran, M. and Shukla, A.A. (2006) Exploitation of the adsorptive properties of depth filters for host cell protein removal during monoclonal antibody purification. *Biotechnol Prog.* **22**, 288–296.

Yu, W.Z., Gregory, J. and Campos, L.C. (2010) Breakage and re-growth of flocs formed by charge neutralization using alum and polydadmac. *Water Res.* **44**, 3959–3965.

Zhao, J.H. and Lai, Y.P. (2010) Experimental analysis of floc formation time and order in coagulation process. *Adv. Mater. Res-Switz.* **113–116**, 1058–1062.

Zhou, Y. and Franks, G.V. (2006) Flocculation mechanism induced by cationic polymers investigated by light scattering. *Langmuir*, **22**, 6775–6786.

PERMISSIONS

LIST OF CONTRIBUTORS

Eltayb Abdellatef, Torsten Will, Aline Koch, Jafargholi Imani and Karl-Heinz Kogel
Centre for BioSystems, Land Use and Nutrition, Institute of Phytopathology and Applied Zoology, Justus Liebig University, Giessen, Germany

Andreas Vilcinskas
Project Group 'Bioresources', Fraunhofer Institute of Molecular Biology and Applied Ecology IME, Giessen, Germany

Jeehye Choi
Department of Biology, University of Western Ontario, London, ON, Canada

Shengwu Ma
Department of Biology, University of Western Ontario, London, ON, Canada
Lawson Health Research Institute, London, ON, Canada
Plantigen Inc., London, ON, Canada

Hong Diao
Lawson Health Research Institute, London, ON, Canada

Zhi-Chao Feng and Rennian Wang
Lawson Health Research Institute, London, ON, Canada
Department of Physiology and Pharmacology, University of Western Ontario, London, ON, Canada

Anthony M. Jevnikar
Lawson Health Research Institute, London, ON, Canada
Plantigen Inc., London, ON, Canada

Arthur Lau
Department of Pathology, University of Western Ontario, London, ON, Canada

Nuri Company, Anna Nadal, Emilio Montesinos and Maria Pla
Institute for Food and Agricultural Technology (INTEA), University of Girona, Girona, Spain

José-Luis La Paz and Sílvia Martínez
Center for Research in Agricultural Genomics (CRAG), Barcelona, Spain

Stefan Rasche and Stefan Schillberg
Fraunhofer Institute for Molecular Biology and Applied Ecology (IME), Aachen, Germany

Feng Gao, Vijaya R. Chitnis and Belay T. Ayele
Department of Plant Science, University of Manitoba, Winnipeg, MB, Canada

Christof Rampitsch, Gavin D. Humphreys and Mark C. Jordan
Cereal Research Centre, Agriculture and Agri-Food Canada, Winnipeg, MB, Canada

Elizabeth E. Hood, Shivakumar P. Devaiah, Keat (Thomas) Teoh and Deborah Vicuna Requesens
Arkansas State University, Jonesboro AR, USA

Gina Fake, Erin Egelkrout, Celine Hayden and John A. Howard
Applied Biotechnology Institute, San Luis Obispo, CA, USA

Kendall R. Hood
Infinite Enzymes, Jonesboro, AR, USA

Kameshwari M. Pappu
University of South Florida, Tampa, FL, USA

Jennifer Carroll
Cal Poly State University, San Luis Obispo, CA, USA

Tejinder Kumar and Ismail Dweikat
Department of Agronomy and Horticulture, University of Nebraska-Lincoln, Lincoln, NE, USA

Tom Elthon
Department of Agronomy and Horticulture, University of Nebraska-Lincoln, Lincoln, NE, USA
Center for Biotechnology, University of Nebraska-Lincoln, Lincoln NE, USA

Tom Clemente
Department of Agronomy and Horticulture, University of Nebraska-Lincoln, Lincoln, NE, USA
Center for Plant Science Innovation, University of Nebraska-Lincoln, Lincoln, NE, USA

Shirley Sato, Zhengxiang Ge, Natalya Nersesian and Han Chen
Center for Biotechnology, University of Nebraska-Lincoln, Lincoln NE, USA

Scott Bean, Brian P. Ioerger and Mike Tilley
Grain Quality and Structure Research Unit, USDA / ARS, Manhattan, KS, USA

Nicholas J. Larkan, Lisong Ma and Mohammad Hossein Borhan
Saskatoon Research Centre, Agriculture and Agri-Food Canada, Saskatoon, SK, Canada

Tarlan Mamedov, Ananya Ghosh, R. Mark Jones, Vadim Mett, Christine E. Farrance, Konstantin Musiychuk, April Horsey and Vidadi Yusibov
Fraunhofer USA Center for Molecular Biotechnology, Newark, DE, USA

Greta Nölke, Marcel Houdelet and Fritz Kreuzaler
Fraunhofer Institute for Molecular Biology and Applied Ecology IME, Aachen, Germany

Christoph Peterhänsel
Institute of Botany, Leibniz-University Hannover, Hannover, Germany

Stefan Schillberg
Fraunhofer Institute for Molecular Biology and Applied Ecology IME, Aachen, Germany
Phytopathology Department, Institute for Phytopathology and Applied Zoology, Justus-Liebig University Giessen, Giessen, Germany

Wei Cheng, Jing-Bo Zhang, Peng Yang and Xian-Wei Kong
Molecular Biotechnology Laboratory of Triticeae Crops, Huazhong Agricultural University, Wuhan, China
College of Plant Science and Technology, Huazhong Agricultural University, Wuhan, China

He-Ping Li, Hong-Jie Du, Qi-Yong Wei and Tao Huang
Molecular Biotechnology Laboratory of Triticeae Crops, Huazhong Agricultural University, Wuhan, China
College of Life Science and Technology, Huazhong Agricultural University, Wuhan, China

Yu-Cai Liao
Molecular Biotechnology Laboratory of Triticeae Crops, Huazhong Agricultural University, Wuhan, China
College of Plant Science and Technology, Huazhong Agricultural University, Wuhan, China
National Center of Plant Gene Research (Wuhan), Huazhong Agricultural University, Wuhan, China

Lauri J. Reuter, Michael J. Bailey, Jussi J. Joensuu and Anneli Ritala
VTT Technical Research Centre of Finland, Espoo, Finland

Jin-Yue Sun, Joe Hammerlindl, Li Forseille, Haixia Zhang and Mark A. Smith
National Research Council Canada, Saskatoon, SK, Canada

Eva C. Thuenemann and George P. Lomonossoff
Department of Biological Chemistry, John Innes Centre, Norwich, UK

Ann E. Meyers and Edward P. Rybicki
Department of Molecular and Cell Biology, University of Cape Town, Rondebosch, South Africa

Jeanette Verwey
Onderstepoort Biological Products SOC Ltd, Onderstepoort, South Africa

Richard L. Tillett, Matthew D. Wheatley, Elizabeth A. R. Tattersall, Karen A. Schlauch, Grant R. Cramer and John C. Cushman
Department of Biochemistry and Molecular Biology, University of Nevada, Reno, NV, USA

Nicola Weichert, Valeska Hauptmann, Kai Schallau, Philip Gunkel and Udo Conrad
Leibniz Institute of Plant Genetics and Crop Plant Research, Stadt Seeland/Ortsteil, Gatersleben, Germany

Matthias Menzel and Uwe Spohn
Fraunhofer Institute for Mechanics of Materials, Halle (Saale), Germany

Thomas C. Hertel and Markus Pietzsch
Institute of Pharmacy, Faculty of Sciences I, Martin Luther University Halle-Wittenberg, Halle (Saale), Germany

Chun-Xiang You, Qiang Zhao, Xiao-Fei Wang, Xing-Bin Xie, Xiao-Ming Feng, Ling-Ling Zhao, Huai-Rui Shu and Yu-Jin Hao
National Key Laboratory of Crop Biology, National Research Center for Apple Engineering and Technology, College of Horticulture Science and Engineering, Shandong Agricultural University, Tai-An, Shandong, China

Jordi Quilis, Belén López-García and Blanca San Segundo
Centre for Research in Agricultural Genomics (Crag), Csic-Irta-Uab-Ub, Edifici Crag, Barcelona, Spain

Donaldo Meynard and Emmanuel Guiderdoni
Cirad, Umr Agap, Montpellier Cedex 5, France

Ling Chen, Brian R. Dempsey, Laszlo Gyenis, Rima Menassa, Jim E. Brandle and Sangeeta Dhaubhadel
Southern Crop Protection and Food Research Centre, Agriculture and Agri-Food Canada, London, ON, Canada

Johannes F. Buyel
Institute for Molecular Biotechnology, RWTH Aachen University, Aachen, Germany

Rainer Fischer
Institute for Molecular Biotechnology, RWTH Aachen University, Aachen, Germany
Fraunhofer Institute for Molecular Biology and Applied Ecology, Aachen, Germany

Index